Electron Distributions and the Chemical Bond

Electron Distributions and the Chemical Bond

Edited by
PHILIP COPPENS

State University of New York
Buffalo, New York

and
MICHAEL B. HALL

Texas A & M University
College Station, Texas

PLENUM PRESS • NEW YORK AND LONDON

Library of Congress Cataloging in Publication Data

Main entry under title:

Electron distributions and the chemical bond.

"Proceedings of a Symposium on Electron Distributions and the Chemical Bond for the national meeting of the American Chemical Society, held March 28–April 2, 1981, in Atlanta, Georgia"—p.
  Bibliography: p.
  Includes index.
  1. Chemical bonds—Congresses. 2. Electron distribution—Congresses. I. Coppens, Philip. II. Hall, Michael, B. III. Symposium on Electron Distributions and the Chemical Bond (1981: Atlanta, Ga.) IV. American Chemical Society.
QD461.E45   541.2′24   82-5390
ISBN-13: 978-1-4613-3469-9   e-ISBN-13: 978-1-4613-3467-5   AACR2
DOI: 10.1007/ 978-1-4613-3467-5

Proceedings of a Symposium on Electron Distributions and the
Chemical Bond held at the National Meeting of the American
Chemical Society, March 28–April 2, 1981, in Atlanta, Georgia

© 1982 Plenum Press, New York
Softcover reprint of the hardcover 1st edition 1982
A Division of Plenum Publishing Corporation
233 Spring Street, New York, N.Y. 10013

PREFACE

This book represents the proceedings of a symposium held at the Spring 1981 ACS meeting in Atlanta. The symposium brought together Theoretical Chemists, Solid State Physicists, Experimental Chemists and Crystallographers. One of its major aims was to increase interaction between these diverse groups which often use very different languages to describe similar concepts. The development of a common language, or at least the acquisition of a multilingual capability, is a necessity if the field is to prosper.

Much depends in this field on the interplay between theory and experiment. Accordingly this volume begins with two introductory chapters, one theoretical and the other experimental, which contain much of the background material needed for a through understanding of the field. The remaining sections describe a wide variety of applications and illustrate, we believe, the central role of charge densities in the understanding of chemical bonding.

We are most indebted to the Divisions of Inorganic and Physical Chemistry of the American Chemical Society, which provided the stimulus for the symposium and gave generous financial support. We also gratefully acknowledge financial support from the Special Educational Opportunities Program of the Petroleum Research Fund administered by the American Chemical Society, which made extensive participation by speakers from abroad possible.

We would like to thank all the authors for their cooperation in preparing the manuscripts, the staff of Plenum Press for the pleasant collaboration which led to the completion of this volume, and Mrs. Carol Reinhardt for the skillful typing of several of the manuscripts.

P.C.
M.B.H.

CONTENTS

Section 1
INTRODUCTION

Section 2
THEORETICAL CONSIDERATIONS

Section 3
EXTENDED SOLIDS: THEORETICAL AND EXPERIMENTAL RESULTS

Section 4
MOLECULAR SOLIDS: THEORETICAL RESULTS

Section 5
MOLECULAR SOLIDS: EXPERIMENTAL RESULTS

CONTENTS

Section 6
ELECTROSTATIC PROPERTIES

Section 1
INTRODUCTION

CONCEPTS OF CHARGE DENSITY ANALYSIS:

THE THEORETICAL APPROACH

Vedene H. Smith, Jr.

Department of Chemistry
Queen's University
Kingston, Ontario K7L 3N6, Canada

1. INTRODUCTION

Over the past twenty years the realization has become more and more wide spread that our understanding of chemical bonding and chemical reactions is based upon the electron density or the one-electron density matrix. This extends beyond the mere interpretation of the phenomena and the provision of a theoretical framework for the analysis of the data, to the fact that theoretical calculations can provide data about the electronic structure and structural properties of systems that are not measured readily.

It is the purpose of the present article to focus on those concepts from the theoretical framework and from computational quantum chemistry and solid-state physics which are needed for electron-density studies. To do this some equations are unfortunately needed and many terms need to be defined, hopefully clearly, to bridge the gap between the experimental scientist, the quantum-chemist and the solid-state physicist. Phrases such as *ab initio* and "first principles" are used to describe calculations* and a veritable alphabet

---

*The expressions *ab initio* and "first principles" are used respectively by quantum chemists and solid-state physicists to describe calculations which are made without using any empirical data save the nuclear charges and geometries. The use of basis set parameters and exchange parameters determined on the basis of computational experience do not cause a calculation to lose its description as *ab initio*. Semi-empirical calculations are those where certain matrix elements of the Fock operator are approximated in terms of experimental or other data and hence sometimes *ab initio* means that all the matrix elements (integrals) are calculated exactly for a given basis set. The origin of the phrase *ab initio* is attributed by Mulliken[1] to Parr, Craig and Ross.[2]

soup of acronyms and abbreviations are used as parts of the standard vocabulary. To emphasize the origins of these latter combination of letters, we shall write them in the following form, the first time they appear, e.g. L(ocalized) M(olecular) O(rbital).

2. WAVEFUNCTIONS AND THE HAMILTONIAN OPERATOR

"The wavefunction is the physicist's description of reality"[3]

The time independent Schrödinger equation

$$\hat{H}\Psi = E\Psi \tag{1}$$

relates the quantized energy values (*eigenvalues*) E to the Hamiltonian operator $\hat{H}$ by means of the wavefunction Ψ. From wavefunctions we can derive theoretical electron densities. In most cases exact solutions to the Schrödinger equation are unobtainable due to the complicated nature of the Hamiltonian. As a result we are forced to rely on approximations to the correct wavefunctions. These are usually obtained by means of the *variational principle* which selects from a set of approximate wavefunctions, the one of lowest energy for the given problem. Unfortunately, it is not necessarily the one of this set which gives the best description of a given physical property, such as the electron density.

The Hamiltonian operator for a system of nuclei $(\alpha, \beta, \ldots)$ with coordinates X, and the electrons $(i, j, \ldots)$ with coordinates x, is

$$\hat{H}(x, X) = - \sum_\alpha \frac{\hbar^2}{2M_\alpha} \nabla_\alpha^2 - \sum_i \frac{\hbar^2}{2m_i} \nabla_i^2 + \hat{V}_{ne}(x, X)$$

$$+ \hat{V}_{ee}(x) + \hat{V}_{nn}(X), \tag{2}$$

where M_α is the mass of nucleus α, m_i is the mass of electron i, $\hat{V}_{ne} = \sum_i \sum_\alpha (-Z_\alpha e^2/4\pi\epsilon_o) r_{i\alpha}$ is the nuclear-electron attraction term, Z_α is the atomic number of nucleus α, e is the electronic charge, and $r_{i\alpha}$ the distance between electron i and nucleus α. Similarly, $\hat{V}_{ee} = \sum_i \sum_j e^2/4\pi\epsilon_o r_{ij}$ and $\hat{V}_{nn} = \sum_\alpha \sum_\beta Z_\beta Z_\alpha e^2/4\pi\epsilon_o r_{\beta\alpha}$.

The Hamiltonian in this form is far too complicated to be solved exactly, except for the simplest systems. An important simplification can be made because of the large difference between the mass (more than three orders of magnitude), and hence velocities, of electrons and of nuclei. During a single vibration of the nuclei the electrons repeat their motions many times; hence, it is usually possible to treat nuclear and electronic motions as virtually independent of each other. The Hamiltonian operator may therefore be decomposed:

$$\hat{H}(x,X) = \hat{H}_n(X) + \hat{H}_e(x,X) , \tag{3}$$

$$\hat{H}_n(X) = - \sum_\alpha \frac{h^2}{2M_\alpha} \nabla_\alpha^2 + \hat{v}_{nn}(X) , \tag{4}$$

$$\hat{H}_e(x,X) = - \sum_i \frac{\hbar^2}{2m_i} \nabla^2 + \hat{v}_{ee}(x) + \hat{v}_{ne}(x,X) . \tag{5}$$

Here $\hat{H}_n(X)$ contains only nuclear coordinates and $\hat{H}_e(x,X)$ contains electronic coordinates as variables and nuclear coordinates as parameters, i.e. the nuclei are assumed fixed in their positions.

In this approximation the total state function Ψ, which is a solution of the time-independent Schrödinger wave equation (1), can be expressed as a product

$$\Psi = \Psi_e(x,X) \cdot \Psi_n(X), \tag{6}$$

where Ψ_e and Ψ_n are solutions of the simpler equations

$$\hat{H}_e(x,X)\Psi_e(x,X) = E_e(X)\Psi_e(x,X) , \tag{7}$$

$$[\hat{H}_n(X) + E_e(X)]\Psi_n(X) = E_n\Psi_n(X) . \tag{8}$$

These equations describe the Born-Oppenheimer[4] approximation and imply that the total wavefunction of a molecule can be separated into an electronic and a nuclear wavefunction. There is clear spectroscopic evidence that this approximation holds quite well in most cases.

Since our primary interest is in electron densities, henceforth any reference to the Hamiltonian H or the wavefunction Ψ will in face refer to $\hat{H}_e$ or Ψ_e, respectively, unless otherwise stated.

We recognize that $\hat{H}_e$ is the quantum-mechanical operator for the motion of the electrons in the field of the nuclei at *fixed positions* and as such is the starting point for nearly all studies of electronic structure. For the specific case of η nuclei and N electrons, it reads

$$\hat{H} = \sum_{i=1}^N (-\hat{h}(i)) + \sum_{i<j} \hat{g}(i,j) \tag{9}$$

where the one-electron Hamiltonian operator for electron i, $\hat{h}(i)$

$$\hat{h}(i) = - \frac{\hbar^2}{2m} \nabla_i^2 - \sum_{\alpha=1}^\eta e^2 z_\alpha/4\pi\varepsilon_o r_{i\alpha}, \tag{10}$$

and the electron-electron repulsion operator,

$$\hat{g}(i,j) = e^2/4\pi\epsilon_o r_{ij} \; . \tag{11}$$

Here the variable i represents the spatial coordinates of electron i, Z_α is the atomic number of nucleus α, $r_{i\alpha}$ is the distance between electron i and nucleus α and r_{ij} is the interelectronic separation between i and j.

However, the Schrödinger equation (8) for the states of nuclear motion is not to be forgotten. It is needed when one considers the effects of nuclear motion on electron densities.

3. ATOMIC UNITS

It is more usual to see the electronic Hamiltonian written in atomic units[5]. The reason for this is that a(tomic) u(nits), in which $\hbar$, m_e, e, and $4\pi\epsilon_o$ are set equal to unity, make the Hamiltonian operator independent of those constants and hence the results of calculations made in atomic units do not depend on the accepted values of the fundamental constants at the time of the calculation. Of course, the ultimate comparison with experiment requires them but the comparison of two calculations made years apart does not.

Thus, in atomic units

$$\hat{h}(i) = -\tfrac{1}{2}\nabla_i^2 - \sum_\alpha Z_\alpha/r_{i\alpha} \tag{12}$$

and

$$\hat{g}(i,j) = 1/r_{ij} \; . \tag{13}$$

It is interesting to note that in this age of the S(ystem) I(nternational), the quantum chemists use a(tomic) u(nits) while the experimentalists use units based on the Å e.g. electron densities are reported in electrons/bohr^3 by the former and in electrons/Å^3 by the latter. It is perhaps worthwhile to compare atomic units with SI and related units.

In Table 1 we have gathered together the definitions of a number of the atomic units in terms of SI units, and in Table 2 some conversion factors which are useful in electron density studies. It should be noted that only the bohr(a_o) and hartree (E_h) are officially sanctioned[6] names for members of the family of atomic units. Otherwise they should be designated specifically in terms of four of the five quantities m_e, e, $\hbar$, E_h and a_o. Thus 2 a.u. of electron density would be tabulated as 2 ea_o^{-3}. It is better to say 2 a.u. of electron density than just the usual 2 a.u.

TABLE 1 - Atomic Versus SI Units

Quantity	A.U. Symbol	A.U. Name or Value*	Expression	SI Value
charge	e	millikan	e	1.602192×10^{-19} C
mass	m_e	thomson	m_e	9.10956×10^{-31} kg
action	$\hbar$	planck	$\hbar$	1.054592×10^{-34} Js
length	a_o	bohr	$\hbar^2 4\pi\varepsilon_o / m_e e^2$	5.291772×10^{-11} m
energy	E_h	hartree	$e^2 / 4\pi\varepsilon_o a_o$	4.359828×10^{-18} J
time	t_o	jiffy	$4\pi\varepsilon_o \hbar a_o / e^2$	2.418885×10^{-17} s
force			$e^2 / 4\pi\varepsilon_o a_o^2$	8.238879×10^{-8} N
velocity	v_o		$a_o t_o^{-1}$	2.187691×10^6 msec^{-1}
momentum	p_o	dumond	$mv_o = \hbar a_o^{-1}$	1.99289×10^{-24} kg msec^{-1}
electron density			ea_o^{-3}	1.08121×10^{12} C m^{-3}
electrostatic potential			$e/4\pi\varepsilon_o a_o$	3.02770×10^{-9} C m^{-1}
velocity of light	c	137.036 bohr jiffy^{-1}	c	2.997925×10^8 msec^{-1}
electric field	E	stark	$e/4\pi\varepsilon_o a_o^2$	5.721526×10^{-21} C m^{-2}
magnetic field strength B	B	zeeman	$\hbar a_o^{-2} e^{-1}$	2.350540×10^5 T
magnetic moment	β	Bohr magneton	$e\hbar/2m_e$	9.27410×10^{-24} JT^{-1}
dipole moment	μ		ea_o	4.48658×10^{-40} C m^2

* If a numerical value is not given, it is equal to unity. Except for bohr, hartree and bohr magneton, most of the names are not commonly used[6,7].

TABLE 2 - Some Conversion Factors Useful in Electron Density Studies.

Energy: 1 a.u. energy = 1 E_h = 1 hartree = $\hbar^2 m_e^{-1} a_o^{-2}$
                = 27.2116 eV = 2 Rydbergs (E_R)
                = 627.509 kcal/mol = 2625.50 kJ/mol
                = 4.35981 x 10^{-11} ergs = 4.35981 x 10^{-18} J

Length: 1 a.u. length = 1 bohr = 1 a_o = 0.529177 Å
                = 0.529177 x 10^{-10} m

Electron
Density :   1 a.u. electron density = ea_o^{-3} = 6.748315 e$Å^{-3}$
                = 1.08121 x 10^{12} C m^{-3}
                1 e$Å^{-3}$ = 0.147424 e(a.u.)$^{-3}$

Electrostatic
Potential   : 1 a.u. electrostatic potential = $e/4\pi\epsilon_o a_o$
                = 27.2116 volts = 3.02770 x 10^{-9} C m^{-1}

Scattering
Variable :   1 a.u. scattering variable = a_o^{-1} = 1.88973 $Å^{-1}$
                = 1.88973 x 10^{10} m^{-1}

4. THE VIRIAL THEOREM AND SCALING OF APPROXIMATE WAVEFUNCTIONS

     The virial theorem has long been recognized as playing a fun-
damental role in the electronic structure of matter[8-11]. Although
it is a necessary condition for a solution to the Schrödinger
equation, many authors have suggested that it may be a better criterion than the energy for testing the goodness of an approximate
wavefunction. As we shall see, a wavefunction may be made to satisfy
the virial theorem through the use of a scaling procedure - which
is different from but similar to the radial κ-scaling employed in
fitting experimental densities.

a) The General Case

     For a collection of N electrons (x) and η nuclei (X) described
by the Hamiltonian operator $\hat{H}(x,X)$ and the exact wavefunction $\Psi(x,X)$
with energy E, the *virial theorem* states that

     T = -1/2V = -E   (14)

where $T = <\Psi|\hat{T}|\Psi>$, and $V = <\Psi|\hat{V}|\Psi>$. The operator $\hat{T}$ represents the
first two terms and V the last three terms in (3). The virial
theorem reflects a *balance* between the kinetic energy contribution
T and the potential energy contribution V to the total energy.

     If we have an approximate wavefunction Φ which does not satisfy
the virial theorem, we may *scale* it to do so and receive the bonus
of lowering the energy in the process. To do this, we introduce a

scale parameter λ such that all N vectors x and all η vectors X are stretched uniformly from the origin by the factor λ and Φ remains normalized, i.e.

$$\Phi_\lambda = \lambda^{3/2\,(N+\eta)}\,\Phi(\lambda x_1, \lambda x_2, \ldots, \lambda x_N; \lambda X_1, \ldots \lambda X_\eta). \tag{15}$$

It is called a *norm-preserving scale transformation*. Substitution of Φ_λ into the variational principle leads to a variationally optimum value λ_{opt},

$$\lambda_{opt} = -\,V/2T \tag{16}$$

where V and T were evaluated using the original unscaled function Φ.

For λ_{opt} given by (16), the virial theorem is satisfied, i.e.

$$-E(\lambda_{opt}) = T(\lambda_{opt}) = -\,1/2V(\lambda_{opt}) \tag{17}$$

$$\text{and}\quad E(\lambda_{opt}) = \frac{-V^2(1)}{4T(1)} \le E(1) \tag{18}$$

where $E(1)$, $V(1)$ and $T(1)$ indicate the expectation values evaluated for $\lambda = 1$, i.e. prior to scaling.

b) The Case of Fixed Nuclear Positions

As we have discussed above, the Hamiltonian operator (9) for the electrons moving in the field due to the fixed nuclei is used for almost all calculations. The virial theorem in this case is the same if the nuclei are at their correct equilibrium locations, and is modified if they are not. For the sake of simplicity, we will state the theorem for the case of a diatomic molecule in a manner similar to (14):

$$T = -1/2V - R\frac{dE}{dR} = -E - 1/2R\frac{dE}{dR}. \tag{19}$$

When the nuclei are in their equilibrium positions or at infinite separation, $\frac{dE}{dR}$ vanishes and the earlier equation is recaptured.

The scaling of an approximate wavefunction $\Phi(\vec{x}_1, \vec{x}_2, \ldots, \vec{x}_N; R)$ which does not satisfy the virial theorem leads to

$$\Phi_\lambda = \lambda^{3N/2}\Phi(\lambda\vec{x}_1, \lambda\vec{x}_2, \ldots, \lambda\vec{x}_N; \rho) \tag{20}$$

where the internuclear distance is stretched by the same amount as the electronic coordinates, i.e. $\rho = \lambda R$.

In this case the optimum scale factor is:

$$\lambda_{opt} = -\,\frac{V(1,\rho) + \rho V_\rho(1,\rho)}{2T(1,\rho) + \rho T_\rho(1,\rho)} \tag{21}$$

where T_ρ and V_ρ denote the derivatives with respect to ρ of $T(1,\rho)$ and $V(1,\rho)$ respectively, the kinetic energy and potential energies for the unscaled electronic coordinates. Since one is usually interested in scaling at the equilibrium point, the condition that $\frac{dE(\lambda,R)}{dR}$ vanishes simplifies the expression for λ_{opt} to

$$\lambda_{opt} = -\frac{V_\rho(1,\rho)}{T_\rho(1,\rho)} = -\frac{V(1,\rho)}{2T(1,\rho)} . \qquad (22)$$

The practical procedure is to minimize $E = E(\rho)$, obtain ρ_e and E_e and then determine $(\lambda_{opt})_e$ from (22) as well as the equilibrium value of R, $R_e = \rho_e/(\lambda_{opt})_e$.

We shall return to this subject later when we discuss the scaling of electron densities and expectation values.

5. REDUCED DENSITY MATRICES, EXPECTATION VALUES AND N-REPRESENTABILITY

We write the electronic wavefunction Ψ for the system of N electrons which was obtained by solution of (9) or by a variational or other approximation to that problem as $\Psi(1,2,\ldots,N)$. The notation $J = (\vec{r}_j, s_j)$ denotes the combined *spin* (s_j) and *space* coordinates of electron j.

In order to account for the indistinguishability of the electrons, it is assumed that Ψ is *antisymmetric*, i.e. that for any permutation P of the coordinates $(1,2,\ldots,N)$

$$\Psi(1,2,\ldots,N) = (-1)^P P \Psi(1,2,\ldots,N) \qquad (23)$$

where p is the parity of P. Unless it is specified otherwise, we shall assume Ψ is normalized, i.e. the normalization integral $\langle \Psi | \Psi \rangle = \int |\psi(1,2,\ldots,N)|^2 d1 \ldots dN$ exists and equals unity.

The expectation value $\langle F \rangle = \langle \Psi | F | \Psi \rangle / \langle \psi | \psi \rangle$ $\qquad (24)$

of a physical quantity F represented by an Hermitian operator

$$\hat{F} = \hat{F}_o + \sum_i \hat{F}_i + \frac{1}{2!} \sum_{i,j}' \hat{F}_{ij} + \frac{1}{3!} \sum_{i,j,k}' \hat{F}_{ijk}$$

$$+ \ldots + \frac{1}{N!} \sum_{i,j,k,l\ldots}' \hat{F}_{ijkl\ldots} \qquad (25)$$

where the terms, symmetrical in the electron indices, successively represent zero, one, two, three, ... and N-particle operators, and the primes denote the omission of all terms with two or more equal indices, may be compactly written in terms of *reduced density matrices,*[12-14] of order p, Γ^P, (p=1,...N) which are obtained by reduction

$$\Gamma^P (1,2,...;P;1',2',...,P')$$

$$= \binom{N}{P} \int \Gamma^N (1,2,...,N;1',2',...,N') d(P+1)...dN \qquad (26)$$

of the N-particle density matrix

$$\Gamma^N (1,2,...,N;1',2',...N')$$

$$= \Psi(1,2,...,N)\Psi^*(1',2',...,N') . \qquad (27)$$

The definition means that the pth reduced density matrix is obtained by eliminating or reducing the coordinates of (N-p) particles.

Then

$$\langle \hat{F} \rangle = \hat{F}_o + \int \hat{F}_1 \Gamma^1 (1;1') d1$$

$$+ \int F_{12} \Gamma^2 (1,2;1',2') d1d2$$

$$+ \int F_{123} \Gamma^3 (1,2,3;1',2',3') d1d2d3$$

$$+ .... \qquad (28)$$

The evaluation of the terms in this equation proceeds operationally in three stages. Consider for the sake of definiteness the two-particle term $\int \hat{F}_{12} \Gamma^2 (1,2;1',2') d1d2$, which involves the two-particle density matrix or 2-matrix Γ^2. Then the three stages are:

a) $\hat{F}_{12}$ operates on only the unprimed coordinates 1,2, i.e.
$\hat{F}_{12} \Gamma^2 (1,2;1',2')$

b) Set 1' = 1 and 2' = 2

c) Integrate over d1 and d2.

Since almost all operators are of either one- or two-particle type or a combination thereof, the one- and two-matrices $\Gamma^1 (1;1')$ and $\Gamma^2 (1,2;1',2')$ are the most important. We shall see later that they are experimentally accessible via x-ray, neutron and electron scattering. For example the energy expectation value for the electronic Hamiltonian (9) may be written in terms of Γ^1 and Γ^2:

$$\langle H \rangle = 2\int \hat{h}_1 \Gamma^1 (1,1') d1 + \int \hat{g}_{12} \Gamma^2 (1,2;1',2') d1d2 \qquad (29)$$

or in terms of Γ^2 alone:

$$\langle H \rangle = \int [\hat{K}^2(1,2)] \Gamma^2(1,2;1',2') d1d2 \tag{30}$$

where $\hat{K}^2(1,2) = (N-1)^{-1} (\hat{h}_1 + \hat{h}_2) + \hat{g}_{12}$ is the *reduced Hamiltonian*.[13-15]

The 4N dimensional (3 space and one spin coordinate for each of the N electrons) integral $\langle \Psi | H | \Psi \rangle$ has been reduced to a sixteen dimensional one (3 space and one spin for *only* four particles: 1, 2, 1' and 2'). It is therefore very tempting to use this lower dimensional integral in the variational principle. However one cannot vary (30) simply with respect to any sixteen dimensional trial Γ^2. If one does so, it is possible to obtain an energy lower than the *true* ground state energy E.[15-17]

The *true* ground state energy corresponds to the minimum of $\int \hat{K}^2 \Gamma^2(1,2;1',2') d1d2$ such that Γ^2 is expressible in the form

$$\Gamma^2(1,2;1',2')$$

$$= \binom{N}{2} \int \Psi^*(1,2,\ldots N) \Psi^*(1',2',3',\ldots N) d3d4\ldots dN \tag{31}$$

i.e. that Γ^2 may be obtained from an antisymmetric wavefunction Ψ by means of (31). Such a Γ^2 is said to be *N-representable*[15] and hence (31) is a necessary and sufficient condition for the N-representability of Γ^2.

A simple mathematical test equivalent to (31) is not known and the search, called the *N-representability problem*,[15] is the subject of much current interest.[18]

Some obvious necessary criteria for the N-representability of Γ^2 include normalization, hermiticity, anti-symmetry and positivity. As defined in (31), Γ^2 is normalized:

$$\int \Gamma^2(1,2;1',2') d1d2 = \binom{N}{2} , \tag{32}$$

hermitian:

$$\Gamma^2(1,2;1',2') = [\Gamma^2(1',2';1,2)]^* , \tag{33}$$

and antisymmetric with respect to the interchange of electrons 1 and 2

$$\Gamma^2(1,2;1',2') = -\Gamma^2(2,1;1',2')$$

$$= \Gamma^2(2,1;2',1')$$

$$= -\Gamma^2(1,2;2',1') , \tag{34}$$

and positive semi-definite, i.e. all of the eigenvalues $\mu_i^{(2)}$ (occupation numbers) in the eigenfunction expansion

$$\Gamma^2(1,2;1',2') = \sum_i \mu_i^{(2)} \phi_i(1,2)\phi_i^*(1',2') \tag{35}$$

must be non-negative. The eigenfunctions $\{\phi_i(1,2)\}$ are usually called *N(atural) S(pin) G(eminals)* and the eigenvalues satisfy the normalization sum rule $\sum_i \mu_i^{(2)} = \binom{N}{2}$.

In a similar manner, a one-matrix $\Gamma^1(1;1')$ is said to be N-representable if and only if

$$\Gamma^1(1;1') = N\int\Psi(1,2,\ldots N)\Psi^*(1',2,\ldots,N)d2\ldots dN \tag{36}$$

i.e. it may be obtained from an anti-symmetric wavefunction Ψ. Similar necessary criteria for its N-representability include its normalization:

$$\int\Gamma^1(1;1')d1 = N, \tag{37}$$

hermiticity:

$$\Gamma^1(1;1') = [\Gamma^1(1';1)]^* \tag{38}$$

and positive semi-definiteness, i.e. all eigenvalues (occupation numbers) $\mu_i^{(1)}$ in the eigenfunction expansion

$$\Gamma^1(1;1') = \sum_i \mu_i^{(1)} \chi_i(1)\chi^*(1') \tag{39}$$

must be non-negative. In fact they are bounded between zero and one, $0 \leq \mu_i^{(1)} \leq 1$ and satisfy the normalization sum rule $\sum_i \mu_i^{(1)} =$ N. The eigenfunctions $\chi_i^{(1)}$ are usually called *N(atural) S(pin) O(rbitals)*.

5. THE INDEPENDENT PARTICLE MODEL

The potential-energy operator in the Hamiltonian for many-electron systems contains electron-repulsion operators of the type e^2/r_{ij}, which make it impossible to separate the coordinate variables of each electron. This means that a many-electron state-function cannot be exactly described as a product of linearly independent one-particle functions ϕ_i. However, one usually assumes as a first approximation the *Hartree product function*[19]

$$\Phi = \phi_i(1) \cdot \phi_2(2) \ldots \cdot \phi_N(N) . \tag{40}$$

Here $\phi_i(J)$ is a function of both spin s_j and space coordinates $\vec{r}_j$ of electron j and is called a *spin-orbital*. In order to account for the indistinguishability of electrons, and the antisymmetry requirement, a normalized antisymmetrized wavefunction based upon the Hartree product (40) may be expressed in the form of the *Slater determinant*,[20]

$$D = \frac{1}{\sqrt{N!}} \begin{vmatrix} \phi_1(1) & \phi_2(1) & \cdots & \phi_N(1) \\ \phi_1(2) & \phi_2(2) & \cdots & \phi_N(2) \\ \phi_1(N) & \phi_2(N) & \cdots & \phi_N(N) \end{vmatrix} \tag{42}$$

where the $\{\phi_i\}$ are an orthonormal set of spin orbitals. The Slater determinant is sometimes written in a more compact notation in terms of its diagonal, i.e.

$$D = \frac{1}{\sqrt{N!}} \det|\phi_i(1)\phi_2(2) \cdots \phi_N(N)| \tag{43}$$

In this I(ndependent) P(article) M(odel), there is a one-to-one correspondence between electrons and spin-orbitals. This leads to statements about core, lone-pair, etc. electrons when one is referring to the core, lone-pair, etc. orbitals.

a) H(artree)-F(ock) Theory

Use of the Slater determinant in the variation principle leads to the *Hartree-Fock equations*,[19-21] which are a set of integro-differential equations for the set of orthonormal spin-orbitals $\{\phi_i(J)\}$ which are optimum in the sense of the energetic criterion of the variation principle. They may be written in the form

$$\hat{F}(1)\phi_i(1) = \sum_{j=1}^{N} \lambda_{ij}\phi_i(1); \quad i = 1, \ldots, N, \tag{44}$$

where $\hat{F}(1)$, called the *Fock operator* (defined below) is a functional of the set $\{\phi_i(1)\}$. It is a one-electron operator which is the effective Hamiltonian operator for an electron (described by ϕ_i) in the attractive field of the nuclei and the average repulsive field of the other N-1 electrons.

The exact solutions of these equations will henceforth be called the HF spin-orbitals. As we shall see, the practical method of solution is the so-called S(elf)-C(onsistent)-F(ield) approach and we shall designate such solutions as SCF spin-orbitals.

We define the Fock one-electron Hamiltonian operator $\hat{F}(1)$ in terms of its matrix elements F_{ij} between spin orbitals ϕ_i and ϕ_j:

$$F_{ij} = \int \phi_i^*(1)\hat{F}(1)\phi_j(1)\,dr_1 \tag{45}$$

$$= H_{ij} + \sum_{k=1}^{N} \{ (1k|\hat{g}|jk) - (1k|\hat{g}|kj) \} \tag{46}$$

where the *one-electron integral* H_{ij} is given by

$$H_{ij} = \int \phi_i^*(1)\hat{h}(1)\phi_j(1)d\vec{r}_1, \tag{47}$$

the *two-electron integrals* are defined by

$$(pq|\hat{g}|rs) = \int \phi_p^*(1)\phi_q^*(2)\frac{1}{r_{12}}\phi_r(1)\phi_s(2)d\vec{r}_1 d\vec{r}_2, \tag{48}$$

and $\hat{h}(i)$ and $\hat{g}(i,j)$ are the one- and two- electron operators respectively in the electronic Hamiltonian operator.

The Slater determinant Φ is determined only to within a unitary transformation amongst the N- members of the set $\{\phi_i(1)\}$. Thus the spin orbitals themselves are not unique and one may obtain different sets related by unitary transformations of the set $\{\phi_i(1)\}$. Quite often, the unitary transformation is chosen such that the matrix of the elements F_{ij} is diagonal, i.e.

$$F_{ij} = \delta_{ij}\varepsilon_i \tag{49}$$

where ε_i are called the *orbital energies* of the HF spin-orbitals:

$$\varepsilon_i = H_{ii} + \sum_{k=1}^{N} (J_{ik} - K_{ik}) \tag{50}$$

where J and K are the two-electron *Coulomb* and *exchange* integrals respectively: $J_{ik} = (ik|\hat{g}|ik)$, $K_{ik} = (ik|\hat{g}|ki)$. The ϕ_i under the requirement (49) are usually called *canonical* HF spin-orbitals. The set of equations (44) may be written for the canonical HF spin-orbitals in the form

$$\hat{F}(1)\phi_i(1) = \varepsilon_i\phi_i(1); \; i = 1, \ldots, N, \tag{51}$$

which justifies the earlier description of $\hat{F}(1)$ as a one-electron effective Hamiltonian operator.

Often the unitary transformation is chosen to *localize* the spin-orbitals according to some criterion.[22,23] Such *localized orbitals* correspond more closely to familiar concepts such as bond and lone pairs than do the canonical orbitals. It should be emphasized that the *total* HF wavefunction and its expectation values (including the electron density) are unchanged by such transformations. Only the individual spin orbitals, -orbital densities and -orbital expectation values are changes.

The total energy, i.e. the expectation value of the Hamiltonian for the Slater determinant D built up from the N-HF orbitals is:

$$E^{HF} = \sum_{i=i}^{N} \varepsilon_i - 1/2 \sum_{i,j}^{N} (J_{ij} - K_{ij}) \tag{52}$$

The HF energy is *not* simply the sum of the orbital energies.

b) Fock-Dirac Density Matrix

The one-(particle density) matrix corresponding to a Slater determinant, $\Gamma_D^1 (1;1')$, is called the *Fock-Dirac density matrix* and may be written in terms of the N occupied spin-orbitals, as

$$\Gamma_D^1 (1;1') = \sum_{i=i}^{N} \phi_i(1)\phi_i^*(1') \tag{53}$$

Comparison with the general expansion (39) of a one-matrix shows that the occupied spin-orbitals have occupation numbers of unity while the unoccupied ones have zero occupation numbers.

It is *idempotent*, i.e. $\int \Gamma_D^1 (1;2) \Gamma_D^1 (2;1') d2 = \Gamma_D^1 (1;1')$, and is invariant under unitary transformations among the N occupied ϕ_i. Since it may be shown that

$$\Gamma_D^2 (1,2;1',2') = 1/2\{ \Gamma_D^1(1;1') \Gamma_D^1(2;2')$$

$$- \Gamma_D^1(1;2') \Gamma_D^1(2;1')\} \tag{54}$$

the energy and all one- and two-particle properties are functionals of Γ_D^1. As a result the Fock-Dirac density matrix, Γ_D^1, is called the *fundamental invariant of the HF scheme.*[12]

c) The Coulomb and Exchange Operators

We note that the second term in (46) gives rise to the so-called Coulomb contribution to the energy while the third is responsible for the exchange term. The Fock operator $\hat{F}$ may be written explicitly in terms of Γ_D^1, i.e. $\hat{F}(1) = h(1) + \hat{C}(1) + \hat{X}(1)$ where the *Coulomb operator*

$$\hat{C}(1) = \int \frac{\Gamma_D^1(1,2) d2}{r_{12}} \tag{55}$$

is a *local* operator and the *exchange* operator

$$\hat{X}(1) = - \int \frac{P_{12}\Gamma_D^1(2,2') d2}{r_{12}} \tag{56}$$

is a *non-local* one. When a local operator acts on a function $f(1)$, the operation is purely multiplicative, i.e. $\hat{C}(1) f(1)$. The non-local operator $\hat{X}$ operates on $f(1)$ such that

$$\hat{X}(1)f(1) = -\int \frac{\Gamma_D^1(1,2)f(2)}{r_{12}} \, d2 = \int x(1,2)f(2) \, dz. \tag{57}$$

$x(1,2)$ is called the kernel of the non-local operator $\hat{X}$. Because of the non-locality of the exchange operator, there has been great interest in replacing it by a local exchange operator for both computational and conceptual reasons.

Most local exchange potentials have as their starting point, the exchange operator in the case of the electron gas: $-4F(k/k_f)$ $(\frac{3}{8\pi} n)^{1/3}$ where n is the constant density of the electron gas and

$$F(k/k_f) = F(\eta) = 1/2 + \frac{1-\eta}{4\eta} \log \frac{1+\eta}{1-\eta} \tag{58}$$

In the original *Slater potential*,[24] the function $F(\eta)$ was replaced by its average over the occupied states, i.e. $F(\eta)_{avg} = 0.75$ and the constant electron density n by $\rho(\vec{r})$. Thus $V_s(\vec{r}) = -3(\frac{3}{8\pi}\rho(\vec{r}))^{1/3}$. Instead of replacing $F(\eta)$ by its average value, one may use $F(1) = 0.5$, the value of $F(\eta)$ at the Fermi surface. This leads to the *Gaspar-Kohn-Sham potential*[25,26] $V_{GKS}(\vec{r}) = -2(3/8\pi\rho(\vec{r}))^{1/3}$ which is just two-thirds of the Slater potential $V_S(\vec{r})$. Introduction of a parameter α defines the Xα exchange potential

$$V_{X\alpha}(\vec{r}) = -3(\frac{3}{8\pi}\rho(\vec{r}))^{1/3} \, . \tag{59}$$

Both the Slater and Gaspar-Kohn-Sham potentials are special cases corresponding to $\alpha = 1$ and $\alpha = 2/3$ respectively. Experience has shown that the use of values of α intermediate to one and two-thirds leads to better results than either of the original values. A number of criteria have been suggested[27-30] for the determination of α and tabulations are available[30] for atomic systems. Most calculations use the atomic α parameters determined by Schwarz[30] to make the Xα and HF total energies the same. The values of α so chosen are generally closer to two-thirds than to one and vary from system to system.

d) Spin Orbitals

So far the discussion of the HF method has involved a single determinant D filled with orthonormal spin-orbitals without speci- fying their form. However, for many systems in particular states it is not possible to restrict the model to a single Slater deter- minant. Several determinants may be degenerate which requires the formation of particular linear combinations of the determinants with the coefficients determined by symmetry. Quite often it is imposs- ible to diagonalize the matrix λ_{ij} in (44). Such cases are "open shell" states. Unless specified otherwise, we shall assume that the state under discussion is a closed-shell one.

In general a spin-orbital $\phi(1)$ may be written as $\phi(1) =$ $A\uparrow(\vec{r})\alpha(s) + B\downarrow(\vec{r})\beta(s)$ where $A\uparrow$ and $B\downarrow$ are spatial functions (orbitals). α and β are the orthonormal spin functions which depend on the spin coordinate s, with values s + 1/2 or -1/2 depending on the direction of spin, i.e. $\alpha(1/2) = 1$, $\alpha(-1/2) = 0$, $\beta(1/2) = 0$, and $\beta(-1/2) = 1$. With such general spin orbitals, D does not necessarily obey the symmetry of the problem. Often *restrictions* are introduced due to symmetry, intuition about the physical situation or because of computational convenience. Typical restrictions of a general spin orbital include $\phi(1) = A(\vec{r})\alpha(s)$ or $B(\vec{r})\beta(s)$, the usual restriction to a definite spin of the spin orbital and (in the atomic case) the separation of the radial and angular coordinates (central-field model), the constraint that $2p_+$, $2p_o$ and $2p_-$ orbitals, for example, all have the same radial part, and the usual restriction to double occupancy of the orbitals.

We shall continue to designate the spin-orbitals with no restrictions imposed as Hartree-Fock. Some authors would call them U(nrestricted) H(artree)-F(ock) -- although other schemes are also called UHF. When the spin orbitals are completely restricted, the method is called R(estricted) HF. The restrictions become especially important in the discussion of systems which are not of closed-shell type and for which in particular, an accurate calculation of the spin density is desired.

e) The Self-Consistent Field Method

Although the HF equations represent a definite simplification of the many-electron problem, their solution is not simple. This is due to the fact the Fock operator $\hat{F}(1)$ depends on the solutions of the equations themselves. The usual procedure, the method of the S(elf)-C(onsistent)-F(ield) is to start with a guess for the solutions, say $\{\phi_i^{(o)}\}$, and use them to construct an initial Fock-Dirac density matrix

$$\Gamma_D^{1(o)}(1,1') = \sum_{i=1}^{N} \phi_i^{(o)}(1)\phi_i^{(o)}(1')^* , \qquad (60)$$

and Fock operator $F^{(o)}$. With $F^{(o)}$ in hand, one solves $F^{(o)}\phi_i^{(1)} = \varepsilon_i^{(1)}\phi_i^{(1)}$ for a new set of spin orbitals $\{\phi_i^{(1)}\}$. The procedure is repeated, the jth step being $F^{(j)}\phi_i^{(j+1)} = \varepsilon_i^{(j+1)}\phi_i^{(j+1)}$, until the desired accuracy is obtained, e.g. $|\rho^{(j+1)} - \rho^{(j)}|$ is sufficiently small. The implementation of this method by the so-called analytical method is the subject of the next section. In the case of atomic systems, the SCF solution of the HF equations by numerical quadrature is now routine[19]. For molecules, this has been done only for a limited number of systems, primarily diatomic ones[31] and the analytical method is the most commonly used procedure.

f) Computational Aspects of the SCF Method

The most commonly used procedure (Roothaan[32]-Hall[33]) for obtaining SCF solutions of the Hartree-Fock problem assumes that the required one-electron orbitals are linear combinations of a set of basis functions which in themselves need have no direct physical interpretation.

In theory the use of a complete set of linearly independent continuous functions, should lead to the least (best) possible value of the energy obtainable in this approximation, which is called the *Hartree-Fock limit*. However, a judicious choice of the basis functions can lead to values close to the Hartree-Fock limit with a finite-sized basis set. Such basis functions are usually obtained by appropriate modifications of the analytic solutions of the one-electron atomic problem (hydrogen atom).

The latter can be obtained in spherical polar coordinates (r,θ,ϕ) and are classified according to their symmetries using a notation based on the natural quantum numbers which result from the classical degrees of freedom of the problem. Therefore, these solutions are designated by $n\ell_m$ (e.g. $1s_0$, $2p_{+1}$, $2p_0$, $2p_{-1}$, $3d_{+2}$, $3d_{+1}$, $3d_0$, etc.), where n is the *principal* quantum number and is restricted to positive integers, ℓ is the *azimuthal* quantum number and for any given n takes integer values from zero to $(n-\ell)$, and m is the *magnetic* quantum number which for a specified ℓ takes integer values from $-\ell$ to $+\ell$. The letters s, p, d, f, g, h, etc., designate $\ell = 0, 1, 2, 3, 4, 5$, etc., respectively. The degenerate sets of orbitals, for example the p type, are more commonly used in their *Cartesian* form, i.e. $2p_x$, $2p_y$, $2p_z$, which are simply appropriately directed linear combinations of the functions $2p_{+1}$, $2p_0$, $2p_{-1}$. The total electronic wavefunction for a state of the electron in the hydrogen atom is a product of a radial function R(r) which depends on n, an angular function $Y_{\ell m}(\theta,\phi)$ (spherical harmonic) which depends on ℓ and m, and a spin function w which depends on the spin quantum number s, with values $+1/2$ or $-1/2$ depending on the direction of spin, i.e.

$$\Psi(1) = R_n(r)Y_{\ell m}(\theta,\phi)w(s) , \tag{61}$$

where $w(1/2) = \alpha$, and $w(-1/2) = \beta$.

In two of the more commonly used families of basis functions, (S(later)-T(ype) and G(aussian)-T(ype) F(unctions) or O(rbitals)), the angular function is held fixed and the radial function varied. Slater-type radial functions are of the form $r^{n-1} e^{-\zeta r}$, where r is the radial distance from the centre (usually a nucleus), n is the principal quantum number and ζ(zeta) is an *orbital exponent para-meter*. Gaussian-type radial functions are of the form $r^{n-1} e^{-\alpha r^2}$, where α is another exponent parameter.

STF's have the advantage that they have the correct form for the one-electron central field problem and are therefore expected to give a better description of the electron density especially in the near-nuclear region where the potential is well approximated by a Coulomb field. Furthermore the electron-nuclear cusp condition

$$\lim_{r_\alpha \to 0} \left[\frac{\partial}{\partial r_\alpha} + Z_\alpha \right] \psi(r_\alpha) = 0 \tag{62}$$

is exactly satisfied by a 1s STF with $\zeta = Z_\alpha$ while a 1s GTF can never satisfy it. At the other extreme in distance wavefunctions should decay exponentially while GTF's decay more rapidly. For these reasons, the use of STF's is more economical than GTF's in the sense that fewer are required to obtain the same degree of accuracy. On the other hand, GTF's have computational advantages because integrations over these functions are much faster due to the fact that the integrals can be evaluated analytically.

Gaussian functions are most commonly used in the so-called Cartesian form where the angular dependence is expressed by powers of x, y, and z. They are of the general form:

$$x^L y^M z^N e^{-\alpha r^2} , \quad L + M + N = \ell \tag{63}$$

and specifically

$$e^{-\alpha r^2} \quad \text{(s-type)} \tag{64}$$

$$xe^{-\alpha r^2} , \quad ye^{-\alpha r^2} , \quad ze^{-\alpha r^2} \quad \text{(p-type)} \tag{65}$$

$$x^2 e^{-\alpha r^2} , \quad xye^{-\alpha r^2} , \quad xze^{-\alpha r^2} ,$$
$$y^2 e^{-\alpha r^2} , \quad yze^{-\alpha r^2} , \quad z^2 e^{-\alpha r^2} , \quad \text{(d-type)} \tag{66}$$

$$x^3 e^{-\alpha r^2} , \quad x^2 ye^{-\alpha r^2} , \quad xyze^{-\alpha r^2} ,$$
$$y^3 e^{-\alpha r^2} , \quad xy^2 e^{-\alpha r^2} , \quad y^2 ze^{-\alpha r^2} ,$$
$$z^3 e^{-\alpha r^2} , \quad xz^2 e^{-\alpha r^2} , \quad yz^2 e^{-\alpha r^2} , \quad x^2 ze^{-\alpha r^2} \quad \text{(f-type)} \tag{67}$$

etc.

It should be noted that there are six functions in the list (66) of d's and ten in the list (67) of f's. If one requires the correct basis functions (real) for the irreducible representations of the rotation group, the proper linear combinations must be formed. Thus for the d's, there are *five* linear combinations which correspond to the familiar d-orbitals (i.e. dyz, dxz, dxy, $d_{x^2-y^2}$ = $(x^2-y^2)e^{-\alpha r^2}$ and $d_{z^2} = (3z^2-r^2)e^{-\alpha r^2}$) and one combination $(x^2+y^2+z^2)e^{-\alpha r^2} = r^2e^{-\alpha r^2}$ which is of s-type (in fact 3s-type). Thus when a set of six Cartesian d-type functions are used, one has added in fact one d-orbital set and one 3s-orbital. If one does not wish the 3s-orbital contamination, then one must use only the proper five linear combinations.

With the angular functions fixed, a finite basis set of r STF's or GTF's is specified by the r exponent parameters (ζ for STF's, α for GTF's). Given such a basis set, one obtains the expansion coefficients of the required atomic or molecular functions which give the least energy for the specified exponent parameters. For atoms, these linear combinations of basis functions are called A(tomic) O(rbitals) and for molecules they are called M(olecular) O(rbitals). They are the SCF solutions to the HF problem for the given set of basis functions.

Usually the MO's are constructed as L(inear) C(ombinations) of A(tomic) O(rbitals) of the appropriate symmetry on each atomic site and are called LCAO-MO to reflect this fact.

It is possible to improve the given basis set by variation of the exponent parameters. The least energy thus obtained is called the *SCF limit* for that basis set and the basis set is said to be atom- (or molecule-) optimized. When limited basis sets are used, it is the usual practice to refer to a basis set with one Slater-type radial function for each occupied orbital in the constituent atoms as the M(inimal) Slater (or S(ingle)-Z(eta)) B(asis) S(et). Thus, a minimal basis for CH_4 would employ $1s_C$, $2s_C$, $2p_C$ (x,y,z) AO's on carbon and $1s_H$ AO's on each of the four hydrogens.

If two or more Slater-type functions are employed per occupied atomic orbital, we speak of D(ouble)- T(riple)- or higher Z(eta) basis sets. A DZ basis set for CH_4 would consist of $1s_C$, $1s_C'$, $2s_C$, $2s_C'$, $2p_C$(x,y,z) and $2p_C'$(x,y,z) STF's on the carbon centre and $1s_H$ and $1s_H'$ STF's on each of the four hydrogen sites. Such an E(xtended) B(asis) set allows the AO to expand or contract from the simpler SZ or MB description. Sometimes the extension of the basis must allow for higher-shell (higher n) AO's, i.e. more diffuse functions which can be critical in treating excited states or anions.

If each Slater function is expanded as a fixed linear combination of Gaussian functions, we say that C(ontracted) GTF's are used and the number of Gaussians per Slater of each symmetry type is specified. Uncontracted Gaussian basis sets use Gaussian functions of appropriate symmetry type, and no Slater functions need to be considered, e.g. a (5,2) Gaussian basis set for carbon means

5 s-type Gaussian and 2 p-type (x,y,z) Gaussian functions are used
to describe the carbon atom, i.e. a total of 5 + 2 x 3 = 11.

When the basis set is augmented with orbitals of higher azimuthal quantum numbers than the ones occupied in the atomic ground
states, these orbitals are called *polarization functions*, because
they can reflect the polarization of the atom in the molecular
situation, e.g. place charge in the bonding region. Since only s-
and p-type orbitals are occupied in carbon or silicon, the d-type
and higher functions are polarization functions. For hydrogen and
chromium p- and f-type functions respectively would be polarization
functions.

If the basis set uses too few s and p functions, the d-functions
may simply be repairing the defects in the basis set rather than contributing to a physical phenomenon. In order to realy distinguish
between these two phenomena, one should study the addition of d
functions at the sp limit. The *sp limit* is the least energy that
may be obtained by the use of s and p functions alone. *spd* and *spdf*
etc. *limits* can be similarly defined.

The inclusion of functions of higher azimuthal quantum numbers
can be a qualitative necessity due to molecular symmetry. If an
occupied MO transforms according to an irreducible representation
of the point group of the molecule for which s and p functions on
first- and second-row atoms (s functions on hydrogen) do not provide
a basis, then higher functions must be included in the basis. As
an example, consider SF_6 where the occupied orbitals belong to a_{1g}, t_{1g}
t_{2g}, e_g, t_{1u}, and t_{2u} irreducible representations. Of these, s and p
functions do not provide a basis for the orbitals of e_g and t_{2g} symmetries. A series of calculations[34] for SF_6, with and without sulfur
d functions revealed that the e_g level is the HOMO if the d orbital
component is omitted, and the fourth highest MO if it is included.
The t_{2g} level is rather constant in its place in the energy order.
The e_g and t_{2g} levels, which belong to the same representations as
the added d basis, were stabilized by the addition of the d component, while all the others were raised in energy.

An alternative to adding polarization functions of higher azimuthal quantum numbers to a basis set is to use s- and p-type functions at off-nuclear points in the bonding regions such as those
along the bond axis.[35,36] Such *bond functions* can be more economical
than polarization functions. Their relative position along the bond
axis will reflect to some degree the electronegativity difference
between the two participating atoms.

In the selection and construction of basis sets one desires a
basis set which is sufficiently *flexible* to describe adequately the
polarization and transfer of charge which accompany molecular formation, is *economical* in size as the computational effort involved in

an SCF calculation is proportional to the fourth power of the number of basis functions employed and is *reliable* in that one can expect similar results when used in similar or related systems, and is *balanced*.

Mulliken[37] defined a *physically balanced* basis set as one which yields "reasonable" calculated values for well-defined experimental quantities and a *formally balanced* basis as one which gives "reasonable" atomic charges. The definitions are arbitrary in the sense that "reasonable" is undefined, atomic charge is an ill-defined concept, and the choice of the experimental quantities is a matter of taste and interest.

The design of a balanced basis set is important as the variation principle used as the only criterion, will place primary emphasis on the regions closer to the nuclei because of the electron-nuclear attraction term in the Hamiltonian operator. This will leave insufficient flexibility in the valence region so critical to the discussion of bonding. One must consider the problem of how to divide a basis of given size amongst the various atomic sites and among the various symmetry types for each atomic site, in order to describe each orbital with the same degree of accuracy. As an example of the latter, in order to describe the ground state of the carbon atom with two electrons in each of the 1s, 2s and 2p orbitals, using sixteen functions, one could use 15 s-type functions and 1 p-type function or 2 s-type function and 14 p-type functions or any intermediate composition. To determine the preferred composition the minimum in total energy fails to remain a good criterion because the 1s (or core) orbitals have considerably lower energy than the 2s or 2p orbitals. One way to decide upon a balanced basis set is to use the criterion that the energy of each orbital should be the same percentage of the energy at the SCF limit. The former balancing is of even more importance in those molecules where electronegativity differences of the atoms should be taken into account. This balancing should be attained initially on the atomic level and then the molecular level.

As mentioned above, one can use exponent parameters which are atom or molecule optimized. Molecular optimization becomes prohibitive when one goes beyond the MB level. On the atomic level one can variationally optimize each exponent. Another approach is to choose the M exponents in a geometric progression $\alpha_k = \sigma \beta^k$ where σ and β are variationally determined. Such a basis is called *even-tempered*.[38]

For GTF's the exponents may be chosen so that the GTF set fits an STO.[39] In that case the basis is called an *STO-MG* set, i.e. one STO is fit by M GTF's. Clearly the weighting function employed in making the fit will make the set better for some uses than for others.

Since polarization and bond functions do not occur by definition in the normal RHF model for the atom, they must be determined for the molecule in some way. Their addition must be done in a *balanced* manner as well. This is because electron density will migrate to the site where the polarization function is added. Thus one must add similar polarization functions at each site.

For electron density studies balanced basis sets are of utmost importance, because a good approximate wavefunction for such purposes would be one where the density in all interesting regions is approximated equally well. Examples are well-known in which wavefunctions of lower or comparable quality give quite inadequate descriptions of the electron density.[13]

Although MB set calculations do give a correct qualitative variation of electron density with distance and in general predict the bonding in molecules, they are inadequate for quantitative description of the density in the important regions. Comparison of the electron density at the HF limit shows that MB sets give too little charge in the bonding region, is inadequate in the nuclear region and overestimates the density in the non-bonding or "lone-pair" regions.

Such errors can become very serious on the scale of the deformation density $\Delta\rho_p$

$$\Delta\rho_p = \rho_{molecule} - \rho_{promolecule} \tag{68}$$

where $\rho_{promolecule}$ is the density of the spherically-averaged free atoms superimposed at the molecular geometry. However, if $\Delta\rho_p$ is calculated using the same basis sets for both the atoms and the molecules, those errors due to atomic features should be eliminated to a large extent.

As an example, consider Figure 1, where $\Delta\rho_p$ along the molecular axis of CO is shown for eight different basis sets identified in Table 3 together with the energies and dipole moments of the SCF wavefunctions.[40] One observes that, in contrast to the larger bases, the SZ deformation density shows an absence of charge in most of the bonding region. For the extended bases (DZ or better quality) the order of decreasing molecular energy is the same as the order of increasing deformation density in the bonding region and decreasing deformation density in the outer regions. Comparison of the DZ (b) and TZ (c) calculations with the others indicates the importance of polarization functions in the description of the deformation density, especially in the bonding region.

Thus it is clear that, although SZ and DZ basis sets are the most commonly used for quantum chemical studies involving the energy, electron density studies require basis sets of at least DZ + DP

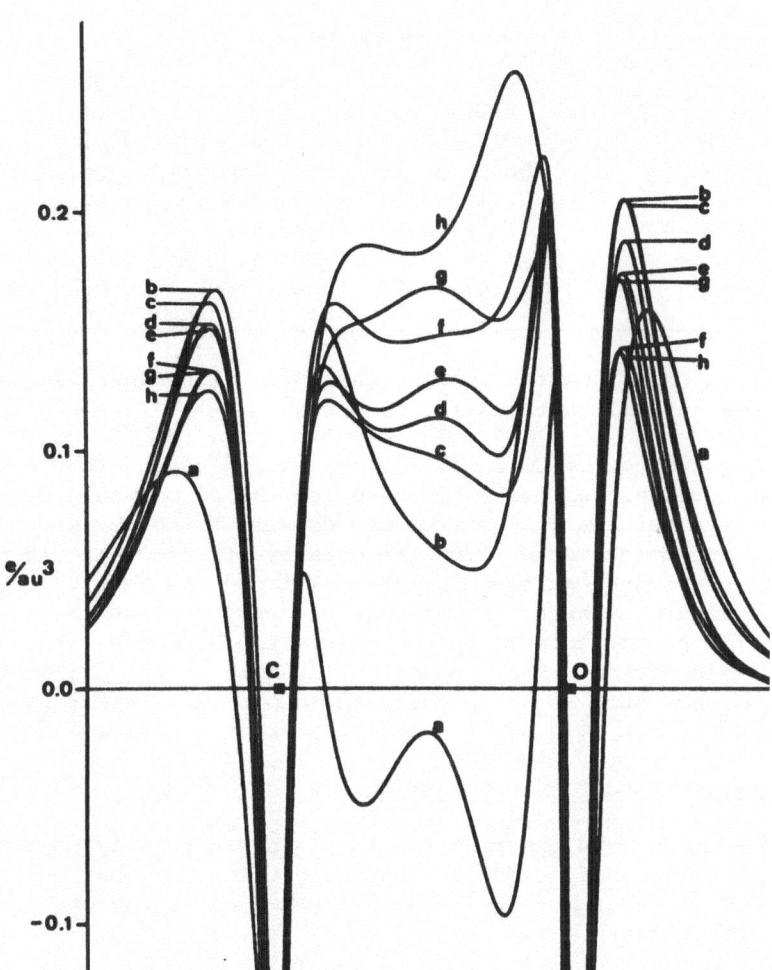

Figure 1: Density difference between the CO molecule and the C and O atoms, as calculated with 8 different basis sets.[40] The bases are specified in Table 3.

Table 3. SCF wavefunctions for CO. The notation n LP
         indicates n L-type P(olarization) functions
         on each centre

Basis No.	Basis Type	$-E$(au)	$\mu_{C^-O^+}$(D)
a	SZ	20.975	0.96
b	DZ	21.348	0.04
c	TZ	21.313	0.01
d	DZ + DP	21.398	0.26
e	TZ + DP	21.423	0.25
f	TZ + 2DP	21.431	0.28
g	TZ + DP + FP	21.459	0.17
h	TZ + 2DP + 2FP	21.469	0.23

quality. As seen from the figure, there is "reasonable" agreement
among those sets of that quality or better.

In the case of Cr(CO)$_6$, the same authors[40] found that a DZ+P
basis was necessary as before to describe the promolecule deformation density, but that the difference density formed by subtracting
atomic Cr and molecular CO from the density of Cr(CO)$_6$ could be
adequately described at the DZ level. This can be seen by examination of Figure 2 where the deformation density formed by subtracting the promolecule is inadequate in the CO region at the DZ
level of calculation. This is another example of the difference
between systems such as TiCl$_4$ where the d orbitals appear as bonding
orbitals and those such as CO where they are polarizing functions.

g) Band or One-Electron Theory of Solids

The one-electron theory of solids is based usually on the IPM
model with the use of a local exchange potential so that the total
potential $V(\vec{r}) = V_{coul} + V_X$ in the effective one-electron
Schrödinger equation

$$\{-\frac{1}{2} \nabla^2 + V(\vec{r})\}\phi(\vec{r}) = \epsilon\phi(\vec{r}) \tag{68}$$

is a local one. With the assumption that $V(\vec{r})$ obeys the symmetry of
the crystal lattice, the single particle solutions $\phi(\vec{k},r)$ satisfy
the Bloch conditions:

$$\phi(\vec{k},\vec{r}+\vec{T}) = \exp(i\vec{k}\cdot\vec{T})\phi(\vec{k},\vec{r}) \tag{69a}$$

where $\vec{T}$ is any primitive translation of the crystal lattice, and
can be written as a product of a plane wave and a periodic function
i.e.

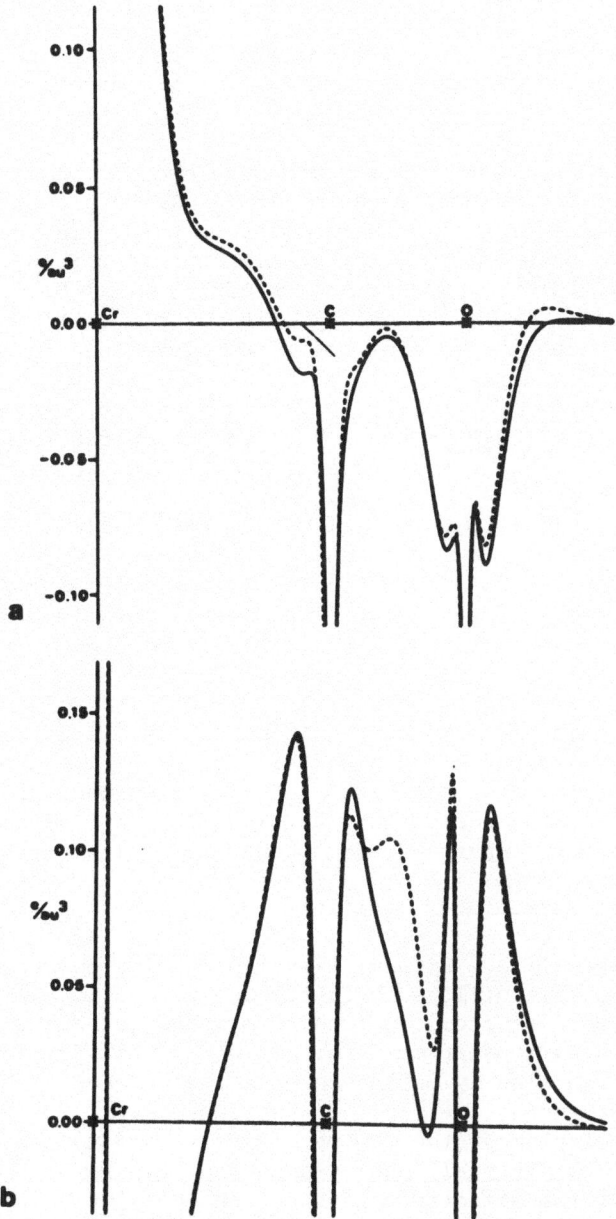

Figure 2: Density difference along a Cr CO axis between Cr(CO)$_6$ and
(a) the Cr atom and the CO molecule;
(b) the promolecule (sum of the spherically
symmetric Cr, C and O atoms.)
Solid lines: DZ basis.
Dashed lines: DZ basis + 3d function on all nuclei.

$$\phi(\vec{k},\vec{r}) = \exp(i\vec{k}\cdot\vec{r})U(\vec{k},\vec{r}) \ . \tag{69b}$$

We observe that these Bloch functions are delocalized in the sense that they exhibit equal probability of being in any of the unit cells.

A number of calculational schemes have been developed to solve the one-electron Schrödinger equation (68). These differ principally in the choice of basis functions to be used. The principal types which have been employed either alone or in combinations are: P(lane) W(aves), A(tomic) O(rbitals), M(uffin)-T(in) O(rbitals) and P(seudo)-O(rbitals). The choice of basis functions and method is usually dependent on the type of solid being examined.

PW's, $\Psi(\vec{k},\vec{r}) = V^{-1/2}\exp(i\vec{k}\cdot\vec{r})$, form a complete, orthogonal set of functions. Thus the PW expansion of the Bloch function is
$$\phi(\vec{k},\vec{r}) = \sum_{\vec{K}}\Psi(\vec{k}+\vec{K},\vec{r})\,A(\vec{k}+\vec{K}) \text{ or equivalently in the form of (69b)}$$
$$\phi(\vec{k},\vec{r}) = e^{i\vec{k}\cdot\vec{r}}U(\vec{k},\vec{r}) \text{ with } U(\vec{k},\vec{r}) = V^{-1/2}\sum_{\vec{K}}e^{i\vec{K}\cdot\vec{r}}A(\vec{k}+\vec{K}).$$

This PW (or weak-binding) expansion is quite slowly convergent due to the presence of the atomic cores and other localized states in the crystal. As a result, PW's are primarily used together with other basis functions.

At the opposite extreme to the PW method (which is best for weakly bound electrons) is the tight-binding or LCAO method (discussed earlier for molecules) based on localized functions (AO's). Since atomic orbitals on different lattice sites are not orthogonal in general, one must construct O(rthogonalized) A(tomic) O(rbitals) and then form Bloch sums for the basis functions in the solution of the one-electron Schrödinger equation.

The O(rthogonalized) P(lane) W(ave) method[41], combines the PW and LCAO methods. It utilizes PW's orthogonalized to the Bloch functions for the core atomic orbitals to solve the Schrödinger equation by a variational procedure. It works best for those sys= tems where one can clearly define core and valence electrons.

One way to implement the OPW method is to add to the crystal potential $V(\vec{r})$ a (non-local, energy-dependent) operator V_{pp} so that the solutions of the effective one-electron equation using $V(\vec{r})$ + $V_{pp}(\vec{r})$ are automatically orthogonal to the core states. One has converted that part of the OPW wavefunction responsible for the orthogonalization to the core into a P(seudo)-P(otential) V_{pp}. This potential is repulsive and cancels to a large extent the attractive core potential leaving a net weak potential.

The pseudo-wavefunctions are relatively smooth in the core regions compared to the oscillatory nature of the original wavefunctions there. This means that we must not use the PP wavefunctions in the core region but rather the correct ones in a study of the electron density.

The PP method is usually used as an empirical method where the PP is parameterized from atomic or experimental data.[42] There are two main variants of the empirical PP method, one local and one-nonlocal. As we have observed above, the PP, by its very definition, should be energy-dependent and non-local. However, a local approximation is simpler to apply and has been used extensively. Nevertheless, the evidence[43,44] is quite convincing that a non-local PP must be used in order to obtain the correct charge density in the bonding region.

One can attempt to solve the effective one-electron equation for the periodic function $U(\vec{k},\vec{r})$ in the representation (69b) of the Bloch function in one unit cell. This involves matching the function and its derivative on the boundary surfaces of the unit cells.[45] Slater[46] simplified the problem by introducing the M(uffin)-T(in) potential model. In the MT model, non-overlapping spheres are centered at the atomic sites and the MT potential is spherically symmetric within the spheres and constant within the intersphere regions.

The Coulombic and exchange contributions to the starting MT potential are generated from the electron density which is the superposition of the free atom densities at the respective atomic sites, i.e. just the procrystal. In this and the other methods, one should proceed to solve the problem in a self-consistent manner, i.e. generating new densities and MT potentials at each step of the cycle.

In Slater's A(ugmented) P(lane) W(ave) method,[46] PW's are used as basis functions in the constant potential region of the MT potential. Within the atomic spheres, the basis functions are expanded in spherical harmonics, $Y_{\ell m}$, i.e.

$$\phi(\vec{k},\vec{r}) = \sum_{\ell=o}^{\infty} \sum_{m=-\ell}^{\ell} C_{\ell m} R_{\ell}(r,\varepsilon) Y_{\ell m}(\theta,\phi) \qquad (70)$$

with the coefficients $C_{\ell m}$ determined by the condition that the expansions for ϕ inside and outside the sphere should match on the surface of the sphere. The radial functions $R_{\ell}(r,\varepsilon)$ are the numerical solutions, regular at the sphere centre, of the radial Schrödinger equation

$$\left[-\frac{1}{2} \frac{d^2}{dr^2} - \frac{1}{r} \frac{d}{dr} + \frac{\ell(\ell+1)}{2r^2} + V(r) - \varepsilon \right] R_{\ell}(r;\varepsilon) = 0. \qquad (71)$$

Since the radial functions $R_\ell(r,\varepsilon)$ are energy dependent, their use in the basis functions ϕ (70) leads to a non-linear eigenvalue problem with roots determined by calculation of the secular determinant for many values of the energy.

It should be noted that the APW's are continuous by their construction but their slopes may be discontinuous at the sphere boundaries. Methods related to the APW scheme have been developed to eliminate this feature.[47]

A method developed in recent years is the L(inear) C(ombination) of M(uffin) T(in) O(rbitals) which uses so-called MTO's as atomic-like basis functions.[48] An MTO for a given sphere is a solution of the radial equation (71) in the sphere, a tail function outside, and an additional term inside designed to cancel the contribution of the tail functions from all the other spheres inside the given sphere.

h) Xα-Methods for Molecules and Clusters

The one-electron effective Schrödinger equations (HF equations) can be applied to molecular problems using the local exchange operator $V_{X\alpha}$ instead of the non-local exchange operator. This approach has been followed with the solution obtained in one of two different ways. In the first of these, the problem is treated by the LCAO-MO method, i.e. a basis of atom centred functions is used. In the LCAO-Xα method[49] a basis of GTO's is employed. In the Xα-D(iscrete) V(ariational) M(ethod),[50] a basis of STO's or numerical functions is employed.

The other approach to the solution of the effective one-electron equations is the S(cattered)-W(ave) or M(ultiple)-S(cattering) Method.[51] In this method, space is first divided into spherical regions around the nuclei (type I regions), the region outside a large sphere surrounding the entire molecule (region III) and the rest of space (the intersphere region II). This partitioning of space is illustrated in Figure 3 for the molecule SF_6. It should be recognized that this is just the MT partitioning for a finite system. A MT potential is constructed, spherically averaged in regions I and III and volume averaged in region II (a constant). In regions I and III the wavefunction is expanded in rapidly convergent series of spherical harmonics and radial wavefunctions which are determined by numerical integration of the radial Schrödinger equation. In region II, a multi centre expansion of the wavefunction is employed. The requirement of continuity of both the wavefunction and its first derivative across the boundaries of the system leads to a set of energy dependent secular equations which are amenable to rapid solution. It is important to appreciate the difference between the basis set in this approach and that in the LCAO method. In the latter, a fixed set of functions is used to construct all of

the MO's whereas in the SW approach the radial functions are determined independently for each level and are numerical.

The SCF-Xα-SW and SCF-Xα-DVM methods have both been used to
calculate electron densities and difference densities.[40,52,53]

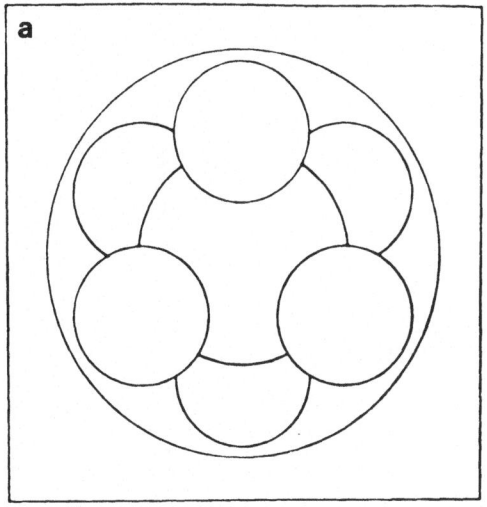

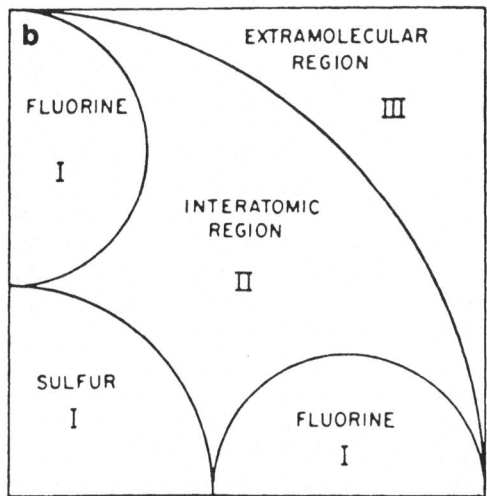

Figure 3: (a) A projected view of the various atomic and the outer
           spheres used for the partitioning the the SF_6 (O_h symmetry)
           molecular space.   (b) A quadrant of the projection of the
           spheres of SF_6 on the xy plane.

6. BEYOND HARTREE-FOCK - THE ROLE OF ELECTRON CORRELATION

The HF model is a relatively accurate description of the electronic structure but it is not the exact solution of the Schrödinger
equation. The error inherent in the HF method is called the
correlation error. For example, the difference between the energy
obtained from the exact solution of the Schrödinger equation and the
HF energy is called the *correlation energy*. Since the HF model is
the optimal IPM from the point of view of the energy, much of the
research devoted to methods are going beyond HF, has focussed on the
correlation energy (see Figure 4).

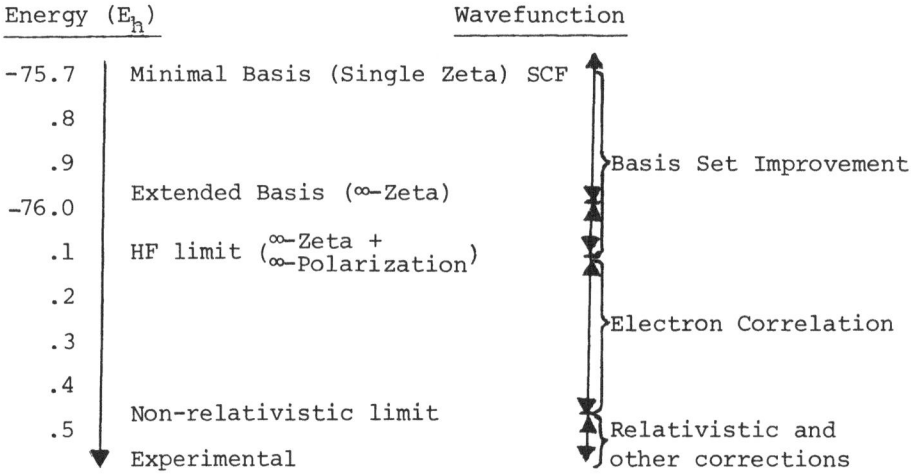

Figure 4: Energetics of the water molecule in ab initio calculations.

It is often said that the HF model offers the same degree of
accuracy for the energy and for the electron density as well as the
expectation values of one-particle operators. We wish to examine
this statement from the point-of-view of electron density studies
and to try to assess how important the role of electron correlation
is in the calculation of accurate electron densities. Our discussion
will be restricted to closed-shell systems.

We expand the exact wavefunction Ψ in terms of a set of normalized Slater determinants over a *complete* set of orthonormalized
spin orbitals $\{\phi_i\}$ whose first N members $\phi_1,\ldots,\phi_N$ are the spin
orbitals occupied in the HF Slater determinant D_{HF},

$$\Psi = a \{D_{HF} + C_s D_s + C_d D_d + C_t D_t + \ldots + C_n D_n\} \tag{72}$$

where $C_s D_s$, $C_d D_d$, $C_t D_t$, $\ldots$, $C_n D_n$ represent respectively, the sums

over the singly, doubly, triply, ..., n-tuptly substituted deter-
minants D_i^a, D_{ij}^{ab}, D_{ijk}^{abc}, D_{ijkl}^{abcd}, etc., with respect to D , and their
respective expansion coefficients C_s, C_d , etc., and a is a norma-
lization factor. The notation D_i^a means simply that spin orbital
ϕ_i occupied in D_{HF} has been replaced by (a virtual) spin orbital
ϕ_a (one not occupied in D_{HF}). The notation for the double, triple,
etc. substitutions has a similar meaning. Such an expansion is
called a S(uperposition) O(f) C(onfigurations) or a C(onfiguration)
I(nteraction) expansion.

The exact expansion (72) is called *complete CI*. When the set
of spin orbitals $\{\phi_i\}$ consists of only a finite number, M > N, the
expansion involves only a finite number of determinants $\binom{M}{N}$. When
all $\binom{M}{N}$ determinants are included, the expansion is called *full CI*
and is the best one can accomplish with the given finite basis of M
spin orbitals. Often the expansion is even more limited. Use of
only D_{HF} and all the singly-substituted determinants $C_s D_s$ is S(ingly)
substituted CI. $D_{HF} + C_s D_s + C_d D_d$, is S(ingly) and D(oubly) substituted CI while $D_{HF} + C_d D_d$ is D(oubly) substituted CI.

The electron density corresponding to the expansion (72) is:

$$\rho(\vec{r}) = (1 + |c_s|^2 + |c_d|^2 + \ldots)^{-1}$$

$$\{\rho_{oo} + 2R_e(C_s\rho_{os}) + |c_s|^2\rho_{ss} + |c_d|^2\rho_{dd} + \ldots\} \quad (73)$$

where ρ_{os}, ρ_{ss}, ρ_{dd}, etc. are matrix elements of the electron density operator with respect to the various determinants in the ex-
pansion of the wavefunction. The energy takes the form

$$E = (1 + |c_s|^2 + |c_d|^2 + \ldots)^{-1}(E_{HF} + 2Re(C_s H_{os}) +$$

$$2Re(C_d H_{od}) + |c_s|^2 H_{ss} + |c_d|^2 H_{dd} + \ldots) \quad (74)$$

where H_{os}, H_{ss}, H_{dd}, etc. are matrix elements of the Hamiltonian
operator H with respect to the same determinants.

Since Brillouin's theorem[54] states that $\langle D_{HF}|H|D_s\rangle = H_{os}$
vanishes, it follows from perturbation theory that only the determinants D_d, doubly substituted with respect to D_{HF}, contribute in
first order to the complete CI expansion (72) of the exact wavefunction. The single, triple and quadruple substitutions appear
in second order through their coupling with the doubles. In terms
of a perturbation theory parameter λ, this means that $C_D = O(\lambda)$,
$C_S = O(\lambda^2)$, $C_T = O(\lambda^2)$, and $C_Q = O(\lambda^2)$. Inspection of (74) reveals
that the energy is correct to second order, i.e. $\frac{\Delta E}{E_{HF}} \sim O(\lambda^2)$. Similarly, it follows from (73) that the electron
density is correct as well to second order (Møller-Plesset theorem[55])
i.e. $\frac{\Delta\rho}{\rho_{HF}} \sim O(\lambda^2)$. This is the basis of the statement made earlier

that the relative errors in the HF energy, electron density, and
one-electron expectation values should be similar.

However, there remains the important question of the magnitude
of the second-order correction, to which both D_s and D_d contribute.
This is in contrast to the energy, where only the latter contributes
in the second and third orders. Consequently, for a long time the
singly substituted terms were neglected because they were not very
important from an energetic point of view. Difficulties encountered
in the prediction by CI of the dipole moments of diatomic molecules
led to the discovery that the singly substituted terms are actually
more important than the doubles in the second-order correction to
the density[56]. This should not be construed to mean that the doubles
should be omitted in correlated studies of the electron density.
The coupling of the singles with D_{HF} through the doubles is quite
important. Hence, one must include the singles and the doubles
simultaneously, i.e. perform for a given basis, SDCI.

Consider the second-order contributions to $\rho(\vec{r})$ in the expansion (73) namely

$$\rho(\vec{r})^{(2)} = 2R_e(C_s\rho_{os}) + |C_d|^2(\rho_{dd} - \rho_{oo}). \qquad (75)$$

A typical term $\rho_{dd}(\vec{r})$ may be written in the form $(\rho_{oo} + \rho_{aa} + \rho_{bb}$
$- \rho_{ii} - \rho_{jj})$ where ρ_{aa}, ρ_{bb}, ρ_{ii} and ρ_{jj} denote orbital densities
corresponding to the virtual orbitals ϕ_a and ϕ_b and the occupied
orbitals ϕ_i and ϕ_j. By elimination of ρ_{oo}, (75) becomes

$$\rho(\vec{r})^{(2)} = 2R_e(C_s\rho_{os}) + |C_d|^2(\rho_{aa} + \rho_{bb} + \rho_{ii} - \rho_{jj}) \qquad (76)$$

Since the most important double substitutions are expected to
involve orbitals localized in the same region as the orbitals for
which they are substituted, the factor $(\rho_{aa} + \rho_{bb} - \rho_{ii} - \rho_{jj})$
should be comparatively small and the first term in (76) should
dominate[57,43]. In the case of the dipole moment operator, the
evidence is that one single substitution is the most important.[57]
For LiH, HF, HCl, and ClF it is that of the L(owest) U(noccupied)
sigma M(olecular) O(rbital) for the H(ighest) O(ccupied) sigma
M(olecular) Orbital. For CO and CS, it is the corresponding LUπMO
substitution for the HOπMO.

How large are these correlation corrections to ρ and do we need
to worry about their effect on $\Delta\rho_p$ in the chemically interesting
regions of the molecule? We can answer this question by examining
recent calculations of the SDCI type. Since they use virtual orbitals generated from the same basis as the SCF determinant, Brillouin's
theorem and the above analysis apply. This means that for a given
basis set, we should be able to determine ρ rather accurately. The
problem of choosing a good basis set for a given system would remain.

In diatomic systems,[43,13] the qualitative effect of electron correlation is to cause charge to accumulate around and behind the nuclei and to be removed from the central region. The net effect for all space would be zero because both the HF and correlated densities are normalized to the same number of electrons, i.e. $\Delta\rho_{corr} = \rho_{SDCI} - \rho_{SCF}$ satisfies $\int \Delta\rho_{corr}d\vec{r} = 0$.

2D contour maps of $\Delta\rho_C$ and the relative density difference function $\Delta\rho_C/\rho_{CI}$ are shown in Figure 5 from our recent calculation[13,43,58] in the case of the water molecule where SDCI and SCF wavefunctions constructed from the same basis set were used. Examination of $\Delta\rho_C$ shows that there is a piling up of charge in regions at and behind each of the nuclei, charge removal from the OH bond regions, and a small depression in the lone-pair region behind the oxygen. These results are consistent with the dipole-moment reduction upon correlation mentioned above. We have calculated $\Delta\rho_C$ and $\Delta\rho_C/\rho_{CI}$ for a number of other molecules $(H_2O)_2$, CH_4, C_2H_2, C_2H_4 wavefunctions of the SDCI type.[58] Becker et al[59] have recently calculated $\Delta\rho_C$ for the CO and N_2 molecules while Lipscomb et al[60] have studies H_2O, H_2S, and BH, with conclusions similar to ours.

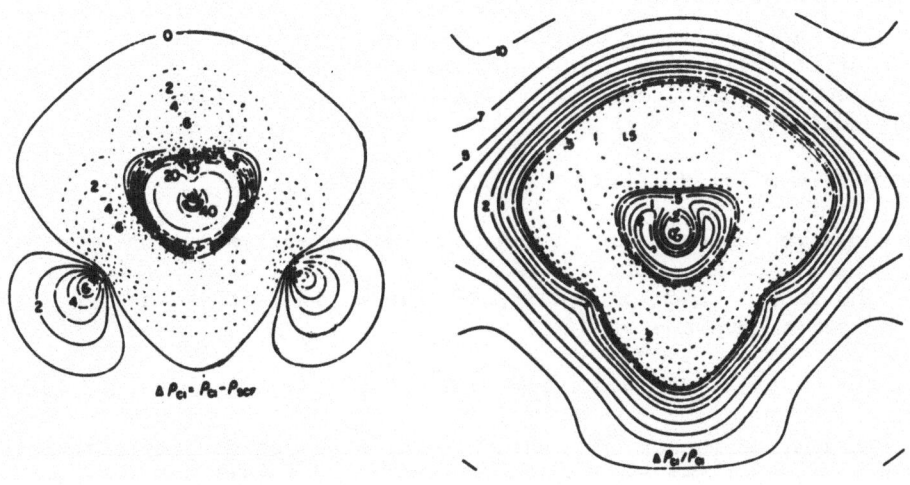

Figure 5: 2D contourmaps of $\Delta\rho_{CI}$ and $\Delta\rho_{CI}/\rho_{CI}$ for H_2O in the molecular plane[13,43,58]. Positive and negative contours are drawn with solid and dashed lines, respectively. The units are: $\Delta\rho_{CI}$ (10^{-3} e/a.u.^{3}) and $\Delta\rho_{CI}/\rho_{CI}$ (%).

Two questions which are often raised are whether one can correct basis-set inadequacy by means of CI or if one could add a basis-set "Error" correction calculated at the SCF level to the CI result. It has been shown[57] that if a small change in the basis set leads to a change in the expectation value of a one-particle operator $<F_1>$ at the SCF level, then it will lead to essentially the same change in $<F_1>$ in a CI calculation constructed from the same basis. This means that the CI calculation cannot correct the error due to basis-set quality in the SCF calculation of $<F_1>$.

It should be pointed out that some of the features of the $\Delta\rho_c$ contour maps reflect the effects of electron correlation at the atomic level[61]. It would be useful, therefore, to calculate second difference densities $\Delta\Delta\rho_c = \Delta\rho_m^{CI} - \Delta\rho_m^{SCF} = (\rho_{molecule}^{CI} - \rho_{SFA}^{CI}) - (\rho_{molecule}^{SCF} - \rho_{SFA}^{SCF}) = \Delta\rho_{c,molecule} - \Delta\rho_{c,SFA}$ where it is presumed that the same atomic basis sets were used in the construction of all of the densities involved.

Electron correlation can play an important role at the atomic level in systems where there are close-lying virtual levels. Examples include beryllium where the 2s-2p near-degeneracy effect is important and transition metals. In Figure 6 HF and CI form factors for the (^4F) ground state of atomic vanadium are compared.[62]

7. ONE- AND TWO-ELECTRON DENSITY FUNCTIONS

The familiar one-electron density $\rho(\vec{r})$ is obtained by the spin-trace of $\Gamma^1(1,1')$, i.e. $\rho(\vec{r}) = \rho(\vec{r},\vec{r})$ where $\rho(\vec{r},\vec{r}') = \int \Gamma^1(1,1')ds$ is called the charge density matrix. This is equivalent to the usual definition

$$\rho(\vec{r}) = N\int \Psi(1,2,\ldots,N)\psi^*(1,2,\ldots,N)ds_1 d_2 \ldots dN \qquad (77)$$

and shows the compactness of the density operator notation.

The radial electron density $D(r)$ is defined in terms of $\rho_o(r)$, the spherical average of $\rho(\vec{r})$:

$$D(r) = 4\pi r^2 \rho_o(r) = \bar{\rho}_o(r) = (4\pi)^{-1}\int_o^{2\pi}\int_o^{\pi}\rho(\vec{r})d\Omega_r \qquad (78)$$

In similar manner, we introduce two-electron density functions which are useful for the interpretation of $\Gamma^2(1,2;1',2')$ and expectation values of two-particle operators, including total x-ray scattering.

Analogous to $\rho(\vec{r},\vec{r}')$ we define the spin-traced two-matrix by

$$\Gamma(\vec{r}_1,\vec{r}_2;\vec{r}_1',\vec{r}_2') = \int\Gamma^2(1,2;1',2')ds_1 ds_2 \qquad (79)$$

and transform it to *extracular* coordinates (electron-electron

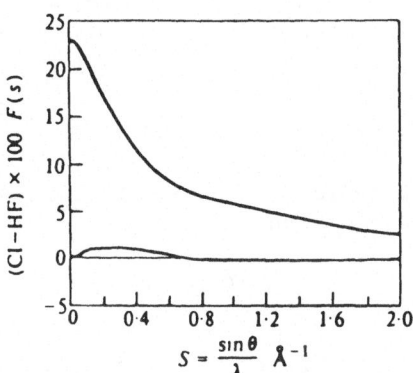

Figure 6: Comparison[62] of HF and CI form factors F(s) for atomic vanadium (^{4}F).

centroid $\vec{R}_o, \vec{R}_o{}'$ and *intracular* coordinates (relative electron-electron) $\vec{r}_{12}, \vec{r}_{12}{}'$

$$\Gamma(\vec{r}_1, \vec{r}_2; \vec{r}_1{}', \vec{r}_2{}') = \overline{\Gamma}(\vec{R}_o, \vec{r}_{12}; \vec{r}_o{}', \vec{r}_{12}{}') \tag{80}$$

where $\vec{R}_o = (\vec{r}_1 + \vec{r}_2)/2$; $\vec{r}_{12} = \vec{r}_1 - \vec{r}_2$ and similarly for the primed coordinates.

Then the intracule matrix $J(\vec{r}_{12}; \vec{r}_{12}{}')$ and the extracule matrix $E(\vec{R}_o; \vec{R}_o{}')$ are obtained by integrating out the extracular coordinates:

$$J(\vec{r}_{12}; \vec{r}_{12}{}') = \int \overline{\Gamma}(\vec{R}_o, \vec{r}_{12}; \vec{R}_o, \vec{r}_{12}{}') d\vec{R}_o \tag{81}$$

and intracular coordinates:

$$E(\vec{R}_o, \vec{R}_o{}') = \int \overline{\Gamma}(\vec{R}_o, \vec{r}_{12}; \vec{R}_o{}', \vec{r}_{12}) d\vec{r}_{12} \tag{82}$$

respectively.[63]

The spherical average of a diagonal element of J,

$$h(r_{12}) = (4\pi)^{-1} \int_o^{2\pi} \int_o^\pi J(\vec{r}_{12},\vec{r}_{12})\ \sin\alpha\ d\alpha d\beta \tag{83}$$

where $\vec{r}_{12} = (r_{12},\alpha,\beta)$, is related in a simple manner to the electron-electron distribution function $P_o(r_{12})$ and weighted pair distribution function $g(r_{12})$:

$$h(r_{12}) = P_o(r_{12})/(4\pi r_{12}^2) = g(r_{12})/(4\pi r_{12}). \tag{84}$$

There is a close parallelism between $h(r_{12})$ and $\rho_o(r)$ expressed by the following set of equations:

$$\rho_o(0) = \sum_i <\delta(\vec{r}_i)>,\quad h(0) = \sum_{i<j} <\delta(\vec{r}_{ij})>, \tag{85}$$

$$4\int_o^\infty r^{k+2}\rho_o(r)dr = \sum_i <r_i^k>,\quad 4\pi\int_o^\infty r_{12}^{k+2} h(r_{12})dr_{12} = \sum_{i<j} <r_{ij}^k>, \tag{86}$$

$$4\pi\int_o^\infty r^2\rho_o(r)dr = N,\quad 4\pi\int_o^\infty r_{12}^2 h(r_{12})dr_{12} = \binom{N}{2}. \tag{87}$$

$$\lim_{r_\alpha \to 0} [1/2\ \rho'_{o\alpha}(r_\alpha)/\rho_{o\alpha}(r_\alpha)]$$
$$= -Z_\alpha,\quad \lim_{r_{12}\to 0} [1/2h'(r_{12})/h(r_{12})] = 1/2 \tag{88}$$

where the last pair of equations (88) reflect the cusp conditions satisfied by the exact $h(r_{12})$ and $P_o(r)$ [$\rho_{o\alpha}$ is the spherical average of $\rho(\vec{r})$ about nucleus α].

It will be shown now that $\Gamma'(1,1')$ provides access to other types of one-electron densities, namely the momentum density and the spin density.

a) Momentum Densities

The Dirac-Fourier transform of $\rho(\vec{r},\vec{r}')$ yields $\Pi(\vec{p},\vec{p}')$, i.e.

$$\Pi(\vec{p},\vec{p}') = (2\pi)^{-3}\int\int\rho(\vec{r},\vec{r}')e^{-i\vec{p}\cdot\vec{r}+i\vec{p}'\cdot\vec{r}'}\ d\vec{r}\ d\vec{r}' \tag{89}$$

and thus the momentum density $\Pi(\vec{p}) = \Pi(\vec{p},\vec{p})$.

Alternatively the Dirac-Fourier transform of the wavefunction Ψ:

$$\hat{\Psi}(\hat{1},\hat{2},\ldots\hat{N}) = (2\Pi)^{-3N/2}\int\Psi(1,2,\ldots N)e^{-i\sum_j \vec{p}_j\cdot\vec{r}_j}$$
$$d\vec{r}_1 d\vec{r}_2 \ldots d\vec{r}_n \tag{90}$$

where $\hat{J} = (\vec{p}_j,s_j)$, followed by the calculation of

$$\Pi^1(\hat{1},\hat{1}') = N\!\int \hat{\psi}(\hat{1},\hat{2},\ldots,\hat{N})\Psi^*(\hat{1},\hat{2},\ldots\hat{N})\,d\hat{2}\ldots d\hat{N} \tag{91}$$

and the spin-trace $\Pi(\vec{p},\vec{p}') = \int \Pi^1(\hat{1},\hat{1}')\,ds_1$ leads to the same result.

It should be noted that $\Pi(\vec{p})$ is not the F(ourier) T(ransform) of $\rho(\vec{r})$. The FT of the latter is the *form factor*

$$F(\vec{\mu}) = \int \exp(i\vec{\mu}\cdot\vec{r})\rho(\vec{r})\,d\vec{r} \tag{92}$$

which may be written[64] as an "A(uto)-C(orrelation)" of $\Pi(\vec{p},\vec{p}')$:

$$F(\vec{\mu}) = \int \Pi(\vec{p},\vec{p} + \vec{\mu})\,d\vec{p} \;. \tag{93}$$

Similarly one may define[65] the *internally-folded density*

$$B(\vec{s}) = \int \Pi(\vec{p})\exp(-i\vec{p}\cdot\vec{s})\,d\vec{p} \tag{94}$$

which is an "AC" of $\rho(\vec{r},\vec{r}')$, i.e. $B(\vec{s}) = \int \rho(\vec{r},\vec{r} + \vec{s})\,d\vec{r}$. The description of these integrals as "AC's" of ρ is by analogy with the usual definition. If we write ρ in terms of its eigenfunction expansions

$$\rho(\vec{r},\vec{r}') = \sum_i \lambda_i \tau_i(\vec{r})\tau_i^*(\vec{r}') \tag{95}$$

then we may write

$$B(\vec{s}) = \sum_i \lambda_i \tau_i(\vec{r})\tau_i^*(\vec{r}+\vec{s})\,d\vec{r} \tag{96}$$

and similarly for $F(\vec{\mu})$.

We may also write

$$F(\vec{p}'-\vec{p}) = \int \Pi(\vec{p},\vec{p}')\,d\vec{P}' \tag{97}$$

and $\quad B(\vec{r}'-\vec{r}) = \int \rho(\vec{r},\vec{r}')\,d\vec{R} \tag{98}$

where $\vec{P} = (\vec{p}'+\vec{p})/2$ and $\vec{R} = (\vec{r}+\vec{r}')/2$ are the extracular coordinates of the one-matrix. These two expressions show that F and B respectively are the relative or intracular coordinate projections of the r-space and p-space one-matrices.[65]

The $B(\vec{r})$ functions are of practical interest because they are related[66] in a simple way to the experimentally measurable Compton profile function:

$$J(\vec{q}) = \int \Pi(\vec{p})\delta(\vec{p}\cdot\vec{q}-|\vec{q}|)\,d\vec{p}, \quad \hat{\vec{q}} = \frac{\vec{q}}{|\vec{q}|} \tag{99}$$

and its isotropic component

$$\overline{J(q)} = \int J(\vec{q})\,d\Omega_q = 2\pi \int_{|q|}^{\infty} p\,\overline{\rho(p)}\,dp \tag{100}$$

By choosing the p_z axis in the $\vec{q}$ direction we may rewrite

$$J(p_z) = \int \rho(\vec{p}) \, dp_x \, dp_y \tag{101}$$

and

$$B(0,0,z) = \int J(p_z) e^{-ip_z z} dp_z \tag{102}$$

Similarly

$$\overline{B(\vec{r})} = \int \overline{J(q)} \, e^{-iqr} dq \ . \tag{103}$$

Therefore, by one-dimensional Fourier transformation of the various $J(\vec{q})$ we can obtain $B(\vec{r})$.

A distribution function which bridges position and momentum space is the Wigner function.[67] If we define intracular coordinates of the one-matrix, $\vec{r}" = \vec{r}'-\vec{r}$ and $\vec{p}" = \vec{p}'-\vec{p}$, then the Wigner function is defined as:

$$(2\pi)^3 W(\vec{P},\vec{R}) = \int \exp(i\vec{p}"\cdot\vec{R}) \, \Pi(\vec{P}-\frac{\vec{P}"}{2},\vec{P}+\frac{\vec{P}"}{2}) \, d\vec{p}" \tag{104}$$

$$= \int \exp(-i\vec{P}\cdot\vec{r}") \rho(\vec{R}-\frac{\vec{r}"}{2},\vec{R}+\frac{\vec{r}"}{2}) \, d\vec{r}" \tag{105}$$

where $\vec{R}$ and $\vec{P}$, the extracular coordinates of the one-matrix were defined earlier.

The position and momentum densities can be recaptured by integrating out respectively the momentum and position extracular coordinates, i.e.

$$\int W(\vec{P},\vec{R}) \, d\vec{P} = \rho(\vec{R}) \text{ and } \int W(\vec{P},\vec{R}) \, d\vec{R} = \Pi(\vec{P}) \ . \tag{106}$$

b) Spin Densities

The spin density distribution $\gamma_s(r)$ may be obtained from $\Gamma^1(1,1')$ by the prescription $\gamma_s(\vec{r}) = \gamma_s(\vec{r},\vec{r})$ where

$$\gamma_s(\vec{r},\vec{r}') = (2M)^{-1} \int s_z(1) \Gamma(1,1') \, ds_1 \ , \tag{107}$$

$s_z(1)$ is the operator for the Z-component of the spin angular momentum, and M is the eigenvalue of the total S_z operator $S_z = \sum_i s_z(i)$.

We decompose $\Gamma^1(1,1')$ in terms of $\rho(\vec{r},\vec{r}')$ and $\gamma_s(\vec{r},\vec{r}')$. We first note that if Ψ is an eigenfunction of an Hermitian operator of the form $\sum_i \hat{F}_i$, then $\hat{F}_1$ commutes with Γ^1. This means that Γ^1 is block-diagonal in any basis of eigenfunctions of $\hat{F}_1$. Since S_z is such an operator, Γ^1 must be block diagonal in α and β, i.e.

$$\Gamma^1(1,1') = \rho_{\alpha\alpha}(\vec{r},\vec{r}')\alpha\alpha'^* + \rho_{\beta\beta}(\vec{r},\vec{r}')\beta\beta'^* \tag{108}$$

$$= \rho(\vec{r},\vec{r}')[1/2(\alpha\alpha'^* + \beta\beta'^*)]$$

$$+ 2M\gamma_s(\vec{r},\vec{r}')[1/2(\alpha\alpha'^* - \beta\beta'^*)] \tag{109}$$

where $\rho(\vec{r},\vec{r}')$ and $\gamma_s(\vec{r},\vec{r}')$ were defined previously.

In terms of $\rho_{\alpha\alpha}$ and $\rho_{\beta\beta}$ they are

$$\rho(\vec{r},\vec{r}') = \rho_{\alpha\alpha}(\vec{r},\vec{r}') + \rho_{\beta\beta}(\vec{r},\vec{r}') \tag{110}$$

$$2M\gamma_s(\vec{r},\vec{r}') = \rho_{\alpha\alpha}(\vec{r},\vec{r}') - \rho_{\beta\beta}(\vec{r},\vec{r}') \tag{111}$$

The latter is just the net density of unpaired spin. If we define N_α, N_β by

$$\int\rho_{\alpha\alpha}(\vec{r},\vec{r})d\vec{r} = N_\alpha, \quad \int\rho_{\beta\beta}(\vec{r},\vec{r})d\vec{r} = N_\beta \tag{112}$$

then

$$\int 2M\gamma_s(1,1')dl = 2M\int s_z(1)\Gamma'(1,1')dl = M = (N_\alpha - N_\beta)/2 . \tag{113}$$

c) Electron Density Functions and X-Ray (and High Energy Electron) Scattering

The intracule matrix $J(\vec{r}_{12},\vec{r}_{12})$ is accessible experimentally by means of total X-ray scattering

$$I_{tot}(\vec{s})/I_{Cl} = N + 2\int J(\vec{r}_{12},\vec{r}_{12})e^{i\vec{s}\cdot\vec{r}}d\vec{r}_{12} . \tag{114}$$

In the gas phase, this becomes

$$I_{tot}(s)/I_{Cl} = N + 2\int h(r_{12})j_o(sr_{12})d\vec{r}_{12} \tag{115}$$

where $s = 4\pi\lambda^{-1}\sin\frac{\theta}{2}$ is the magnitude of the momentum transfer $\vec{s}$, λ is the wavelength of the incident radiation, θ is the scattering angle, and $j_o(x) = x^{-1}\sin x$ is a zeroth order spherical Bessel function. Similarly the intensity for coherent X-ray scattering $I_C(\vec{s})$ provides access to the diagonal portion of the one-electron density matrix $I_C(\vec{s}) = |F(\vec{s})|^2$ where $F(\vec{s}) = \int\rho(\vec{r},\vec{r})e^{i\vec{s}\cdot\vec{r}}d\vec{r}$. In the gas phase this simplifies to $F(\vec{s}) = \int\rho_o(\vec{r})j_o(s\vec{r})d\vec{r}$.

The incoherent scattering function $S(\vec{s})$ is defined as the difference between the total and incoherent scattering intensities $S(\vec{s}) = I_t(\vec{s})/I_{Cl} - I_c(\vec{s})/I_{Cl}$.

In the scattering of electrons,[68] the situation becomes somewhat more complicated because the incident electron interacts not only with the electronic charge distribution but with the nuclei as well. Within the first Born approximation the total, elastic, and inelastic scattering intensities for electron scattering from neutral free atoms are related to the corresponding X-ray quantities defined above:

$$I_t^e(s)/I_{Cl}(s) = 4s^{-4}((N-F(s))^2 + S(s)) \tag{116}$$

$$I_{el}^e(s)/I_{Cl}(s) = 4s^{-4}(N-2F(s)) + F^2(s) \tag{117}$$

$$I_{inel}^e(s)/I_{Cl}(s) = 4s^{-4}S(s) \tag{118}$$

In the case of scattering from neutral free molecules, the form factors involved have to be calculated with respect to fixed centres (indicated by quotation marks in the equations below). Thus

$$I_{el}(\vec{s}) = 4s^{-4}\{ \sum_{\alpha=1}^{\eta} z_\alpha^2 + \sum_{\alpha=1}^{\eta}\sum_{\beta=1}^{\eta} z_\alpha z_\beta \exp(i\vec{s}\cdot\vec{R}_{\alpha\beta})$$

$$+ 2 \sum_{\alpha=1}^{\eta} z_\alpha "F"_\alpha + "F^2" \tag{119}$$

where

$$"F_\alpha" = \int d\vec{r}\rho(\vec{r}) j_o(s|\Delta\vec{r}_\alpha - \vec{r}|) \tag{120}$$

and

$$"F^2" = \int d\vec{r}\rho(\vec{r}) \int d\vec{r}'\rho(\vec{r}') j_o(s|\vec{r}-\vec{r}'|) . \tag{121}$$

and $\Delta\vec{r}_\alpha$ indicates the vector from the molecular centre of mass to nucleus α.

The first and second terms in (119) describe the coulombic and interference scattering from the bare nuclei of the molecule. The third term represents the interference of the nuclear coulomb scattering with the scattering due to the electron distribution while the last term describes the electron-electron interference scattering.

The inelastic intensity

$$I_{inel}^e(\vec{s}) = 4s^{-4}\{\sum_{\alpha=1}^{\eta}z_\alpha - "F"^2 + \int d\vec{r}_{12} h(\vec{r}_{12}) j_o(sr_{12})\} \tag{122}$$

contains the intracule function discussed previously.

The sum of I_{el}^e and I_{inel}^e, the total intensity

$$I_t^e(s) = 4s^{-4}\{N + \sum_{\alpha=1}^{\eta} z_\alpha^2 + \sum_{\alpha}\sum_{\beta} z_\alpha z_\beta \exp(i\vec{s}\cdot\vec{R}_{\alpha\beta})$$

$$-2 \sum_{\alpha=1}^{\eta} z_\alpha "F"_\alpha + \int d\vec{r}_{12} J(\vec{r}_{12},\vec{r}_{12}) j_o(sr_{12})\} \tag{123}$$

does not contain the electron-electron interference term, $"F^2"$.

d) Rôle of the Promolecule

As we have commented earlier, it is often advantageous to remove the promolecule density, the density of the superposition of the spherically averaged atoms at the nuclear positions from the true molecular density ρ i.e. $\Delta\rho_p = \rho - \rho_{pro}$. In the case of the electron scattering expressions, a simplification in the equations occurs with the result that both calculations and analysis are simplified, whereas in the X-ray case, $\Delta F(\vec{S}) = \int \Delta\rho_p(\vec{r}) e^{i\vec{S}\cdot\vec{r}} d\vec{r}$ and $F(\vec{S})$ present similar problems for the inversion.

This is due to the fact that some of the terms in the scattering expressions are described correctly by the promolecule. Thus

$$\Delta I_t^e(s) = I_t^e - I_{t,p}^e = -2\Sigma_\alpha z_\alpha \int d\vec{r} \Delta\rho_p(\vec{r}) j_o(s|\Delta\vec{r}_\alpha - \vec{r}_\alpha|)$$

$$+ \int d\vec{r} \Delta_p J(\vec{r},\vec{r}) j_o(sr) \tag{124}$$

where $\Delta\rho_p$ was defined above and $\Delta_p J(\vec{r},\vec{r}) = J_{mol}(\vec{r},\vec{r}) - \Sigma_\alpha J_\alpha^{atom}(\vec{r},\vec{r})$. For the elastic intensity, we have

$$\Delta I_{el} = 4s^{-4}\{-2\Sigma_\alpha \int d\vec{r}\Delta_p\rho(\vec{r}) j_o(s|\Delta\vec{r}_\alpha - \vec{r}|$$

$$+ \int d\vec{r} \int d\vec{r} \Delta_p[\rho(\vec{r})\rho(\vec{r}')] j_o(s|\vec{r} - \vec{r}'|) \tag{125}$$

or

$$\Delta I_{el} = rs^{-4}\{-2\Sigma_\alpha[Z_\alpha - F_\alpha(s)] \int d\vec{r}\Delta_p\rho(\vec{r}) j_o(s|\Delta\vec{r}_\alpha - \vec{r}_\alpha|)$$

$$+ \int d\vec{r} \int d\vec{r}' \Delta_p\rho(\vec{r})\Delta_p\rho(\vec{r}') j_o(s|\vec{r}-\vec{r}'|) \quad . \tag{126}$$

The inversion to obtain $\Delta_p\rho(\vec{r})$ is not as straightforward as in the X-ray case where $\Delta F(\vec{s}) = \int \Delta_p\rho(\vec{r}) e^{i\vec{S}\cdot\vec{r}} d\vec{r}$ presents in principle the problem of direct Fourier inversion. If measurements of ΔI_t^e are in hand, (124) leaves the additional contribution $\int d\vec{r}\Delta_p J(\vec{r},\vec{r})$ $j_o(sr)$ to contend with. Fink et al[69] have assumed that it is negligible and have proceeded to invert the first term $-2\Sigma_\alpha z_\alpha \int d\vec{r}\Delta_p\rho(\vec{r}) j_o(s|\Delta\vec{r}_\alpha - \vec{r}_i|)$ by means of a multipole model model for $\Delta_p\rho(\vec{r})$.

8. SOME PROPERTIES OF THE ELECTRON DENSITY

a) Density Functional Theory

We have seen above that the energy in the HF scheme is a functional of the Fock-Dirac density matrix $\Gamma_D^{(1)}$ (1;1') while the exact energy is a functional of $\Gamma_\psi^{(2)}$ (1,2;1',2'). However, one can show (*Hohenberg-Kohn theorem*) that *a non-degenerate ground state is a unique functional of $\rho(\vec{r})$*, i.e. the energy E, the local potential

$\hat{V}$ in the electronic Hamiltonian, and the wavefunction Ψ are all unique functionals of $\rho(\vec{r})$. It should be noted that Hohenberg and Kohn[70] established only the *existence* of such functionals. The problem of their construction remains. Density functional theory provides a theoretical framework and justification for the use of local approximations to the non-local exchange operator in the HF theory.

It is important to note that although the original proof of the Hohenberg-Kohn theorem does not explicitly state so, it was required that $\rho(\vec{r})$ be derivable from an admissible wavefunction Ψ, i.e. be N-representable.

b) N-Representability of the Electron Density

One of the basic assumptions of density functional theory is that the electron density $\rho(\vec{r})$ be derivable from an anti-symmetric wavefunction and in some applications it is further assumed that the wavefunction is a Slater determinant. This leads to the question of which conditions $\rho(\vec{r})$ must satisfy for this to be true.

If $\rho(\vec{r})$ is finite, non-negative, differentiable, and is normalized to the number of electrons N, i.e. $\int \rho(\vec{r}) d\vec{r} = N$ then it may be obtained from an antisymmetric wavefunction, i.e. it is *N-representable*.

The proof of this statement is trivial in the case of a density due to one or two single electrons. One need only take the square root of $\rho(\vec{r})$ to define an orbital, i.e. $\phi(\vec{r}) = \sqrt{\rho(\vec{r})}/N$ and hence the wavefunction. In the case of the single-electron system H_2^+, the orbital corresponding to the promolecule $\rho(\vec{r})$ was examined by Pearson and Palke[71] for its contribution to molecular bonding.

The proofs in the literature[72,13] for more than two electrons are by means of algorithms for the construction of a set of M orthonormal orbitals $\{\phi_k\}$ such that the one-density matrix $\rho(\vec{r},\vec{r}')$ corresponding to $\rho(\vec{r})$ may be written

$$\rho(\vec{r},\vec{r}') = 2 \sum_{k=1}^{M} \phi_k(\vec{r}) \mu_k \phi_k^*(\vec{r}') \tag{127}$$

where $o \le \mu_k \le 1$, $\sum_{k=1}^{M} \mu_k = N/2$, and $\rho(\vec{r},\vec{r}')$ satisfies the conditions for N-representability*. By their very nature these proofs bring out the point that there are many possible wavefunctions corresponding to a given $\rho(\vec{r})$. They do not solve the problem which remains of determining that wavefunction which satisfies as many as possible of those
* For the sake of simplicity, we discuss the case where N is even, $\gamma_s(r)$ vanishes and $M = N/2$.

conditions which characterize the system under consideration.

The most recent construction is due to Harriman[73] and it is perhaps the most physically satisfying of them all. The earlier ones[72] divided space into $M(\geq N/2)$ regions R_k $(1 \leq k \leq M)$ and identified $\sqrt{\rho(\vec{r})}$ in each such region with a different spin orbital which vanished outside its particular region, i.e. $\phi_k(\vec{r}) = \sqrt{\rho(\vec{r})}/\int \rho(\vec{r})d\vec{r}$ for r in R_k and zero otherwise. By this construction, the ϕ_k orbitals are orthogonal and normalizable but suffer from discontinuities in both themselves and their derivatives at the region boundaries.

Harriman[73] defines each orbital as the product of an amplitude $\sqrt{\rho(\vec{r})}$ and a phase factor chosen to ensure orthonormality, i.e. for $k=1, \ldots, M$

$$\phi_k = \sqrt{M^{-1}\rho(\vec{r})} \, \exp[ikf(\vec{r})] \tag{128}$$

and

$$|\phi_k(\vec{r})|^2 = M^{-1}\rho(\vec{r}). \tag{129}$$

The Harriman $\{\phi_k\}$ are a great improvement conceptually in that for $\rho(\vec{r})$ smooth and continuous they are as well. However, it should not be forgotten that they are only a means toward establishing the N-representability of $\rho(\vec{r})$ and are probably not physically significant themselves.

Because the Hohenberg-Kohn theorem established a one-to-one correspondence between the local potential $\hat{V}$ of the Hamiltonian $\hat{H}$ and $\rho(\vec{r})$, there is still to be answered the related problem of the $\hat{V}$-representability[74] of $\rho(\vec{r})$, i.e.

Given $\rho(\vec{r})$, can a local potential $\hat{V}$ be found such that the ground state of the Hamiltonian with this $\hat{V}$ has the stipulated density $\rho(\vec{r})$?

Of course, if we know that $\rho(\vec{r})$ correctly describes an electronic system, we can construct the local potential $\hat{V}$ in the form $\hat{V} = \sum_\alpha Z_\alpha/r_\alpha$ by searching for the cusps of $\rho(\vec{r})$. This equation would yield the nuclear locations R_α and the atomic numbers Z_α.

The work of Massa and collaborators for the analysis of experimental electron densities (Chapter 2, section 2) is based upon the assumption that $\rho(\vec{r})$ may be represented exactly by an idempotent one-density matrix, i.e. one obtained from a Slater determinant. If M is restricted to be equal to N/2 in the above discussion, it follows that any finite, non-negative, differentiable and normalizable $\rho(\vec{r})$ may be derived from a Slater determinant.

c) Cusp and Asymptotic Conditions on the Electron Density

Knowledge of the short and long-range behaviour of the electron density is useful for assessing the quality of experimental and calculated electron densities.

At each of the nuclear positions α, the exact electron density $\rho(r)$ (and the exact $\rho_{HF}(r)$) satisfy a Kato-type electronuclear *cusp condition*

$$\lim_{r_\alpha \to 0} (\frac{\partial}{\partial r_\alpha} + 2Z_\alpha) \rho_{o\alpha}(r_\alpha) = 0 \qquad (130)$$

where $\rho_{o\alpha}(r_\alpha)$ is the spherical average of $\rho(\vec{r})$ about α.[75]

The long-range behaviour of the exact HF $\rho(r)$ is an exponential decay given by $\rho_{HF}(r) = \exp[-2(-2\varepsilon_{max})^{1/2}r]$.

d) Some Theorems Involving the Electron Density

There are a number of theorems which involve the electron density in addition to those concerning the expectation values of one-electron operators.

1) Linear response theorem: The *response* of a system when a local one-particle potential $\hat{U} = \sum_i u(\vec{r}_i)$ is turned on is to lowest order $\Delta E = \int u(\vec{r})\rho(\vec{r})d\vec{r}$.

2) Hellman-Feynman theorem: A special case of the Hellman-Feynman theorem is of great importance in the interpretation of the nature of bonding from a knowledge of $\rho(\vec{r})$. The general *Hellman-Feynman theorem*[76] states that for the exact Ψ (or the Hartree-Fock)

$$\frac{\partial E}{\partial \lambda} = \int \Psi * \frac{\partial \hat{H}}{\partial \lambda} \Psi \, dz = \langle \frac{\partial \hat{H}}{\partial \lambda} \rangle, \qquad (131)$$

where λ is a parameter in $\hat{H}$. If λ is chosen to be an internuclear coordinate (say R), then only the electron-nuclear interaction term $\hat{V}_{ne}$ in $\hat{H}$ gives a non-vanishing contribution to $\langle \partial \hat{H}/\partial R \rangle$. Since the change in the electronic energy due to a small displacement of nucleus α in the x-direction is the negative of the force on α in that direction, we have

$$-F_{\alpha x} = \frac{\partial E}{\partial x_\alpha} = \int \frac{\partial \hat{V}_{ne}}{\partial x_\alpha} (\vec{r})\rho(\vec{r})d\vec{r}. \qquad (132)$$

This means that the entire x-component of force exerted on this
nucleus is calculable from $\rho(\vec{r})$ and the configuration of the other
nuclei. As a result, the forces holding the nuclei together
(balanced by the nuclear repulsions at equilibrium) can be discussed classically using $\rho(\vec{r})$.

   3) <u>Wilson-Frost Charging Process:</u>[77,78] If a scaling parameter
λ is introduced such that each atomic number in the molecule is λz_α,
and $\partial E/\partial\lambda$ is calculated by the Hellman-Feynman theorem:

$$\frac{\partial E}{\partial \lambda} = \int [\hat{V}_{ne}(\lambda=1,\vec{r},\{R\})\rho(\vec{r},\lambda)]_N d\vec{r} \tag{133}$$

then the energy of the original, unscaled molecule, $E(\lambda=1)$ can be
obtained from the knowledge of the electron density as a function
of λ, $\rho(\vec{r},\lambda)$ by integration of (133).

$$E(\lambda=1) = E(\lambda=\lambda_o) + \int_{\lambda_o}^{1} d\lambda \int_N [\hat{V}_{ne}(\lambda=1,\vec{r},\{R\})\rho(\vec{r},\lambda)]_N d\vec{r} \tag{134}$$

Here λ_o is greater than or equal to λ_c where λ_c is defined as the
critical value of λ such that the system is bound. It should be
noted that usually in the literature (133) and (134) are written
with λ_c set incorrectly equal to zero. This error in the general
case occurred probably because it is correct in the case of the
hydrogen atom. The subscript N indicates that the number of electrons is fixed during the process.

   Frost[78] introduced F, an average electron density function:

$$F(\vec{r}) = \int_{\lambda_o}^{1} [\rho(\vec{r},\lambda)]_N d\lambda \tag{135}$$

Then

$$E(\lambda=1) - E(\lambda=\lambda_o) = \int \hat{V}_{ne}(\lambda=1,\vec{r},\{\vec{R}\})F(\vec{r})d\vec{r} \tag{136}$$

   4) <u>Schwendeman's theorem:</u>[79] The total electronic contribution
to a force constant may be obtained from $\rho(\vec{r})$ by

$$-K_A^e = \int \frac{\partial \rho(a,r_A)}{\partial x_A} \frac{\cos\theta_A}{r_A^2} d\vec{r}_A \tag{137}$$

where $\rho(a,\vec{r}_A)$ is $\rho(\vec{r})$ expanded about site A. It is to be differ-
entiated with respect to the nuclear coordinate x_A only through the
dependence of $\rho(a,r_A)$ on the parameters a in ρ and the wavefunction
which are functions of the nuclear coordinate x_A because of the
Born-Oppenheimer approximation.

5) <u>Electrostatic Potential</u>: The *electrostatic potential* $U(r)$ due to the η nuclei and N electrons of a system is $U(r) = U_N(r) + U_e(r)$ where the nuclear and electronic contributions are respectively

$$U_N(\vec{r}) = \sum_\alpha \frac{Z_\alpha}{|\vec{R}_\alpha - \vec{r}|} \tag{138}$$

and

$$U_e(\vec{r}) = -\int \frac{\rho(\vec{r}')}{|\vec{r}' - \vec{r}|} \, d\vec{r}'. \tag{139}$$

It is the solution of Poisson's equation $\nabla^2 U(\vec{r}) = 4\pi\rho(\vec{r})$. Although $U(\vec{r})$ is not an interaction energy, $qU(\vec{r})$ is the Coulombic inter-action energy between the unperturbed (static) electron density of the system and a positive unit (point) charge at $\vec{r}$.

The electrostatic potential $U(\vec{r})$, calculated from the quantum mechanically calculated or experimental $\rho(\vec{r})$ can be used to study by electrostatics the interaction between two distributions A and B, i.e.

$$E(\text{lectrostatic}) \ (A,B) = \int \frac{\rho_A(\vec{r}_1, \vec{R}_\alpha) \ \rho_B(\vec{r}_2, \vec{R}_\beta)}{|\vec{r}_1 - \vec{r}_2|} \, d\vec{r}_1 d\vec{r}_2 \tag{140}$$

or equivalently

$$= \int U_A(\vec{r}_2, \vec{R}_\alpha) \rho_B(\vec{r}_2, \vec{R}_\beta) \, d\vec{r}_2$$
$$= \int U_B(\vec{r}_2, \vec{R}_\beta) \rho_A(\vec{r}_2, \vec{R}_\alpha) \, d\vec{r}_2 \tag{141}$$

where U_A and U_B are the electrostatic potentials due to ρ_A and ρ_B. Of course, on the classical level, this neglects the contributions due to the mutual polarization of ρ_A and ρ_B while on the quantum mechanical, the exchange effects are not included.

The electrostatic potential of a given molecule A, U_A, is an index of reactivity of molecule A which gives a global picture of its interaction with reactant point charges over the entire space external to A. There is a rich literature on the subject of the molecular electrostatic potential and chemical interactions and reactivity.[80]

6) <u>Electrostatic potential at the nucleus</u>:[80] Evaluation of (138) and (139) for the electrostatic potential at one of the nuclei, say nucleus β, leads to

$$U(\vec{R}_\beta) = \sum_{\alpha \neq \beta} \frac{Z_\alpha}{|\vec{R}_\alpha - \vec{R}_\beta|} - \int \frac{\rho(\vec{r}) d\vec{r}}{|\vec{r} - \vec{R}_\beta|} \tag{142}$$

for a molecule. For an atom, this reduces to

$$U(0) = - Z\int \frac{\rho(\vec{r})}{r} d\vec{r} = -<V_{en}>$$
(143)

These expressions lead to the following expressions for the total energy of an atom:

$$E^{at} = \frac{1}{2} (Z-Z_C)U(0) - \frac{1}{2} \int_{Z_C}^{Z} [Z' \frac{\partial U(0)}{\partial Z'} - U_0]_N dZ'$$

and for molecules

$$E^{mol} = \sum_\beta \frac{1}{2}(Z_\beta-Z_C)U(\vec{R}_\beta)$$
$$- \sum_\beta \frac{1}{2} \int_{Z_C}^{Z_\beta} [Z'_\beta (\frac{\partial U(\vec{R}_\beta)}{\partial Z_\beta} - U(\vec{R}_\beta)]dZ_\beta$$
(144)

7) Scaling of the Electron Density and Single-Particle Expectation Values: As discussed earlier, the scaling of an approximate wavefunction $\Phi(\vec{x}^N;\vec{R})$ utilizes

$$\Phi_\lambda = \lambda^{3N/2} \Phi(\lambda\vec{x}_1,\lambda\vec{x}_2,\ldots,\lambda\vec{x}_N;\lambda R)$$
(145)

This in turn leads to a scaling relation for the electron density,

$$\rho(\vec{r},\vec{R},\lambda) = \lambda^3 \rho(\lambda\vec{r},\lambda\vec{R},1).$$
(146)

Expectation values of single-particle operators, $\Omega_1 = \sum_i \Omega(i)$ which are of the power form $\Omega(1) = r_1^j$, scale according to $<\Omega_1>_\lambda = \lambda^{-j}<\Omega_1>_1$.

We note that in momentum space that the momentum density scales as

$$\pi(\vec{p};\vec{R},\lambda) = \lambda^{-3}\pi(\lambda^{-1}\vec{p},\lambda\vec{R},1)$$
(147)

while for those Ω_1 for which $\Omega(1) = p_1^k$, the relation is $<\Omega_1>_\lambda = \lambda^k<\Omega_1>_1$.

e) Representations of the Electron Density

Although the electron density has reduced the information in the wavefunction from a function of 3N variables to one of 3 variables, it still contains more information than required for many purposes. As a scalar function of three variables, four dimensions are required for its complete graphical description. For obvious reasons it is usually desirable to summarize this information in a digestible and useful form. We distinguish between *"pictorial"* and *"tabular"* representations of density features.

Pictorial representations can be classified by dimensionality
(2, 3 or 4) and by type (graphs, contours, etc.). The current media
of scientific communication force us to concentrate on representations which are two-dimensional. Three-dimensional representations
can be made using models, or using time as the third dimension, as
in a movie. Similarly, colour, density of dots or intensity of
light can be used as the ρ dimension in representations of $\rho(\vec{r})$.

We recall how a two-dimensional contour map arises from a
function defined on a plane. The graph of the function forms a
two-dimensional surface in three-dimensional space, where we imagine
the "f-axis" is upwards. This is a *3D graph*. Its contour map
arises from the intersection of this surface with a set of horizontal
planes, $f = k_i$. We call this a *2D contour map*. A cross-section of
this 3D graph is then a *2D graph*.

These considerations lead to the following general definitions.[61] The graph of a scalar function f of n variables $(x_1, x_2,
..., x_n)$ is an n-dimensional surface in n + 1 dimensions. This is
an *(N+1)D graph*. An *mD graph* (m = 1,2,...,n+1) can be defined as
the intersection of the surface with an m-dimensional "vertical"
surface specified by the n - (m-1) independent constraint equations
on the x_i, $g_i(x_1, x_2, ..., x_n) = 0$; j=1,...n-(m-1). Usually these
vertical surfaces would be planar, which requires that the constraint equations be linear.

An *mD contour* (m=1,2,...n) is defined as the intersection of
the functional surface with an m-dimensional "horizontal" surface
f=k, where k is a constant. These contours (if they exist) are
either (m-1)-dimensional closed surfaces in m dimensions or isolated points. There are, of course, many equivalent definitions
of these mD graphs and contours. The intersection of an mD graph
with the plane f=k results in an (m-1)D contour. The information
contained in the (uncountably infinite) set of all such (m-1)D
contours is equal to that contained in the mD graph. In practice,
an mD graph is evaluated only at a finite set of points and thus a
finite set of (m-1)D contours contains the same amount of
information.

With the electron density, we are primarily interested in 2,
3 and 4D graphs, and 2 and 3D contours. The 2D graph and the 2D
contour map (where a finite number of 2D contours are drawn) are
familiar. The 2D contour map allows one to visualize an approximation to the 3D graph. The accuracy of this visualization depends
on how smooth the function is between the contour values chosen for
the map. While the 3D graph requires three dimensions, its perspective projection, also known as "cross-sectional electron-

density diagrams" can be drawn on a two-dimensional sheet; i.e. a
printed page. Similarly, the perspective projection of a 3D contour
(sometimes called a "shape plot") can be drawn as well. These also
allow a visualization of an approximation to the 3D surface, except
for those parts of the surface hidden from view. If the 3D surface
is reasonably smooth, all of it can be viewed by means of a few
perspectives at suitably chosen viewing angles.

A single 3D (or 2D) contour conveys almost no information. A
suitable set of such contours (together with our "smoothing by eye"
between contour values) provides us with a complete description of
the electron density. Unfortunately, unless the 3D contours are
"transparent", the outer contour hides the others from view. Even
if they were transparent, only the outermost few would be readily
discernible. Differential colouring might improve the situation a
little. An alternative is to show perspectives of "cut-away" 3D
contour maps. A number of these with different parts of the map
cut away would enable the reader to visualize the 3D contour map.
The numerical information content is probably maximized by the
separate presentation of a few perspectives of each of a suitable
set of 3D contours, but the visualization is made more difficult.
(Cross sections of 3D contour maps, are, of course, merely 2D
contour maps.)

Illustrations of some of the types of contour maps and graphs
are given in Figures 7 and 8. In the former, the 2D contour map
for C_2H_4 in the molecular (H_2CCH_2) plane displayed together with a
corresponding 3D graph at the orientation ($\theta = 70°$, $\phi = 330°$) and
a 3D contour map of $\rho(\vec{r})$ for the same choice of orientation of the
observer and the defining electron density value of 0.2 e/(a.u.)^3.

In Figure 8 the effect of different choices of the defining
electron density is illustrated. It is interesting to observe that
various of the prevalent notions about molecules and molecular
structure correspond to different choices of the defining density,
e.g. diagram (e) looks like the usual ball-and-stick models,
diagrams (f) and (g) represent the space-filling picture, while
(h) and (i) show the shapes appropriate for crystall packing of
the molecules. At large defining densities (a) and (b) the molecule is obviously inadequately described by only the carbon balls.
As the defining density decreases, diagrams (c) and (d) show the
C-C bond as well as the hydrogen atoms. The full bonding picture
is represented in diagrams (e) through (i), and at extremely low
electron density one would expect the C_2H_4 molecule to be simply
an ellipsoidal ball.

In Figure 9, the change in the electron density as one proceeds

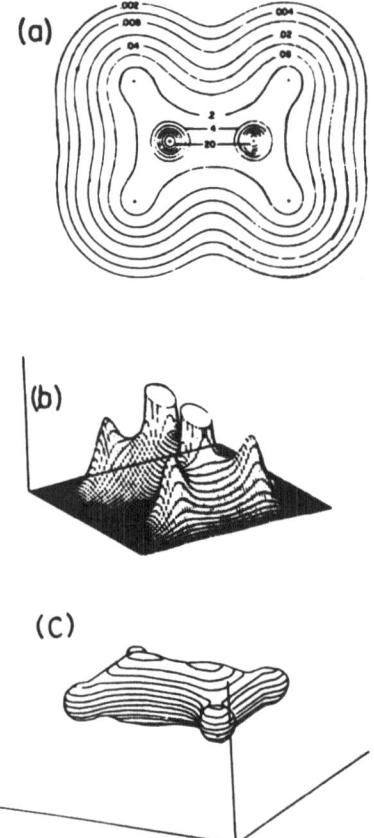

Figure 7: (a) 2D contour map; (b) a 3D graph (at orientation
($\theta = 70°$, $\phi = 330°$) of the electron density of C_2H_4 in the
molecular (H_2CCH_2) plane; (c) a 3D contour map ($\rho = 0.2e/$
$(a.u.)^3$) of the electron density for the same orientation
of the observer.

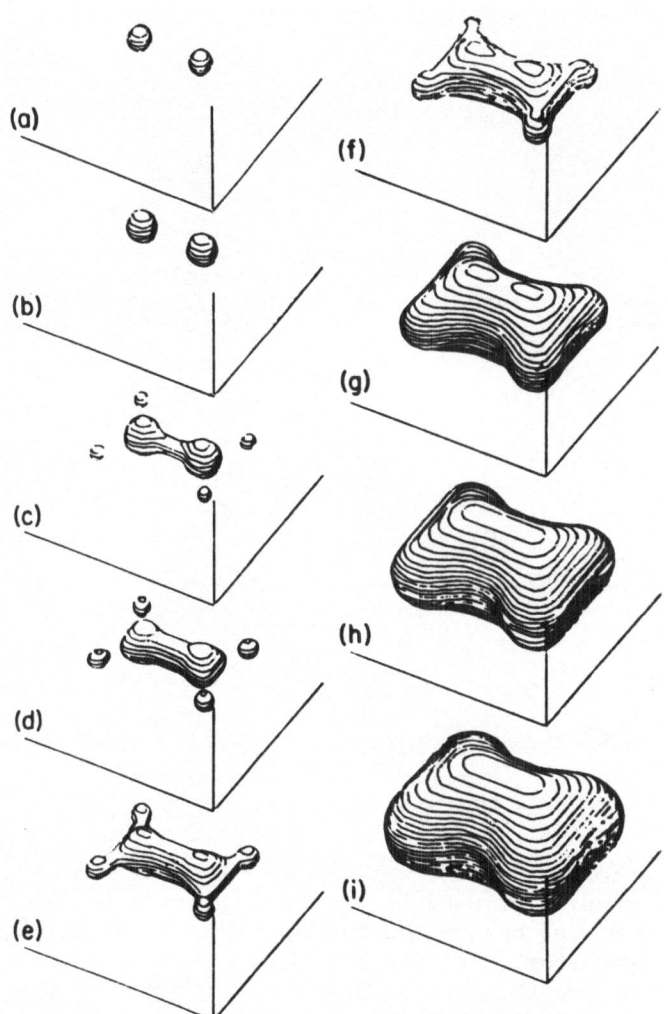

Figure 8: 3D contour maps of the electron density of C_2H_4 for ρ = 0.8, 0.4, 0.3, 0.27, 0.25, 0.2, 0.08, 0.04, and 0.02 e/(a.u.)^3.

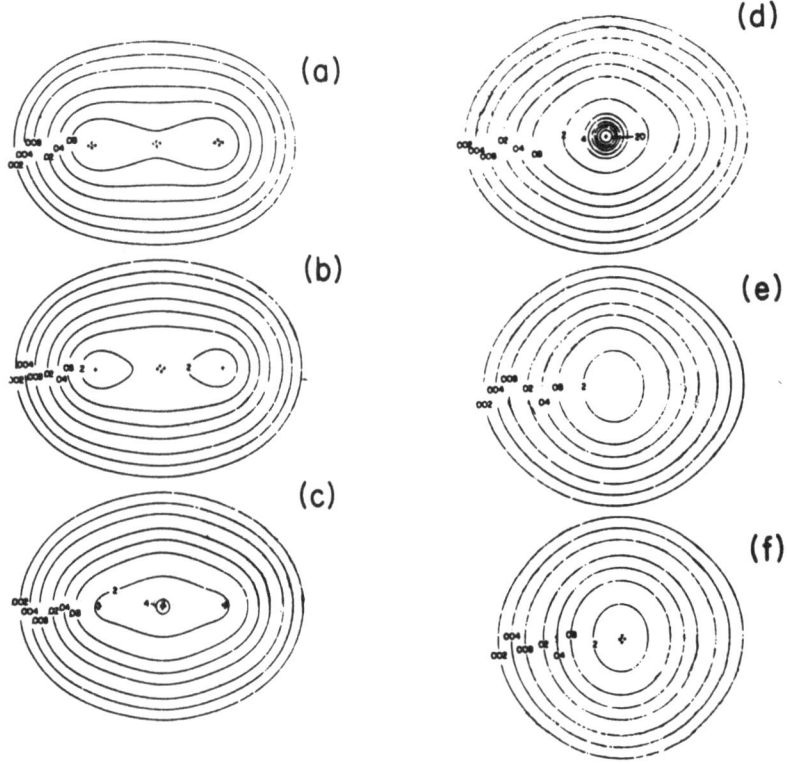

Figure 9: 2D contour maps of a walk along the CC axis of the elec-
           tron density of C_2H_4 in planes orthogonal to the CC axis.
           (a) 0.5 a.u. before the hydrogen atoms; (b) through the
           hydrogen atom;   (c) through the midpoints of the C-H bonds;
           (d) through the carbon atom; (e) a quarter of the way along
           the CC bond; (f) midway along the CC bond. Note that the
           symbol + indicates a nucleus in the plane under considera-
           tion while $\frac{1}{1}$ indicates the projected position of an
           out-of-plane nucleus.

along the CC axis is presented by means of 2D contour maps for
various planes orthogonal to the axis. Since the 2D contour maps
are an order lower than the 3D contour maps, the contours of the
2D maps in this figure can be considered as cross sections orthogonal to the C-C axis through the 3D contour maps of the type shown
in Figure 8. However the 3D graphs of this figure represent electron densities in planes orthogonal to the molecular plane in which
the electron densities of the 3D graphs of Figure 7 are shown, i.e.
the intersection of the 3D graph of this figure with appropriate 3D
graphs of Figure 7 (wherever possible) is a line curve. The features of diagrams (a) through (f) are self-explanatory and agree
well with the previous figures.

f) Tabular Representations

Tabular representations are of two types: The density can be
approximated by some functional fit such as a point-charge or multipole model, and the parameters of the fit listed. Three-dimensional
space can be searched for specific topographical features and their
properties listed. The most interesting of these for any scalar
function are those where the various-order derivatives vanish or
are discontinuous. As noted above, the electron density ("at rest")
is not everywhere smooth as it has cusps at the various nuclei.
These electron-nuclear cusps are maxima, but the gradient of $\rho(\vec{r})$ is
discontinuous there and thus the second derivatives as well as the
gradient itself are not defined there. At non-nuclear points, it is
believed that $\rho(\vec{r})$ is a smooth function and thus can be characterized
by its critical points where $\nabla\rho(\vec{r})$ vanishes. Critical points
$\bar{x} = (\bar{x}_1,\ldots,\bar{x}_n)$ can be classified in terms of the *invariants* of the
Hessian or curvature matrix (h_{ij}) where

$$H_{ij} = \frac{\partial^2 f}{\partial x_i \partial x_j}\bigg|_{\bar{x}}$$

These are its *rank* r (the number of non-zero eigenvalues), *index*
i (the number of negative eigenvalues) and *signature* p = r-2i (the
difference between the numbers of positive and negative eigenvalues).
Non-degenerate critical points have rank n. We list such critical
points in Table 4 for functions of 1, 2 and 3 variables. If we
assume that there are no degenerate critical points, then

$$\chi_f(m) = \sum_{i=0}^{n} (-1)^i \nu_i$$

where $\chi_f(m)$ is the homological Euler characteristic of the domain
m (of dimension n) of f and ν_i is the number of critical points of
index i. For three-dimensional real space, $\chi = -1$, and for the
infinite periodic lattice, $\chi = 0$.

TABLE 4 - Non-degenerate Critical Points for Functions of 1, 2 and
3 Variables

Dimension	(r,i)	(r,p)	Type	Name[81]
1	(1,0)	(1,1)	minimum	pit
	(1,1)	(1,-1)	maximum	peak
2	(2,0)	(2,2)	minimum	pit
	(2,1)	(2,0)	saddle point	pass
	(2,2)	(2,-2)	maximum	peak
3	(3,0)	(3,3)	minimum	pit
	(3,1)	(3,1)	saddle point	pale
	(3,2)	(3,-1)	saddle point	pass
	(3,3)	(3,-3)	maximum	peak

The requirement of non-degenerate critical points is not
practically restricting, as degenerate ones can be removed by infinitesimal changes of the function. In the case of a molecular
electron density for a given nuclear configuration, an infinitesimal
change in the coordinates of any one of the nuclei would be expected
to remove a degeneracy. This does not mean that they can be ignored
as they can be vital in some applications such as the representation
of bonding in terms of the catastrophe points[82,83] of the electron
density. The number of maxima (they are not true critical points
since $\nabla\rho$ is not defined there) of $\rho(\vec{r})$ is equivalent to the number
of nuclei, the number of passes with the number of bond paths or
bonds, the number of pales with the number of rings and the number
of pits with the number of cages.

The studies of Bader and co-workers[83] on the virial partition-
ing of $\rho(\vec{r})$ are based on the topological properties of $\nabla\rho(\vec{r})$ and lead
to definitions of concepts such as atoms in a molecule and bond paths.
Zero flux surfaces in three-dimensional real space are defined as
those closed surfaces through which the flux of $\nabla\rho(\vec{r})$ is everywhere
zero. The condition of closure together with the nuclear cusp condition eliminates nuclei from any of the surfaces and, if it is
further assumed that the nuclei are the only maxima, reduces the
number of partitioning surfaces to a finite number. A *partitioning
surface* is formed by all the paths of steepest descent from an internuclear pass (since the index of a pass is two, these form a two-
dimensional surface) and by the (one-dimensional) surface of the
paths of steepest descent from a pale. The surfaces close at minima,
either by a point or by a surface at infinity. The path of steepest
ascent to the nuclei (along the charge ridge) defines the "bond path".

1. R. S. Mulliken, "Selected Papers of Robert S. Mulliken", University of Chicago Press, Chicago (1975).
2. R. G. Parr, D. P. Craig, and I. Ross, J. Chem. Phys. 18: 1561 (1950).
3. G. Zukav, "The Dancing Wu Li Masters", Wm. Morrow, New York (1979), p. 80.
4. M. Born and J. R. Oppenheimer, Ann. Phys. 84: 457 (1927).
5. D. R. Hartree, Proc. Cambridge Phil. Soc. 24: 89 (1928).
6. D. H. Whiffen, Pure Appl. Chem. 50: 75 (1978).
7. J. E. Harriman, "Theoretical Foundations of Electron Spin Resonance", Academic Press, New York (1978).
8. E. A. Hylleraas, Z. Physik 54: 347 (1929).
9. V. Fock, Z. Physik 63: 855 (1930).
10. J. C. Slater, J. Chem. Phys. 1: 687 (1933).
11. P. O. Löwdin, J. Mol. Spectr. 3: 46 (1959).
12. P. O. Löwdin, Phys. Rev. 97: 1474 (1955).
13. V. H. Smith, Jr. and I. Absar, Isr. J. Chem. 16: 87 (1977).
14. V. H. Smith, Jr., NATO Adv. Study Inst. Series B48: 3 (1980).
15. A. J. Coleman, Rev. Mod. Phys. 35: 668 (1963).
16. C. Elson and V. H. Smith, Jr. (to be published).
17. C. Elson, M.Sc. Thesis, Queen's University at Kingston (1969).
18. H. Kummer and I. Absar, J. Math. Phys. 18: 335 (1977).
19. D. R. Hartree, "Calculation of Atomic Structures", Wiley, New York (1957).
20. J. C. Slater, Phys. Rev. 35: 210 (1930).
21. V. Fock, Z. Phys. 61: 126 (1930).
22. C. Edmiston and K. Ruedenberg, Rev. Mod. Phys. 35: 457 (1963).
23. H. Weinstein, R. Pauncz, and M. Cohen, Adv. At. Mol. Phys. 7: 97 (1971).
24. J. C. Slater, Phys. Rev. 81: 385 (1951)
25. R. Gaspar, Acta Phys. Acad. Sci. Hung. 3: 263 (1954).
26. W. Kohn and L. J. Sham, Phys. Rev. 140: A1133 (1965).
27. J. C. Slater, T. M. Wilson and J. H. Wood, Phys. Rev. 179: 28 (1969).
28. D. J. McNaughton and V. H. Smith, Jr., Int. J. Quantum Chem. 3S: 775 (1970).
29. D. J. Stukel, R. N. Euwema, T. C. Collins, and V. H. Smith, Jr. Phys. Rev. B 1: 779 (1970).
30. K. Schwarz, Phys. Rev. B 5: 2466 (1972).
31. P. A. Christiansen and E. A. McCullough, Jr., J. Chem. Phys. 67: 1877 (1977).
32. C. C. J. Roothaan, Rev. Mod. Phys. 23: 69 (1951); 32: 179 (1960).
33. G. G. Hall, Proc. Roy. Soc. A205: 541 (1951).
34. N. Rösch, V. H. Smith, Jr., and M. H. Whangbo, J. Amer. Chem. Soc. 96: 5984 (1974).
35. S. Rothenberg and H. F. Schaeffer III, J. Chem. Phys. 54: 2764 (1971).

36. H. L. Hase and A. Schweig, Angew. Chem. 89: 264 (1977).
37. R. S. Mulliken, J. Chem. Phys. 36: 3428 (1962).
38. R. C. Raffenetti, J. Chem. Phys. 59: 5936 (1973).
39. W. J. Hehre, R. F. Stewart, and J. A. Pople, J. Chem. Phys. 51: 2657 (1969).
40. W. Heijser, E. J. Baerends, and P. Ros, J. Mol. Struct. 63: 109 (1980).
41. C. Herring, Phys. Rev. 57: 1169 (1940).
42. M. L. Cohen, Science 179: 1189 (1973).
43. V. H. Smith, Jr., Phys. Scripta 15: 147 (1977).
44. Y. W. Yang and P. Coppens, Sol. St. Commun. 15: 1555 (1974).
45. E. P. Wigner and F. Seitz, Phys. Rev. 43: 804 (1933).
46. J. C. Slater, Phys. Rev. 51: 846 (1937).
47. D. D. Koelling, Rep. Prog. Phys. 44: 139 (1981).
48. O. K. Andersen and R. V. Kasowski, Phys. Rev. B4: 1063 (1971).
49. H. Sambe and R. H. Felton, J. Chem. Phys. 62: 1122 (1975).
50. E. J. Baerends and P. Ros, Chem. Phys. 8: 412 (1975).
51. K. H. Johnson, Adv. Quantum Chem. 7: 143 (1973).
52. D. R. Salahub, A. E. Foti, and V. H. Smith, Jr., J. Amer. Chem. Soc. 99: 8067 (1977).
53. J. Mrozek, V. H. Smith, Jr., D. R. Salahub, P. Ros, and A. Rozendaal, Mol. Phys. 41: 509 (1980).
54. L. Brillouin, Actual. Sci. Ind. 71: 1 (1933); 159: 1 (1934).
55. C. Møller and M. S. Plesset, Phys. Rev. 46: 618 (1934).
56. C. F. Bender and E. R. Davidson, J. Chem. Phys. 49: 4222 (1968).
57. S. Green, Adv. Chem. Phys. 25: 179 (1974).
58. V. H. Smith, Jr., NATO Adv. Study Inst. Series B 48: 27 (1980).
59. P. Becker, M. E. Stephens, and F. Grimaldi (to be published).
60. J. Bicerno, D. S. Marynick, and W. N. Lipscomb, J. Amer. Chem. Soc. 100: 732 (1978).
61. V. H. Smith, Jr., P. F. Price, and I. Absar, Isr. J. Chem. 16: 187 (1977).
62. D. Munch and E. R. Davidson, J. Chem. Phys. 63: 980 (1975).
63. A. J. Thakkar and V. H. Smith, Jr., J. Chem. Phys. 67: 1191 (1977).
64. R. Benesch, S. R. Singh, and V. H. Smith, Jr., Chem. Phys. Letters 10: 151 (1971).
65. A. J. Thakkar, A. M. Simas and V. H. Smith, Jr., Chem. Phys. 63: 175 (1981).
66. W. Weyrich, P. Pattison and B. G. Williams, Chem. Phys. 41: 271 (1979).
67. E. Wigner, Phys. Rev. 40: 749 (1932).
68. R. A. Bonham and M. Fink, "High Energy Electron Scattering", Von Nostrand Reinhold, New York (1974).
69. M. Fink, D. Gregory and P. G. Moore, Phys. Rev. Letters 37: 15 (1976).
70. P. Hohenberg and W. Kohn, Phys. Rev. 136B: 864 (1964).

71. R. G. Pearson and W. E. Palke, Proc. Natl. Acad. Sci. USA 77:
     1725 (1980).
72. T. L. Gilbert, Phys. Rev. B 12: 2111 (1975).
73. J. E. Harriman, Phys. Rev. A 24: 680 (1981).
74. E. Larsson (unpublished).
75. E. Steiner, J. Chem. Phys. 39: 2365 (1961).
76. S. T. Epstein, A. C. Hurley, R. E. Wyatt and R. G. Parr,
     J.Chem. Phys. 47: 1275 (1967).
77. E. B. Wilson, Jr., J. Chem. Phys. 36: 2232 (1962).
78. A. A. Frost, J. Chem. Phys. 37: 1147 (1962).
79. R. Schwendeman, J.Chem. Phys. 44: 556 (1966).
80. P. Politzer and D. G. Truhlar, eds., "Chemical Applications
     of Atomic and Molecular Electrostatic Potentials", Plenum,
     N. Y. (1981).
81. M. Morse, "Pits, Peaks and Passes", Math. Assoc. America (1966).
82. K. Collard and G. G. Hall, Int. J. Quantum Chem. 12: 623 (1977).
83. S. Srebrenik, R. F. W. Bader, and T. Tung Nguyen-Dang,
     J. Chem. Phys. 68: 3667 (1978).

CONCEPTS OF CHARGE DENSITY ANALYSIS: THE EXPERIMENTAL APPROACH

Philip Coppens

Department of Chemistry
State University of New York at Buffalo
Buffalo, New York 14214

INTRODUCTION

Though the potential for deriving electron distributions from
elastic diffraction data was recognized in the decade following
Von Laue's discovery of X-ray diffraction, its practical application was delayed by many unrecognized problems of both experimental and theoretical nature. Only with the advent of four-
circle diffractometers and high-speed computing could sufficiently
accurate data be collected, corrected for physical effects like
extinction, absorption and thermal diffuse scattering, and interpreted with a scattering factor formalism based on a more
realistic description of bonded atoms.

As with any other area of science the development of the
field has been accompanied by the genesis of a small flood of new
words, a peculiar jargon, readily understood by participants in
the venture, but a secret language to those with less direct
involvement and interest.

It is the purpose of this article to explain some of these
terms and the concepts they represent, and where appropriate to
give examples taken from recent studies. I will briefly treat
some experimental concepts which are needed for the subsequent
discourse, continue with a discussion of formalisms used to
describe the electron density and its Fourier transform, the X-ray
structure factors, and proceed with treatments of electron density
mapping, of space partitioning and the propagation of experimental
errors into the experimental charge density. Wherever possible
this the emphasis will be on the linguistic rather than the

mathematical aspects which are treated in detail
in the referenced publications.

FOURIER TRANSFORMS AND VALENCE SCATTERING

When X-rays are diffracted by a crystal with its regular
periodicity the scattered beams are observed in specific well
defined directions which make an angle described as $2\theta_{hk\ell}$ with the
incident beam, h, k and ℓ being the Miller indices of the diffracting planes. Scattering theory shows that the amplitude F of
a beam diffracted in the direction described by the scattering
vector $\underset{\sim}{S}$ with length $2\sin\theta/\lambda$ (Fig. 1) is given by the Fourier

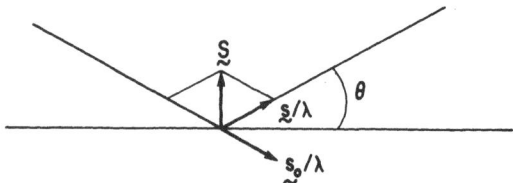

Fig 1. Definition of the scattering vector S. The vectors s_o
and s are unit vectors in the directions of the incident
and diffracted beams respectively. S is perpendicular to
the diffracting plane and has a magnitude $2\sin\theta/\lambda$.

transform relationship

$$F(\underset{\sim}{S}) = \int_{\substack{unit \\ cell}} \rho(\underset{\sim}{r}) \exp 2\pi i \ \underset{\sim}{S}.\underset{\sim}{r} \ d\underset{\sim}{r} \tag{1}$$

where $\rho(\underset{\sim}{r})$ is the electron density at the point defined by the
vector $\underset{\sim}{r}$.

It is a well known property of such a Fourier transform
relation that compact functions transform into diffuse ones and
vice versa. In our case this means that the compact inner electron shells of the atomic "core" are diffuse in scattering space,
i.e. have relatively large values of F at high values of $|S|$,
while valence electrons scatter mainly in the low order region.
This is the distinction between <u>high-order</u> (= large angle) data
and <u>low-order</u> data which are much more susceptible to bonding
effects concentrated in the valence electrons. We define:

High-order data

> "The data above a value of $\sin\theta/\lambda$ at which
> valence scattering is estimated to be small"

How can we define something as subjective as "small" in this context? For a free nitrogen atom in its ground state the valence scattering becomes insignificant above about 0.5 Å^{-1} (Fig. 2). But when the Fourier transform of a nitrogen molecule is considered a different result emerges. Fig. 3 shows the difference $\Delta F(S)$ between $F(S)$ of the nitrogen molecule[1] and the transform of $\tilde{t}$wo free nitrogen atoms at bonding distance. It illustrates that the effect of bonding is anisotropic, i.e. depends on the direction of S [which is a logical result of bonding depending on (r)], and $\tilde{a}$lso that the effect of bonding may extend beyond the $\tilde{v}$alence

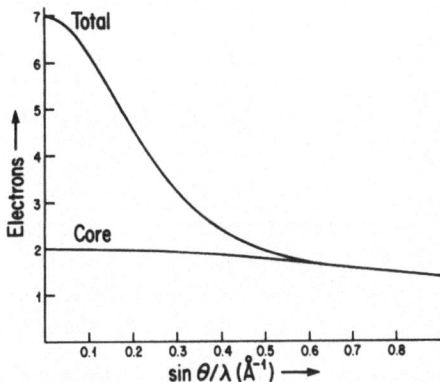

Fig. 2.   Total and core scattering of the free nitrogen atom.

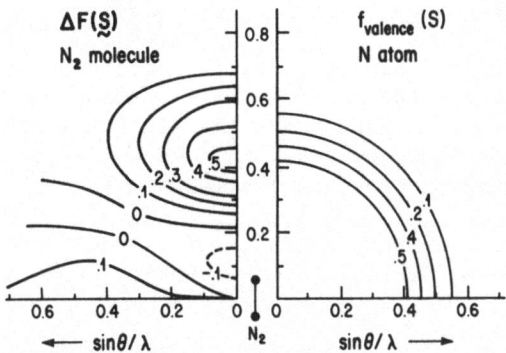

Fig. 3.   Transform of the theoretical deformation density of the nitrogen molecule (Ref. 1), compared with the valence shell scattering of an isolated nitrogen atom.

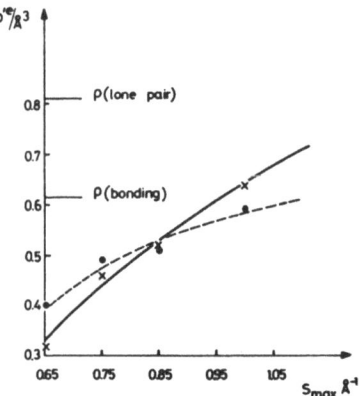

Fig. 4.   The peak value of the X-N deformation density in the
bonding and lone pair regions of p-nitropyridine N-oxide
as a function of the cut-off value in sin θ/λ. Circles
are average values for the maximum in the C-C and C-N
bonds, crosses are average peak values in the oxygen
lone pair regions. The limits for infinite resolution,
as well as the line describing the best least-squares
fit are indicated (Refs. 2, 18).

scattering limit of the isolated atom. This is because electron
density accumulates in a <u>compact</u> region of the covalent bond.
Thus its Fourier transform is more diffuse and extends to higher
values of $\sin\theta/\lambda$.

For more compact parts of the valence density the scattering
extends even further as illustrated in Fig. 4 for a lone-pair peak
on the oxygen atoms of the nitro-group in p-nitropyridine N-
oxide,[2] which increases in height as diffraction data of higher
and higher sin θ/λ are added in the calculation.

Thus, the appropriate choice of what constitutes high-order
depends on the material, the nature of its bonding, and the kind
of atoms present. Over the last years the sin θ/λ limit has
perceptibly moved from about 0.55-0.65 $\mathring{A}^{-1}$ to 0.75-0.85 $\mathring{A}^{-1}$ as
more extended data sets have become available.

THE CALCULATION OF X-RAY SCATTERING AMPLITUDES: MODELING THE
CHARGE DENSITY

Most commonly experimental measurements of scattering amplitudes are made on a relative rather than an absolute scale. Thus
there is a need for the introduction of a <u>scale factor</u> k defined
here as

$$F_{observed} (\underset{\sim}{S}) = k\, F_{calc} (\underset{\sim}{S}) \qquad (2)$$

where $F_{calc}(\underset{\sim}{S})$ as defined in (1) is to be evaluated at the points H of the reciprocal lattice corresponding to the periodicity of the crystal under consideration:

$$F_{calc} (\underset{\sim}{H}) = \int_{\substack{unit \\ cell}} \rho(\underset{\sim}{r}) \exp 2\pi i\; \underset{\sim}{H}.\underset{\sim}{r}\; d\underset{\sim}{r} \qquad (3)$$

It can be shown in a straightforward manner[3] that the integral in (3) may be replaced by a sum over the scattering of individual fragments of the charge density in the unit cell.

In the most common application of this principle the fragments are atoms with scattering amplitude $f_i(H)$ at the positions $\underset{\sim}{r}_i$ which gives

$$F_{calc} (\underset{\sim}{H}) = \sum_{\substack{all \\ atoms}} f_i (\underset{\sim}{H}) \exp 2\pi i\; \underset{\sim}{H}.\underset{\sim}{r}_i\; d\underset{\sim}{r} \qquad (4)$$

When f_i is the free-atom scattering factor equation (4) represents the free-atom model. As charge density analysis is concerned with the effects of bonding on the electron density, the free-atom model is clearly inadequate. The reasons for this inadequacy can be loosely divided into three parts (a) atoms are not exactly neutral but participate in charge transfer; (b) they show deviations from spherical symmetry because of preferential occupation of non bonding orbitals containing "lone pairs" of electrons, and (c) they may form covalent bonds in which electron density is shared by two or more atoms.

When these effects are neglected as in the free-atom model the atomic positions obtained by least-squares adjustment of the F_{calc}'s to the set of F_{obs} may be biased; they may be moved for example towards the direction of the lone pair electrons[4] by a small amount (< 0.01 Å), the size of which is dependent not only on bonding but also on the $\sin\theta/\lambda$ cut-off of the data set.[5] We define

The asphericity shift

"The difference between the atomic positions derived by least-squares refinement of X-ray data using the free-atom model and the unbiased position from, for example, neutron diffraction or refinement of very high-order X-ray data".

A similar model-dependent bias may occur in the experimental temperature parameters. The most straightforward improvement beyond the free-atom model is incorporation of charge transfer between the valence shells of the atoms in the unit cell. Since a change in electron population affects the electron-electron repulsions, a change in the radial dependence of the valence shell density must also be incorporated in such a model. This may be done by means of a parameter kappa (κ), such that the modified valence density ρ' is given by[6]

$$\rho'_{valence}(r) = \rho_{valence}(\kappa r) \tag{5a}$$

Thus if kappa is larger than one, the "true" valence density at distance r from the nucleus corresponds to the valence density at a larger distance in the free atom, or in other words, the atom is contracted. Kappa can therefore be described as an expansion-contraction parameter which may be derived together with the valence shell population parameter P in the atomic density expression:

$$\rho_{atom}(r) = \rho_{core}(r) + P_{valence}\,\rho_{valence}(\kappa r) \tag{5b}$$

The Fourier transform of (5b) which is used in the scattering amplitude calculation is

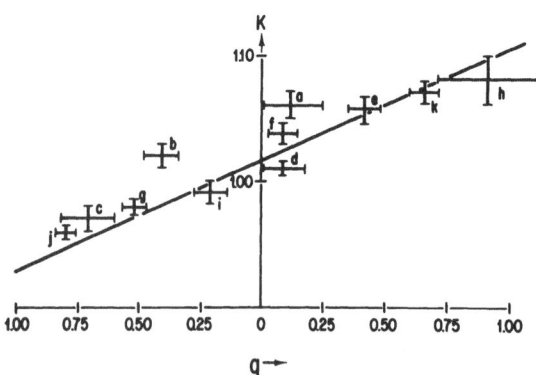

Fig. 5   Relation between κ and q (net charge) for N atoms in a number of structures (a), (b) glycylglycine, (c) formamide, (d), (e) p-nitropyridine N-oxide, (f) sulfamic acid, (g), (h) NH_4SCN, (i) NaSCN, (j), (k) KN_3. Bars indicate estimated standard deviations. The full line is the relation predicted by Slater's recipe for atomic orbital exponents. (Ref. 6).

$$f_{atom}(S) = f_{core}(S) + P_{valence} \kappa^3 f_{valence}(S/\kappa) \quad (6)$$

which shows again that the scattering of a more compact atom ($\kappa > 1$) extends to larger values of $S = 2\sin\theta/\lambda$.

Since the introduction of kappa was justified by the variation in electron-electron repulsion with electron population, a relationship must exist between κ and P. This is nicely demonstrated in Fig. 5 for a series of compounds in which nitrogen atoms bear both positive and negative charges.[6] The full line in the figure is the correlation predicted by Slater's recipe for the relation between atomic charge and orbital exponent,[7] a very clear experimental demonstration of the validity of Slater's rules!

To recap:

The kappa parameter

"A factor multiplying the radial coordinate of the atomic valence density, allowing expansion or contraction of the valence-shell with variation of atomic charge".

The refinement including the kappa parameter is often referred to in the local slang as a radial refinement. The logical second step past this spherical atom model is the introduction of aspherical atom-centered density functions. Application of such formalisms has been pioneered by Dawson,[8] by Stewart[9] and by Hirshfeld.[10] The aspherical density fragments are functionaly identical to the p, d, f, g----hydrogen atom orbitals, with the important distinction that the electron probability distribution, rather than its wave function is described. The angular functions are the well known spherical harmonics $y_{\ell m}$ labelled with the quantum numbers ℓ and m, with $\ell = 1,2,3,4,5$--- representing dipolar, quadrupolar, octopolar, hexadecapolar, tridecadipolar etc. functions (Fig. 6).

Expression 5(b) for the atomic density is thus generalized to give:

$$\rho_{atom}(\underset{\sim}{r}) = \rho_{core}(r) + P_{valence}\, \rho_{valence}(\kappa r)$$

$$+ \sum_{\ell} \sum_{m} P_{\ell m}\, y_{\ell m}(\theta,\Phi)\, R_{\ell}(r) \quad (7)$$

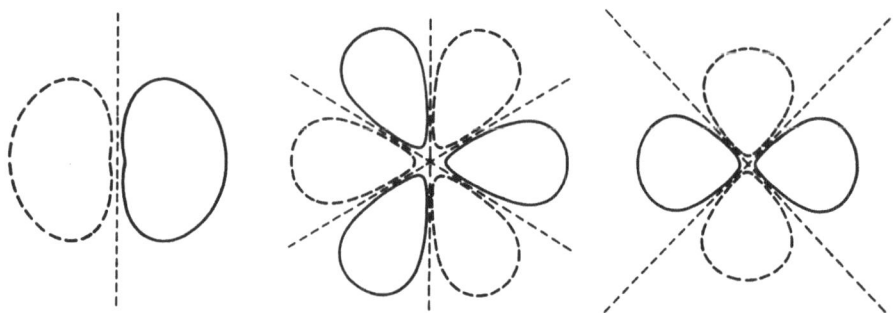

Fig. 6 Graphic representation of some multipolar functions:
        (from left to right) a dipole, an octopole and a quad-
        rupole function. Functions are negative within dotted
        areas. If the x axis is horizontal and the y axis is
        vertical in the figure, the functions are x, x^3-3xy^2 and
        x^2-y^2 respectively.

in which θ, Φ and r are the spherical polar coordinates, R_ℓ is a
radial function, which may be different for different ℓ, and the
$P_{\ell m}$ represent the population coefficients. In the structure
factor equation (7) is replaced by its reciprocal space equivalent:

$$f_{atom} (\underset{\sim}{S}) = f_{core} (\underset{\sim}{S}) + P_{valence} \kappa^3 f_{valence} (S/\kappa)$$

$$+ \sum_\ell \sum_m P_{\ell m} f_{\ell m} (\underset{\sim}{S}) \qquad\qquad (8)$$

where $f_{\ell m} (\underset{\sim}{S})$ is the transform of $y_{\ell m} (\theta,\Phi)R_\ell(r)$, the evaluation
of which is well documented.[11] Note that in (7) $\rho_{atom}(r)$ has been
replaced by $\rho_{atom} (r)$ to indicate the aspherical nature of the
atom. The radial function $R_\ell(r)$ is usually of the form $r^n\ell$
exp-ζr, as described in more detail in the literature.[12]

The formalism developed by Hirshfeld[10] can be shown to be
equivalent to the expansion of spherical harmonic functions. It
uses what we will describe as "Hirshfeld functions" of the form
$r^n cos^n \theta_K$ where the θ_K are the angles with well defined directions,
which in the case of n = 1 are the three edges of a cube. For n
= 1 the Hirshfeld angular functions are the three dipolar functions of the spherical harmonic expansion; for higher values of
n, however, the relationship is somewhat more complicated. For n
=2, for example, there are six functions which are equivalent to

Table 1   Direction cosines in Hirshfeld formalism and relation
           the spherical harmonics

n	directions	corresponding spherical harmonics and their number
0		s(1)
1	100,010,00$\bar{1}$	p(3)
2	110,1$\bar{1}$0,101,10$\bar{1}$,011,01$\bar{1}$	d(5)+s(1)
3	110,1$\bar{1}$0,101,10$\bar{1}$,011,01$\bar{1}$,111, 1$\bar{1}\bar{1}$,$\bar{1}$1$\bar{1}$,$\bar{1}\bar{1}$1	f(7)+p(3)
4	100,010,001,α11,1α1,11α,$\bar{\alpha}$11 1$\bar{\alpha}$1,11$\bar{\alpha}$,$\alpha\bar{1}$1,1$\alpha\bar{1}$,$\bar{1}$1α,α1$\bar{1}$,$\bar{1}\alpha$1 1$\bar{1}\alpha$†	g(9)+d(5)+s(1)

† $\alpha = \sqrt{2} - 1$

the five quadrupoles plus an $\ell = 0$ monopolar term. The relationships are summarized in Table 1. As we will see in some examples
in section 6 both formalisms lead to very similar results when
applied in a least-squares adjustment of the charge density
parameters $P_{valence}$, $P_{\ell m}$, κ and ζ together with the conventional
structure parameters.

In principle the aspherical atom formalism, which accounts
for both charge transfer and atomic asphericity, should be comple-
mented with the addition of functions allowing for the overlap
density represented by products $\Phi_A \Phi_B$ of atomic orbitals Φ on atoms
A and B.

The product $\Phi_A \Phi_B$ is referred to as a two-center term or bond
term. It may have large values in the bond region between adjacent atoms. It's Fourier transform is rather complicated but has
been described in the literature so that incorporation in a
structure amplitude formalism is straightforward.[13,14] The
problem encountered in practice is that the scattering of $\Phi_A \Phi_B$
closely resembles the scattering of a combination of multipoles
centered on atoms A and B.[15] Such redundancy in the formalism
introduces a large correlation in a least-squares refinement and
results in lack of reliability of the populations of the individual terms as obtained from the refinement.

Other bond-centered functions which have been proposed are simply Gaussian density functions located at or near the bond centers.[16] These "charge-clouds" (Ladungswolken), as they have often been referred to, fall off more rapidly towards the atom centers than the two-center terms $\Phi_A \Phi_B$. Thus correlation with one-center terms may be less serious. But agreement between observed structure factors and those calculated using a charge cloud formalism with a small set of one-center functions is usually not better than what may be obtained with a full multipole expansion without bond terms.

The "charge-cloud" model has also been applied successfully to describe spin density localized in Fe-O bonds in yttrium iron garnet.[17]

When the formalisms described are used in a least squares refinement the corresponding labels are attached to describe the refinement. Thus:

Kappa refinement

"Least-squares refinement of all structural parameters and
kappa and spherical valence shell population parameters on
all atoms"

Multipole refinement

"Least-squares refinement of all structural parameters,
kappa, and spherical valence shell and multipole population
parameters on all atoms."

Hirshfeld refinement

"Least-squares refinement of all structural parameters,
exponents and population parameters of Hirshfeld-type
$r^n \cos \theta_K \exp(-\zeta r)$ functions on all atoms"

An important difference between the multipole and the Hirshfeld refinement is that the latter describes the deformation from the free-atom density which is the leading term in the expansion, while the former allows the free-atom valence shell itself to be modified through the use of the kappa parameters. A comparison of the two refinements is discussed in section 6.

ELECTRON DENSITY MAPS: ρ_{TOTAL}, $\rho_{VALENCE}$ AND $\rho_{DEFORMATION}$

The total charge density $\rho_{total}(\tilde{r})$ is dominated by the spherical atom densities. When the latter are subtracted the deviations from the free-atom model become more obvious. The subtracted part and the difference are referred to as

The promolecule density

"The charge density prior to the formation of bonds, i.e. the superposition of spherical free-atoms each centered at its position in the molecule"

The procrystal density

"The charge density prior to the formation of bonds, i.e. the superposition of spherical free-atoms each centered at its position in the crystal"

and the

The deformation density

"The difference between the total electron density and the promolecule (or procrystal) density"

The deformation density represents the rearrangement of the electron density due to interatomic bonding.

The more generally used valence density is just the probability distribution of the valence electrons. In the context of an experimental study it is defined as follows:

The valence density

"The difference between the total electron density and the superposition of free-atom atomic core densities"

This working definition assumes that the atomic cores are not affected by chemical bonding. Though this is not strictly correct the deviations are not observable in an X-ray experiment of the resolution and accuracy obtained at present.

While the deformation density gives a good demonstration of

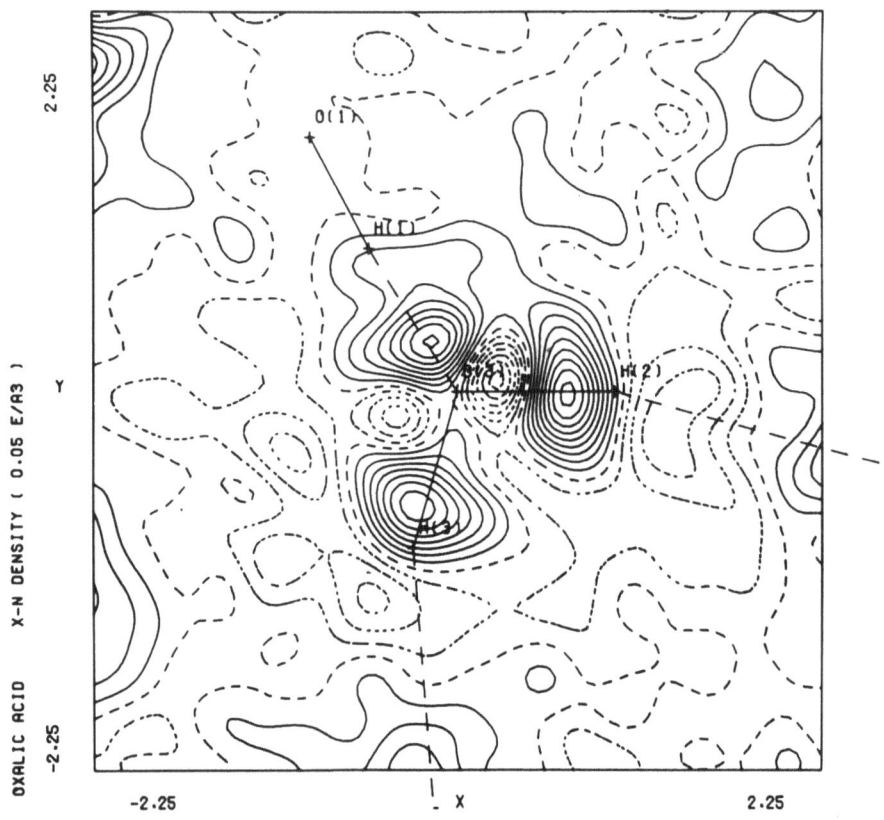

Fig. 7.   X-N deformation density in the plane of the water molecule
in α-oxalic acid dihydrate.   Contours 0.05eÅ⁻³, zero
and negative contours broken (Ref. 23).

covalent bonding and lone pair density (Fig. 7), the valence
density has been used by Streitwieser and collaborators in their
analysis of bonding in organic molecules (Chapter 6).   It is also
obtained directly in band structure calculations of metals and
semiconductors in which the core electrons are represented by a
pseudo potential (Chapter 3, part 1).

To illustrate the difference between the valence and the
deformation densities let us consider a simple model for a $(He)_2$
molecule.   Such a molecule has 2 molecular orbitals formed by the
positive and negative combinations (bonding and antibonding
respectively) of the 1s orbitals on the two He atoms which we
label A and B.   The charge density is given by

$$\rho_{bonding} = \frac{2}{2+2S} (\Phi_A^2 + 2\Phi_A\Phi_B + \Phi_B^2)$$

$$\rho_{antibonding} = \frac{2}{2-2S} (\Phi_A^2 - 2\Phi_A\Phi_B + \Phi_B^2) \qquad (9a)$$

where S is the overlap integral $\int \Phi_A\Phi_B$ dv. Each orbital is populated by 2 electrons, while the denominator in (9a) respresents the normalization factor.

Summing the two terms gives for the total density (which is equal to valence density in this example):

$$\rho_{total} = \rho_{valence} = \frac{2}{1-S^2} (\Phi_A^2 + \Phi_B^2) - \frac{4S}{1-S^2} \Phi_A\Phi_B \qquad (9b)$$

We are interested in the density at the bond midpoint where Φ_A and Φ_B have equal values of, say, Φ_{mid}. Thus

$$\rho_{valence} = \frac{4}{1-S^2} \Phi_{mid}^2 - \frac{4S}{1-S^2} \Phi_{mid}^2 = \frac{4}{1+S} \Phi_{mid}^2$$

while

$$\rho_{deformation} = (\frac{4}{1+S}) \Phi_{mid}^2 - 4\Phi_{mid}^2 = -\frac{4S}{1+S} \Phi_{mid}^2 \qquad (9c)$$

Thus for this unstable molecule which has no net covalent bonding, the deformation density at the bond midpoint is always negative (S is positive), but the valence density at bond midpoint is positive since the overlap integral S has values between zero and one.

While the presence of density accumulation in the bonds in a deformation map is indicative of covalent bonding this is not the case for the valence density. But the opposite statement cannot be made: the absence of a bond peak in the deformation density does not prove the absence of covalent contributions to bonding, because neutral spherical atoms are subtracted which, for elements with more than half-filled shells, have more than one electron per orbital. When, for example, a spherical oxygen atom is subtracted in an N-O or O-O bond, 6/4 electrons are removed per valence hybrid. But the atomic bonding orbitals Φ in expression (9a) must be singly occupied so that valence bond "pairing" of electrons can be accomplished. So the bonded atom configuration for an oxygen atom may be written as $(1s)^2(2s)^2(2p_x)^2(2p_y)(2p_z)$

plus the two-center term in (9a) (z axis along the bond direction). But when we subtract out 6/4 electrons per oxygen orbital, the extra half electron subtracted out in the bonding region may more than compensate for the bond peak, thus giving a negative deformation density.

This aspect of the deformation density has led to considerable confusion in the literature as N-O bonds in nitropyridine N-oxide,[18] the O-O bond in hydrogen peroxide,[19] the Cr-Cr bond in chromous acetate[20] and the Mn-Mn bond in $Mn_2(CO)_{10}$, dimanganese decacarbonyl,[21] showed little or no bonding density.

Instead of spherical atoms we may subtract out prepared fragments, like the $Mn(CO)_5$ fragment in the case of $Mn_2(CO)_{10}$. Such a "procomplex" or "procluster" density corresponds to a prepared fragment prior to formation of the final bond under examination.

The difference between the theoretical total and "procluster" densities of $Mn_2(CO)_{10}$ shows a small but significant amount of accumulation in the Mn-Mn bond.[22]

ELECTRON DENSITY MAPPING WITH EXPERIMENTAL DATA

Since the structure factors F_{obs} are the Fourier transform of the charge density (expression 3), the charge density is the inverse Fourier transform of the F_{obs}:

$$\rho(\underset{\sim}{r}) = \int \{F_{obs}(\underset{\sim}{H})/k\}exp-2\pi i\underset{\sim}{H}.\underset{\sim}{r}d\underset{\sim}{r} = \frac{1}{V}\Sigma\{F_{obs}(\underset{\sim}{H})/k\}exp-2\pi i\underset{\sim}{H}.\underset{\sim}{r} \quad (10a)$$

where the integral is replaced by a summation over the amplitudes of all diffracted beams allowed by Bragg's law; k is the scale factor defined by expression (2).

For the application of expression (10a) we must know not only the magnitude of F(H), but also its phase, which is not observed. This offers no handicap in the case of a centrosymmetric structure for which the phases must be either 0 or π, and are almost always correctly calculated with the free-atom model (an exception occurs in very simple structures where weak reflections sometimes are due only to deviations from the free-atom model, the 222 reflection in silicon being the best known example). When the structure is acentric, however, the phases can have any value between 0 and 2π, and the calculated values are model dependent. The best approach is to calculate the phases with the most flexible

model available such as the multipole expansion; but to some
extent the phases and therefore the map of $\rho(\underset{\sim}{r})$ remain model
dependent in the acentric case.

To calculate the deformation density from the experimental
data the structure factors as calculated with the free-atom model
must be subtracted from the summation in (10a)

$$\rho_{deformation} \, (\underset{\sim}{r}) = \frac{1}{V} \, \Sigma [F_{obs} (\underset{\sim}{H})/k - F_{calc, \, free \, atom} (\underset{\sim}{H})]$$

$$exp-2\pi iH.r \qquad\qquad\qquad\qquad\qquad (10b)$$

The $F_{calc, \, free \, atom}$ values in (10b) must be calculated with
unbiased atomic parameters, free from asphericity shifts, otherwise the deformation density features will be sharply reduced and
only the features not accounted for in the least-squares refinement remain.

The map obtained with the least-squares parameters allows us
to judge the success of the fitting procedure. It contains
features not accounted for by the model and is referred to as
the:

Residual map

"The difference between the total charge density and the
charge density calculated with the scattering factors and
parameters of a least squares refinement, i.e. the remnant
of the density not described by the refinement model"

A completely independent set of parameters may be obtained
from a neutron diffraction experiment. The deformation density
map calculated with these parameters is often called an X-N map:

X-N map

"The difference between the experimental total charge density
and the charge density calculated from Hartree-Fock atomic
X-ray form factors and neutron positional and thermal
parameters".

A drawback of the X-N method is that the temperature parameters of the neutron experiment are sometimes systematically
different from the X-ray values for reasons which are not well
understood, but are due to unrecognized systematic errors in
either data set. It becomes then necessary to scale the average
values of the neutron thermal parameters $(U_{ii})_N$ to the X-ray
X-N map.

An alternative to the X-N method is the use of high order X-ray rather than neutron parameters; referred to as:

X-X map

"The difference between the experimental total charge density and the charge density calculated from Hartree-Fock atomic X-ray form factors and high-order X-ray positional and thermal parameters".

This "X-X" procedure is quicker, as it eliminates the neutron experiment, and therefore has become quite popular. But it is not satisfactory when regions around hydrogen atoms are to be studied. A compromise which combines some of the advantages of both methods, uses parameters from a joint X-ray-multipole/ neutron refinement. In this refinement structural parameters depend on both data sets while charge density parameters are, of course, only dependent on the X-ray data.[23]

The corresponding deformation map, with F_{calc}'s from the free atom model, is referred to as the X-(X+N) map, in analogy with the X-X and X-N designations.

Examples of X-N and X-X maps can be found in Chapter 5, part 2 by Goddard and Kruger. An X-(X+N) map on oxalic acid dihydrate shows a slight increase in lone pair peak heights in comparison with an X-X map based on the same X-ray data. This is to be expected if the high-order X-ray data results were not entirely free of bias due to the free-atom model.

After successful completion of an aspherical atom refinement the density due to the multipole functions may be plotted; giving a

Multipole deformation map

"The difference between the total charge density calculated with the multipole model parameters and the charge density calculated from Hartree-Fock free atom X-ray form factors and multipole model positional and thermal parameters."

If the multipole model functions are plotted directly in the unit cell the multipole deformation map refers to a (pseudo)- static density because the thermal parameters are not included. It also eliminates the effects of truncation of the summation over the structure factors in (10) beyond the values of F which have been derived from the experiment. Such an extrapolation to infinite resolution introduces detail which depends on the multi-

pole functions selected, which are not unique (because of the
finite size of the expansion, the choice of the exponent n_ℓ in
the radial function etc). A dynamic multipole deformation density
based more directly on the experimental data is obtained by
Fourier transforming

$$\Sigma(F_{calc,multipole\ model} - F_{calc,free\ atom})$$

analogous to (1). Since the calculated structure factors include
the effect of thermal motion, the density is smeared by the
molecular vibrations and therefore referred to as a

Dynamic density:

"The charge density smeared by the vibrational motions in
 the crystal:

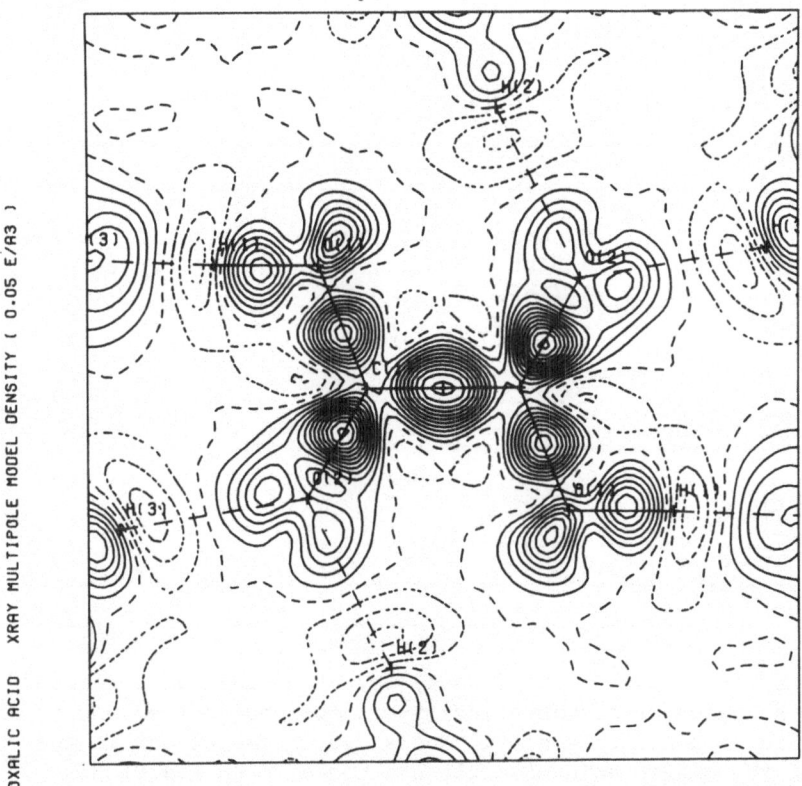

Fig. 8a. Multipole model maps in the plane of the oxalic acid
         molecule (a) dynamic density contours 0.05 e$\overset{o}{A}^{-3}$. Zero
         and negative contours broken.

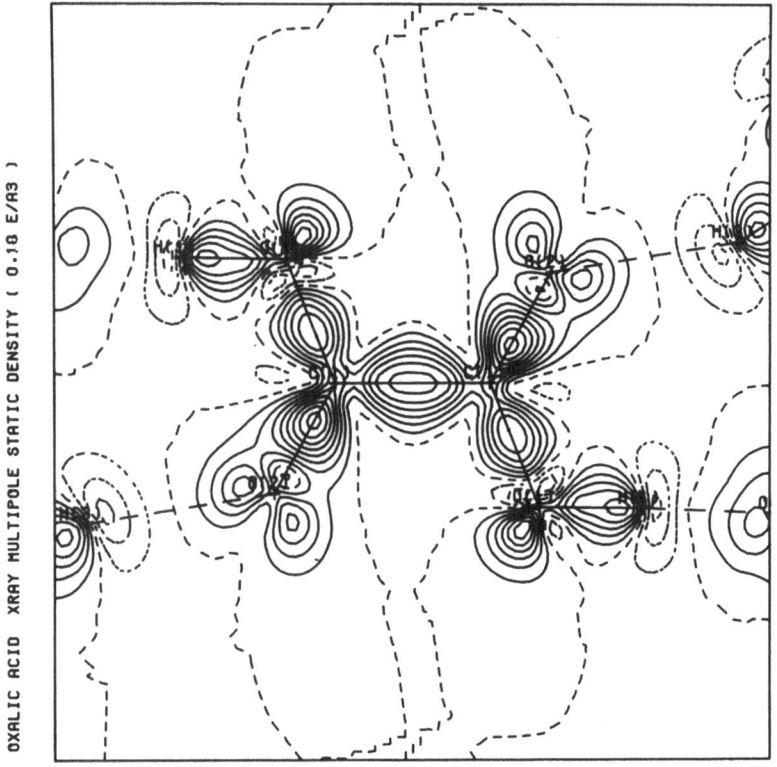

Fig. 8b. Multipole model maps in the plane of the oxalic acid
         molecule. (b) Static density at twice the contour
         interval of a) (Ref. 23).

compared with the:

Static density

    "The charge density of an atom, promolecule or molecule at
     rest"

Note that though the static density may be derived as described,
as well as calculated theoretically, it does not correspond to an
observable in nature. A static and a dynamic model map in the
plane of the oxalic acid molecule are compared in Figure 8.

    Table 2 provides a summary of the density functions discussed
in this section.

Table 2. Definition of Fourier difference maps

General Expression:

$$\Delta\rho = \frac{1}{V} \Sigma(F_1 - F_2) \exp{-2\pi i H \cdot r}$$

a) Residual map

$F_1 = F_{obs}$
$F_2 = F_{calc,model}$.

b) X-N deformation map

$F_1 = F_{obs}$
$F_2 = F_{calc,neutron\ parameters,\ free-atom\ model}$

c) X-X deformation map

$F_1 = F_{obs}$
$F_2 = F_{calc,high\ order\ parameters,\ free-atom\ model}$

d) X-(X+N) deformation map

$F_1 = F_{obs}$
$F_2 = F_{calc,joint\ X-ray\ neutron\ parameters,\ free-atom\ model}$.

e) X-N valence map
    X-X valence map
    X-(X+N) valence map

as b,c,d but F_2 calculated with core
electron contribution only

f) dynamic multipole
    deformation map

$F_1 = F_{model}$
$F_2 = F_{calc,multipole\ model\ structural\ parameters,\ free-atom\ model}$.

g) static multipole
    deformation map

$\rho_{valence,multipole\ model} - \rho_{valence,free-atom\ model}$

Table 3   Aspherical atom refinements of pyridinium dicyano
methylide

	Hirshfeld formalism	Multipole formalism
Number of variables (v)	190	162
Number of reflections (n)	2391	2390
$R = \{\dfrac{\Sigma|F_o^2 - k^2 F_c^2|}{\Sigma F_o^4}\}^{1/2}$	0.028	0.029
$R_w = \{\dfrac{\Sigma w|(F_o^2 - k^2 F_c^2)|}{\Sigma w F_o^4}\}^{1/2}$	0.035	0.039
$S = \{\dfrac{\Sigma w|(F_o^2 - k^2 F_c^2)|}{n-v}\}^{1/2}$	1.71	1.85

AN EXAMPLE: THE COMPARISON OF ELECTRON DENSITY MAPS FROM
ALTERNATIVE LEAST SQUARES REFINEMENTS.

        The density functions described above may be used to judge
the different least squares models.   In a study by Baert, Coppens,
Stevens and Devos, both the multipole and the Hirshfeld refinements were applied to a set of diffraction data on pyridiniumdi-
cyanomethylide.[24]   The data extend to $\sin\theta/\lambda = 1.06\text{Å}^{-1}$ and show
good agreement between symmetry-related reflections, the internal
consistency R factor ($\frac{\Sigma|<F^2>-F^2|}{\Sigma F^2}$) being equal to 0.005, 0.013
and 0.4 for $\sin\theta/\lambda < 0.4$ Å^{-1}, $0.4 < \sin\theta/\lambda < 0.9$ Å^{-1} and $\sin\theta/\lambda$
> 0.9 Å^{-1} respectively.   Some details of the two refinements are
summarized in Table 3.   The number of parameters is larger in the
Hirshfeld refinement.   This is due to the additional monopolar,
dipolar and quadrupolar functions implied in the Hirshfeld
formalism (Table 1) and also to a slightly more flexible model
for the hydrogen atoms used in this particular study.   As may be
expected the final R factors and the goodness of fit are slightly
lower for the more flexible model.

        How similar are the results of the two refinements?   The
least-squares parameters give very little insight into the
question.   Rather, one has to resort to electron density maps or
to physical properties such as electrostatic moments which are
implied in the results.

        Residual maps, and static model maps in the plane of the
pyridinium ring are shown in figs. 9.   A crystallographic mirror
plane, oriented perpendicular to the ring, passes through the
N(1) and C(1) atoms, so that the left and right hand sides of the
figures are identical.

        Examination of the residual maps (fig. 9a) shows a striking
similarity, with slightly higher features in the map based on the
less flexible multipole refinement.   The dynamic model maps are
also very similar, though the peaks in several of the ring bonds
are somewhat more elongated along the bond in the Hirshfeld maps.
This difference is accentuated in the static model maps (fig.
9b), which are plotted at infinite resolution and zero thermal
motion.   The conclusion of this study is that the static maps are
quite dependent on the functions which make up the model.   This
is not unexpected as the maps represent an extension of the image
beyond the experimental resolution.   This extension relies completely on the correctness of the fitting functions.

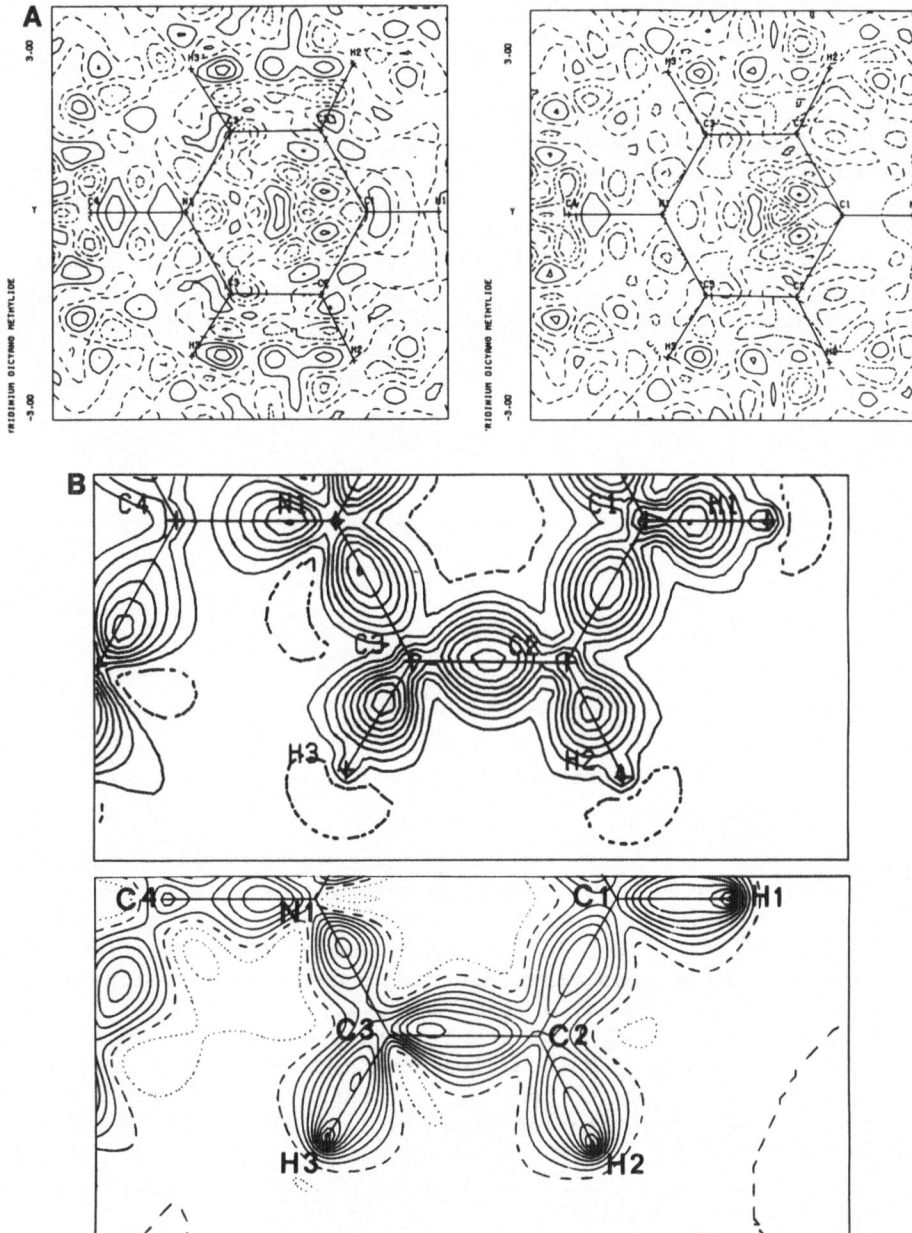

Fig. 9.   Residual and static model maps in the plane of the
          pyridine ring of pyridiniumdicyanomethylide, $C_5H_5NC(CN)_2$

          a) residual map left: from multipole refinement;
             right: from Hirshfeld refinement, contours 0.05 eÅ^{-3}.

          b) static model maps: top from multipole refinement;
             bottom: from Hirshfeld refinement. Contours 0.1eÅ^{-3}.
             (Ref. 24).

DEFINITION OF ATOMS AND MOLECULES: THE PARTITIONING OF CRYSTAL
SPACE.

The multipole refinement using atom-centered functions
implicitly divides the charge density into atomic fragments.
But, since the multipole functions and especially those with
large ℓ extend quite far from the atom they are centered on, they
also represent some density localized near neighboring nuclei.
There is a basic arbitrariness in the partitioning of a continuous
function such as the charge density, in particular when the
density does not decrease to a very small value in the region
between the centers of the fragments.   Stewart[25] has used the
term pseudoatom rather than atom to emphasize the ambiguity:

Pseudoatom

"All the charge density described by functions centered at
 the position of the nucleus."

Similarly we have

Pseudomolecule

"All the density attributed to the pseudoatoms which constitute the molecule."

Since the density between molecules in a crystal usually
decreases to very small values, choice of a pseudomolecular
fragment will be not as crucial, and pseudomolecular properties
may be expected to be less dependent on the definition selected.

The multipoles on adjacent atoms overlap each other, much
like atomic orbitals, so that no clear boundary between fragments
can be traced. This type of boundary is described as "fuzzy";
it is functional rather than localized.

Fuzzy boundary

"The separation between overlapping functions of neighboring
 fragments of the charge density".

In an attempt to define better localized fuzzy atoms Hirshfeld[26] has developed the:

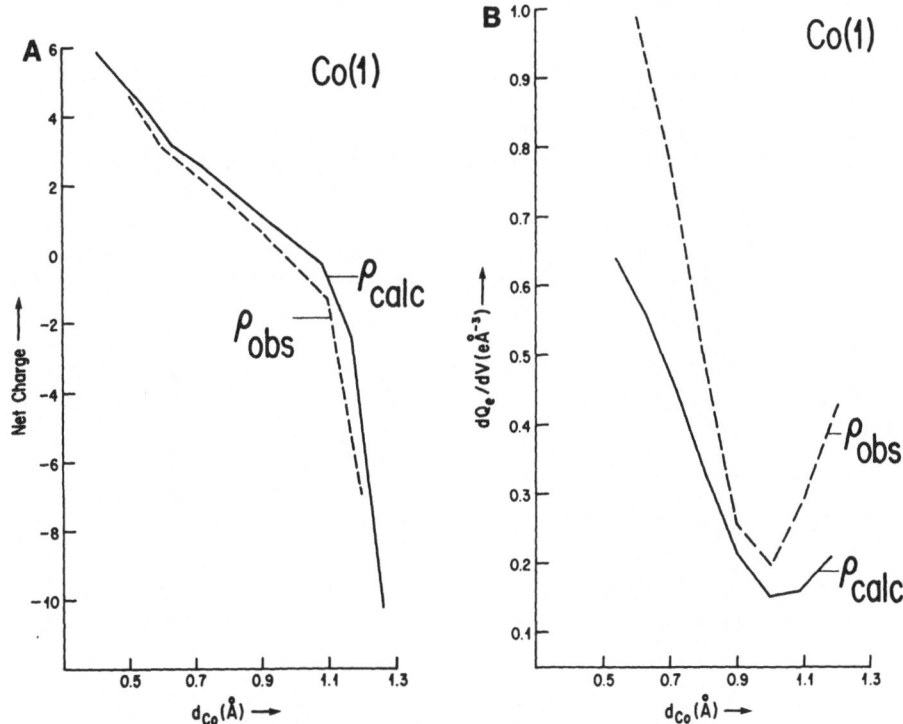

Fig. 10 Charge integration over the volume of one of the cobalt
atoms in methylidyne tricobalt nonacarbonyl, $Co_3CH(CO)_9$.

a) net charge as a function of the distance of the boun
dary planes between Co and C from the cobalt atom, d_{Co}

b) the average electron density (in $e\text{Å}^{-3}$) on the surface
of the boundary polyhedron as a function of d_{Co}.
The calculated value is based on spherical isolated
atoms, while the observed value is derived from the
experimental structure factors. Note that the bonded
cobalt atom is negative relative to the cobalt atom
in the promolecule (Ref. 29).

Stockholder concept

"A way of partitioning the charge density such that each
atom is assigned a fraction of the charge at a point in the
crystal proportional to its contribution to the procrystal
density at that point".

This stockholder concept, described in more detail in Chapter
6, part 1, can be applied both to the static model density and
the dynamic experimental density. In the latter case the free-
atom densities, which are the basis for the partitioning, must be
thermally smeared with thermal motion as determined by the experiment. It is often found that the net atomic charges obtained in
this way are smaller than those from the kappa refinement (see
Table 4 in Chapter 6, part 1), but they show good transferability
between identical groups in different molecules.

Instead of using a fuzzy boundary a fence can be drawn, the
inside of which defines the volume assigned to the atom. This
border has become known as a

Discrete boundary

"The surface defining an atom's (or molecule's) closed
volume, such that all space within the surface is assigned
to that atom".

Discrete boundary partitioning is used in the definition of
the "virial fragments" by Bader and coworkers[27]. These fragments
have non-planar surfaces and individually satisfy the virial
theorem. This partitioning has never been applied to experimental
results, though its use in subdividing the static model density
should be no different from the partitioning of a theoretical
charge density.

A discrete boundary partitioning leading to polyhedral
volumes for each atom has been defined by the condition:[27]

$$\frac{(\underset{\sim}{r}-\underset{\sim}{r}_A)\cdot \underset{\sim}{r}_{AB}}{R_A} = \frac{(\underset{\sim}{r}-\underset{\sim}{r}_B)\cdot -\underset{\sim}{r}_{AB}}{R_B} \tag{11}$$

where $\underset{\sim}{r}_A$ and $\underset{\sim}{r}_B$ are the vectors from a point in space to the
adjacent atoms A and B, $\underset{\sim}{r}_{AB}$ is the interatomic vector from A to
B, and R_A and R_B are the radii of A and B. The volume is a gener-
alization of the Wigner–Seitz cell, used in the theory of metals,

which has boundary planes perpendicular to, and bisecting the
vectors between adjacent atoms.   Thus:

Generalized Wigner-Seitz cell

"A polyhedral volume defined by planes perpendicular to
vectors between the atom(s) in a molecule (group) and its
neighbors".

The Generalized Wigner-Seitz cell may be used to define an
atom or a molecule. When only two types of atoms are involved,
as is the case in binary compounds such as V_3Si or CaF_2, the
ratio of the radii R_A and R_B may be treated as a variable and the
desired property plotted as a function of the position of the
boundary plane between adjacent atoms.[28] Similarly in the tri-
nuclear cobalt complex $Co_3CH(CO)_9$ each cobalt atom has two other
cobalt and three carbon atom neighbors, so that the charge in the
cobalt atom region may be plotted as a function of the position
of the boundary plane, as illustrated in Fig. 10.[29] The average
density at the surface of the polyhedron is also plotted in this
figure. There is clearly an area of lower density between the
atoms, and the position of the minimum may be defined as the
"best" discrete boundary of the cobalt atom.

An illustration of the Wigner-Seitz cell for the molecular
crystal of formamide is found in Chapter 6.1 of this book.

ELECTROSTATIC PROPERTIES DERIVED FROM THE ELECTRON DENSITY

Though the electron densities are unusually informative and
have the intuitive advantage that "a picture is better than a
thousand words", there is a need for comparison of numerical
properties with those from theory and other experimental techniques.

In general a property χ of the electron density may be
obtained by applying the corresponding operator $\hat{\gamma}$ to the electron
density and subsequently integrating over the volume V of interest

$$\chi = \int_V \hat{\gamma} \, \rho(\underset{\sim}{r}) \, d\underset{\sim}{r} \qquad (12)$$

When $\hat{\gamma} = 1$, this expression represents the total electronic
charge; for $\hat{\gamma} = x$, y or z, χ is the dipole moment component in
the x, y, z direction respectively.

For the nth moment $\rho(r)$ is multiplied by n components of r:

$$\chi = \int_V r_\alpha r_\beta r_\gamma \cdots r_\eta \rho(\underset{\sim}{r}) \; d\underset{\sim}{r} \tag{13}$$

with $\alpha, \beta, \gamma \cdots \eta$ being either 1, 2 or 3 and $r_1 = x$, $r_2 = y$ and $r_3 = z$.

Moments for which $\hat{\gamma}$ is a positive power of r_α are often called <u>outer moments</u>, because they depend more on the peripheral parts of the integration volume where r_α is large. They are contrasted with <u>inner moments</u> such as the electric field gradient at a nucleus which depend on negative powers of r_α:

$$\chi = \int r_\alpha^{-1} \, r_\beta^{-1} \, r_\gamma^{-1} \cdots r_\eta^{-1} \; \rho(\underset{\sim}{r}) \; dr \tag{14}$$

From the experimental viewpoint there are two important distinctions between the inner and the outer moments:

1.   The outer moments depend on the inner part of the diffraction pattern while the inner moments are much more affected by the intensities of the high order reflections and thus are sensitive to truncation errors.

2.   The outer moments depend on a definition of the volume of integration. The inner moments do not suffer from such an ambiguity since the contribution of remote parts of the electron density is small, so that the difference between integration over a pseudo-molecular volume or all of crystal space is negligable.

Examples of the calculation of outer moments and of the electrostatic potential are given by Moss in Chapter 6, part 1, while Schwarzenbach and Lewis (Chapter 6, part 2) and Stevens (Chapter 5, part 3) discuss the calculation of electric field gradients from X-ray data and the complementary nature of field gradients from spectroscopic experiments and X-ray determined charge densities.

SOURCES OF EXPERIMENTAL ERROR AND THEIR EFFECT ON THE ELECTRON DENSITY MAPS [30-33]

> "When an experimentalist obtains a result everybody believes it, except the experimentalist himself"

The above observation based on many repeated experiments underlines the multitude of sources of error which may affect an experiment. Some of the errors are <u>random</u> and can be reduced by repeated experiments. They include counting statistics and short term fluctuations in source intensity and temperature. Other errors are <u>systematic</u> in the sense that their effect can be reproduced by an adjustment in the parameters of the least squares model used to refine the data. When for example some intensity of the high order measurements is lost because of an insufficiently wide counter aperture (the wavelength spread in the beam broadens high-order reflections more severely), a positive bias will be introduced in the thermal parameters of the atoms. Other sources of bias are diffractometer positioning and scan speed errors, and uncorrected long range instabilities in the radiation source.

A second type of systematic error is introduced by secondary interactions between the radiation and the crystal. Accurate correction for effects like absorption, extinction, multiple reflection and thermal diffuse scattering are the subject of an extensive literature. Like the systematic measurement errors they affect primarily the scale and temperature factors. But sometimes errors are introduced in the attempt to apply a correction. For example, Schwarzenbach and Lewis describe how an anisotropic extinction treatment of data on Al_2O_3 mimicked electron density features near the Al nuclei which before correction agreed with spectroscopic electric field gradient measurements (Chapter 6, part 2).

It is instructive to follow the effect of random errors through the theory of propagation of errors: Since

$$\Delta\rho = \rho'_{obs}/k - \rho_{calc} = \rho_{obs} - \rho_{calc}$$

(ρ'_{obs} being the unscaled density) the variance in $\Delta\rho$ has contributions from both the observed and the calculated densities:

$$var\ (\Delta\rho) = var(\rho_{obs}) + var(\rho_{calc}) - 2\ cov(\rho_{obs}, \rho_{calc}) \quad (15)$$

with

$$\text{var}(\rho_{obs}) = \text{var}(\rho'_{obs}) + \rho'^2_{obs} \text{ var}(k)/k^4 = \text{var}(\rho'_{obs}) +$$

$$\rho^2_{obs}/k^2\sigma^2(k) \tag{15a}$$

The covariance term (cov) arises because a least squares determined scale factor is commonly used to obtain the scaled density ρ_{obs}. Thus the error in ρ_{obs} will be correlated with the error in ρ_{calc}:

$$\text{cov}(\rho_{obs}, \rho_{calc}) = \frac{d(\rho'_{obs}/k)}{d\ k} \text{ cov}(k, \rho_{calc}) = -\frac{\rho_{obs}}{k} \text{ cov}(k, \rho_{calc})$$

$$\tag{16}$$

An additional model dependence of ρ_{obs} exists for acentric structures for which the structure factor phases must be calculated before ρ_{obs} can be obtained. For this reason most electron density studies have been done on centrosymmetric crystals.

In the centrosymmetric case, neglecting possible covariances between the intensities of different reflections, the effect of the errors in the measurements is given by:

$$\text{var}(\rho'_{obs}) = \frac{4}{V^2} \sum_{1/2} \sigma^2(F_{obs}) \cos^2 2\pi(hx+ky+\ell z)$$

$$= \frac{2}{V^2} \sum_{1/2} \sigma^2(F_{obs}) [1 + \cos 4\pi(hx+ky+\ell z)] \tag{17}$$

For a reasonably complicated structure the second term in (17) averages to zero, except at or near special positions in the unit cell. The expression for $\sigma^2(\rho'_{obs})$ thus becomes

$$\text{var}(\rho'_{obs}) = \sigma^2(\rho'_{obs}) = \frac{2}{V^2} \sum_{1/2} \sigma^2(F_{obs}) \tag{18}$$

This expression, originally derived by Cruickshank,[34] implies that the error in ρ'_{obs} is constant over the unit cell. In contrast the scale factor dependent term in (15a) and the second and third terms in (15) are strongly dependent on position. The last two are given by

$$\text{var}(\rho_{calc}) = \sum_m \sum_n \left(\frac{\partial \rho_{calc}}{\partial U_m}\right) \left(\frac{\partial \rho_{calc}}{\partial U_n}\right) \sigma(U_m) \sigma(U_n) \gamma(U_m, U_n) \qquad (19)$$

and

$$-2\text{cov}(\rho_{obs}, \rho_{calc}) = 2\rho_{obs}/k \sum_m \left(\frac{\partial \rho_{calc}}{\partial U_m}\right) \sigma(U_m) \sigma(k) \gamma(U_m, k) \qquad (20)$$

in which the set of U_m are the refined positional and thermal parameters and the γ's are the correlation coefficients. In a well designed experiment the covariances among the scale factor and temperature parameters are often small, in which case $\sigma^2(\rho_{calc})$ is dominated by the variances in the parameters:

$$\text{var}(\rho_{calc}) = \sum \left(\frac{\partial \rho_{calc}}{\partial U_m}\right)^2 \sigma^2(U_m) \qquad (19a)$$

The derivatives in this expression tend to be large wherever ρ_{calc} is large, i.e. in the vicinity of the atomic positions. For the temperature parameters, for example, $(\partial \rho_{calc}/\partial U_m)^2$ has a large value at the nuclear position the magnitude of which increases with increasing atomic number Z.

Similarly the scale factor dependent term in (15a):

$\rho_{obs}^2 \frac{\sigma^2(k)}{k^2}$ is also large near the nuclear position. Though the combined contribution of (19a) and the scale factor error is reduced somewhat by the covariance term (20), the error function still shows pronounced peaks near the nuclear positions of the heavier atoms in the crystal, as illustrated in Fig. 11.

As a result the experimental maps are unreliable in the vicinity of the nuclear positions. But this is the region where information can be gained from measurement of the electric field gradient by magnetic resonance spectroscopy, so that a combination of the different techniques can lead to a much more complete representation of the charge density.

One should note that when a molybdenum atom complex is compared with an isomorphous chromium atom complex, the errors near the molybdenum atom are not increased proportional to the increase in atomic number, because the $\sigma(U_m)$ are usually smaller

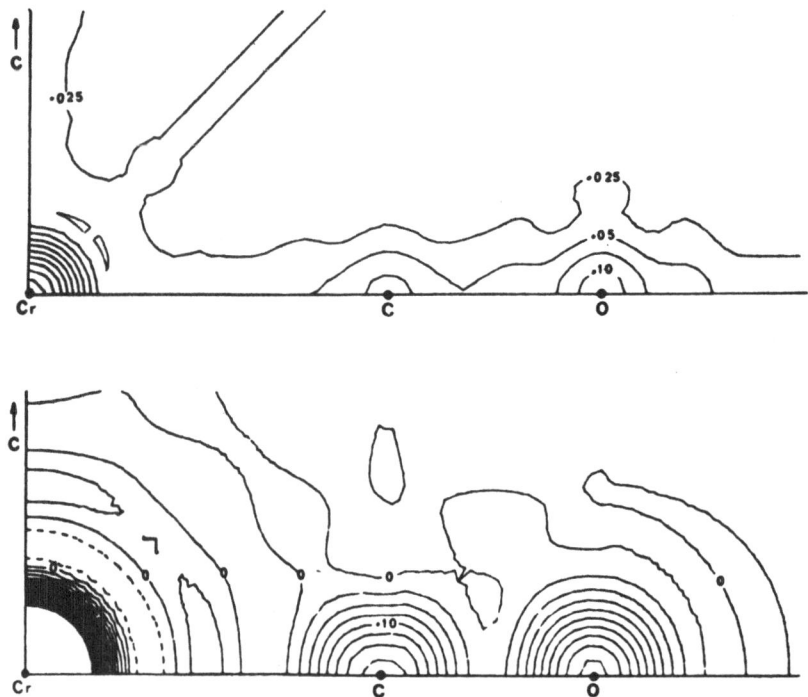

Fig. 11. Standard deviation of the deformation density of $Cr(CO)_6$
after averaging over chemically equivalent planes.
Top: Excluding the error in the scale factor. Bottom:
Effect of the scale factor error for $\sigma(k)/k = 1/100$.
Contours 0.025 $e\text{Å}^{-3}$. Contours above 0.40 $e\text{Å}^{-3}$ omitted
in bottom picture. The peaks at the nuclear positions
will be somewhat less than the sum of the values illustrated here because of the correlation between scale
and temperature parameters (Ref. 30).

for a heavier atom, while the scale factor is often more precisely
determined when stronger scatterers are present.

The above discussion does not take into account the bias
introduced by systematic errors in the data. Since a large part
of these errors can be accounted for by a change in the thermal
parameters, ρ_{calc} will be most affected in the vicinity of the
atomic positions. The same is true for ρ_{obs} to which ρ_{calc} is a

best lease squares fit.† Thus when both functions are based on the same data set as in the X-X technique, the effects on $\Delta\rho$ cancel in first approximation.

This is confirmed in studies by Helmholdt and Vos[35] on the effect of TDS on $\Delta\rho$. For ammonium hydrogen oxalate at 110°K these authors calculated errors in the carbon-carbon bonds of 0-0.03 eÅ^{-3} while at the atomic positions errors of 0.05-0.08 eÅ^{-3} were found.

Approximate cancellation of errors will not occur when ρ_{obs} and ρ_{calc} are obtained from different experiments as in the X-N method. In fact, differences between least squares X-ray and neutron thermal parameters are one of the main disadvantages of the X-N method, and often require rescaling of neutron parameters to X-ray values. In sulfamic acid for example scale factors $U_{ii}(X)/U_{ii}(N)$ were 1.01, 1.14 and 1.03 for i = 1, 2, 3 respectively.[51]

The effect of systematic errors can be analyzed by comparing parallel experiments on different specimen and different instruments. Such measurements on oxalic acid dihydrate have been performed in at least four X-ray and three neutron diffraction laboratories under a project sponsored by the Commission on Charge, Spin and Momentum densities of the I.U.Cr. Preliminary results show that densities are reliable to within about 0.10-0.15 eÅ^{-3}, and indicate that systematic errors cannot be neglected.

Acknowledgements. Support of this work by the National Science Foundation is gratefully acknowledged. I would like to thank Dr. Grant Moss for his critical reading of the manuscript.

References

1. D. Feil, Isr. J. of Chem. 16:103 (1977).
2. M. S. Lehmann and P. Coppens, Acta Chem. Scand. A31:530 (1977).

---

†The effect of systematic errors is to add terms representing the covariance cov(F_H,F_H') to expression (17). If for example all strong reflections are underestimated peak heights will be too low since such reflections give positive contributions to the electron density near the nuclear position.

3. D. Sherwood, Crystals, X-rays and Proteins, John Wiley, New York, p. 342.
4. P. Coppens and C. A. Coulson, Acta Cryst. A23:718 (1967).
5. J. W. Bats, P. Coppens and T. F. Koetzle, Acta Cryst. B33:37 (1977).
6. P. Coppens, T. N. Guru Row, P. Leung, E. D. Stevens, P. J. Becker, and Y. W. Yang, Acta Cryst. A35:63 (1979).
7. C. A. Coulson "Valence" Oxford University Press, Oxford (1961).
8. B. Dawson, Proc. Roy. Soc. (London) A298:255 (1967).
9. R. F. Stewart, J. Chem. Phys. 51:4569 (1969).
10. F. L. Hirshfeld, Isr. J. of Chem. 16:226 (1977).
11. R. F. Stewart Nato Adv. Study Inst. Series B48:439 (1977).
12. N. K. Hansen and P. Coppens, Acta Cryst. A34:909 (1978).
13. J. Bentley and R. F. Stewart, J. Comp. Physics 11, 127 (1973).
14. R. F. Stewart, J. Bentley and B. Goodman, J. Chem. Phys. 63:3786 (1975).
15. D. S. Jones, D. Pautler and P. Coppens, Acta Cryst. A28:635 (1972).
16. E. Hellner, Acta Cryst. B33:3813 (1977).
17. M. Bonnet, A. Delapalme, P. Becker and H. Fuess, J. Magn. and Magn. Materials 7:23 (1978).
18. P. Coppens and M. S. Lehmann Acta Cryst. B32:1777 (1976).
19. J.-M. Savariault and M. S. Lehmann J. Am. Chem. Soc. 102:1298 (1980).
20. M. Benard, P. Coppens, M. L. DeLucia and E. D. Stevens, Inorg. Chem. 19:1924 (1980).
21. M. Martin, B. Rees and A. Mitschler, Acta Cryst. B38:6 (1982).
22. W. Heijser, Ph.D. Thesis, Free University, Amsterdam, Holland, 1979.
23. P. Coppens, R. Boehme, P. Price and E. D. Stevens Acta Cryst. A37:857 (1981).
24. F. Baert, P. Coppens, E. D. Stevens and L. Devos, Acta Cryst. A38:143 (1982).
25. R. F. Stewart, Acta Cryst. A32:565 (1976).
26. F. L. Hirshfeld, Theor. Chim. Acta 44:129 (1977).
27. P. Coppens and T. N. Guru Row, Ann. NY Acad. Sc. 313:244 (1978).
28. J. L. Staudenmann, Ph.D. Thesis, University of Geneva, Switzerland (1975).
29. A. Holladay and P. Coppens, to be published.
30. B. Rees, Acta Cryst. A32:483 (1976).
31. E. D. Stevens and P. Coppens, Acta Cryst. A32:915 (1976).
32. B. Rees, Acta Cryst. A34:254 (1978).
33. M. S. Lehmann, Nato Adv. Study Inst. Series B48:355 (1979).
34. D.W.J. Cruickshank, Acta Cryst. 9:754 (1956).
35. R. B. Helmholdt and A. Vos, Acta Cryst. A33:38 (1977).

Section 2
THEORETICAL CONSIDERATIONS

DENSITY FUNCTIONAL THEORY

Robert G. Parr

Department of Chemistry
University of North Carolina
Chapel Hill, North Carolina 27514

Cards suffuse,
         displays abound.
How can they
         be turned to swans?

I. PREFACE

Contemporary quantum chemistry presents a quandry. Solution
of the Schrodinger equation proceeds to higher and higher accuracy,
and molecular properties are computed ever more precisely. But the
understanding of molecular properties seems thereby not enhanced.
The beauty that is chemistry eludes us.

Density functional theory[1] offers hope of freedom from this
curse. This theory is in its infancy. It is far from completely
understood, and it is far from completely developed. At this time
its domain of applicability is limited. Nevertheless it is rigorous, equivalent to the usual wavefunction theory, and enticing.

I shall here try to convey to you the flavor of this theory,
and I shall here try to convince you that it is enticing. The
specific result on which I shall focus is that the theory validates
and enlarges the concept of electronegativity.[2]

II. DENSITY FUNCTIONAL THEORY

The conventional treatment of a problem in molecular electronic
structure proceeds as follows. For a given N-electron system with

electron i moving in the field of a nuclear (or other) potential
v(i), determine the 4N-dimensional wavefunction $\psi(1,2...,N)$ by
solving the variational equation

$$\delta(<\psi|\hat{H}|\psi> - E <\psi|\psi>) = 0, \tag{1}$$

where

$$H = -\frac{1}{2} \sum_i \nabla_i^2 + \sum_i v(i) + \sum_{i<j} (1/r_{ij}), \tag{2}$$

and the total electronic energy E is the Lagrange multiplier for
the normalization constraint. From ψ, compute properties of interest as expectation values of appropriate operators, for example the
several components of $E = <\psi|\hat{H}|\psi> = E[N,v]$: the electronic ki-
netic energy $T = <\psi|\hat{T}|\psi>$, the electron-nuclear attraction energy
$V_{ne} = <\psi|\hat{V}_{ne}|\psi>$, and the electron-electron repulsion energy $V_{ee} =
<\psi|V_{ee}|\psi>$. Obtain the three-dimensional electron density from
the formula

$$\rho(1) = N \int \cdot \cdot \int |\psi|^2 ds_1 dv_2 dv_3 \cdot \cdot dv_N . \tag{3}$$

The integration is over space and spin coordinates of all electrons
but the first, and over spin coordinates of the first. Note that
$N = \int \rho(1) d\tau_1$.

The elementary properties[3] of the electron density of Eq. (3)
are well understood and experimentally accessible; it is the subject of the present symposium.

Density functional theory derives from the amazing fact that
for a ground state[4] the electron density determines all properties,
even ψ.[1] This means that we can write

$$E[\rho] = T[\rho] + V_{ne}[\rho] + V_{ee}[\rho], \tag{4}$$

where each quantity is a "functional" of ρ--a number the value of
which is determined by the function ρ. Furthermore, if a wrong ρ
is inserted in Eq. (4), the resulting E is above the true E provided ρ is a density corresponding to the right number of particles.[1,5,6] That is, one has the variational principle

$$\delta(E[\rho] - \mu N[\rho]) = 0, \tag{5}$$

where μ is a Lagrange multiplier, called the chemical potential.
Note the formal similarities between Eqs. (1) and (5).

III. CHEMICAL POTENTIAL   .

From Eq. (5) follows $\mu = (\partial E/\partial N)_v$, which means that μ is the slope of the E versus N curve for constant nuclear potential. Up to a sign, this property is just the electronegativity χ,[2]

$$\mu = -\chi \approx - \left(\frac{I+A}{2}\right). \tag{6}$$

Here we include the classical Mulliken formula for electronegativity,[7] which (as is well known) is what one gets from fitting a quadratic in N to the energies of the system of interest and its monopositive and mononegative ions.

For a given system of interest, just as the energy E in Eq. (1) is a property determinable in principle from the fundamental constants of physics, so the chemical potential μ in Eq. (5) is a property determinable in principle from the fundamental constants of physics.[8] Electronegativity is a precise concept.

When atoms come together to form a molecule, differing chemical potentials (electronegativities) become equal by electron transfer. Electrons tend to flow from a region of high chemical potential (low electronegativity) to a region of low chemical potential (high electronegativity). Sanderson's Principle of Electronegativity Neutralization (Equalization) is validated.[9]

Proceeding with care, the state of an atom in a molecule can be defined (it is a perturbed atomic ground state),[2] and for cases in which the atomic chemical potential is exponentially decreasing with N, the geometric mean law for electronegativity equalization can be shown to hold.[10] To first order, charge transfer on molecular formation is proportional to electronegativity difference.[2] As one goes down a column in the periodic table, the limiting law $\mu \sim Z^{-1/3}$ should hold.[11]

The chemical potential of classical chemical thermodynamics has an exact meaning, with exact consequences which are important for understanding macroscopic phenomena. So also with the chemical potential of density functional theory for individual atoms and molecules. It has exact consequences for microscopic behavior. It is a property of an electronic distribution which is constant throughout a molecule in its ground state; it is a property any difference of which drives electron transfer. One may call it electronegativity, or one may call it chemical potential.[12]

IV. IMPLEMENTATION

The practical implementation of the density functional method presents great difficulties because the exact functional $E[\rho]$ is not yet known.

Briefly, the situation may be described as follows. Equation (4) contains three functionals, $T[\rho]$, $V_{ne}[\rho]$ and $V_{ee}[\rho]$. $T[\rho]$ is not known, although various approximations to it are known; among the more accurate I may mention one recently proposed in our laboratory,[13]

$$T[\rho] \approx \frac{1}{8} \int \frac{\nabla\rho \cdot \nabla\rho}{\rho} \, d\tau + (1 - \frac{C}{N^{1/3}}) C_F \int \rho^{5/3} d\tau, \tag{7}$$

where C_F is the Thomas-Fermi coefficient and C is another constant. V_{ne} is known; it is $\int \rho v d\tau$. $V_{ee}[\rho]$ is not known, although if it is rewritten in the form

$$V_{ee}[\rho] = J[\rho] - K[\rho], \tag{8}$$

with

$$J[\rho] = \frac{1}{2} \int\int \frac{\rho(1)\rho(2)}{r_{12}} \, d\tau_1 d\tau_2, \tag{9}$$

J is known and various approximations to $K[\rho]$ are known. Note here that Eq. (8) itself does not imply an approximation; $K[\rho]$ is the entire exact difference between V_{ee} and J, containing correlation as well as exchange effects. In a word, all contemporary calculations entail approximations for two components of $E[\rho]$: $T[\rho]$ and $K[\rho]$.

The situation is more favorable if one goes to another functional theory, the first-order density matrix functional theory.[14] Then only K needs to be approximated. The so-called Xα method is an example of such a theory, and values of μ for the elements have already been calculated by this method.[15]

The old Thomas-Fermi-Dirac approximations,

$$T[\rho] \approx C_F \int \rho^{5/3} d\tau, \quad K[\rho] \approx C_X \int \rho^{4/3} d\tau, \tag{10}$$

are merely the natural "local" approximations to $T[\rho]$ and $K[\rho]$. Adding a corresponding local approximation to $J[\rho]$,

$$J[\rho] = B_0 N^{2/3} \int \rho^{4/3} d\tau, \tag{11}$$

has recently been shown to give an interesting completely local theory.[16]

I expect information on $T[\rho]$ and $K[\rho]$ to continue to accumulate until one can deal with each of them to arbitrary accuracy. Progress may be slow, however, and complete success is not assured.

Meanwhile, of course, the exact consequences of the theory can be developed and applied, thereby exploring the whole conceptual description implied by Eq. (5). Among the interesting results obtained so far are an equation of Gibbs–Duhem type giving the change of chemical potential from one situation to another[2] and a definition of the pressure at a point in the electronic distribution (It varies from point to point, just as does the pressure in a star).[17] Inserting the latter into the former gives the interesting, perhaps highly useful equation, reminiscent of classical thermodynamics,

$$N\delta\mu = \int \delta P d\tau + \int \rho \delta v^* d\tau + \delta X, \qquad (12)$$

where v^* is the classical electrostatic potential and X is a correction for nonlocal effects which is, for example, identically zero in the approximations of Eq. (10) above.

V. CONCLUSION

At the University of North Carolina we are trying to develop more and more accurate density functional calculational methods, while at the same time exploring the formal consequences of the exact theory. We use elaborate wavefunction calculations [Eq. (1)] as needs be to test presumptions about density functional calculations [Eq. (5)]. We seek in the first instance a precise first-order density matrix functional theory, and we anticipate that the final elegant theory, if ever found, will in its orbital representation be a theory of natural orbitals,[18] not Hartree–Fock or Kohn–Sham orbitals.[19] Such a theory would indeed constitute relief from the predicament described at the outset of this talk.

REFERENCES

1. P. Hohenberg and W. Kohn, Phys. Rev. 136:B864 (1964).

2. R. G. Parr, R. A. Donnelly, M. Levy and W. E. Palke, J. Chem. Phys. 68:3801 (1978).

3. For an excellent summary, see V. H. Smith, Jr. and I. Absar, Israel J. Chem. 16:87 (1977). See also Professor Smith's article in the present volume, Chapter 1.

4. But see S. Valone and J. Capitani, Phys. Rev. A 23:2127 (1981).

5. M. Levy, Proc. Natl. Acad. Sci. USA 76:6062 (1979).

6. S. M. Valone, J. Chem. Phys. 73:1344:4653 (1980).

7. R. S. Mulliken, J. Chem. Phys. 2:782 (1934); 3:573 (1975).

8. The criticism that E is defined for only integral values of N can be dealt with. See J. Katriel, R. G. Parr and M. R. Nyden, J. Chem. Phys. 74:2397 (1981).

9. R. T. Sanderson, Science 121:207 (1955).

10. R. G. Parr and L. J. Bartolotti, unpublished. The exponential fall-off parameter must be the same in the two atoms.

11. N. H. March and R. G. Parr, Proc. Natl. Acad. Sci. USA 77:6285 (1980).

12. Professor Harry B. Gray suggests the name "chemipotential" for this quantity. I prefer "chemical potential" because the analogy with the macroscopic chemical potential is strict and revealing and the term is already in common use in the physics literature. Note that in the present discussion nothing whatever has been said about orbital chemical potential or orbital electronegativity, the introductions of which obscure the essence of density functional theory, and the properties of which obscure the very chemical potential concept. See references 2 and 14.

13. P. K. Acharya, L. J. Bartolotti, S. B. Sears and R. G. Parr, Proc. Natl. Acad. Sci. USA 77:6978 (1980).

14. T. L. Gilbert, Phys. Rev. B 12:2111 (1975); R. A. Donnelly and R. G. Parr, J. Chem. Phys. 69:4431 (1978).

15. L. J. Bartolotti, S. R. Gadre and R. G. Parr, J. Am. Chem. Soc. 102:2945 (1980).

16. R. G. Parr, S. R. Gadre and L. J. Bartolotti, Proc. Natl. Acad. Sci. USA 76:2522 (1979). See also R. G. Parr and A. Berk, pp. 51-62 in Chemical Applications of Atomic and Molecular Electrostatic Potentials (Plenum Press, 1981), P. Politzer and D. G. Truhlar, eds.

17. L. J. Bartolotti and R. G. Parr, J. Chem. Phys. 72:1593 (1980).

18. P.-O. Löwdin, Phys. Rev. 97:1474 (1955).

19. W. Kohn and L. J. Sham, Phys. Rev. 140:A1133 (1965).

QUANTUM MODEL OF THE COHERENT DIFFRACTION

EXPERIMENT: RECENT GENERALIZATIONS AND APPLICATIONS

C.A. Frishberg
School of Theoretical and Applied Science
Ramapo College, Mahwah, New Jersey

M.J. Goldberg and L.J. Massa
Graduate School and Hunter College, City University of
of New York, New York 10021

ABSTRACT

The scattering power for the X-ray crystallographic experiment is the Fourier transform of the electron density. This can be cast into a quantum formalism by representing the density as the elements of the 1-electron density matrix. The condition that a 1-density matrix come from a wave function is ensured by requiring that its eigen-values fall in the range between zero and one. A particular case we have studied because it is a most important first approximation is that of a single determinant of molecular orbitals. This case is completely characterized by idempotency of the density matrix. Measured X-ray intensities may be used to fix the elements of the density matrix. Specific applications of this idea are presented. In particular we consider (1) open shell systems and (2) Bloch orbital representations.

INTRODUCTION

We wish to discuss the use of quantum mechanical ideas in modeling the coherent X-ray diffraction experiment. A theme which underlies our view is that exactly the same language and conceptual structure ought to be used to describe theory and experiment. Consider Figure 1 where we sketch the rough outlines of an X-ray crystallographic experiment. A particular unit cell of a crystal contains a set of molecular orbitals. These orbitals give rise to the electron density of the unit cell. The electron density in turn yields ideal structure factors associated with the X-ray scattering. The quantities actually measured are intensities

$I(\vec{K})$ along the directions of the scattering vectors $\vec{K}$ which satisfy the conditions of Von-Laue and Bragg for coherent diffraction.
The experimental intensities are contaminated by effects of polarization, extinction, thermal motions, etc.[1], but if one has a sufficiently good model of these effects it is possible to reduce
the intensities to a set of ideal structure factors $F(\vec{K})$. These
are just the Fourier transform of the electron density, including
phases

$$F(\vec{K}) = \int e^{i\vec{K}\cdot r} \rho(r) d^3 r \tag{1}$$

Symbolically the crystallographic experiment may be thought of as
the sequence

$$\{\phi\} \Rightarrow \rho(r) \Rightarrow \{F(\vec{K})\} \tag{2}$$

where the $\{\phi\}$ is a set of orbitals, and $\rho(r)$ is the related density. But now consider the reverse process, viz,

$$\{F(\vec{K})\} \Rightarrow \rho(r) \Rightarrow \{\phi\} \tag{3}$$

i.e., given the set of structure factors $F(\vec{K})$, is it possible to
recover the associated electron density and orbitals? That is the
problem addressed in this paper.

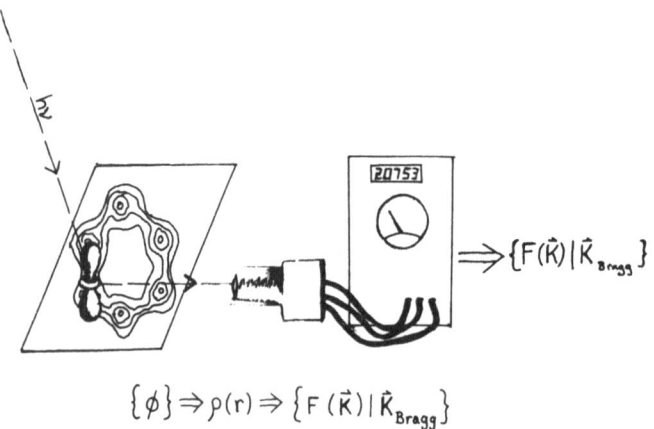

$$\{\phi\} \Rightarrow \rho(r) \Rightarrow \left\{F(\vec{K}) | \hat{K}_{Bragg}\right\}$$

Figure 1.   Idealized sketch of X-ray coherent diffraction

One may represent[2] the molecular orbitals $\vec{\phi}$ in an atomic orbital basis $\vec{\psi}$ as

$$\vec{\phi} = \vec{\vec{C}}\vec{\psi} \tag{4}$$

where $\vec{\vec{C}}$ is a matrix collection of LCAO coefficients. The density matrix is

$$\rho_1 = \text{tr}\,\vec{\phi}\vec{\phi}^+ = \text{tr}\,\vec{\vec{C}}^+\vec{\vec{C}}\vec{\psi}\vec{\psi}^+ \tag{5}$$

and the electron density ρ which gives rise to the structure factor is the diagonal of the density matrix, i.e.,

$$\rho(r) = \rho_1(r,r')\Big|_{r'=r} \tag{6}$$

It is convenient to define a matrix

$$\vec{\vec{P}} = \vec{\vec{C}}^+\vec{\vec{C}} \tag{7}$$

which reduces the density matrix to the compact form

$$\rho_1 = \text{tr}\,\vec{\vec{P}}\vec{\psi}\vec{\psi}^+ \tag{8}$$

And now consider the following program. Treat the elements of the matrix $\vec{\vec{P}}$ as experimental parameters. Hence, given structure factors $F(\vec{K})$, least squares fit the elements of $\vec{\vec{P}}$ to the data $F(\vec{K})$. Conclude that since $\vec{\vec{P}}$ determines the density matrix one could recover through the above equations the density matrix and orbitals of the problem. In practice, this program is fallacious.[3] The reason is that a least squares fitting procedure in itself does not incorporate certain of the requirements imposed upon the density matrix by quantum mechanics. As a result not every matrix $\vec{\vec{P}}$ implies a valid density matrix.

Certain of the restrictions upon the matrix $\vec{\vec{P}}$ which follow from the Schrödinger equation

$$\hat{H}\Psi(1,\ldots N) = E\Psi(1,\ldots N) \tag{9}$$

may be considered in reference to Figure 2. Every solution or approximate solution to the Schrödinger equation must be antisymmetric because of the indistinguishability of Fermions. The figure indicates every element of the set of all antisymmetric N-body wavefunctions Ψ can be mapped into the set of all valid density matrices ρ_1 through the rule

$$\rho_1 = \int\Psi\Psi^*d2\ldots dN \tag{10}$$

So given Ψ one can generate ρ_1 and if ρ_1 is valid, it must be related to some antisymmetric Ψ. But problems arise with a simple
experimental fitting of the elements of $\vec{P}$ becuase the set of all
possible matrices $\vec{f}(1,1')$ is wider than the set $\vec{\rho}_1$ $(1,1')$ mappable
into wavefunctions. This is indicated in the figure by the arrow
from the wider set $\{f\}$ not reaching the set $\{\Psi\}$. Hence any experimental fitting procedure based upon functions of the wider set $\{f\}$
will in general sacrifice the relationship to $\{\Psi\}$ in favor of
"goodness-of-fit". In practical situations involving an approximate representation of the density matrix the crystallographer's
least square "program" left to its own devices can arrive at a
representation in the wider space $\{f\}$ not connected to wavefunctions.

Fortunately, it's easy to characterize the mapping between
wavefunctions and density matrices.[4] In particular

$$0 \le w_i \le 1 \quad \text{and} \quad \sum_i w_i = N \tag{11}$$

That is, the density matrices related to wavefunctions have
eigenvalues w_i in the range zero to one and sum to the number of
occupied molecular orbitals. The extreme eigenvalues zero and
one characterize single Slater determinant representability which
encompasses the Hartree-Fock approximation. An equivalent condition ensuring single Slater determinant representability is, that
the density matrix be idempotent or a projector. This occurs if

$$\vec{P}^2 = \vec{P} \tag{12}$$

A very good representation of the density is reachable from
a single Slater determinant. In that case modeling the crystallographic density is reduced to the problem of finding normalized
projector density matrices which satisfy the X-ray scattering
factors of the problem. The mathematical structure which accom-
plishes this is embodied in the Lagrangian minimization

$$\delta \, \text{tr} \, \{(\vec{P}^2 - \vec{P}) + \lambda_k \vec{P}\vec{f}(\vec{K}) + \lambda_N \vec{P}\} = 0 \tag{13}$$

In this expression $\vec{P}$ is the "experimental" population matrix, $\vec{f}(\vec{K})$
is the matrix of Fourier transforms of atomic orbital products
evaluated along a scattering direction $\vec{K}$, and λ_K, and λ_N are un-
determined Lagrangian multipliers whose values are fixed by the
equations of constraint

$$F(\vec{K}) = \text{tr} \, \vec{P}\vec{f}(\vec{K}) \quad \text{and} \quad N = \text{tr}\vec{P} \tag{14}$$

where F(K) is an X-ray structure factor and N is the number of occupied molecular orbitals. To carry out the variation indicated
above is to arrive at an equation for the matrix $\vec{P}$, which can be
written in iterative form as

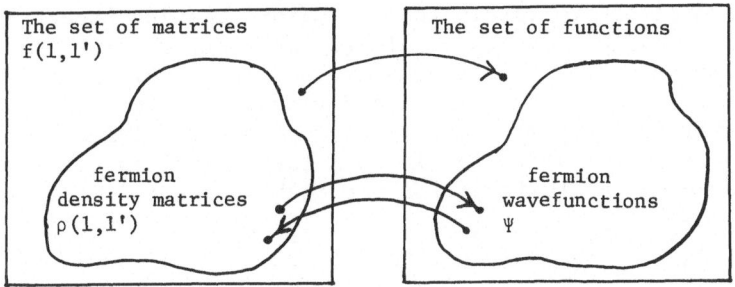

Figure 2. Sketch indicating the mapping problem associated with wave function representability of density matrices.

$$\vec{P}_{n+1} = 3\vec{P}_n^2 - 2\vec{P}_n^3 + \lambda_N\vec{1} + \lambda_K\vec{f}(\vec{K}) \qquad (15)$$

The equation is solved by selecting an iterate $\vec{P}$, solving for λ_K and λ_N using the equations of constraint, evaluating the right hand side of the above equation obtaining P_{n+1} for the next matrix iterate, and repeating the algorithm to solution characterized by

$$\vec{P}_{n+1} = \vec{P}_n \qquad \text{and} \quad \lambda_N \rightarrow 0 \qquad \text{and} \qquad \lambda_K \rightarrow 0 \qquad (16)$$

The equation delivers solutions which are normalized projectors satisfying the scattering conditions of an experiment. One obtains as it were a density matrix consistent with the requirements of theory and experiment. Moreover (1) in a projector matrix the number of free parameters is in general much smaller than in a non-projector of the same dimensions and (2) the serious difficulty of correlation of parameters is alleviated.

In the following two sections we consider generalizations of the ideas associated with the above $\vec{P}$ -equations along two directions. Firstly, we examine the case of open-shell density matrices.[5] Numerical results for the cases of lithium and beryllium atoms are discussed. Secondly, we consider an improved way of accounting for interactions among neighboring unit cells by using Bloch orbitals.[16]

Projector Equations for Open Shells

If we consider an open shell system, i.e., one for which we do not assume doubly occupied orbitals, then we may associate different projectors with the spatial orbitals containing particles of different spin. Symbolically we have

$$\vec{\phi}^\alpha \Rightarrow \vec{P}^\alpha \qquad \text{and} \qquad \vec{\phi}^\beta \Rightarrow \vec{P}^\beta \qquad (17)$$

where the symbols above retain their previous meaning except that the appropriate spin is noted explicitly as a superscript. The Lagrangian minimization condition discussed previously applied to the case of determining two normalized projectors simultaneously that also satisfy X-ray structure factor conditions leads to two coupled equations each of which is term for term of exactly the same form previously displayed for the doubly occupied case, i.e.,

$$\vec{\rho}^{\alpha}_{n+1} = 3\vec{\rho}^{\alpha 2}_n - 2\vec{\rho}^{\alpha 3}_n + \lambda_k \vec{g}_{\alpha,\beta}(\vec{K}) + \lambda_{N\alpha}\vec{I} \tag{18}$$

$$\vec{\rho}^{\beta}_{n+1} = 3\vec{\rho}^{\beta 2}_n - 2\vec{\rho}^{\beta 3}_n + \lambda_k \vec{g}_{\alpha,\beta}(\vec{K}) + \lambda_{N\beta}\vec{I} \tag{19}$$

The equations are coupled by the matrix $\vec{g}(\vec{K})$ which plays the role of the Fourier transform matrix in the doubly occupied case and here depends upon the relative contribution of α and β type particles to the total scattering. The numerics of solving the coupled equations are entirely analogous to those for the doubly occupied case.

We discuss the application of the open-shell projector equations to examples of lithium and beryllium atoms. We utilize structure factors as data which are essentially exact for these systems. In the case of lithium Larsson[6] has computed a Hylleraas wavefunction which includes 99% of the correlation energy, and for beryllium Weiss[7] configuration interaction wavefunction accounts for 93% of the correlation energy. Benesch and Smith[8] have obtained a natural orbital expansion of these two functions and calculated structure factors for 108 values of $\sin\theta/\lambda$ in the range $0-4\text{Å}^{-1}$. These structure factors provided our data. The basis sets used for representation of the density matrices were those of Clementi[9] for the essentially exact Hartree-Fack calculations of lithium and beryllium.

On taking as an initial guess the atomic Hartree-Fock density matrices for the two cases considered, the coupled open shell equations iterate smoothly to solution. We now discuss a few of the results obtained from these solutions. For both lithium and beryllium, structure factors calculated over a wide range of $\vec{K}$ using our X-ray fitted density matrices are much closer to the exact structure factors than those obtained from a Hartree-Fock wavefunction in the same basis. If one recollects that the Hartree-Fock density is thought of as a fairly accurate density[10] then this is a striking result. One may quantify this by calculating the magnitude of

$$\varepsilon = \sum_K |F_{exact}(\vec{K}) - F_{approx}(\vec{K})| \tag{20}$$

where F_{exact} $(\vec{K})$ is the exact structure factor calculated from the essentially exact wavefunction, and F_{approx} $(\vec{K})$ is the structure factor calculated from an approximate density matrix.

We calculated this goodness of fit parameter over a wide range of scattering factors for lithium and beryllium using our X-ray density matrices (ε^{xr}) as well as Hartree-Fock density matrices (ε^{HF}). For the case of lithium ε^{xr} = .0020, and ε^{HF} = .1506, while for the beryllium ε^{xr} = .0093, and ε^{HF} = .9945. As may be seen, the goodness of fit parameter ε is reduced by a factor of about 100 for the case ε^{xr} versus ε^{HF} in both cases.

Of course by definition the total energy must be less for the Hartree-Fock case than for the X-ray case, but interestingly, the increase in energy for the latter is very small. The increases amount to only .000055 a.u.& .001526 a.u. for lithium and beryllium respectively. These small increases are consistent with slightly less good X-ray values which occur for electron - electron repulsion terms compared to Hartree-Fock values. However the one electron properties $\langle \hat{T} \rangle$, $\langle \frac{1}{r} \rangle$, and $\langle r^2 \rangle$ are all better calculated by X-ray than Hartree-Fock for both atoms. We draw attention to the fact that not only is the kinetic energy calculable from the X-ray density matrix, it is in fact a better value than from the Hartree-Fock wavefunction. This contrasts markedly to fitting procedures involving multipole models[11] or Hirshfeld models[12] of the density because, in those cases, from the density representations alone, it is not now possible to recover the expectation values of general quantum mechanical observables such as kinetic energy. Of course every observable is obtainable given density matrices.[13]

The details of the numerical results refered to above will be published in a separate paper.[5]

Bloch Orbital Representation of the Density Matrix

We have discussed the density matrix as represented in an atomic orbital basis. This is a practical representation for an isolated molecule or for a molecular crystal involving weak interactions among neighboring unit cells. Organic molecules often crystallize under conditions involving such weak interactions among neighbors. However, more generally it is desirable to have the density matrix in a form which incorporates interactions, weak or strong, among neighboring unit cells. This may be accomplished by writing the density matrix in a basis of Bloch orbitals.[14] In this way contact is also made with the band theory of crystals so useful in discussing a range of properties whether of insulators, semiconductors, or metals.

Bloch functions are of the   form

$$\psi_k = e^{i\vec{k}\cdot\vec{r}} \, U_k(\vec{r}) \tag{21}$$

where $\vec{k}$ is a wave vector in the reciprocal space of the crystal and varies in that space over a volume of one unit cell also called a Brillouin zone. The function U_k has by definition the periodicity of the crystal lattice. The Bloch functions extend over the whole space of the crystal, and for that reason provide a natural way of accounting for interactions among neighboring unit cells. In fact, Bloch orbitals are a symmetry basis which span the representations of the space group of the crystal. A well defined prescription for constructing the periodic functions U_k out of non-periodic functions ψ centered at each of the lattice sites is the following

$$U_k(\vec{r}) = \sum_1 e^{-i\vec{k}\cdot(\vec{r}-\vec{R}_1)} \, \psi(\vec{r}-\vec{R}_1) \tag{22}$$

where the summation is over lattice sites and the $\vec{R}$ are lattice vectors. Figure 3 displays over a few unit cells, a portion of a Bloch function constructed out of an atomic orbital centered at successive lattice sites. The illustration indicates the concept that a Bloch orbital is a wave of the orbitals out of which it is constructed, in the sense that the orbitals are simply multiplied by a phase in going from site to site. The crystal orbitals may be taken as linear combinations of a Bloch orbital basis, i.e.,

$$\vec{\phi}_{\vec{k}} = \overset{\Rightarrow}{C}\vec{\psi}_{\vec{k}} \tag{23}$$

where $\vec{\phi}$ is a column vector of the crystal orbitals, $\vec{\psi}_k$ is a column vector of Bloch orbital basis functions, and the matrix $\overset{\Rightarrow}{C}$ contains

Figure 3. A portion of a Bloch orbital pictured as a wave of atomic orbitals

the coefficients of linear combination. The density matrix becomes

$$\rho_{1k} = \text{tr } \vec{\phi}_k \ \vec{\phi}_k^+ = \text{tr } \vec{C}^+\vec{C} \ \vec{\psi}_k \vec{\psi}_k^+ \tag{24}$$

But this is exactly the form of the density matrix we've used
previously with the exception that the basis is made of Bloch
orbitals. Hence, as one may guess, the iterative equations previously displayed for closed and open shell cases will apply in the
same form here so long as the basis used is a basis of Bloch orbitals. We view this as having special significance because it allows
one to envision the use of X-ray scattering data to determine the
density matrix for the widest possible class of crystals ranging
through insulators, semiconductors, and metals. One's optimism
in this is fostered by the successes of band theory in the description of such a range of solids.

An indication of how the interactions of neighboring unit
cells enter the formalism may be obtained by examining the structure factors for a particular $\vec{K}$ expanded in a Bloch orbital basis,[15]
i.e.,

$$F(\vec{K}) = \text{tr } \vec{C}^+\vec{C} \ \{\vec{f}^{00}(\vec{K}) + 2\vec{f}^{01}(\vec{K})e^{-i\vec{k}\cdot(\vec{R}_0 - \vec{R}_1)} + \ldots \} \tag{25}$$

where the various symbols retain their previous meaning but the
Fourier transform matrices are labelled with superscripts indicating the lattice sites from which the lattice centered functions
are taken. If only the leading term in curly brackets, $\vec{f}^{00}(\vec{K})$, is
retained one recovers the previous formalism for non-interacting
unit cells, but including the following term, $2\vec{f}^{01}(\vec{K})\exp(-i\vec{k}\cdot(\vec{R}_0-\vec{R}_1))$
incorporates interactions between all nearest neighbor unit cells.
Including successive terms accounts for unit cells still further
removed, etc.

Summary and Concluding Remarks

The density matrix is an appropriate object with which to match
the requirements of quantum mechanics and of crystallography. In the
single Slater determinant approximation, the projector equations
may be used to accomplish this. The projector condition, $\vec{P}^2 = \vec{P}$
also serves to reduce the number of parameters otherwise required
and alleviates correlation of parameters which otherwise occurs.
Open shell systems may also be treated with coupled projector
equations and we have discussed their numerical application to
model problems of lithium and beryllium atoms. The numerical results are consistent with a conjecture of Gilbert[16] to the effect
that even an exact density is N-representable by a single Slater
determinant, a point that has previously been investigated by
Henderson.[17] The density matrix written in a Bloch orbital representation is a suitable way to model the X-ray scattering for a

diversity of crystal types ranging through insulators, semiconductors and metals.

REFERENCES

1. G.H. Stout and L.H. Jensen, "X-ray Structure Determination," MacMillan, NY (1968).
2. E.R. Davidson, "Reduced Density Matrices in Quantum Chemistry," Academic Press, NY (1976).
3. R.F. Stewart, J. Chem. Phys. 58:1668 (1973).
4. W.L. Clinton, C.A. Frishberg, L.J. Massa, and P.A. Oldfield, Int. J. Quantum Chem. Symp. 7:505 (1973).
5. C.A. Frishberg and L.J. Massa, submitted to Phys. Rev.
6. S. Larsson, Phys. Rev. 169:49 (1968).
7. A.W. Weiss, Phys. Rev. 122:1326 (1961).
8. R. Benesch and V.H. Smith, Jr., Acta Cryst. A26:586 (1970), J. Chem. Phys. 53:1466 (1970).
9. E. Clementi, IBM J. Res. Develop. 9 (1965).
10. P.O. Löwdin, Phys. Rev. 97:1490 (1955).
    W.L. Clinton, J. Nakhleh, and F. Wunderlich, Phys. Rev. 177:1 (1969).
    C. Möller and M.S. Plesset, Phys. Rev. 46:618 (1934).
    W. Kutzelnigg and V.H. Smith, Jr., J. Chem. Phys. 42:2791 (1965).
    J. Goodisman and W.J. Klemperer, J. Chem. Phys. 38:721 (1963).
    I. Shavitt, in: "Modern Theoretical Chemistry", vol. III, H.F. Schaefer III, ed., Plenum Press, NY (1976).
11. J.N. Varghese and R. Mason, Proc. R. Soc. Lond. A372:1 (1980)
12. F.L. Hirshfeld, Acta Cryst. A32:239 (1976).
13. P.O. Löwdin, Advances in Quantum Chem. 12:263 (1980).
14. C.M. Quinn, "An Introduction to the Quantum Chemistry of Solids", Clarendon Press, Oxford (1973).
15. M.J. Goldberg and L.J. Massa, submitted to Int. J. Quantum Chem.
16. T.L. Gilbert, Phys. Rev. B12:2111 (1975).
17. G.A. Henderson, Int. J. Quantum Chem. 13:143 (1978).

THE INFLUENCE OF RELATIVITY ON MOLECULAR PROPERTIES:

A REVIEW OF THE RELATIVISTIC HARTREE-FOCK-SLATER METHOD

J.G. Snijders and E.J. Baerends

Scheikundig Laboratorium der Vrije Universiteit
De Boelelaan 1083
1081 HV Amsterdam, The Netherlands

ABSTRACT

A review of the relativistic Hartree-Fock-Slater method is given. The accuracy of the method is assessed by comparison with fully relativistic atomic calculations. It is applied to the bond properties of some diatomics where relativity plays a major role, among other things giving a new explanation for relativistic bondlength contractions. Relativistic effects in photoelectron spectra are studied. Finally the changes in the molecular electron density due to relativity are investigated.

INTRODUCTION

It has become apparent in recent years that the properties of molecular systems containing heavy elements can be substantially influenced by the requirements of the special theory of relativity. For electrons in deep lying core orbitals this is immediately clear, if one considers that an atomic 1s electron moves approximately with a velocity equal to the nuclear charge in atomic units while the velocity of light in the same units is approximately 137. Consequently in heavy elements these electrons have a velocity that is a substantial fraction of the speed of light and relativistic effects are to be expected.

On the other hand it is not immediately obvious that valence electrons, moving with much smaller velocities, will be influenced at

all and it moght be expected that chemical effects, which are
largely determined by the dynamics of these valence electrons, are
comparatively unaffected by relativity.

It turns out, however that this is not the case and that on the
contrary relativistic effects on valence properties can be substantial.
There are basically two reasons for this situation. In the first
place valence orbitals, especially those with considerable atomic
s character, do have tails that penetrate the core deeply, hence
these electrons spend a small portion of the time near the nucleus
where relativistic effects are huge, which can have a substantial
influence in the energetics of these electrons. Secondly the core
orbitals are of course heavily affected by relativity, so that there
can be an appreciable indirect effect on the valence electrons through
core-valence interaction. The purpose of this paper is to give a
review of these effects   studied by the relativistic Hartree-Fock-
Slater method developed by us. For other methods we refer to the
literature [1,2].

In section I we give a summary of the Hartree-Fock-Slater LCAO
method and the perturbative scheme by which we treat relativity. In
section II we evaluate the reliability of the method by comparing with
fully relativistic Dirac-Fock-Slater results, which are available
for atoms. In section III we study the effects of relativity on the
bonding parameters of some molecules.

In section IV we apply the method to the interpretation of photoelectron spectra of some systems that exhibit relativistic effects.
Finally in section V we look at the relativistic changes in electron
densities.

I. Method

The nonrelativistic starting point used in this work is the
Hartree-Fock-Slater (HFS) equation, which reads

$$(-\frac{1}{2}\nabla^2 + V_N + V_C + V_X)\phi_i = \varepsilon_i\phi_i$$

where V_N is the nuclear potential, V_C the Coulomb potential and
V_X the exchange potential. The difference with the Hartree-Fock
equations consists of the replacement of the non local exchange
operator by the local exchange potential proposed by Slater [3]

$$V_X = -3\alpha \left[(3/8\pi)\rho \right]^{1/3} ,$$

where ρ is the electronic density and α is a constant which will always be taken as $\alpha=0.7$. The equation is solved by the method of Baerends et al. [4], which involves expansion of the orbitals in an accurate Slater-function basis. The matrix elements of the Fock operator, needed to set up the resulting secular equation, are calculated by 3-dimensional numerical integration. The solutions are then iterated to self-consistency in the usual way.

It should be emphasized that this method does not include un-warranted approximations of the potential like the muffin-tin ap-proach. In fact the method is asymptotically exact in the sense that arbitrarily accurate solutions of the HFS equations can be obtained by using large enough basis sets and a large enough num-ber of integration points. .

To include the effects of relativity a perturbational scheme was developed by Snijders et al. [5,6]. This method takes advantage of the fact that the largest influence of relativity in molecules is to be found among the core orbitals, which are very similar to their atomic counterparts. The core orbitals are therefore kept frozen and are transferred from fully relativistic Dirac-Fock-Slater calculations on the constituent atoms which can be routinely performed. The smaller relativistic effects on the valence orbitals can then be treated by perturbation theory to first order in the square of the fine structure constant α^2 $(\alpha=e^2/4\pi\varepsilon_0\hbar c)$.

We can distinguish four perturbations due to relativity:

a) The mass-velocity correction due to the relativistic increase of the mass of an electron with its velocity:

$$h_{MV}^1 = -\frac{1}{8}\alpha^2 p^4 = -\frac{1}{8}\alpha^2 v^4 ,$$

where $\vec{p}$ is the linear momentum operator.

b) The Darwin correction. This term has no classical relativistic analogue and is due to the small scale irregular motion of an electron about its mean position, the so-called "Zitterbewegung" (see e.g. ref. [7]).

$$h_D^1 = \frac{1}{8}\alpha^2 \nabla^2 V_T^0 ,$$

where V_T^0 is the total (non-relativistic) potential. The main contribution to this operator comes from the nuclear potential

$$V_N = \sum_\mu z_\mu / r_{1\mu} ,$$

so

$$h_D^1 \simeq \frac{1}{8} \alpha^2 \nabla^2 V_N = \frac{1}{8} \alpha^2 \sum_\mu - 4\pi Z_\mu \delta (r_1 - r_\mu)$$

is almost a sum of Dirac δ-functions on the nuclei. It is therefore only important for orbitals that have a non-zero value at
the nuclei, i.e. orbitals that contain atomic s-orbital contributions.

c) Spin-orbit coupling. This is the well-known coupling of the
electron-spin to the magnetic field produced by its own orbital
motion,

$$h_{SO}^1 = \frac{1}{2} \alpha^2 \vec{s} \cdot (\vec{\nabla} V_T^0 \times \vec{p})$$

where $\vec{s}$ is the spin angular momentum operator.

d) The change in electronic potential induced by relativity. This
is the indirect effect mentioned in the introduction. The core
contribution to this term can be written as just the difference
of the relativistic and non-relativistic atomic core potentials.
For the valence contribution we need the first order change in
the orbitals ϕ_i^1, which are calculated by perturbation theory,
to construct the first order change in the valence density ρ_{val}^1

$$h_{POT}^1 = \sum_\mu (V_{core,\mu}^{rel} - V_{core,\mu}^{nrel}) + \int \frac{\rho_{val}^1}{r_{12}} d\tau_2 + \text{exchange terms.}$$

Since $h^1 = h_{MV}^1 + h_D^1 + h_{SO}^1 + h_{POT}^1$ depends on ρ^1 and hence the ϕ_i^1 which
are determined through the perturbation equation

$$(h^1 - \varepsilon_i^1) \phi_i^0 = (\varepsilon_i^0 - h_i^0) \phi_i^1$$

which in turn contains h^1, one has to solve this equation iteratively
until self-consistency is reached.

The correction to the orbital energies can now be calculated from

$$\varepsilon_i^1 = <\phi_i^0|h^1|\phi_i^0> = \varepsilon_{MV}^1 + \varepsilon_D^1 + \varepsilon_{SO}^1 + \varepsilon_{POT}^1.$$

Apart from computational advantages, one of the important assets
of a perturbational approach is the clear cut separation of the
relativistic corrections of different physical origin (h_{MV}^1, h_D^1, h_{SO}^1
h_{POT}^1), which allows a more illuminating interpretation of the
results.

II.   Comparison with atomic DFS calculations

     In order to assess the adequacy of the proposed perturbation
scheme, we have compared it with full Dirac-Fock-Slater calculations
on several atomic systems [5].   In table I we show the valence
orbital energies for three elements (Au, Hg and Rn) for which
relativistic effects are particularly large. In the first two
columns we compare the non-relativistic orbital energies from
numerical calculation and HFS-LCAO calculation showing that the
basis set used [8] is large enough that we have essentially
reached the basis set limit.   In the third column we show the effect
of the direct relativistic corrections (h^1_{MV}, h^1_D, h^1_{SO}). In all these
cases there is a sizable stabilization by relativity. In column
four we add the indirect potential effect which partly counteracts
the direct stabilization of orbital energies. It is quite clear
that this effect is large and should be included to reach any kind
of agreement with the full relativistic result in the last column.
Any remaining discrepancies with the full relativistic result must
be due to effects of higher order than α^2. They can come from two
sources, however, namely higher order corrections to the relativistic
perturbation operators (i.e. h^2 etc.) and energetic effects of h^1
in higher order. The importance of the last effect can be taken
into account by solving the so-called quasi relativistic equation:

$$(h^0 + h^1)\, \phi_i = \varepsilon_i\, \phi_i$$

effectively taking h^1 into account up to infinite order while
neglecting h^2 [9].

     The results of this procedure are shown in the fifth column of
table I. In Au and Hg the influence on the d levels is negligible
but the 6s orbital energies are brought into much closer agreement
with the full relativistic result. The remaining discrepancy in the
Rn $6p\frac{1}{2}$ energy must be due to h^2 which is difficult to calculate.
It can thus be concluded that first order perturbation theory,
possibly improved by the quasi-relativistic procedure can give an
accurate description of relativistic effects on orbital energies
except in cases where relativistic effects are extremely large as
in the spin-orbit splitting of Rn.

     Another criterion to assess the accuracy of the method is to com-
pare relativistic changes in the orbitals themselves as given by ϕ^1 in
first order, with the changes determined from the difference of a
non-relativistic and a fully relativistic numerical calculation.
As an example we show in fig. 1 the orbital densities of the $6s\frac{1}{2}$
orbital of Gd^+. It can be seen that the first order result is quite
satisfactory. For other orbitals and other elements comparable
results have been obtained [5].

Table I. Orbital energies in eV.

	nrel(num.)	nrel(anal.)	pert.(dir.)	pert.(dir.+indir.)	quasirel.	rel.
Au						
5d 3/2	7.47	7.48	10.17	7.20	7.24	7.24
5d 5/2			8.67	5.69	5.81	5.70
6s 1/2	3.67	3.68	5.36	4.72	5.13	5.28
Hg						
5d 3/2	11.50	11.52	14.73	10.40	10.37	10.38
5d 5/2			12.93	8.61	8.67	8.52
6s 1/2	4.80	4.80	6.88	5.75	6.16	6.30
Rn						
6s 1/2	16.19	16.21	22.70	19.51	20.90	21.16
6p 1/2	7.12	7.13	10.97	8.70	9.16	9.73
6p 3/2			8.55	6.28	6.27	6.11

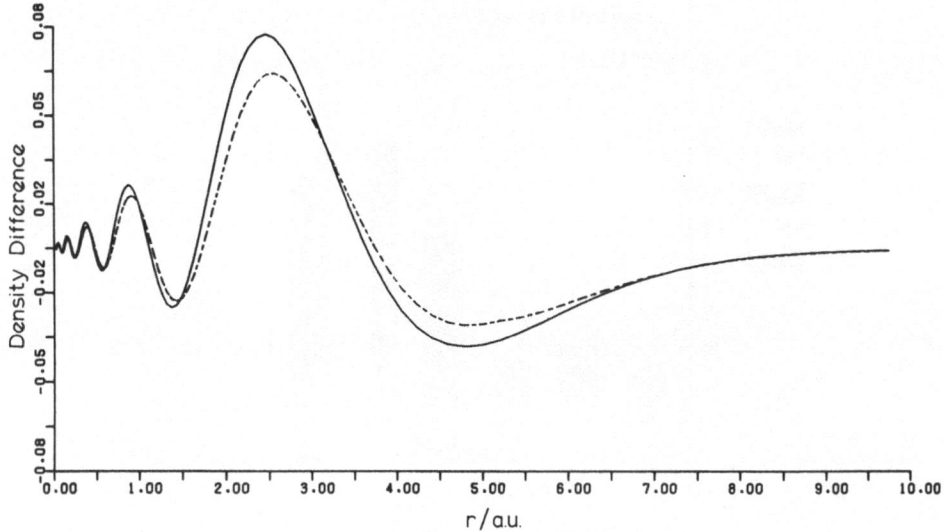

Fig. 1.   Relativistic correction to the $6s\frac{1}{2}$ density in Gd^{+}, exact (———) and by perturbation theory (— — —).

III. Relativistic effects on bond properties

To get an idea about the importance of relativity for the properties of the chemical bond, our method has been applied to a large number of small molecules [10,11]. In fig. 2-4 we show the effect of relativity on bondlengths (R_e), dissociation energies (D_e) and vibrational frequencies (ω_e) for the Cu, Ag and Au hydrides and dimers. It is apparent that while relativity plays a minor role for the Cu compounds, its importance increases as expected, when one goes down a column of the periodic table. For the Au compounds it is seen that relativity is of paramount importance, the relativistic results being in much closer agreement with experiment. AuH and Au_2 are almost twice as stable relativistically as was predicted by the non-relativistic theory. In fact even the trends are reversed by relativity: the Au compounds are more strongly bound than the corresponding Ag and even Cu compounds, in agreement with experiment and in disagreement with the non-relativistic prediction. Similar results are obtained for the vibration frequencies, which for Au_2 is doubled by relativity, bringing it much closer to experiment. The bonds are contracted by relativity in all cases studied, for Au_2 by as much as 0.5 Å .

INFLUENCE OF RELATIVITY ON BONDLENGTHS

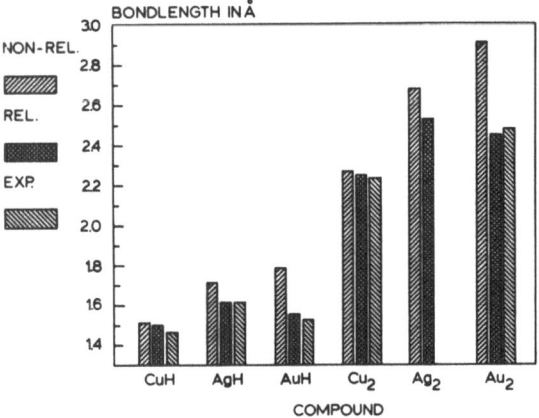

Fig. 2

INFLUENCE OF RELATIVITY ON BONDENERGIES

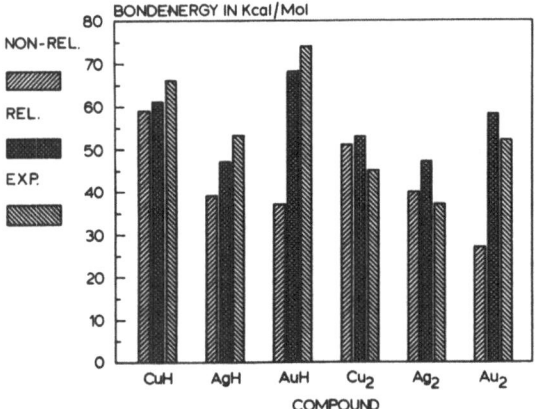

Fig. 3

INFLUENCE RELATIVITY ON VIBR. FREQ.

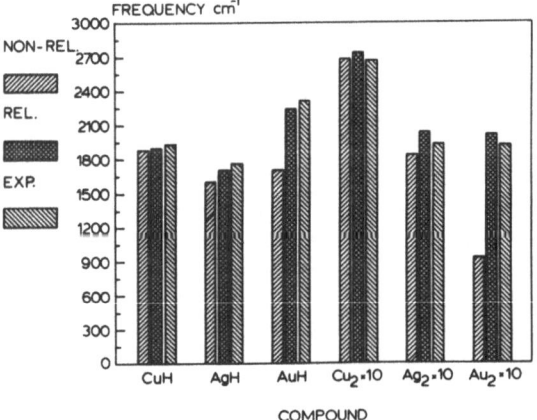

It has been known for some time from relativistic work on atoms that direct relativistic effects cause s and p orbitals to contract, while the d orbitals expand somewhat due to increased shielding of the nucleus by s and p electrons [12,14]. This circumstance has led various authors to the conclusion that in s or p bonded diatomics the relativistic bondlength contraction is a consequence of the atomic orbital contraction, the reason being that in order to achieve efficient overlap of the contracted atomic orbitals the nuclei have to be closer to each other.

The present method lends itself very well to analyze the causes of the bond contraction more closely as we shall show now. For the first order correction to the total energy one can derive:

$$E^1 = \int_{X=X'} (h_{MV} + h_D + h_{SO}) \, \rho^0 (X,X') \, dX$$

where ρ^0 is the zeroth-order (i.e. non-relativistic) electron density. Note that the indirect potential correction h_{POT}^1 is absent in this expression, a consequence of the zeroth-order energy being stationary against first order changes in the density (for proof see ref. [11]). It should be emphasized that since E^1 does not contain the first order density corrections ρ^1 in any way, it is totally independent of any relativistic contractions of the orbitals. This effect enters only in second order, for which one obtains:

$$E^2 = \frac{1}{2} \int_{X=X'} (h_{MV} + h_D + h_{SO}) \, \rho^1 (X,X') \, dX$$

It should be mentioned that for closed-shell systems, such as those studied here, the contribution of the spin-orbit term h_{SO} actually vanishes.

If we now study the binding energy of a molecule AB as a function of internuclear distance R, we can write up to second order in relativity:

$$\Delta E_{AB}^{rel} (R) = \Delta E_{AB}^{nrel} + \Delta E_{AB}^1 (R) + \Delta E_{AB}^2 (R)$$

where $\Delta E_{AB}^{nrel} (R) = E_{AB}^{nrel} (R) - E_A^{nrel} - E_B^{nrel}$ etc.

The minimum of this function of course determines the equilibrium bondlength. In fig. 5 we show the curves ΔE^1 and ΔE^2 in the case of Au_2 and AuH.

It is seen that in both cases ΔE^2 is a very flat function of R and hence has little influence on the position of the minimum of ΔE^{rel} even though its contribution to the dissociation energy is not quite negigible, at least in AuH. The first order correction ΔE^1 on the other hand is a steeply rising function of R and can therefore be identified as the main factor in the contraction of bondlengths due to relativity. We have seen, however, that ΔE^1 is completely

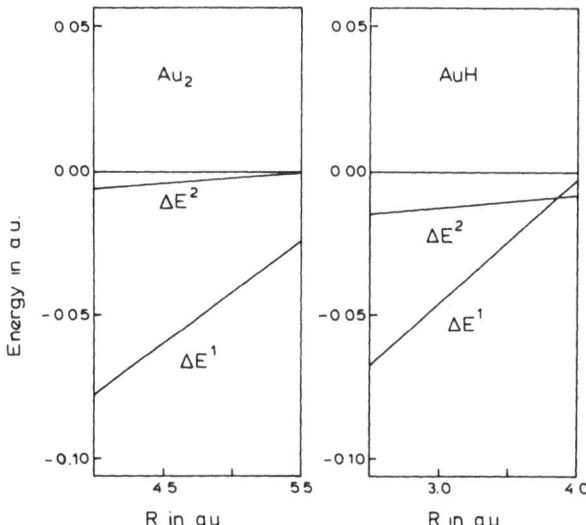

Fig. 5. First and second order relativistic corrections
        to the bonding energies of Au_2 and AuH as a
        function of internuclear distance.

independent of orbital contractions and we can therefore conclude
that contrary to what was generally believed, bond contractions
are not due to orbital contractions (this has in the meantime been
confirmed by a completely different method, see ref. [15]).
Actually it is not difficult to understand the physical reason
behind the contraction caused by ΔE^1. Non-relativistically the
kinetic energy rises as the atoms are pushed together, in accordance with the uncertainty principle since the electrons are
forced to occupy a smaller volume (of course the potential energy
falls at first leading to a minimum in the energy curve).
The most important contribution to ΔE^1 is the mass-velocity
correction to the kinetic energy, $-\alpha^2/8p^4$, a negative definite
quantity which will tend to become more and more negative as the
positive non-relativistic kinetic energy goes up. It therefore
partly relaxes the kinetic repulsion and will cause the minimum
in the binding energy curve to shift to smaller internuclear
distances, thus explaining the relativistic bondlength contractions.

   Using another method [16], it has been shown that in compounds like CsH and BaH^+ the bonding is not adequately described
using metal s functions but that in these compounds the d polarization functions are very important. Moreover the relativistic
contraction was much smaller if one included these d functions.
This was then explained by noting that atomic d functions in fact
expand due to relativity. Recently we applied our method to these

compounds to see if maybe here we had a case where ΔE^2 does play
a role in the contraction [17]. Although we did find important d
orbital participation and also a contraction which was only half
that found with s orbitals alone, these contractions were again
found in first order, thus ruling out the explanation through d
function expansion. To understand why in these cases the d functions
nevertheless diminish the contraction, we have to look a little
closer into the contraction mechanism.

The bonding σ molecular orbital in these hydrides can be written
approximately as:

$$\phi_\sigma = a\,\phi_{M6s} + b\,\phi_{H1s} + c\,\phi_{Mcore}$$

i.e. as a linear combination of metal 6s, hydrogen 1s and some
metal core orbitals, which are needed to keep this orbital orthogonal to the metal core, with which the H1s function overlaps.
In this case $E^1 = 2<\phi_\sigma|h^1|\phi_\sigma>$. It turns out that the most important
contribution in E^1 comes in fact from the core part in ϕ_σ
$E^1 \sim |c|^2$.

The coefficient c increases at shorter bond distances due to
larger H1s-Mcore overlap. The slope of E^1 which determines the con-
traction is thus proportional to $|c|$:

$$\frac{d\Delta E^1}{dR} \sim |c|\,\frac{d|c|}{dR}$$

If we now include d functions in these compounds part of the charge
flows back from hydrogen to the metal so that the coefficient b
in the bonding MO decreases. Because of the orthogonality constraint
to the core this means also that less core orbital is mixed into
the MO, i.e. c is smaller. Therefore the slope ΔE^1 diminishes and
the contraction is smaller than without d functions. Thus the less
the charge is polarized to hydrogen the smaller the bond contractions
are.

IV. Relativistic effects in photoelectron spectra

To get a feeling of the accuracy that can be achieved in
predicting ionization potentials (I.P) by the HFS method, we present
in fig. 6 the I.P. of the rare gases and the Hg atom and compare
with experiment. The I.P. have been calculated using Slater's
transition state concept to account for relaxation [18]. In general
the agreement between theory and experiment is quite reasonable
and in particular the spin-orbit splittings seen in the heavier
elements are reproduced quite well, a relativistic treatment being
of course essential in these cases.

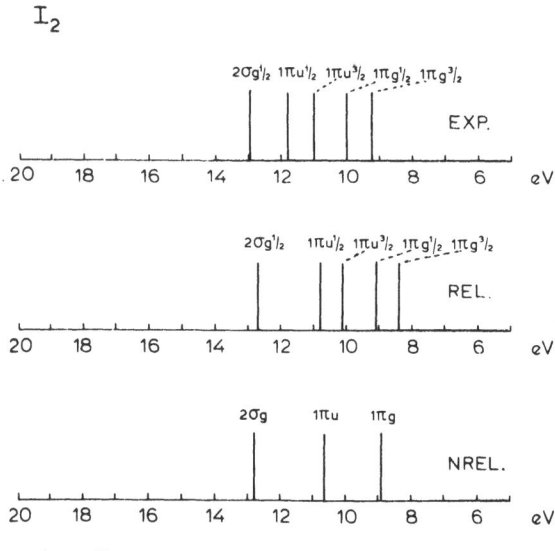

Fig. 7. Experimental and theoretical
photoelectron spectrum of I_2

In fig. 9 we show how the various relativistic effects shift
the I.P. in the case $TeBr_2$. It is seen that there is a considerable
stabilization of all levels through the mass-velocity correction,
while those orbitals having considerable Te character are affected
most, Te being the heaviest centre. The Darwin term is only important for those orbitals that have some Te or Br s character ($1a_1$,
$1b_1$) as expected from the considerations in section II.

The potential corrections, which have to be taken into account
here in contrast with the total energy considered in the previous
section, overcompensate the relativistic stabilization so that
the relativistic I.P. are actually somewhat smaller than the non-
relativistic ones. Among the three non-bonding MO's an interesting
additional effect appears. Since all three are of different non-
degenerate symmetry there is no direct spin-orbit splitting (the
diagonal elements of h_{SO}^1 identically vanish), but since these
levels are so close in energy, they can in fact be coupled due to
off-diagonal elements of the spin-orbit operator between them (they
are all of the same double group symmetry $e_{\frac{1}{2}}$ in C_{2v}^*). This separates
one of the nonbonding MO's from the other two as found experiment-
ally in $TeBr_2$ (fig. 10). The effect is calculated to be very small
in $TeCl_2$, which is not surprising since the relevant orbitals are
almost entirely of halogen character and relativity is expected
to be of less importance for Cl than for Br and indeed experimentally the splitting is not resolved in $TeCl_2$.

It seems therefore rather well established that we are dealing

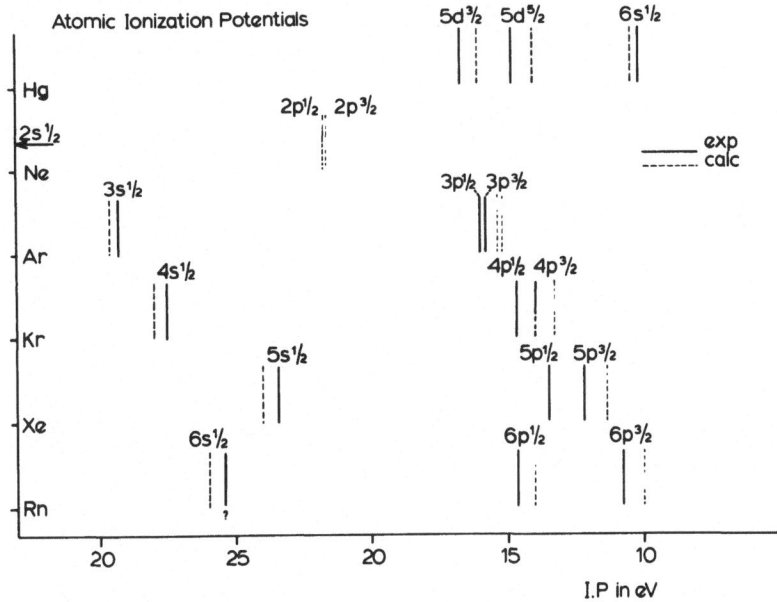

Fig. 6.   Experimental and theoretical ionization potentials for
          various atoms.

As an example of a simple molecule we show in fig. 7 the non-
relativistic, relativistic and experimental photoelectron spectrum
of I_2 [6] . Again a similar agreement with experiment is reached
as in the rare gases. The main effect of relativity is a spin-
orbit splitting of the π_u and π_g I.P.

An interesting case are the telluriumdihalides [19,20] $TeCl_2$,
$TeBr_2$, TeI_2. The general features of the spectrum of these bent
systems are readily understood by only taking into account the
interaction of the p orbitals of the three centres (see fig. 8).
In the plane of the molecule we have two bonding orbitals ($1a_1$ and
$1b_1$) made up of halogen p orbitals directed along the Te-halogen
bonds interacting strongly with the in-plane p-orbitals on Te.
There are also two non-bonding in plane orbitals ($2a_1$, $2b_1$)
consisting mainly of halogen p orbitals at roughly right angles
to the bond axes. Perpendicular to the molecular plane we have
two π type MO's ($1b_2$ and $2b_2$) one being a bonding combination of
halogen p_z and Te p_z orbitals, the other, also filled in these
species, its antibonding counterpart. Finally there is a third
non-bonding MO ($1a_2$) consisting almost entirely of halogen p
orbitals, since there is no valence orbital of the right symmetry
on Te.

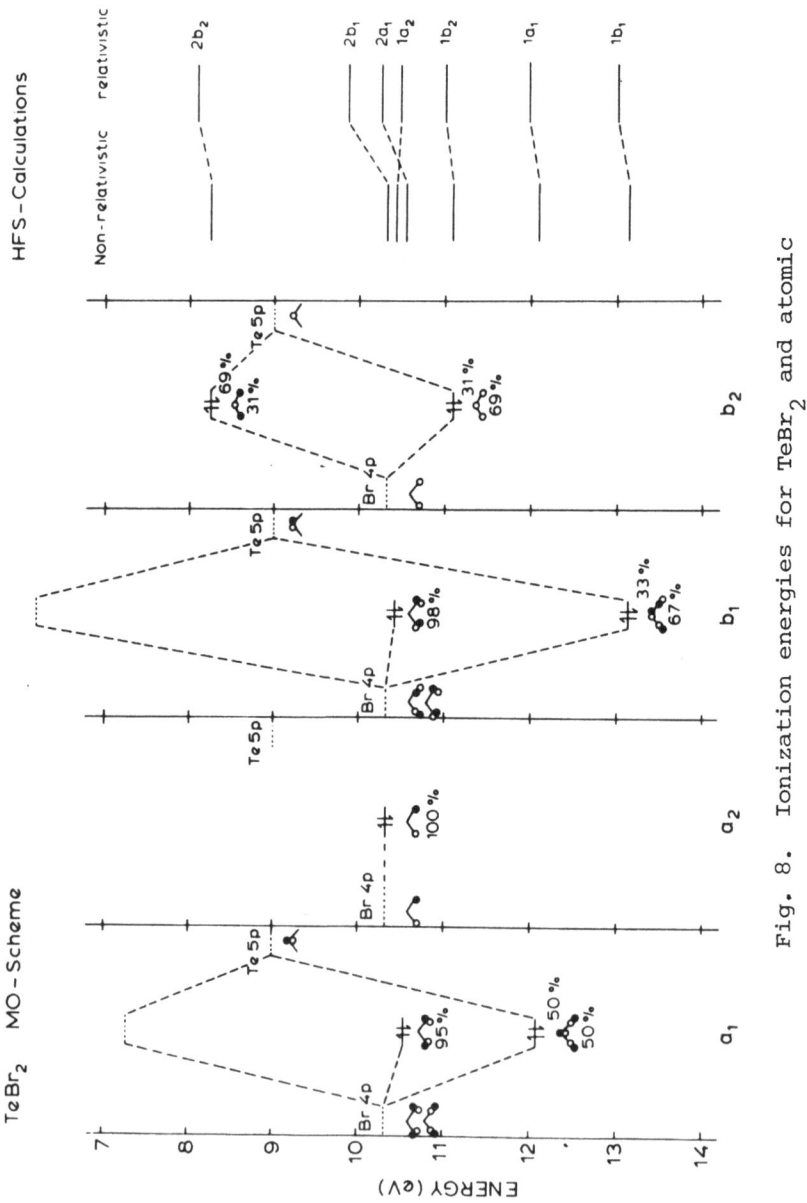

Fig. 8.   Ionization energies for TeBr$_2$ and atomic character of the valence MO's.

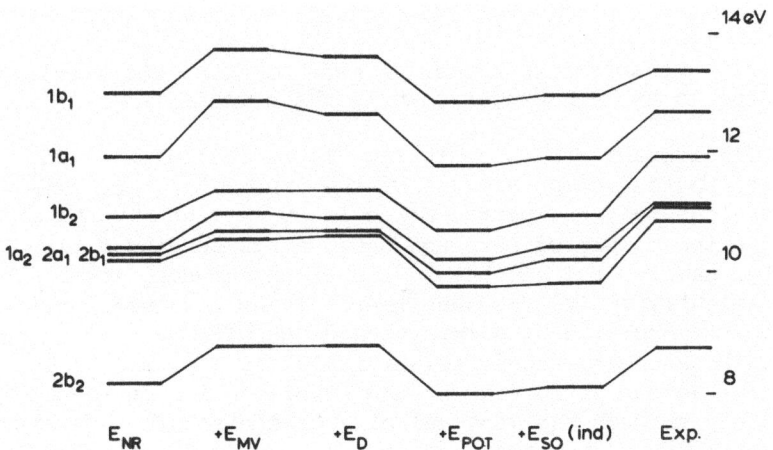

Fig. 9.   Relativistic corrections to the I.P. of $TeBr_2$.

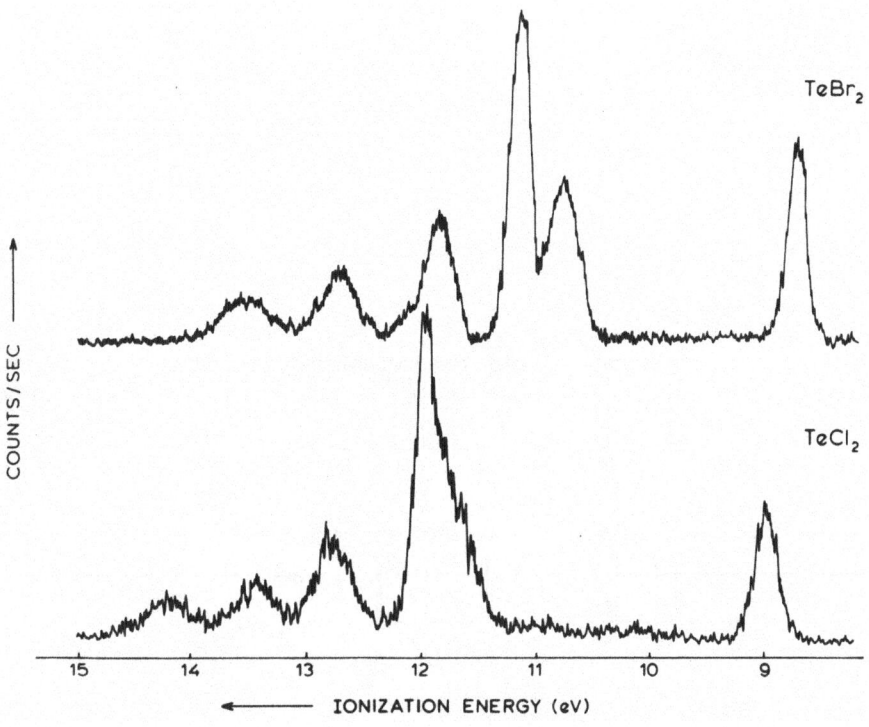

Fig. 10. Experimental photoelectron spectra of $TeCl_2$ and $TeBr_2$.

here with a spin-orbit splitting that is not predicted by symmetry
as such.

Currently measurements are being made on TeI_2, where this effect
is calculated to be even bigger than in the bromide [20].

As a final example we show in fig. 11 the calculated and
experimental photoelectron spectra of the linear HgI_2 [6]. The
spectrum divides into two parts. Below 14 eV we find a σ_g bonding
level consisting mainly of $I5p_z$ and Hg6s, a non-bonding σ_u level
concentrated mainly on I and the π_u and π_g non-bonding levels which
are only slightly split, because there is no low-lying orbital
of the proper symmetry on the intervening Hg with which they could
interact in this energy range. Relativity stabilizes the σ_g MO,
which is not unexpected in view of the stabilization of Hg6s
orbital mainly through the mass-velocity mechanism. The π orbitals
are spin-orbit split by an amount that is actually larger than the
π_u/π_g separation so that the two level systems interpretate as is
found experimentally.

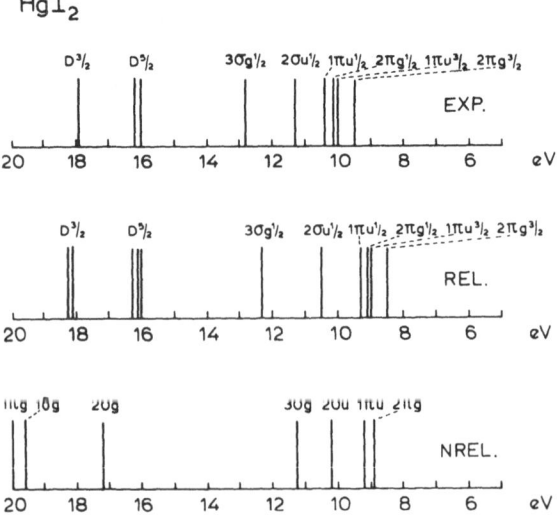

Fig. 11. Experimental and theoretical photoelectron
spectrum of HgI_2 .

Above 14 eV we find the ionizations due to the Hg5d electrons. The experimental spectrum shows essentially two bands separated by an amount close to the spin-orbit splitting of the 5d level in the Hg atom. This has been interpreted at first as an indication that the d electrons in HgI_2 are essentially atomic in character, the levels being shifted down by the increased charge on Hg and slightly split by a weak ligand field of axial symmetry. Calculations, however, give a somewhat different interpretation. The non-relativistic treatment shows that in fact the ligand field is quite strong, separating a level of σ_g symmetry $(5d_{z^2})$, from the two remaining ones of π_g and δ_g symmetry by a considerable amount, mainly due to (antibonding) interaction with the I5s orbitals. The direct relativistic effects, however, partly cancel this interaction and bring the levels much closer together again, while off-diagonal spin-orbit interactions between them separate two groups in an atomic-like pattern, in agreement with experiment. Again it is seen that it is quite important to take into account off-diagonal spin-orbit interactions between levels that are closely spaced as in the telluriumhalide case treated above.

Other systems for which photoelectron spectra have been interpreted by this method include SnO, ThF_4 and UF_4 [21].

V. Influence of relativity on electron densities

Another aspect that can be studied by the present method are the effects of relativity on electron density maps. It has been shown earlier [22,23] that the LCAO-HFS method can be used to provide density difference maps of a quality comparable to Hartree-Fock maps and that these maps compare well with experiment. Although relativistic effects on the total electron density are very large, most of course comes from the core electrons, so that the effects on deformation densities are much smaller. Nevertheless even the valence orbitals (at least of s and p type) can experience a substantial contraction in the heavier elements, so that for molecules containing these elements the difference maps are modified by relativity.

As an example we show in fig. 12a and 12b the non-relativistic and relativistic deformation densities for $HgCl_2$ [24]. The general features of these maps are readily understood. In this molecule there are 10σ, 12π and 4δ type valence electrons, while in the sum of the spherically averaged atoms, with $d^{10}s^2$ configuration for Hg and s^2p^5 for Cl, we have 11 1/3 σ, 10 2/3 π and 4 δ type electrons. As a consequence in the deformation maps we always find a transfer of 4/3 of an electron from the σ to the π system and therefore a depletion of charge in the the σ region and an excess of charge in the π region, mainly in Clpπ orbitals. The ono-relativistic and relativistic maps are similar, but differ in detail. In particular, since the Hg6s orbital is considerably stabilized by relativity, charge flows back from Cl to Hg, giving rise to a depletion of

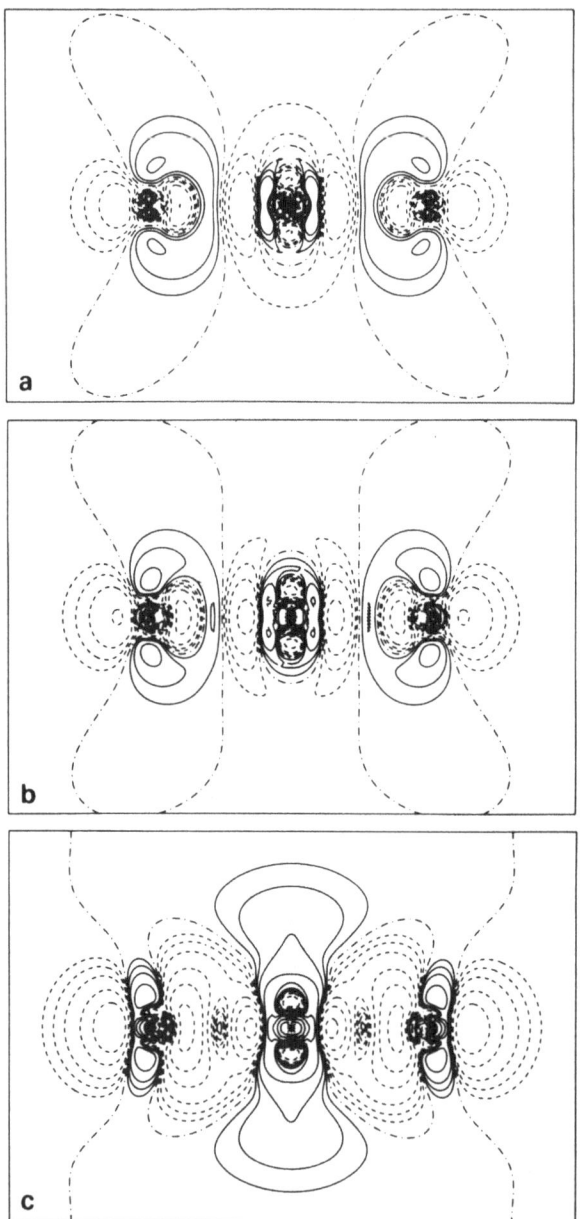

Fig. 12. Nonrel. deformation density (a), rel. deformation density
         (b) and difference of rel. and nonrel. deformation densi-
         tics (c) for HgCl$_2$.
         Contour values: for (a) and (b): 0, ±0.0025, ±0.005,
         ±0.01, ±0.025, ±0.05, ±0.1; for (c): 0, ±0.00025, ±0.0005,
         ±0.001, ±0.0025, ±0.005, ±0.01, ±0.025. ——— : positive,
         - - -: negative, -·-· : zero contour.

charge in the Clpσ region and an enhanced density in the Hg6s region. This can be most clearly seen in fig. 12c, where we plot the difference between relativistic and non-relativistic deformation densities.

The largest differences that we observe in these maps are approximately 0.02 e/au^3. The corresponding figures for the first and second row analogues $ZnCl_2$ and $CdCl_2$ are 0.0025 and 0.005 e/au^3 respectively. Since the experimental accuracy can be assumed to be of the order of 0.01 e/au^3, relativistic effect might become important in comparing with experiment for third row transition metal compounds.

Other molecules that have been investigated [11] are AuH, AuCl and Au_2, which show very similar effects, the dominant contribution coming from the stabilization of the Au6s orbital.

References

1. K.S. Pitzer, Acc. Chem. Res. 12, 271 (1979).
2. P. Pyykkö and J.P. Desclaux, Acc. Chem. Res. 12, 276 (1979).
3. J.C. Slater, Phys. Rev. 81, 385 (1951); 91, 528 (1953).
4. E.J. Baerends, D.E. Ellis and P. Ros, Chem. Phys. 2, 41 (1973).
5. J.G. Snijders and E.J. Baerends, Mol. Phys. 36, 1789 (1978).
6. J.G. Snijders, E.J. Baerends and P. Ros, Mol. Phys. 38, 1909 (1979).
7. R.E. Moss, Advanced Molecular Quantum Mechanics (Chapman and Hall, London, 1973).
8. J.G. Snijders, E.J. Baerends and P. Vernooijs, At. and Nuc. Data Tables, to be published.
9. J.G. Snijders, E.J. Baerends and P. Vernooijs, to be published.
10. T. Ziegler, J.G. Snijders and E.J. Baerends, Chem. Phys. Lett. 75, 1 (1980).
11. T. Ziegler, J.G. Snijders and E.J. Baerends, J. Chem. Phys. 74, 1271 (1981).
12. R.G. Boyd, A.C. Larson and J.T. Waber, Phys. Rev. 129, 1629 (1963).
13. D.F. Mayers, Proc. Roy. Soc. A241, 93 (1957).
14. S.J. Rose, I.P. Grant and N.C. Pyper, J. Phys. B11, 1171 (1978).
15. J.G. Snijders and P. Pyykkö, Chem. Phys. Lett. 75, 5 (1980).
16. P. Pyykkö, J. Chem. Soc. Faraday II, 75, 1256 (1979).
17. P. Pyykkö, J.G. Snijders and E.J. Baerends, Chem. Phys. Lett., to be published.
18. J.C. Slater, The Self Consistent Field for Molecules and Solids, Vol. 4 (McGraw-Hill, New York, 1974).
19. G. Jonkers, C.A. de Lange and J.G. Snijders, Chem. Phys. 50, 11 (1980).
20. G. Jonkers, C.A. de Lange and J.G. Snijders, to be published.
21. J. Dyke and J.G. Snijders, to be published.

22. W. Heijser, A.Th. van Kessel and E.J. Baerends, Chem. Phys.
     16, 371 (1976).
23. W. Heijser, E.J. Baerends an P. Ros, J. Mol. Struct. 63, 109,
     (1980).
24. P. Ros, J.G. Snijders and T. Ziegler, Chem. Phys. Lett. 69,
     297 (1980).

Section 3
EXTENDED SOLIDS: THEORETICAL AND EXPERIMENTAL RESULTS

THEORETICAL DETERMINATION OF ELECTRONIC CHARGE

DENSITIES IN COVALENTLY BONDED SEMICONDUCTORS

C. S. Wang

University of Maryland
College Park, MD 20742
and
Naval Research Laboratory
Washington, D.C. 20375

B. M. Klein

Naval Research Laboratory
Washington, D.C. 20375

INTRODUCTION

In this paper we present theoretical results of electronic charge densities and x-ray structure factors of diamond structure Si and Ge, and zinc-blende structure GaP, GaAs, ZnS and ZnSe. The calculated charge densities result from self-consistent energy band calculations for these crystals using the method of linear combination of Gaussian orbitals (LCGO), with a local density form of exchange-correlation potential. The calculations are fully *ab initio* with the only input being assumed crystal structures and lattice constants. Further details regarding the method may be found in the next section; and the interested reader can find a full discussion and results for the energy bands, densities of states, effective masses and optical properties in Refs. 1 and 2.

Our results discussed in Section III are in good agreement with the experimental structure factor measurements and indicate that local density theory gives a good quantitative representation of the crystal charge density for these, and presumably other materials. This is to be expected as the ground state charge density in local density theory results from a variational principle, with the major uncertainty being the choice of the exchange-correlation functional. It appears that first-principles calculations are capable of yielding results of sufficient accuracy so that reliable interpretations of chemical bonding in materials may be made. Where theory and experiment deviate, principly in the bond density strength and asphericity, improvements may be possible by using more realistic (non-local) exchange-correlation functionals.

METHODOLOGY

An effective one-particle equation for wave vector $\vec{k}$ and band index n can be obtained within the Hohenberg-Kohn-Sham local density functional formalism:

$$\left[-\frac{\hbar^2}{2m}\nabla^2 - \sum_{\mu,\tau} \frac{Z_\tau}{|\vec{r}-\vec{R}_\mu-\vec{r}_\tau|} + \int \frac{\rho(\vec{r}')}{|\vec{r}-\vec{r}'|}\, d^3r' + V_{xc}(\rho(\vec{r})) \right] \psi_n(\vec{k},\vec{r})$$

$$= E_n(\vec{k})\psi_n(\vec{k},\vec{r}). \tag{1}$$

Here the first term is the kinetic energy, the second the electron-nuclei Coulomb interaction (nuclei charge Z_τ), the third the electron-electron Coulomb potential, and the fourth, the local density exchange-correlation potential. The charge density $\rho(\vec{r})$ is related to the occupied one-particle wave function $\psi_n(\vec{k},\vec{r})$ by

$$\rho(\vec{r}) = \sum_{nk}^{\text{occupied}} \psi_n^*(\vec{k},\vec{r})\psi_n(\vec{k},\vec{r}) \tag{2}$$

which in turn determines the electron-electron Coulomb and exchange-correlation potentials. Thus Eqs. (1) and (2) need to be solved self-consistently. We have used the self-consistent linear combination of Gaussian orbitals (LCGO) method[3-5] which has been applied successfully to study the electronic properties of a wide range of simple[6,7] and transition metals,[8-10] and a few covalently bonded materials.[11,12] Here we will give a brief discussion of the procedure that we have adopted to calculate the self-consistent charge density of representative semiconductors.

Linear Combination of Gaussian Orbitals Basis

As in the usual variational treatment, the crystal wave functions $\psi_n(\vec{k},\vec{r})$ are expanded in terms of a Bloch basis set $\phi_{i\tau}(\vec{k},\vec{r})$

$$\psi_n(\vec{k},\vec{r}) = \sum_{i,\tau} C_{i\tau,n}(\vec{k})\phi_{i\tau}(\vec{k},\vec{r}) \tag{3}$$

where the standard LCAO form was used for the basis function i centered at the sublattice site τ,

$$\phi_{i\tau}(\vec{k},\vec{r}) = \frac{1}{\sqrt{N}} \sum_{\vec{R}_\mu} e^{i\vec{k}\cdot\vec{R}_\mu} u_{i\tau}(\vec{r}-\vec{R}_\mu-\vec{r}_\tau). \tag{4}$$

An optimum choice of a basis set $[u_{i\tau}(\vec{r})]$ which provides maximum variational freedom yet still maintains a manageable size of the secular equation has been the subject of numerous discussions.[13] In our study for covalently bonded diamond and zinc-blende materials we have chosen a linear combination of atomic orbitals *minimum* basis set with an additional shell of s, p, and d virtual Gaussian type orbitals (GTO) for added variational flexibility. The atomic orbitals were solutions to the atomic secular equation constructed with the self-consistent Herman-Skillman atomic potential[14] and a basis set of GTO. The Gaussian exponents were varied non-linearly to minimize the atomic energies as follows: For each orbital quantum number we choose a set of even-tempered Gaussians that satisfy

$$\alpha_l = \alpha_1 \left| \frac{\alpha_2}{\alpha_1} \right|^{l-1}, \tag{5}$$

where the two most localized Gaussian exponents α_1 and α_2 were varied non-linearly to minimize the corresponding lowest state atomic energy. Typically 8, 5, and 4 GTO were used to describe the atomic 1s, 2p and 3d states, respectively. Two more extended GTO were added with their exponents varied to minimize the energies of the next lowest states. This procedure was continued until all valence states of the atom are completed. The atomic orbitals were then augmented by an additional s, p, and d shell of independent GTO for the variational freedom needed to describe the wave functions in a solid. The presence of these diffuse GTO often leads to approximate linear dependence of the basis set, hence the overlap matrices must be checked carefully against negative or unphysically small eigenvalues throughout the Brillouin zone. We found it helpful to truncate the tail of the highest valence atomic orbitals by setting minimum values for the corresponding Gaussian exponents. Variationally this has little effect because the long tails of the atomic orbitals are strongly modified in a solid by the overlap of the wave functions on the neighboring atomic sites. These changes can only be fitted with the aid of the additional virtual orbitals. The resulting basis set is not complete, but does overlap both the bound and low excited subspaces of the wave functions. It leads to a dimension of 36 × 36 for the secular equation for Si, 45 × 45 for GaP and ZnS, and 54× 54 for Ge, GaAs, and ZnSe.

Crystal Potential

The initial Coulomb potential is constructed as a superposition of overlapping spherically symmetric atomic potentials. The corresponding overlapping atomic charge density is used to evaluate the local density exchange-correlation potential which we chose to be a Wigner interpolation formula[15] of the form (energies are in Ry):

$$V_{xc}(\vec{r}) = -2(3/\pi)^{1/3}\rho^{1/3}\left\{1.0 + \frac{(0.0569\,\rho^{1/3} + 0.0060)}{(\rho^{1/3} + 0.079)^2}\right\}, \tag{6}$$

where the charge density ρ is evaluated at the point $\vec{r}$.

Due to the analytic properties of the Gaussian orbitals it is customary to expand the crystal potential in a set of either (1) symmetrized plane waves (SPW)

$$V(\vec{r}) = \sum_s V_s \sum_{\{R/\vec{d}_R\}} e^{iR\vec{K}_s \cdot (\vec{r} - \vec{d}_R)}, \tag{7}$$

where the sum runs over all group operations $\{R\,|\vec{d}_R\}$ and stars of reciprocal lattice vectors $\vec{K}_s$; or (2) overlapping GTO,

$$V(\vec{r}) = \sum_{\tau,l,\mu} V_{l,\tau}\, e^{-|\alpha_l|(\vec{r} - \vec{R}_\mu - \vec{r}_\tau)|^2} \tag{8}$$

centered at each atomic site.

The plane wave basis has the advantage of being completely general, but the disadvantage of slowly converging near the nuclei due to the Coulomb singularity. For example, even with the aid of an Ewald type procedure, it is necessary to include 4000 stars of reciprocal lattice vectors to describe the crystal potential for Ni[8] and Fe.[9] For a more complex structure with less symmetry, such a procedure can be prohibitively expensive. The GTO on the other hand, converge rather quickly (less than 20 GTO are needed to describe an atomic potential to good accuracy). However, overlapping spherical GTO are too spherical around each atom to describe the directional bond in a covalent material. Although results can be improved by including higher angular momentum terms around each atom and/or additional GTO centered at the interstitial tetrahedral sites, such a procedure is somewhat arbitrary compared to the Fourier series expansion where the reciprocal lattice vectors are rigorously defined. Thus we chose to expand our potential in a mixed basis of GTO and SPW as proposed by Euwema.[16] The self-consistent Herman-Skillman neutral atomic Coulomb and exchange-correlation potentials were each fitted with 18 even-tempered GTO. The four longest range GTO which can be represented by rapidly converging Fourier series were deleted to avoid excessive lattice sums in constructing the overlapping crystal potential. The difference between the exact atomic Coulomb potential and the truncated GTO series was tabulated over a logarithmic radial mesh and their Fourier coefficients calculated numerically. In general the crystal $V_{xc}(\vec{r})$ cannot be expressed as a superposition of atomic $V_{xc}(\vec{r})$. However, subtracting out the contribution from the overlaping atomic GTO series helps to eliminate the cusp behavior near the nuclei to yield a rapidly converging Fourier expansion for the remainder. Typically 25 independent Fourier coefficients were evaluated via a three dimensional least-square fitting procedure based on a sampling of 400 random points in the unit cell. Having defined the initial potential and the basis functions, it is straightforward to evaluate the usual linear variational secular equation:

$$\sum_{j\sigma} [H_{i\tau,j\sigma}(\vec{k}) - S_{i\tau,j\sigma}(\vec{k}) E_n(\vec{k})] C_{j\sigma,n}(\vec{k}) = 0. \qquad (9)$$

When expanded in terms of the basis orbitals $u_{i\tau}(\vec{r})$, the Hamiltonian and overlap matrix elements are

$$H_{i\tau,j\sigma}(\vec{k}) = \sum_{\vec{R}_\mu} e^{-i\vec{k}\cdot\vec{R}_\mu} < u_{i\tau}(\vec{r} - \vec{R}_\mu - \vec{r}_\tau) |\hat{H}| u_{j\sigma}(\vec{r} - \vec{r}_\sigma) > \qquad (10)$$

$$S_{i\tau,j\sigma}(\vec{k}) = \sum_{\vec{R}_\mu} e^{-i\vec{k}\cdot\vec{R}_\mu} < u_{i\tau}(\vec{r} - \vec{R}_\mu - \vec{r}_\tau) | u_{j\sigma}(\vec{r} - \vec{r}_\sigma) > . \qquad (11)$$

The above two-and three-center integrals for a Gaussian type basis can be expressed in closed form. Results have been published elsewhere.[5,17] The secular equation was solved at a set of $\vec{k}$ points in Brillouin zone and a new charge density was calculated from Eq. (2).

Self-consistency

The major modification to the charge density due to self-consistency (charge transfer, change in valence electron s and p occupation, etc.) are expected to occur in the interstitial region, particularly along the tetrahedral bond. Therefore the Gaussian expansion describing the atomic potential near the nucleus was kept frozen at their starting values; only the Fourier coefficients were varied following the procedure of Callaway and Fry.[18] At each iteration, the Fourier coefficients of the input Coulomb potential $V_c(\vec{K}_s)$ were evaluated from the output charge density $\rho(\vec{K})$ of the previous iteration, with

$$V_c(\vec{K}_s) = \frac{8\pi}{K_s^2} \rho(\vec{K}_s) \qquad (12)$$

and $\rho(\vec{K}_s)$ calculated analytically from the wave functions:

$$\rho(\vec{K}_s) = \sum_{n\vec{k}} C^*_{i\tau,n}(\vec{k}) C_{j\sigma,n}(\vec{k}) S_{i\tau,j\sigma}(\vec{k}, -\vec{K}_s). \tag{13}$$

Here the sum runs over all occupied bands (index n) and wave vectors $\vec{k}$, $C_{i\tau,n}(\vec{k})$ are the eigenvectors, and the generalized overlap matrices

$$S_{i\tau,j\sigma}(\vec{k}, \vec{K}_s) = \int \phi^*_{i\tau}(\vec{k}, \vec{r}) e^{i\vec{K}_s \cdot \vec{r}} \phi_{j\sigma}(\vec{k}, \vec{r}) d^3 r \tag{14}$$

are simply the matrix elements of the plane waves calculated between the basis functions $\phi_{i\tau}(\vec{k}, \vec{r})$. Ten special $\vec{k}$-points[19] in the 1/48'th of the irreducible Brillouin zone were used in our iterations for a self-consistent potential. In order to speed up the convergence, $\rho(\vec{K}_s)$ was relaxed by mixing with 50% of the input $\rho(\vec{K}_s)$ from the previous iteration before it was substituted into Eq. (2). Once $\rho(\vec{K}_s)$ has been determined it is straightforward to calculate $\rho(\vec{r})$ and hence the exchange-correlation potential at the selected points in the unit cell. The contribution from the overlapping GTO are subtracted-out before the new Fourier coefficients $V_{xc}(\vec{K}_s)$ are determined by a 3-dimensional least-square fitting procedure. Finally the plane wave contributions of both the Coulomb and the exchange-correlation potential to the new Hamiltonian were calculated via

$$H_{i\tau,j\sigma}(\vec{k}) = \sum_{\vec{K}_s} (V_c(\vec{K}_s) + V_{xc}(\vec{K}_s)) S_{i\tau,j\sigma}(\vec{k}, \vec{K}_s), \tag{15}$$

where $S_{i\tau,j\sigma}(\vec{k}, \vec{K}_s)$, which needs to be calculated only once, is the same matrix used in Eq. (13) to evaluate $\rho(\vec{K}_s)$. We found that $\rho(\vec{K}_s)$ converges to within 10^{-4} a.u., and the energies to 0.02 eV, at the end of 5 iterations.

In the following section we compare our self-consistent $\rho(\vec{K}_s)$ (Eq. (13)) with experimentally measured structure factors. The *valence* electron charge density maps also shown were calculated with a mixed basis of GTO and SPW. The GTO were obtained by fitting the atomic valence electron charge density $\rho(\vec{r})$, while the effect of self-consistency was included through Eq. (13). The sum over bands was limited to the valence electrons only. Deformation densities, $\delta\rho(\vec{r})$, were calculated by subtracting a superposition of spherical neutral-atom valence densities used as starting values in our SC calculations from the final SC valence charge densities.

RESULTS AND DISCUSSION

In this section we present our results for the valence and deformation charge densities, and the x-ray structure factors calculated according to the procedures described in the preceding section. All of the experimental structure factors used for comparison were reduced to zero temperature and corrected for anomalous dispersion in order to facilitate a meaningful correlation between theory and experiment.

<u>Si and Ge</u>

In Figs. 1-3 we show our calculated valence and deformation density maps and bond axis densities, respectively, for Si and Ge. Tables I and II compares our calculated structure factors, F(hkl), with experiment and other *ab initio* theoretical results.

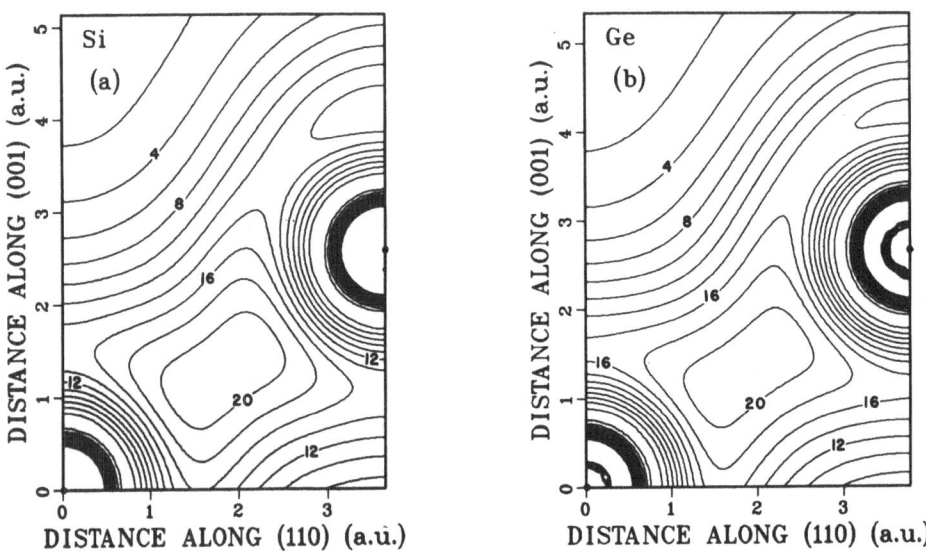

Fig. 1. Calculated self-consistent valence charge density in a portion of the $(1\bar{1}0)$ crystal plane for: a) Si, and b) Ge. The contours are in units of electrons/unit cell, and the contour interval is 2 electrons/unit cell.

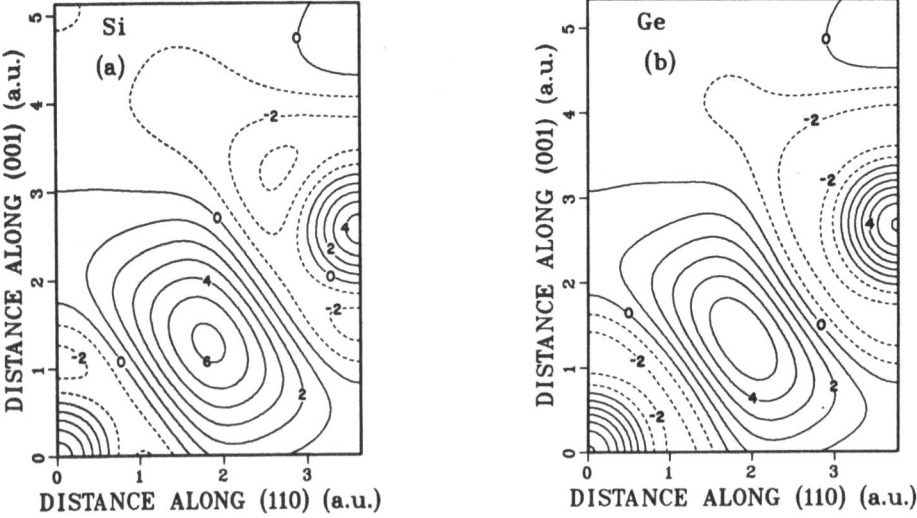

Fig. 2. Calculated valence deformation charge density in a portion of the $(1\bar{1}0)$ crystal plane for: a) Si, and b) Ge. The contours are in units of electrons/unit cell, and the contour intervals is 1 electron/unit cell.

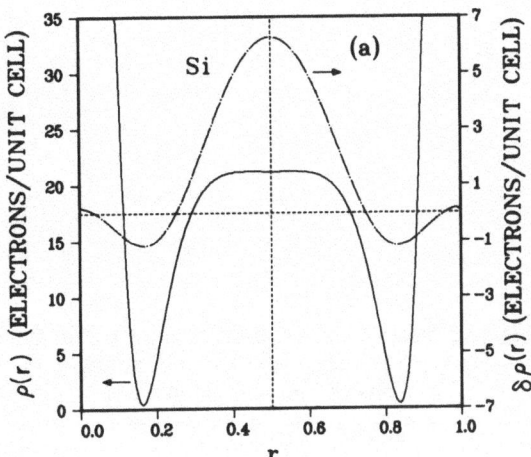

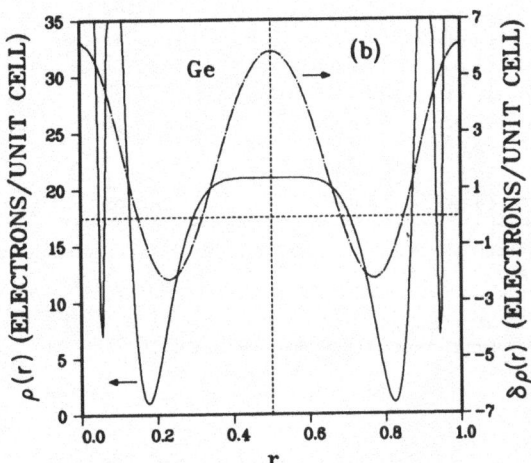

Fig. 3. Calculated $\rho(r)$ (solid lines) and $\delta\rho(r)$ (chain-dotted lines) along the (111) bonding direction for: a) Si, and b) Ge. $\rho(r)$ is the self-consistent valence charge density, $\delta\rho(r)$ is the valence deformation charge density (self-consistent minus starting overlapping atomic charge densities), and r is the fraction of the nearest neighbor (bond) distance.

Table I. Comparison of Theoretical and Experimental
X-ray Structure Factors $F(hkl)$ for Si
(in Units of Electrons/Unit Cell)

hkl	Expt.[a]	Present Results	Theory[b]	Theory[c]	Theory[d]
000	(28.0)	(28.0)	(28.0)	(28.0)	(28.0)
111	15.19	15.11	15.07	15.12	15.13
220	17.30	17.26	17.31	17.28	17.14
311	11.35	11.37	11.41	11.33	11.02
222	0.38	0.25	—	0.34	0.38
400	14.89	14.92	14.92	14.88	14.70
331	10.25	10.17	10.15	10.20	9.94
422	13.42	13.37	13.37	13.36	13.30
333	9.09	9.07	9.07	9.02	8.92
511	9.11	9.08	9.08	9.08	8.98
440	12.08	12.04	—	12.04	12.00

[a]Experiment: Aldred and Hart, Ref. 27., except for (222) which is from Ref. 28.
[b]LCAO theory: Heaton and Lafon, Ref. 12.
[c]SCOPW theory: Raccah, *et al.*, Ref. 29.
[d]Pseudopotential theory: Zunger and Cohen, Ref. 21.

Table II. Comparison of Theoretical and
Experimental X-ray Structure
Factors $F(hkl)$ for Ge
(in Units of Electrons/Unit Cell)

hkl	Expt.[a]	Present Results	SCOPW[b]
000	(62.0)	(62.0)	(62.0)
111	39.42	38.83	38.95
220	47.44	47.23	47.26
311	31.37	31.29	31.21
222	0.26	0.22	0.48
400	40.50	40.56	40.54
331	27.72	27.39	27.53
422	36.10	35.91	36.00
333	24.50	24.35	24.34
511	—	24.35	24.38
440	32.34	32.23	32.23

[a]Matsushita and Kohra, Ref. 24.
[b]Raccah, et al., Ref. 29.

Figures 1-3 clearly show the covalent nature of the bonding in Si and Ge, with the characteristic charge pile-up near the bond center for both materials. It is further interesting to note from the deformation densities shown in Figs. 2 and 3 how charge is pulled into the bond compared with the situation for overlapping charge densities— much of the covalency results from charge redistribution, not merely from wave function overlap.

The valence and deformation maps are in good agreement with previous *ab initio* calculations[20,12] and the experimental densities constructed by Yang and Coppens[22] for Si. The characteristic features of the valence density being elongated along the bond, and opposite for the deformation density, are reproduced in the calculated maps. However, Yang and Coppens[22] find a valence bond maximum of 27.6, while we find 21.3 (electrons/unit cell) for Si. Our value is very close to the theoretical value found by Hamann[20] (22.2 electrons/unit cell) but somewhat smaller than that found by Zunger and Cohen[21] (24.0 electrons/unit cell). Recently, Scheringer[23] has pointed out that the Yang and Coppens[22] value may be overestimated by as much as 3.6 electrons/unit cell due to inaccuracies in four of the measured high momentum structure factors. For Ge, where experimental structure factor data exists for only a relatively few number of reflections[24,25] ($\sin \theta/\lambda < 0.50$ A^{-1}), quantitative results for the bonding have been limited.[26]

Calculated and measured structure factors for Si and Ge are compared in Tables I and II, respectively. For silicon, except for the (222) forbidden reflection, the experimental results of Aldred and Hart[27] who used the Pendellösung fringe method are shown in Table I. For the (222) reflection in Si the value shown was chosen from the work of Roberto, Batterman and Keating,[28] corrected to zero temperature (see Ref. 28). The experimental results of Matsushita and Kohra[24] for Ge were chosen for comparison in Table II. These measurements were made using the half-width of the Bragg-case diffraction curves measured in the triple crystal arrangement using Cu-Kα

radiation. Tables I and II also show the present theoretical structure factor results, together with calculated values of: Heaton and Lafon[12] (for Si) who used a self-consistent LCAO method similar to ours but with a Kohn-Sham ($\alpha = 2/3$) exchange-correlation potential and a smaller basis set; Raccah *et al.*[29] (for Si and Ge) who used a self-consistent orthogonalized plane wave (SCOPW) approach also with $\alpha = 2/3$; and Zunger and Cohen[21] (for Si) who used the first-principles self-consistent pseudopotential method with the Hedin-Lundqvist[30] form of local density exchange-correlation potential.

Our structure factor results for Si compare extremely well with the measurements of Aldred and Hart.[27] Most of the theoretical results also agree very well, especially the present results and the other SC-LCAO[12] and the SCOPW[29] calculations. There are more substantial disagreements between the SC-pseudopotential[21] results and the other calculations, and with the measurements. The exception is for the (222) reflection where the present results are considerably smaller than experiment, while the SC-OPW and SC-pseudopotential results are much closer. Since the calculations of Heaton and Lafon[12] and our own were done in a similar manner except for the choice of exchange-correlation potential and basis functions, the very close agreement between the two sets of results seems to indicate that (at least for Si) the self-consistent charge density is rather insensitive to these choices.

Our structure factor results for Ge also compare very well with the measurements of Matsushita and Kohra[24] and with the SCOPW[29] calculations. Here too our calculated value of F(222) is smaller than the measured value, though closer than was the case for Si. The SCOPW value of F(222) for Ge is almost a factor of two larger than experiment.

GaP, GaAs, ZnS and ZnSe

Figures 4-9 show our calculated valence and deformation maps and bond axis densities for the zinc-blende compounds. Tables III-V compare the absolute values of our calculated structure factors with experiment and with other *ab initio* results, while in Table VI we present our results for both the real and imaginary parts of $F(hkl)$ for all four compounds. It should be noted that the density maps for ZnS and ZnSe include the 3d-states of Zn and Se, since the Zn 3d-states fall in the upper valence band region (see Ref. 1), while for the gallium compounds, the much lower-lying Ga and As 3d-states were kept in the core. Density maps with the Zn and Se 3d-states kept in the core may be found in Ref. 1.

The valence density maps for all four materials in Figs. 4 and 5 show a charge build-up towards the anion sites (P, As, S, or Se). However, the deformation maps and bond axis densities shown in Figs. 6-9 are even more instructive. The deformation maps clearly show that the bonding induced by the crystalline environment is both ionic and covalent. In fact, the deformation maps are qualitatively similar to Si and Ge insofar as they show a bond charge buildup elongated perpendicular to the bond axis, but with the important difference that the deformation bonding charge is shifted towards the anion site. From Figs. 8 and 9 it can be seen that the bond axis deformation charge densities peak towards the anion site, but the bond axis valence densities are shifted even further towards the anion, indicating that a large part of the valence charge asymmetry is due to atomic wave function overlap effects.

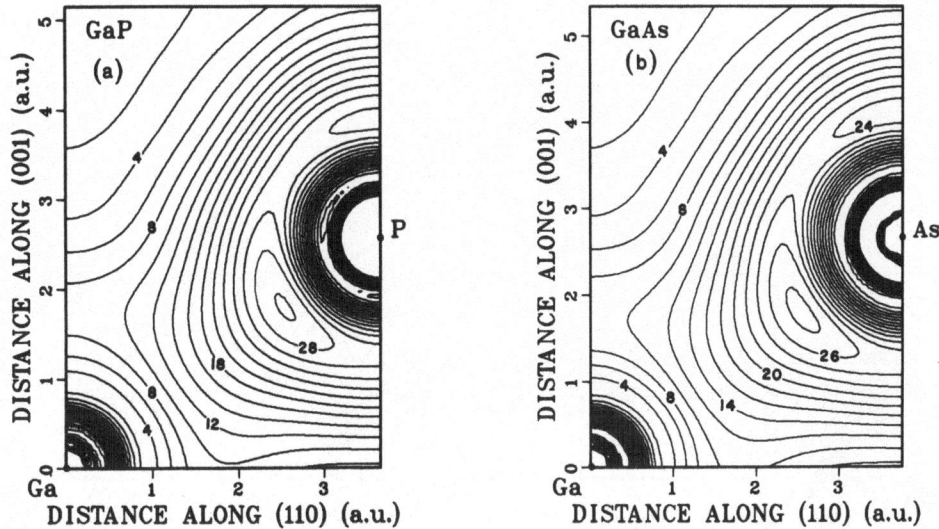

Fig. 4. Calculated self-consistent valence charge density in a portion of the ($1\bar{1}0$) crystal plane for: a) GaP, and b) GaAs. The contours are in units of electrons/unit cell, and the contour interval is 2 electrons/unit cell.

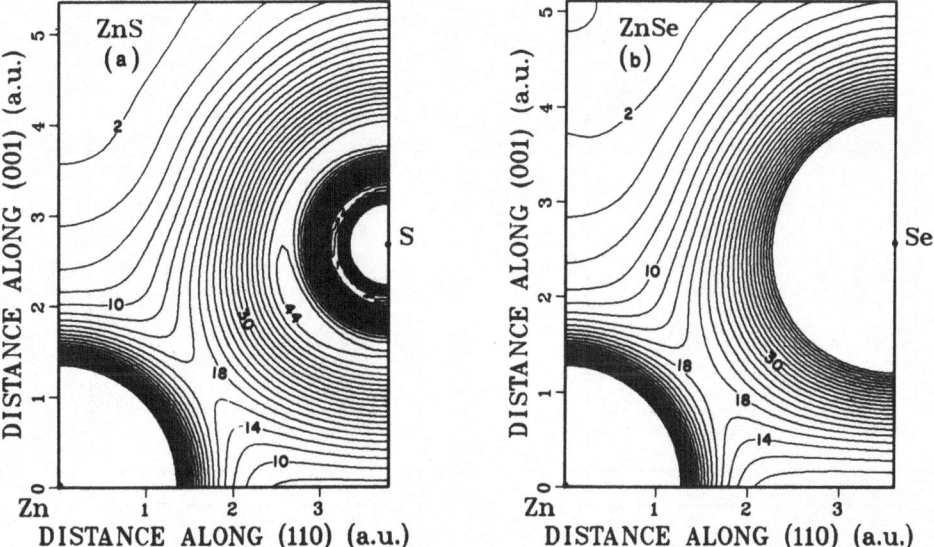

Fig. 5. Calculated self-consistent valence charge density in a portion of the ($1\bar{1}0$) crystal plane for: a) ZnS, and b) ZnSe. The contours are in units of electrons/unit cell, and the contour interval is 2 electrons/unit cell. For ZnSe we only show contours of 50 or less.

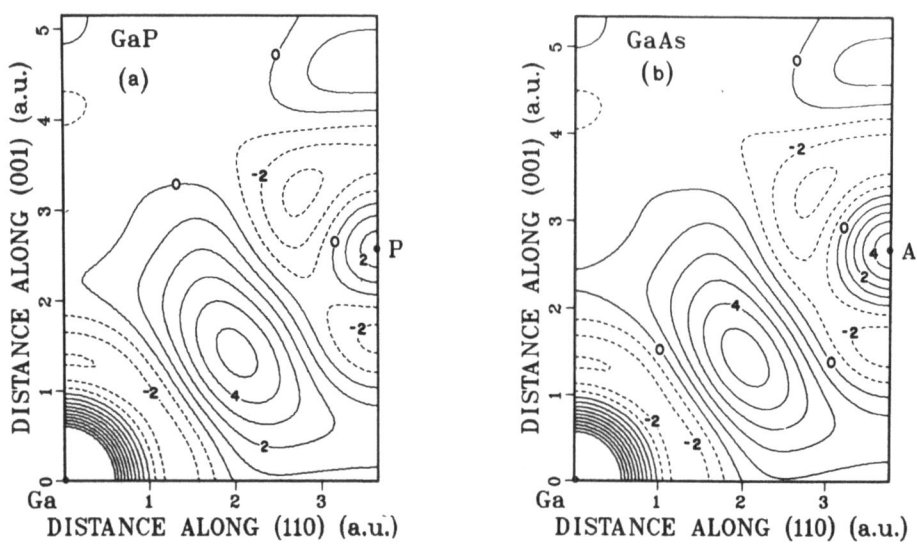

Fig. 6. Calculated valence deformation charge density in a portion of the $(1\bar{1}0)$ crystal plane for: a) GaP, and b) GaAs. The contours are in units of electrons/unit cell, and the contour interval is 1 electron/unit cell.

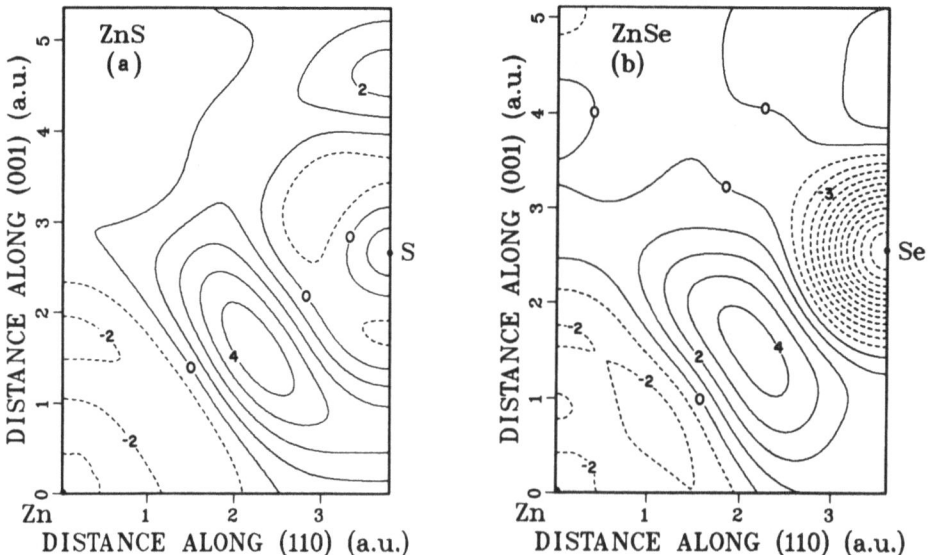

Fig. 7. Calculated valence deformation charge density in a portion of the $(1\bar{1}0)$ crystal plane for: a) ZnS, and b) ZnSe. The contours are in units of electrons/unit cell, and the contour interval is 1 electron/unit cell.

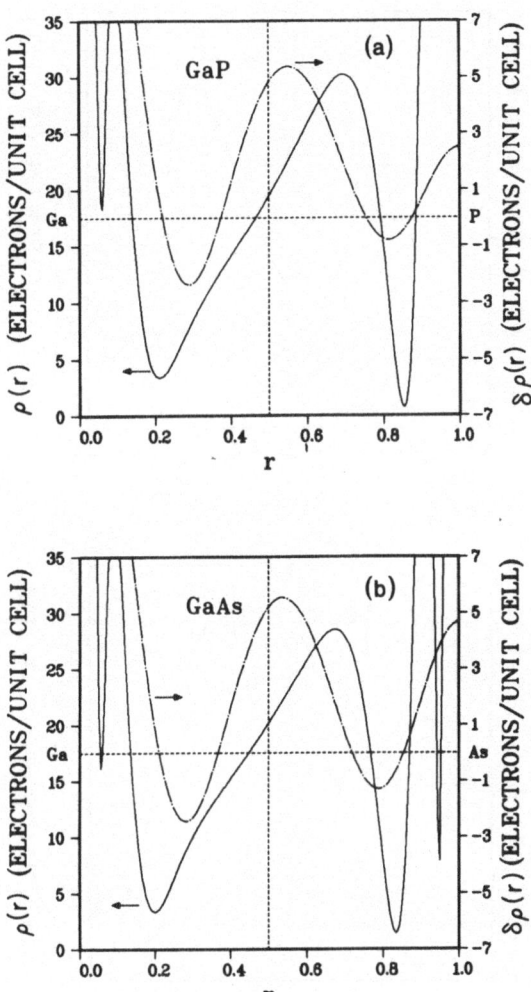

Fig. 8. Calculated $\rho(r)$ (solid lines) and $\delta\rho(r)$ (chain-dotted lines) along the (111) bonding direction for: a) GaP; and b) GaAs. $\rho(r)$ is the self-consistent valence charge density, $\delta\rho(r)$ is the valence deformation charge density (self-consistent minus starting overlapping atomic charge densities), and r is the fraction of the nearest neighbor (bond) distance.

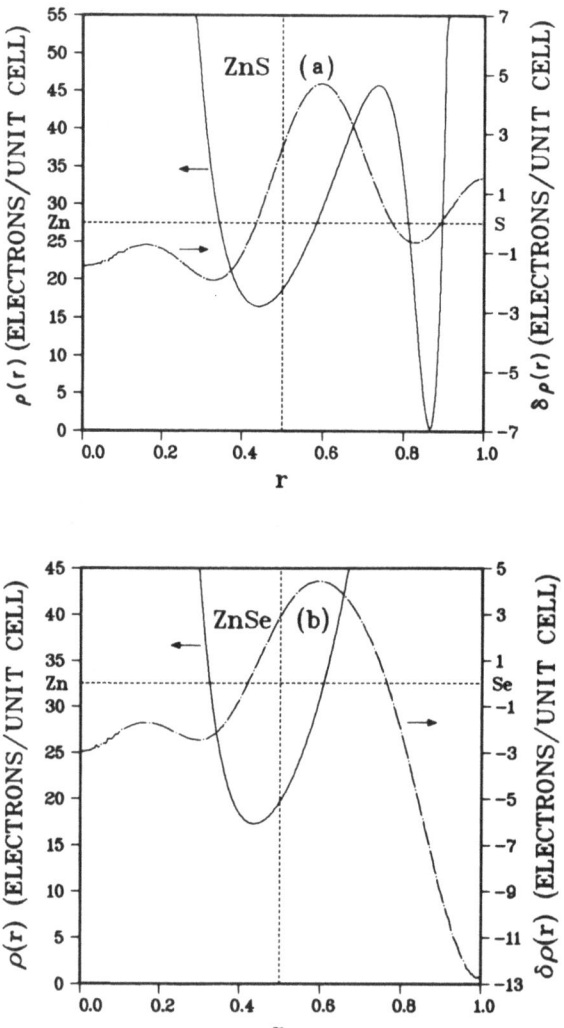

Fig. 9. Calculated $\rho(r)$ (solid lines) and $\delta\rho(r)$ (chain-dotted lines) along the (111) bonding direction for: a) ZnS; and b) ZnSe. $\rho(r)$ is the self-consistent valence charge density, $\delta\rho(r)$ is the valence deformation charge density (self-consistent minus starting overlapping atomic charge densities), and r is the fraction of the nearest neighbor (bond) distance.

Table III. Comparison of Theoretical
and Experimental Absolute Values
of The X-ray Structure Factors
$|F(hkl)|$ for GaAs
(in Units of Electrons/Unit Cell)

hkl	Expt.[a]	Expt.[b]	Present Results
000	(64.0)	(64.0)	(64.0)
111	39.44	39.06	38.85
220	47.18	46.66	47.18
311	31.37	31.04	31.25
400	40.84	40.40	40.44
331	27.75	27.45	27.32
422	36.22	35.84	35.80
333	24.37	24.10	24.30
511	24.54	24.27	24.30
440	32.38	32.04	32.15
444	26.94	26.66	26.77

[a]Experimental results of Matsushita and Hayashi (Ref. 31) corrected for anomalous dispersion $(\bar{f}' = -1.31)$ and temperature $(\bar{B} = 0.595 \ A^2)$.
[b]Experimental results of Matsushita and Hayashi (Ref. 31) corrected for anomalous dispersion $(\bar{f}' = -1.01)$ and temperature $(\bar{B} = 0.629 \ A^2)$.

Table IV. Comparison of
Theoretical and Experimental
Absolute Values of The
X-ray Structure Factors
$|F(hkl)|$ for GaP
(in Units of Electrons/Unit Cell)

hkl	Expt.[a]	Present Results
000	(46.0)	(46.0)
111	28.83	28.84
200	14.40	14.63
220	32.19	31.77
311	22.92	22.89
222	12.79	12.45
400	26.19	26.95
331	19.43	19.69
420	10.48	10.44
422	23.86	23.79
333	17.24	17.34
511	17.24	17.35
440	21.01	21.32

[a]Uno and Ishigaki, Ref. 34.

Table V. Comparison of Theoretical and
Experimental Absolute Values of the
X-ray Structure Factors $|F(hkl)|$ for ZnSe
(in Units of Electrons/Unit Cell)

hkl	Experiment[a]	Present	SCOPW[a]
000	(64.0)	(64.0)	(64.0)
111	39.64	38.99	39.06
200	3.72	2.99	2.79
220	47.45	47.16	47.37
311	32.28	31.23	31.38
222	2.91	2.55	2.38
400	40.58	40.27	40.63
331	27.39	27.23	27.59
420	3.02	2.82	2.77
422	35.14	35.63	36.11
440	32.19	32.06	32.55

[a]Ref. 29.

Table VI. Calculated Real and Imaginary Parts of the X-ray
Structure Factors $F(hkl)$ for the Zinc-Blende
Compounds. The origin is at a cation site, and the
units are electrons/unit cell.

hkl	GaAs		GaP		ZnS		ZnSe	
	ReF	ImF	ReF	ImF	ReF	ImF	ReF	ImF
111	26.65	28.27	26.42	11.57	25.62	12.44	25.91	29.13
200	-1.48	0.00	14.63	0.00	13.06	0.00	-2.99	0.00
220	47.18	0.00	31.77	0.00	31.54	0.00	47.16	0.00
311	21.48	-22.70	21.12	-8.83	20.31	-8.86	20.76	-23.33
222	-1.20	0.19	12.45	0.19	11.21	0.10	-2.55	0.13
400	40.44	0.00	26.95	0.00	26.35	0.00	40.27	0.00
331	18.64	19.98	18.20	7.51	17.26	7.77	17.83	20.58
420	-1.39	0.00	10.44	0.00	9.23	0.00	-2.82	0.00
422	35.80	-0.02	23.79	-.03	23.11	-0.04	35.63	-0.04
333	16.40	-17.93	15.92	-6.86	14.98	-7.16	15.57	-18.62
511	16.42	17.91	15.95	6.84	15.00	7.13	15.60	18.59
440	32.15	0.00	21.32	0.00	20.72	0.00	32.06	0.00

In Table III we compare our calculated values of $|F(h\,k\,l)|$ for GaAs with the experimental results of Matsushita and Hayashi[31] corrected for thermal vibrations and anomalous dispersion. The experimental data were determined using the half-width of the Bragg-case diffraction curves measured in the triple crystal arrangement using Cu-Kα radiation (similar to Matsushita and Hayashi's experiments[28] on Ge described above). For the case of polyatomic materials it is considerably more uncertain how to correct for anomalous dispersion (f'_j) and determine the Debye-Waller correction (e^{-M}). Different values of f'_j exist in the literature for Ga and As, and the value of B $(M = B(\sin\theta/\lambda)^2)$ depends on the choice of f'_j. Matsushita and Hayashi[31] discuss their results in terms of two sets of f'_j, from Cromer[32] and from Cromer and Liberman,[33] respectively, without making a clear-cut choice between the sets. We have therefore reduced their experimental data by considering $\overline{f'}$ as the Ga and As mean for both sets: a) $\overline{f'} = -1.31$, $\overline{B} = 0.595\ A^2$, and b) $\overline{f'} = -1.01$, $\overline{B} = 0.629\ A^2$. Both sets of reduced experimental data are shown in Table III and compared with our theoretical results. Given the uncertainty in the $\overline{f'}$ and $\overline{B}$ corrections to the experimental data, and the quoted $\sim 1\%$ accuracy in the measurements, the agreement between theory and experiment is remarkably good.

Uno and Ishigaki[34] have measured the structure factors of GaP using two different methods: a) the angle dispersive method using monochromatized Cu-Kα x-rays and, b) the white x-ray diffraction method, both performed on powder samples. The angle dispersive method results, being more accurate (see Ref. 34), are compared with our calculations in Table IV. The experimental results shown have been corrected to zero temperature using a value of $\overline{B} = 0.84$ obtained from Fig. 3 of Ref. 34 [a $\log(|F_{\text{exp}}|/|F_{\text{theory}}|)$ plot using Cromer's[32] values for f'_{Ga} and f'_P]. As for the GaAs case, there is some uncertainty in the f' and $\overline{B}$ values, and the quoted experimental accuracy is $\sim 1-3\%$. Within these experimental uncertainties our results for GaP again agree very well with the measurements.

For ZnSe we compare our results with the experiments and SCOPW calculations of $|F(hkl)|$ of Raccah et al. in Table V. The measurements were done on a powder sample with the thermal and anomalous diffraction corrections taken into account by Raccah et al.[29] (they present the corrected results), with a quoted experimental uncertainty of several percent. Our calculations agree quite well with the measurements, with the most substantial disagreement being for $|F(222)|$. Our results also agree very well with those resulting from the SCOPW calculations.

CONCLUSIONS

We can conclude from the theoretical-experimental comparisons made in this paper that local density theory gives a good description of charge densities in semiconductor systems. The major discrepancy between theory and experiment appears to be that the theory underestimates the charge density asphericity, as evidenced by the too small vlaues of F(222) and bond charge for Si and Ge. This seems to be a general result of many SC calculations for charge and spin densities and is due to the underlying local density approximation. Some examples for metals are shown in Table VII. Improvements in the exchange-correlation functional (e.g., non-local corrections) may resolve these subtle but interesting disagreements between theory and experiment.

Table VII. Directional Anisotropy of Charge and Spin Densities
for Transition Metals by Comparing the Ratios of Structure
Factors for Wave Vectors of the Same Magnitude.

Element	Structure Factors	Wave Vectors	Theory	Experiment
Ni	Spin	$\dfrac{(3,3,3)}{(5,1,1)}$	2.179^a	3.028^b
Fe	Spin	$\dfrac{(3,3,0)}{(4,1,1)}$	2.351^c	2.848^d
Fe	Charge	$\dfrac{(3,3,0)}{(4,1,1)}$	1.0025^c	1.010^e
				1.011^f
V	Charge	$\dfrac{(3,3,0)}{(4,1,1)}$	1.0039^g	1.024^h
				1.0085^i

^{a}Ref. 8, ^{b}Ref. 35, ^{c}Ref. 9, ^{d}Ref. 36,
^{e}Ref. 37, ^{f}Ref. 38, ^{g}Ref. 10, ^{h}Ref. 39,
^{i}Ref. 40.

However, it is clear from the present study that good quality *ab initio* calculations are capable of giving useful and reliable results for the bonding and chemistry of covalently bonded semiconductors.

ACKNOWLEDGMENTS

The authors are grateful to L. L. Boyer, D. A. Papaconstantopoulos, and W. E. Pickett for comments on the manuscript; and to A. Koppenhaver for technical assistance. This work was supported by the Office of Naval Research, contract no. N00014-79-WR-90028.

REFERENCES

(1) Wang, C.S.,; and Klein, B.M. *Phys. Rev.* **1981**, *B24*, in press.
(2) Wang, C.S.; and Klein, B.M. *Phys. Rev.* **1981**, *B24*, in press.
(3) Lafon, E.E.; Lin, C.C. *Phys. Rev.* **1966**, *152*, 579.
(4) Chaney, R.C.; Tung, T.K.; Lin, C.C.; Lafon, E. *J. Chem. Phys.* **1970**, *52*, 361.
(5) Wang, C.S.; Callaway, J. *Comput. Phys. Commun.* **1978**, *14*, 327.
(6) Ching, W.Y.; Calloway, J. *Phys. Rev.* **1975**, *B11*, 1324; **1974**, *B9*, 5115.
(7) Singhal, S.P.; Callaway, J. *Phys. Rev.* **1977**, *B16*, 1744.
(8) Wang, C.S.; Callaway, J. *Phys. Rev.* **1977**, *B15*, 298; Callaway, J,; Wang, C.S. *Phys. Rev.* **1973**, *B7*, 1096.
(9) Callaway, J.; Wang, C.S. *Phys. Rev.* **1977**, *B16*, 2095; Tawil, R.A.; Callaway, J. *Phys. Rev.* **1973**, *B7*, 4252.
(10) Laurent, D.G.; Wang, C.S.; Callaway, J. *Phys. Rev.* **1978**, *B17*, 455.

(11) Heaton, R.; Lafon, E. *Phys. Rev.* **1978**, *B17*, 1958.

(12) Heaton, R.; Lafon, E. *J. Phys.* **1981**, *C14*, 347.

(13) Simmons, J.E., Lin, C.C.; Fouquet, D.F.; Lafon, E.E.; Chaney, R.C. *J. Phys.* **1975**, *C8*, 1549.

(14) Herman, F.; Skillman, S. "Atomic Structure Calculations," Prentice-Hall: New York, 1963.

(15) Wigner, E. *Phys Rev.* **1934** *46*, 1002.

(16) Euwema, R.N. *Phys. Rev.* **1971**, B4, 4322.

(17) Chaney, R.C.; Norman, F. *Int. J. Quantum Chem.* **1974**, *8*, 465.

(18) Callaway, J.; Fry, J.L. in "Computational Methods in Band Theory"; Marcus, P.M.; Janak, J.F.; Williams, A.R., Ed.; Plenum: New York, 1972; 512.

(19) Chadi, D.J.; Cohen, M.L. *Phys. Rev.* **1973** *B8*, 5747.

(20) Hamann, D.R. *Phys. Rev. Lett.* **1979** , *42*, 662.

(21) Zunger, A.; Cohen, M.L. *Phys. Rev.* **1979**, *B20*, 4082.

(22) Yang, Y.W.; Coppens, P. *Solid State Commun.* **1974**, *15*, 1555.

(23) Scheringer, C. *Acta. Cryst.* **1980**, *A36*, 205.

(24) Matsushita, T.; Kohra, K. *Phys. Stat. Sol.* **1974**, *24*, 531.

(25) DeMarco, J.J.; Weiss, R.J. *Phys. Rev.* **1965**, *137*, A1869.

(26) Yang, Y.W. "Ph.D. Thesis, State Univesity of New York at Buffalo" unpublished, 1976.

(27) Aldred, P.J.E.; Hart, M. *Proc. Roy. Soc. London* **1973**, *A332*, 223.

(28) Roberto, J.B.; Batterman, B.W.; Keating, K.T. *Phys. Rev.* **1974**, *B9*, 2590.

(29) Raccah, P.M.; Euwema, R.N.; Stukel, D.J.; Collins, T.C. *Phys. Rev.* **1970**, *B1*, 756.

(30) Hedin, L.; Lundqvist *J. Phys.* **1971**, *C4*, 2064.

(31) Matsushita, T.; Hayashi, J. *Phys. Stat. Sol.* **1977**, A41, 139.

(32) Cromer, D.T. *Acta. Cryst.* **1965**, *18*, 17.

(33) Cromer, D.T.; Liberman, D. *J. Chem. Phys.* **1970**, *53*, 1891.

(34) Uno, R.; Ishigaki, J. *J. Appl. Cryst.* **1975**, *8*, 578.

(35) Mook, H.A. *Phys. Rev.* **1966**, *148*, 495.

(36) Terasaki, D.; Uchida, Y.; Watanabe, D. *J. Phys. Soc. Jpn.* **1975**, *39*, 1277.

(37) Diana, M.; Mazzone, G. *Phys. Rev.* **1974**, *B9*, 3898.

(38) DeMarco, J.J.; Weiss, R.J. *Phys. Lett.* **1965**, *18*, 92.

(39) Weiss, R.J.; DeMarco, J.J. *Phys. Rev.* **1965**, *140*, A1223.

(40) Diana, M.; Mazzone, G. *Philos. Mag.* **1975**, *32*, 1227.

FERMI GAS APPROACH TO X-RAY SCATTERING BY METALLIC SOLIDS

Pierre Becker

Laboratoire de Mineralogie-Cristallographie
Universite Nancy I, co 140, 54037 NANCY Cedex, France

INTRODUCTION

Most methods of analyzing electron change density derive
from localized schemes (such as multipolar expansions), where the
atomic origin of valence electrons is retained as a leading fact.
If the ground state wavefunction of a system were known, an
infinite number of representations might be chosen; in other
words, it would be equally possible to use a localized or a
delocalized picture to describe this system. Practically this is
not the case and one has to proceed through successive approximations in order to describe the system: the starting point is
thus crucial.

Consider an ionic solid: a rather good approximation consists
in assuming the electrons to be localized in "ionic orbitals"
that obey a pseudo-atomic hamiltonian which takes the crystal
field into account. It is afterwards equivalent to describe the
solid by use of translational invariance, using Bloch functions.
The simple crystal of LiH has been recently studied following
such schemes.[1,2] The fundamental parameters are the screened
nuclear charges of the ions and the overlap integrals: they
correspond by construction to a localized description.

Consider the other extreme: metallic materials. Most
physical properties can be explained on the basis of the electron
gas model, rather than using free atoms. Afterwards, it is also
possible to localize the basis of representation (Wannier functions). The fundamental parameters are the Fermi wave number and
the Fourier coefficients of the crystal potential.

Extensive studies have been done on metallic Beryllium,[3-10]
from structure factors or Compton profiles. Be is a rather

complex case since it is not an ideal metal and all experiments
confirm the existence of covalent interactions (described through
directional effects). No simple model exists which explains the
features of observed structure factors or Compton profiles.
Structure factors have been "filtered" nicely by Stewart[4] through
a multipolar expansion (localized). But the populations that are
obtained cannot be interpreted easily, nor can they be transferred to the analysis of Compton experiments.

This leads to the simple question raised by S. Berko in a
recent conference: "Can one tell from an electron density function whether a solid is a conductor or an insulator?" The answer
is not simple and one should keep in mind that for example
elastic scattering is only one experiment among many others: in
order to understand the experiment in physical terms, one should
include in the model previous knowledge of the material. In
particular, many basic properties of a solid involve a partial
knowledge of excited states and not only of the ground state.

Complementarity of scattering experiments is essential. If
$\rho(\vec{r})$ and $\bar{\rho}(\vec{p})$ are the charge and momentum densities, one can
write[11]

$$\rho(\vec{r}) = \gamma^1(\vec{r},\vec{r}) \tag{1}$$

where $\gamma^1(\vec{r},\vec{r}')$ is the first order spinless density matrix. The
structure factor $F(\vec{k})$ is:

$$F(\vec{k}) = \int \rho(\vec{r}) \exp(i\vec{k}.\vec{r}) \, d\vec{r} \tag{2}$$

If one defines $B(\vec{t})$ as the one particle auto-correlation function[27]

$$B(\vec{t}) = \int \gamma^1(\vec{r},\vec{r}+\vec{t}) \, d\vec{r} \tag{3}$$

then

$$\bar{\rho}(\vec{p}) = \int B(\vec{t}) \exp(-i \frac{\vec{p}.\vec{t}}{\hbar}) \, d\vec{t} \tag{4}$$

Compton profiles are proportional to the projection of $\bar{\rho}(\vec{p})$
on the scattering vector direction. Thus, by measuring various
Compton profiles, one can sample $B(\vec{t})$ in various directions, and
reconstruct three dimensional momentum distributions.[12]

It is well established that elastic scattering differentiates
the electronic density around various sites in the unit cell,
while Compton scattering is essentially sensitive to the dynamics
of electrons.

The purpose of the present notes is simply to show that, starting from a Fermi gas model, one can describe charge and momentum densities with a unique set of parameters. Moreover, it is possible to cover with such an approach the range of materials from ideal metals to covalent semi-conductors. We shall use over simplified models. This type of description is familiar to solid state physicists.[13]

FREE ELECTRON GAS MODEL

The simplest model consists in describing valence electrons as free electrons (experiencing a constant average potential). Let Ω be the volume of the crystal, and $\vec{k}$ the wave vector of the electrons:

$$\psi^0_{\vec{k}}(\vec{r}) = \langle\vec{r}|\vec{k}\rangle = \Omega^{-1/2} \exp(i\vec{k}.\vec{r}) \tag{5}$$

$$\gamma^1_0(\vec{r},\vec{r}\,') = 2\,\Omega^{-1} \sum_{k<k_f} \exp\{i\vec{k}.(\vec{r}-\vec{r}\,')\} \tag{6}$$

where k_f is the Fermi wave number. The density $\vec{k}$ vectors is $(\Omega/4\pi^3)$, leading to a uniform charge density:

$$\rho_0(\vec{r}) = \frac{N}{\Omega} = \frac{k_f^3}{3\pi^2} \tag{7}$$

With such a model, valence electrons should not contribute to the diffraction. This does not agree with experimental observations on Al^{14} or Be^3 where, if f is the effective form factor

$$f < f_{atom} < f_{core} \tag{8}$$

for low order reflections (f_{atom} is the free atom form factor, and f_{core} corresponds to core electrons). It should be noted that use of an improper model for low order reflections leads to a biased scale factor (too small if defined as F_{obs}/F_{calc}) and biased thermal parameters (too small).

From (6) we get:

$$B(\vec{t}) = \frac{k_f^3}{\pi^2} \frac{j_1(k_f t)}{k_f t} \tag{9}$$

Therefore, $B(\vec{t})$ and $\bar{\rho}(\vec{p})$ are isotropic, in contrast with the observations on Compton profiles. The non-periodic zeros of $B(\vec{t})$ are characteristic of systems with a Fermi surface and can be used in general to reconstruct the Fermi surface.[15] The

Compton profile is also isotropic and its valence part is the
inverted parabola:

$$J(p) = \frac{3N}{4p_f^3} (p_f^2 - p^2) \qquad p < p_f$$

$$= 0 \qquad\qquad\qquad p > p_f$$

leading to a discontinuity at p_f for the radial momentum density.

One should point out that this model is improper since free
electron plane waves are not orthogonal to core states.[17] Thus
the 1-particle density matrix is not the simple sum of core and
valence densities. The effect of this nonorthogonality is the
alteration of the density in the core region (orthogonalisation
hole): some electron density is pulled out of the core region
towards the interatomic domain.[16] This leads to complications in
the expressions of structure factors or Compton profiles. But
this alteration is small compared to the effects of potential
that have been neglected so far. The use of OPW method does not
improve very much the structure factors of Be.

If $|\vec{k}>$ is the plane wave, one can construct an OPW $|0_{\vec{k}}>$
as being:

$$|0_{\vec{k}}> = [1 - \hat{P}] \; |\vec{k}> \qquad\qquad (10)$$

where $\hat{P}$ is the projector on core states. If we denote the core
wavefunctions by the symbol ψ_c:

$$\hat{P} = \sum_c |\psi_c><\psi_c| \qquad\qquad (11)$$

$|0_{\vec{k}}>$, as defined by (10) are neither normalized nor orthogonal to
each other. The effect of the operation $\hat{P}$ is to create nodes in
the OPW:

$$< 0_{\vec{k}} | 0_{\vec{k}} > = 1 - <\vec{k}|\hat{P}|k> \qquad\qquad (10')$$

Significant charge buildup has been observed in metals at
interstitial sites.[4,5,18] This fact has been interpreted as an
evidence for covalency. Let us discuss this point in the case of
Beryllium, as a transition to the next paragraph.

First we assume a free electron gas: the deformation density
is:

$$\delta\rho_o = \rho_o - \rho_{2s} \qquad (12)$$

where ρ_{2s} stands for free atom valence density. One expects holes in the cores and positive peaks in interstitial regions. Note that ρ_o equals 0.246 e$\overset{\circ}{A}^{-3}$ in Be. Using for 2s electrons simple Slater orbitals, one gets for $\delta\rho_o$ 0.05 at tetrahedral sites, 0.06 at octahedral sites, and 0.02 between neighboring atoms. This does not agree with observations, from which $\delta\rho(\text{tet.}) \simeq 3\delta\rho(\text{oct.}) \simeq 0.09e\overset{\circ}{A}^{-3}$. We may now take the core valence non-orthogonality into account. Generally, if electrons are described by orbitals ϕ_i, and if S_{ij} is the overlap integral

$$S_{ij} = <\phi_i|\phi_j> = \delta_{ij} + \Delta_{ij}$$

the charge density is given by:

$$\rho(\vec{r}) = \sum_{ij} \phi_i \phi_j^* \{S^{-1}\}_{ij}$$

Up to second order in overlap, this leads to:

$$\rho(\vec{r}) = \sum_i |\phi_i|^2 \{1 + \sum_j |\Delta_{ij}|^2\} - \sum_{i\neq j} \phi_i \phi_j^* \Delta_{ij}$$

In the case of a metal, one gets:

$$\rho(\vec{r}) = 2 \sum_k |\psi_k^0|^2 \{1 + 2 \sum_c |<\vec{k}|c>|^2\}$$

$$+ 2 \sum_c |\psi_c|^2 \{1 + 2 \sum_k |<c|\vec{k}>|^2\}$$

$$- 4 \sum_{ck} \{\psi_k^0 \psi_c <\vec{k}|c> + c.c.\} \qquad (13)$$

where ψ_c stands for a core 1s orbital at site c. The first term shows the enhancement of the density due to overlap. Using Slater orbitals, one gets a factor 1.14 in front of the free electron contribution. This increases the deformation density by 0.03e$\overset{\circ}{A}^{-3}$, leading to a buildup similar to observation. The second and third terms give contributions in the core and can be calculated quite simply by analytical expressions. Note that adding the core density and the squares of OPWs (after renormalizing those), leads to the same density as (13).

Suppose now that we turn on the potential $V(\vec{r})$, assumed to be small. Considering the equation for the Fermi energy, and assuming V to vary slowly, we can write locally:

$$\rho(\vec{r}) = \frac{k_f(\vec{r})^3}{3\pi^2} \qquad E_f = \mathcal{H}^2 k_f(\vec{r})^2/2m + V(\vec{r})$$

$$\rho(\vec{r}) = a\{E_f - V(\vec{r})\}^{3/2} \tag{14}$$

with $\qquad a = (2m/\mathcal{H}^2)^{3/2} \cdot \dfrac{1}{3\pi^2}$

This is the Thomas-Fermi model. The effect of V is to enhance the charge density in regions of attractive potential, at the expense of interstitial regions. Therefore V makes the density look more atomic-like. One should also point out that any "real" theory tends to (14) when the potential becomes slowly varying. If one develops (14) in series of the perturbating potential V, one obtains:

$$\rho(\vec{r}) = \rho_o \left\{1 - \frac{3}{2}(V(\vec{r})/E_f) + \frac{3}{8}(V(\vec{r})/E_f)^2 + \cdots\right\} \tag{15}$$

We shall see in the next paragraph that to first order, V can be considered as a superposition of spherically symmetric screened ion potentials. Therefore, to first order, one can define spheri-cal pseudoatoms and anisotropy is only a quadratic correction. We may define covalency as the part of charge density beyond linear response. Its occurence depends on the value of the ratio V/E_f with respect to one. Notice also that to first order, $\rho(\vec{r})$ remains normalized. Finally, we shall see that momentum density is only affected by the potential to second order.

The major point of this qualitative discussion is that the potential creates pseudo-atoms that follow its fluctuations. The problem is thus to define properly a self consistent potential for valence electrons. Conversely, it appears that the measure-ment of charge density can be very helpful in describing that potential.

LINEAR SCREENING

Let $V(\vec{r})$ be the potential acting on the electrons:

$$\hat{H}\,|\psi_k\rangle = (\hat{T} + V)\,|\psi_k\rangle = E_k|\psi_k\rangle \tag{16}$$

The true wavefunction ψ_k can be expanded in OPWs:

$$|\psi_k> = (1 - \hat{P}) \, |\phi_k> \tag{17}$$

with $\qquad \hat{P} = \sum_c |\psi_c><\psi_c|$

and $\qquad |\phi_k> = \sum_{\vec{q}} a_q(\vec{k}) \, |\vec{k}+\vec{q}> \tag{18}$

$\vec{q}$ being a reciprocal vector. ϕ_k is the pseudo wavefunction, which satisfies the pseudo-hamiltonian:

$$(\hat{T} + \hat{U}) \, |\phi_k> = E_k \, |\phi_k> \tag{19}$$

The pseudo-potential $\hat{U}$ can be written as:

$$\hat{U} = V + (E - \hat{H}) \, \hat{P}$$

$$= V + \sum_c (E_k - E_c) |\psi_c><\psi_c| \tag{20}$$

The second term in (20) is positive and partially cancels the strong attractive character of V inside the ionic cores. Thus the variation of $\hat{U}$ is slow compared to V, which allows one to develop a perturbation theory, starting from the free electron gas, though it would have been impossible with the true potential. Notice that the second term in (20) pushes the electrons out of the cores toward interatomic regions.

$\hat{U}$ is non local and energy dependent, thus complicated to deal with exactly. It is not unique: one can change ϕ_k by any combination of core functions without modifying the true wavefunction or the energy E_k. It is generally chosen[16-19] in such a way to give a very smooth pseudowavefunction and pseudopotential inside the cores. Energy dependence and nonlocality are very important in the accurate description of electronic properties of some solids (particularly semi-conductors). However, most features can be explained on the basis of a local approximation for the pseudo potential, $U(\vec{r})$. We shall adopt this attitude here.

If we do so, the pseudo-hamiltonian becomes an hermitian operator, the solutions of which are orthonormalized. But, from (10') and (17), we know that ϕ_k and ψ_k cannot be simultaneously orthonormalized: thus locality of U leads to some contradiction. For instance, the true charge density is $\{\rho_{core} + \sum_k |\psi_k|^2\}$, and not $\{\rho_{core} + \sum_k |\phi_k|^2\}$. ϕ_k is not orthogonal to the core and at

first sight, it seems adequate to take this fact into account, by
a procedure similar to that used in the previous paragraph (13).
However, if the energy dependence of $\hat{U}$ had been considered, from
(20):

$$< \phi_k | \frac{\partial \hat{U}}{\partial E_k} | \phi_k> = <\phi_k | \hat{P} | \phi_k>$$

which equals the orthogonalization hole itself. Harrison[20]
concluded that if one does not take this energy dependence into
account, one should forget about the orthogonalisation of pseudo-
wavefunctions to the core, thus assuming:

$$\rho(\vec{r}) = \rho_{core}(\vec{r}) + 2 \sum_k |\phi_k|^2 \tag{21}$$

This point is not so clear in the literature on pseudopotentials,
and may lead to appreciable variations: fortunately it does only
affect the core region.

Now assume that U is the sum of the potentials from ions U_i
and a potential due to the valence electron sea (screening
potential) U_s:

$$U(\vec{r}) = U_i(\vec{r}) + U_s(\vec{r})$$

$$U_i(\vec{r}) = \sum_c u_i(|\vec{r}-\vec{R}_c|) \tag{22}$$

U_s is related to the charge density coming from the varying
potential, ρ_1, through the Laplace equation:

$$\nabla^2 U_s = -4\pi e^2 \rho_1 \tag{23}$$

$$\rho_1 = 2 \sum_k |\phi_k|^2$$

Developing ρ_1 in Fourier series:

$$\rho_1 = \frac{1}{\Omega_o} \sum_q F_1(\vec{q}) \, e^{i\vec{q}\cdot\vec{r}}$$

and

$$U_s(\vec{q}) = \frac{1}{\Omega} \int e^{-i\vec{q}\cdot\vec{r}} U_s(\vec{r}) \, d\vec{r} \tag{24}$$

one obtains:

$$U_s(\vec{q}) = \frac{4\Pi e^2}{\Omega_o q^2} F_1(\vec{q}) \tag{25}$$

$F_1(\vec{q})$ is the valence structure factor, and Ω_o is the volume of the unit cell: $\Omega = M\Omega_o$.

The coefficients $a_q(\vec{k})$ can be obtained from perturbation to first order:

$$a_q(\vec{k}) = \frac{\langle \vec{k}+\vec{q}|U|\vec{k}\rangle}{T_{\vec{k}}-T_{\vec{k}+\vec{q}}} \tag{26}$$

$$T_{\vec{k}} = \hbar^2 k^2/2m$$

For a general, non local potential, $\langle \vec{k}+\vec{q}|\hat{U}|\vec{k}\rangle$ depends on $\vec{k},\vec{q}$ and E_k. But, for a local approximation:

$$\langle \vec{k}+\vec{q}|U|\vec{k}\rangle = \frac{1}{\Omega} \int e^{-i\vec{q}\cdot\vec{r}} U(\vec{r}) \; d\vec{r} = U(\vec{q}) \tag{27}$$

Obviously, from (26) singularities will occur when $T_{\vec{k}} \sim T_{\vec{k}+\vec{q}}$, i.e. when $\vec{k}$ is on the edge of a Brillouin zone. Thus, we expect singularities from scattering when the Fermi surface cuts the Brillouin zone. The local approximation can be formulated in another way. Since large effects of the potential occur at the Fermi surface, one may consider essentially the neighbourhood of it, and make the "on Fermi surface approximation": $k = |\vec{k}+\vec{q}| \sim k_f$ and $E_k \sim E_f$, in which case $\langle \vec{k}+\vec{q}|U|\vec{k}\rangle$ does depend only on q, as in (27).

To first order, we get:

$$F_1(\vec{q}) = \frac{4}{M} U(\vec{q}) \sum_{\vec{k}} \frac{1}{T_{\vec{k}} - T_{\vec{k}+\vec{q}}} \tag{28}$$

The integration leads to the well known screening function:

$$F_1(\vec{q}) = -\Omega_o \frac{m \; k_f}{\Pi^2 \hbar^2} U(\vec{q}) \; \chi(q/2k_f) \tag{29}$$

$$\chi(x) = \frac{1}{2} + \frac{(1 - x^2)}{4 \; x} \ln \left|\frac{1 + x}{1 - x}\right| \tag{30}$$

From this expression, we deduce:

$$U_s(\vec{q}) = -\frac{\lambda^2}{q^2} U(\vec{q}) \chi(q/2k_f) \tag{31}$$

$$U(\vec{q}) = U_i(\vec{q}) / \epsilon(q) \tag{32}$$

where the dielectric function $\epsilon(q)$ is given by

$$\epsilon(q) = 1 + \frac{\lambda^2}{q^2} \chi(q/2k_f) \tag{33}$$

λ^{-1} is the screening length.

$$\lambda^2 = 4k_f me^2/\pi\hbar^2 \quad \text{and} \quad \lambda^2/q^2 = (2k_f/q)^2/(\pi a_o k_f)$$

Finally, before a general discussion, we may write the structure factor:

$$F_1(\vec{q}) = \frac{-\Omega_o q^2}{4\pi e^2} U_i(\vec{q}) [1 - \epsilon^{-1}(q)] \tag{34}$$

If $\vec{R}_n$ is the position of the nth ion in the cell, and if

$$u_1(q) = \frac{1}{\Omega_o} \int u_i(r) e^{-i\vec{q}\cdot\vec{r}} d\vec{r} \tag{35}$$

$$U_i(\vec{q}) = u_i(q) \sum_n e^{-i\vec{q}\cdot\vec{R}_n}$$

Therefore, we define, at Bragg vectors, a pseudo-form factor

$$f_1(q) = \frac{-\Omega_o q^2}{4\pi e^2} u_i(q) [1 - \epsilon^{-1}(q)] \tag{36}$$

Within the linear response model, this pseudo-form factor (and thus the pseudoatom) is spherically symmetric. Notice that it cannot be defined without going through the collective response of the Fermi gas to the potential. Finally, if core-valence overlap would be included, the procedure of eqn (13) should be followed, replacing ψ_k^0 by ϕ_k.

Let us now discuss the general features of the model we have just developed. The whole model depends on the form of $u_i(r)$ and on the Fermi momentum. Any property, to first order, must depend linearly on $u_i(q)$: for instance the energy.[16] Thus the

whole physics of the system under study depends on a unique set
of parameters, which allows one to see which experiment needs a
more sophisticated theory to be interpreted. This simple linear
response model has been successful in explaining basic properties
on non transition metals and alloys.[16,19,20]

Let us focus attention on the pseudo form factor and the
total potential which can be decomposed in a sum of spherically
symmetrical atomic contributions u(r) or u(q). The screening
function χ varies abruptly from 1 to 0 for $q/2k_f \simeq 1$ (logarithmic
singularity):   this corresponds to intersections of the Fermi
surface with the Brillouin zone edges. Taking the example of a
f.c.c. structure with z valence electrons per atom:

$$x = q/2k_f = (\frac{\Pi}{12z})^{1/2} (h^2+k^2+1^2)^{1/2}$$

For monovalent metals, x > 1 for all Bragg reflections. The
higher z, the more important screening effects are. For divalent
and trivalent metals, one expects significant features for the
first Bragg peaks.

The charge density being normalized to first order in the
perturbation, $F_1(0)$ has to be equal to the total number of electrons in the cell. This implies, from (29) that:

$$U(q) \to -\frac{2}{3} E_f \qquad \text{if } q \to 0$$

From (36), we also conclude that:

$$u_i(q) \to -\frac{4\pi ze^2}{q^2\Omega_o} \qquad \text{if } q \to 0$$

which means, at large distances, that $u_i(r) \sim -ze^2/r$. Very
generally, the logarithmic discontinuity in χ and ε leads to
oscillations in the pseudoatom charge density (Friedel oscillations)
proportional to $\cos 2k_f r/(2k_f r)^3$.[16]

If $U(\vec{r})$ varies very slowly, $U(\vec{q})$ is very sharp and thus χ
can be replaced by 1. In that case the limit of (29) becomes
identical to the term linear in V in (15), the Thomas-Fermi
model. If the T-F model is correct, it means that U(q) is negligible when $q/2k_f$ is of the order of 1. Since $q/2k_f$ is larger
than 1 for most reflections it means that the electronic structure
cannot be explained from elastic scattering experiments. In
order to be able to detect interesting features, U(q) and $\chi(q/2k_f)$
should be both appreciable for the first Bragg peaks.

The simplest model for $u_i(r)$ is the "empty core model" which assumes a total cancellation of the ionic potential by the electrons inside the core:

$$u_i(r) = 0 \qquad \text{if } r < R_c$$
$$= -ze^2/r \text{ if } r > R_c \tag{37}$$

where R_c is a characteristic core radius.

It leads to:

$$u_i(q) = \frac{4\pi ze^2}{q^2\Omega_o} \cos(qR)_c$$

$$u(q) = u_i(q)/\varepsilon(q)$$

$$f_1(q) = z \cos(qR_c)(\frac{\varepsilon - 1}{\varepsilon}) \tag{38}$$

The choice of R_c is crucial since it defines the regions where $u_i(q)$ is negative. Since it also fixes the height of secondary extrema in he potential, it is clear that the model is too crude. Generally, R_c is fixed by making the first zero of $u(q)$ coincide with the correct one. The width of $u(q)$ is approximately $\pi/2R_c$: it should be of the same order as $2k_f$ in order to see some features in the structure factors. From the values given by Harrison,[20] this is the case for most metals.

In reality, the cancellation is not perfect inside the core. A two parameter potential can be used, where:

$$u_i(r) = -A \qquad \text{if } r < R_c \tag{39}$$

The derivations of $u(q)$ and $f(q)$ are straightforward and the model works correctly for Al.[14]

Finally, if one projects out of the pseudo wavefunction angular components (through the projector $P_\ell = \sum_m |Y_{\ell m}> < Y_{\ell m}|$) about one center, it can be shown that for the values of ℓ which are present in the core functions the cancellation is good; however it is bad for those ℓ-components that are not present in the core (for instance p-symmetry in Be is not cancelled). One can define the "model potential", which is non local and energy dependent:

$$u_i = -\sum_\ell A_\ell(E) P_\ell \tag{40}$$

where A_ℓ is important for the ℓ that are not present in the core.

In the discussion, we did not account for exchange and correlation. In the local exchange approximation,[19] one can write:

$$U_s(\vec{q}) = \{\frac{4\pi e^2}{q^2} + X(q)\} \, F_1(\vec{q}) \tag{41}$$

with the exchange-correlation term $X(q)$. $X(0)$ must fit the free electron gas case, and for large q, $X(q) = - 2\pi e^2/q^2$. A reasonable, though not unique choice is:

$$X(q) = - \frac{3\pi e^2}{2k_f^2} \{1 + \frac{3q^2}{4k_f^2}\}^{-1} \tag{42}$$

Equation (29) is still valid, but in (32), ε should be replaced by:

$$\varepsilon^*(q) = 1 + \frac{\lambda^2}{q^2} \, \chi(q/2k_f) \, \{1 + X(q) \, q^2/(4\pi e^2)\} \tag{43}$$

From the wave function (16) it is easy to see that $B(\vec{r})$ does not contain terms of first order in $U(\vec{q})$ or $a_q(\vec{k})$. It means that Compton profiles are not modified in first order. However, we shall see that the second order modification, not being screened as for the first order contribution in the charge density, can be very appreciable.

Finally, let us say a few words about alloys. Under alloying, the Fermi wave number is changed, and thus the position of the Fermi surface with respect to the BZ. There will be a collective effect associated with this, and the more the Fermi surface will cross the BZ, the more important will be the effects of the potential and may be covalent effects (next paragraph). Besides this collective effect, resulting in χ or ε, there will be a localized effect, due to the ionic potential. If we take a binary ordered alloy, with sites a and b, the whole theory remains valid, except that we can define two pseudoatoms (equation 36) $f_1^a(q)$ and $f_1^b(q)$, where u_i is replaced by $u_i^a(q)$ or $u_i^b(q)$. Therefore, the model is very general and can be very helpful in describing properties of metallic systems, and the change of metallicity under alloying.

In order to end this paragraph, we give some numbers related to this theory, in the case of Beryllium. Some basic data are summarized in Table 1.

Table 1. Basic data on Beryllium metal

Space group: $P6_3/mmc$

$a = 2.2858\text{\AA}$ $c = 3.5843\ \text{\AA}$

$\Omega_o = 16.21\ \text{\AA}^3$

$\rho_o = .246\ e\text{\AA}^{-3}$

$k_f = 1.94\ \text{\AA}^{-1}$

The analysis of interesting structure factors is summarized in Table 2, together with the screening functions.

It is clear, from Table 2, taken from ref. 3, that the free atom does not fit the experiment. There is some contest in the literature about those structure factors.[6,7] However, absolute scaling is very important in such a problem and only Brown's data are scaled. Besides that, the new data[6] fit free atoms and we shall see in the following lines that it does not agree with the basic physics of this metal. Let us point out that the Wannier function calculation of March and coworkers[9] does not fit either the data at lowest Bragg angle, which is not surprising since it starts from a localized atom-like calculation.

In Fig. 1, we present the screening function χ with the reflections of Table 2 specified. We see that the critical region does correspond to those Bragg peaks.

Table 2. Some low-order structure factors and screening functions for Be.

h k l	$q/2k_f$	χ	ε	$f_1(obs)$	$f_1(\text{free atom})$
1 0 0	.818	.73	1.33	-.022	.019
0 0 2	.90	.65	1.25	-.196	-.017
1 0 1	.93	.62	1.23	-.162	-.030
1 0 2	1.21	.27	1.05	-.072	-.030
1 1 0	1.41	.19	1.03	-.083	-.033

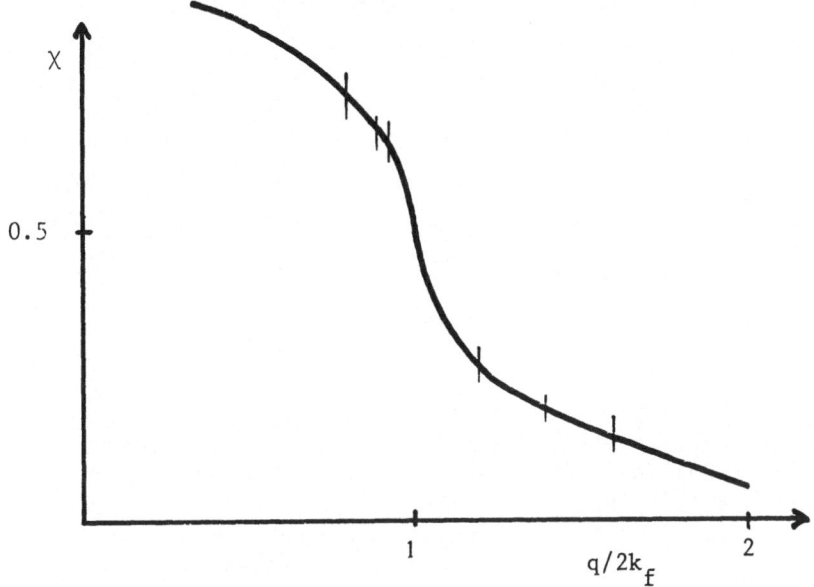

Fig. 1. The function $\chi(q/2k_f)$. First Bragg reflections of Be
are marked on the curve.

In Fig. 2, we present the potential $u(q)$ deduced from a
model non-local pseudo-potential.[16] We see that the first zero
does correspond approximately to: $(q/2k_f)$ = 0.7. Using an empty
core model would give: $q/2k_f = \pi/(4R_c k_f)$, and thus R_c = .57 Å,
in agreement with the literature. In Fig. 2 we also present the
curve $u(q)$ corresponding to this empty core model. For the first
Bragg reflections, we see that the agreement is fair, but that
there is a complete misfit concerning the oscillations of $u(q)$.
Thus the one parameter "empty core model potential" is too crude.
Since the 2p symmetry is not screened, we need at least a two
parameter model, as given in equation (39).

In order to draw the curves of Fig. 2, we notice that:

$$u(q) = (-2/3\ E_f)\ \lambda^2/q^2\ \cos(qR_c)/\varepsilon \qquad (44)$$

for the empty core model. The form factor can also be obtained
from:

$$f_1(q) = z\ (2/3\ E_f)^{-1}\ u(q)\ \chi(q/2k_f) \qquad (45)$$

Table 3.   Form factors from various models

h k l	Model Pot.	Empty core	March[9]	OPW[3]	Tight binding[3]	Exp[3]
1 0 0	-.135	-.118	+.141	-.06	-.017	-.022
0 0 2	-.173	-.173	-.200	-.05	-.16	-.196
1 0 1	-.166	-.180	-.05	-.06	-.05	-.162
1 0 2	-.060	-.090	-.100	-.020	-.07	-.072
1 1 0	-.020	-.060	-.035	-.019	-.02	-.083

Finally, we summarize the values of form factors in Table 3.   We
observe that (100) and (110) are not well reproduced, but that
the reflections (101) and (002), the most affected ones, are
correctly described (they were poorly reproduced by OPW or tight
binding).   We see that the model is fairly successful in describing structure factors of Be.   It is certainly worth being investigated more for interpreting x-ray data.

Tight binding results for 100, compared with other models,
seem to indicate that there is covalency for this reflection,
cancelling the Fermi gas effect.

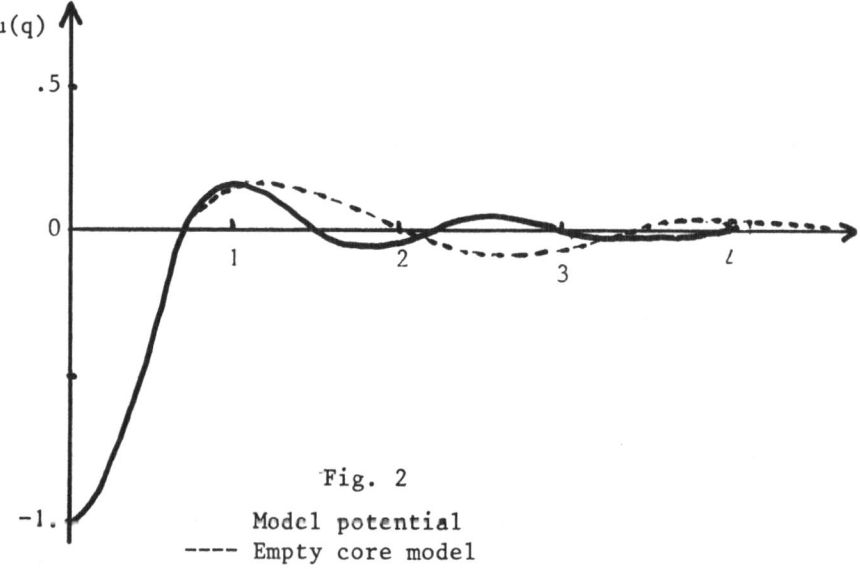

Fig. 2

—— Model potential
---- Empty core model

Fig. 2.   The function u(q).

HIGH ORDER RESPONSE: COVALENCY

The calculation of the response of the electrons to the potential, beyond the linear response, is very complicated and will only be sketched briefly here.

We have seen that the first order perturbation gives very basic trends in metal physics and leads to spherical pseudoatoms: therefore, besides the screening effect, no prefered direction appears from this model. Besides that, singularities are expected when the Fermi surface cuts the first BZ. We might say that the linear response is a kind of "reference state for solids with unfilled bands". Covalency may be defined as the result of contributions from second, third ... order terms in the potential.[21]

The effect of high order perturbations can be calculated using the perturbation expansion of the first order density matrix proposed by March and coworkers.[22] An exact calculation has been performed to second order;[23] the second order contribution to the structure factor $F_2(\vec{q})$ is:

$$F_2(\vec{q}) = \frac{m^2 \Omega_o}{2\pi^2 \hbar^4 k_f} \sum_{\vec{q}'} U(\vec{q}') U(\vec{q}-\vec{q}') \, \Phi(\vec{q},\vec{q}') \qquad (46)$$

where $\Phi(\vec{q},\vec{q}')$ is a quite complicated screening function. It can be approximated by the simple function:

$$\Phi(\vec{q},\vec{q}') \sim \frac{1}{2} x^{*-1} \ln\left|\frac{1 + x^*}{1 - x^*}\right| \qquad (47)$$

with

$$x^* = \{\vec{q}^2 + (\vec{q}-\vec{q}')^2 + \vec{q}.(\vec{q}-\vec{q}')\}^{-1/2}/(2k_f) \qquad (48)$$

The singularity of x^* occurs at 1, where Φ goes to infinity (Fig. 3). Thus, if a pair of reflections $(\vec{q},\vec{q}')$ approximately satisfy (48) the second order effect is important, and leads to selective fluctuation of the charge density, and thus to direction effects. When z=1, one can see that x^* is always much larger than 1, and therefore first order theory is basically correct. However, when k_f decreases, by an increase of z or by alloying, one can have appreciable covalent contributions; in fact, second order theory is sufficient to qualitatively describe the variation of properties from metallic systems to semiconductors.[23] More elaborate theories can be applied[24] and lead to similar results. In the case of Be, several triplets $(\vec{q},\vec{q}-\vec{q}',\vec{q}')$ lead to quasi critical second order effects: for example (100, 002, 102) (010, 1$\bar{1}$0, 100). They involve 100, which resists the most first order treatment.

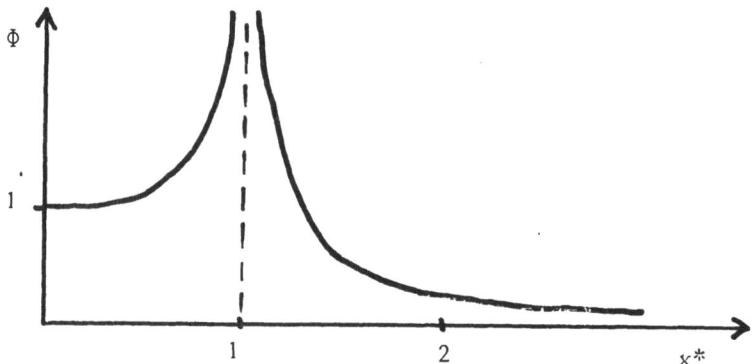

Fig. 3.   Φ as a function of x* (equations 47, 48)

We will end this paper by some comments on the momentum density. Starting from the pseudo wavefunction (18), the auto-correlation function $B(\vec{t})$ has the form:

$$B(\vec{t}) = B_o(t) + \Omega^{-1} \sum_{\vec{q}} \sum_{\vec{k}} |a_q(\vec{k})|^2 \; e^{i(\vec{k}+\vec{q})\cdot\vec{t}} \qquad (49)$$

As said earlier, it shows that there is no contribution in first order in U. But, compared to the charge density, $B(\vec{t})$ has more pronounced singularities when $q=2k_f$, because of the square of the diverging term. $B_o(t)$ is the free electron function (eqn. 9). The second part of (49) allows for anisotropy, because of the exponential term. The effect has been calculated precisely to first order, taking also exchange and correlation into account.[25] Notice finally that orthogonalisation to the core does not play a very important part in Compton profiles.[26]

CONCLUSION

In this paper, we have sketched a method of investigating charge and momentum densities that starts from the crystal poten-tial and involves a unique set of parameters, valid for many experiments. We hope that it can be used also for accurate x-ray data.

REFERENCES

1. G. Grosso & G. Pastori Parravicini, Phys. Rev. B8:3421 (1978).

2. P. Pattison & W. Weyrich, J. Phys. and Chem. Solids, 40:213 (1979).

3. P. J. Brown, Phil. Mag., 26:1372 (1972).

4. R. F. Stewart, Acta Cryst., A34:33 (1977).

5. Y. W. Yang and P. Coppens, Acta Cryst., A34:61 (1978).

6. F. K. Larsen, M. S. Lehmann, M. Merisalo, Acta Cryst., A36:159 (1980).

7. S. Manninen and P. Suortti, Phil. Mag., B40:199 (1979).

8. N. K. Hansen, P. Pattison, J. R. Schneider, Physik B35:215 (1979).

9. C. C. Matthai, P. J. Grout and N. H. March, J. Phys., F10:1621 (1980).

10. G. Loupias, J. Petiau, A. Issolah, M. Schneider, Phys. Stat. Solidi B: (1981) in the press.

11. "Electron and Magnetization Distributions in molecules and crystals" 908 pp, P. Becker, ed. Plenum (1980).

12. W. Schulke, Phys. Stat. Solidi, B82:229 (1977).
    N. K. Hansen and P. Pattison, H.M.I. Report 77 (1980).

13. D. Pines and P. Nozieres. "The Theory of Quantum Fluids" Benjamin (1966).

14. B. Dawson. "Studies of Atomic Charge Densities by X-Ray and Neutron Diffraction". Advances in Struct. Research by Diffraction Methods. Pergamon (1975).

15. W. Schulke. Phys. Stat. Solidi, B80:K67 (1977).
    P. Pattison and B. Williams, Solid State Comm., 20:585 (1976).

16. W. A. Harrison. "Pseudo-potentials in the Theory of Metals" Benjamin (1966).

17. T. O. Woodruff, Sol. State Phys., 4:367 (1957).

18. J. Hafner, Solid State Comm., 27:263 (1978).

19. a. V. Heine, b. M. L. Cohen and V. Heine, c. V. Heine and
    D. Weaire, Solid State Physics 24: Academic Press (1970).

20. a. R. W. Shaw and W. Harrison, Phys. Rev., 163:604 (1967).
    b. W. Harrison, "Electronic Structure and Properties of
    Solids" Freeman (1980).

21. a. G. L. Krasko and A. B. Makhnovetskii, Phys. Stat.
    Solidi, B80:713 (1977).
    b. A. B. Makhnovetskii and G. L. Krasko, Phys. Stat.
    Solidi, B81:263 (1977).

22. N. H. March, W. H. Young and S. Sampanthar, The Many Body
    Problem in Quantum Mechanics; University Press (1967).

23. a. R. Pickenhain and A. Milchev, Phys. Stat. Solidi, B77:571
    (1976).
    b. A. Milchev and R. Pickenhain, Phys. Stat. Solidi, B79:549
    (1977).
    c. G. Paasch and A. Heinrich, Phys. Stat. Solidi, B102:323
    (1980).
    d. R. Pickenhain and K. Unger, Phys. Stat. Solidi, B102:611
    (1980).

24. a. M. Rasolt and R. Taylor, Phys. Rev. B11:2717 (1975).
    b. L. Dagens, M. Rasolt and R. Taylor, Phys. Rev. B11:2726
    (1975).

25. E. Daniel and S. H. Vosko, Phys. Rev. 120:2041 (1960).
    P. Eisenberger, L. Lam, P. M. Platzman and P. Schmidt, Phys.
    Rev. B6:3671 (1972).

26. W. Schulke, Phys. Stat. Solidi, B62:453 (1974).

27. R. Benesch, S. R. Singh and V. H. Smith, Jr., Chem. Phys.
    Letters 10:151 (1971).

THEORETICAL AND EXPERIMENTAL CHARGE DISTRIBUTIONS IN EUCLASE

AND STISHOVITE

J.W. Downs

Department of Geological Sciences
Virginia Polytechnic Institute and State University
Blacksburg, VA 24061-0796

R.J. Hill

CSIRO, Division of Mineral Chemistry
Port Melbourne, Victoria 3207, Australia

M.D. Newton

Chemistry Department
Brookhaven National Laboratories
Upton, NY 11973

J.A. Tossell

Department of Chemistry
University of Maryland
College Park, MD 20742

G.V. Gibbs

Department of Geological Sciences
Virginia Polytechnic Institute and State University
Blacksburg, VA 24061-0796

INTRODUCTION

The crust and mantle of the earth are composed predominantly
of silicates. Thus, a knowledge of silicate crystal chemistry is
of primary importance if we are to thoroughly understand the nature
of the earth's materials and the physical and chemical state of
the earth's interior. From a chemical point of view, perhaps the

173

most basic property of a silicate mineral is its charge density distribution. An accurate knowledge of this distribution may provide insight into observed bond length and angle variations, bond type variations, phase transformations, and various physical properties. Furthermore, accurate determinations of experimental electron distributions in silicates may also serve as a test for the modelling of charge density distributions with theoretical wave functions calculated for various silicate molecules.

CHARGE DENSITY ANALYSIS OF SILICATES

The study of silicate minerals offers certain advantages as well as disadvantages to the charge density analyst. Usually, silicates tend to be hard, offering the advantage of increased specimen stability during the collection of large data sets and relatively low thermal vibration amplitudes compared with molecular systems. However, the inherent hardness of most silicates also tends to increase extinction effects which must be recognized, and then properly corrected or minimized for successful charge density analysis. The low thermal vibration amplitudes in silicates also allows scattering to persist at high angles, indicating that correction for TDS (thermal diffuse scattering) may be important for silicates. Extinction effects, coupled with the somewhat diffuse valence density and the more ionic character of the bonding (compared to say, organic systems) pose the most serious challenge to the charge density analyst who wishes to study silicate minerals. The paucity of multiple bonds and purely covalent bonds in silicates suggests that the spherical approximation (i.e., modelling the re-distribution of valence density using generalized scattering factors of spherical symmetry) may be more valid for silicate minerals than for many organic or even inorganic molecules.[1] Unfortunately many minerals are not suitable for charge density analyses due to varying degrees of order-disorder (e.g., Al/Si in feldspars), solid solution, and twinning. These problems may be effectively eliminated by careful choice of the crystal to be studied.

The purpose of this paper is to report the results of two recent charge density analyses of euclase,[2] $AlBeSiO_4(OH)$, and stishovite, the high pressure (90 kb) polymorph of SiO_2 which features six-coordinate Si and which has a density some 60 percent greater than that of α-quartz.[3]

EUCLASE, AlBeSiO$_4$(OH)

Crystal Structure

The structure of euclase, AlBeSiO$_4$(OH), is composed of chains of interconnected three-membered rings of BeO$_4$ and SiO$_4$ tetrahedra (Fig. 1) cross-linked by zig-zag chains of Al-containing octahedra. Each oxygen atom in the structure is three-coordinate and receives a formal Pauling bond strength sum of exactly 2.0.

Method

The low order X-ray data ($\sin\theta/\lambda < 0.70$Å^{-1}) for euclase, treated for isotropic and anisotropic (Type I, Gaussian) extinction during a conventional least-squares refinement (using neutral atom gas phase form factors), resulted in two sets of F_{obs}^2. The data treated for anisotropic extinction were then averaged from P$\bar{1}$ to P2$_1$/a symmetry consistent with the space group of the mineral. Positional and thermal parameters were obtained from separate conventional refinements of high order X-ray and neutron data.[4] With positional and thermal parameters fixed at the neutron or high-order refined values, the scale factor was refined using the low order data. At this point, $\Delta\rho_{X-X}$ and $\Delta\rho_{X-N}$ maps were calculated. With the introduction of separate core and valence form factors on all atoms and with positional and thermal parameters fixed at the HO or neutron values, the core/valence populations were varied. This was followed by refinement of the radial parameters (κ). In the final stages of the refinement, the population and kappa

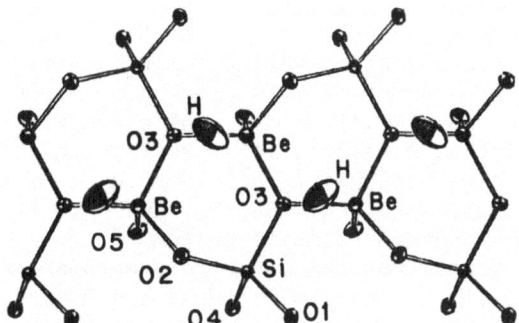

Figure 1. The tetrahedral chain of the euclase structure. The hydrogen atoms, bonded to 05, are represented by large ellipsoids located 2.17 A from 03 and 2.35 A from the midpoint of the horizontal Be03 bond. These chains are cross-linked by Al atoms (not shown).

parameters were varied simultaneously, the calculation being completed with the computer program RADIEL.[5] During these refinements the atomic scattering factor was defined as:

$$f_{atomic}(S,\kappa) = P_{core} f(S)_{core} + P_{valence} f(S/\kappa)_{valence} \qquad (1)$$

where $S = 2\sin\theta/\lambda$, P_{core} and $P_{valence}$ are the core and valence populations, respectively, and f_{core} and $f_{valence}$ are the appropriate form factors normalized to one electron. Kappa (κ) is a parameter which expands or contracts the valence scattering function in reciprocal space causing a concomitant contraction or expansion of the atom in direct space.[6] As the population parameters were varied, the total charge on the system was constrained to be neutral.

Atomic Charges and Kappas

The results of the population and kappa refinements for euclase are given in Table 1. Their comparison shows the effect of different models for extinction and thermal motion. In nearly every case, the temperature factors obtained from the neutron refinement of euclase are larger than those from the high order X-ray refinement. It appears that larger thermal parameters induce an increase in the cation kappas and charges, and in the anion charges, whereas the kappa parameters for oxygen are nearly unchanged. The cation charges and kappas obtained using isotropic extinction corrected F_{obs}^2 are slightly lower than those obtained with an anisotropic extinction model.

Population parameters are highly dependent upon the nature of the distribution function over which the charge is integrated.[7] Therefore, although we do not attach any absolute significance to the charges, we may nonetheless qualitatively compare the charges and kappas for different atoms refined using identical models.

In all three radial refinements, the estimated net atomic charges are consistent with chemical theory in that the Al, Be, Si, and H have a positive net charge while oxygen atoms are negative. Furthermore, the kappas show Al, Be, Si, and H to be contracted and O atoms to be expanded relative to the neutral free atoms. The charges and kappas from the three refinements are considered to be statistically identical. Although this holds true for the oxygen charges in each individual refinement, the O3 oxygen has the smallest charge and the largest kappa in each case. In euclase each oxygen is three coordinate with O1 and O4 bonded to two Al and one Si; O2 is linked to Be, Si, and Al; O5 to Be, H, and Al; and O3 to two Be and one Si. As expected, the least expanded oxygen with the smallest charge is bonded to the most electronegative atoms. To the extent that the ratio of the refined net

Table 1. Net atomic charges and kappa values
for euclase from kappa refinements

Atom	Neutron Anisotropic Extinction	Neutron Isotropic Extinction	H.O. X-ray Anisotropic Extinction
		Charges	
Al	+1.7(2)	+1.6(3)	+2.3(2)
Si	+1.9(2)	+1.8(2)	+2.1(2)
Be	+0.9(2)	+0.8(2)	+1.1(2)
H	+0.4(1)	+0.4(1)	+0.4(1)
O1	-1.0(1)	-1.2(1)	-1.2(1)
O2	-1.2(1)	-1.0(1)	-1.3(1)
O3	-0.8(1)	-0.5(1)	-1.0(2)
O4	-0.9(1)	-1.0(1)	-1.2(1)
O5	-1.1(1)	-0.8(1)	-1.3(1)
		Kappas	
Al	1.16(7)	1.07(6)	1.7(1)
Si	1.13(3)	1.11(3)	1.31(5)
Be	1.22(8)	1.09(8)	1.4(1)
H	1.3(1)	1.3(1)	1.6(3)
O1	0.92(1)	0.90(1)	0.92(1)
O2	0.91(1)	0.91(1)	0.92(1)
O3	0.93(1)	0.95(1)	0.93(1)
O4	0.92(1)	0.91(1)	0.92(1)
O5	0.90(1)	0.93(1)	0.91(1)

atomic charge to the formal charge may be a measure of the
"ionicity" of the bond, it is seen that the AlO bond has nearly
10 percent more ionicity than the SiO or BeO bonds.

Electron Density Maps

The deformation electron density is given by,

$$\Delta\rho = \rho_{obs}/k - \rho_{procrystal} \qquad (2)$$

where k is the scale factor and $\rho_{procrystal}$ is the thermally
smeared density of isolated, neutral atoms located at the positions
determined by refinement of neutron or high order X-ray data. When
positional and thermal parameters from the neutron refinement are
used, we obtain the deformation density $\Delta\rho_{X-N}$, and when parameters
from refinement of high order X-ray data are used, we have $\Delta\rho_{X-X}$.[8]

Charge deformation maps were calculated through several bonds in
euclase and are shown in Figures 3-8. The contour interval on all
experimental maps is 0.1 e/Å³ with an estimated standard error between atoms of 0.1 e/Å³, based on the variance in ΔF.

Figure 2a is a $\Delta\rho_{X-N}$ section through the Be–O5–H linkage
showing significant build-up of electron density in the BeO and
OH bonds. The bond peak is situated closer to the oxygen atom in
the BeO than in the OH bond indicating more charge transfer to
oxygen for the more ionic BeO interaction. The peak in the BeO
bond also appears to be off the bond axis. The density in the
lone pair region of oxygen is ascribed to bond density associated
with an Al atom located somewhat out of the plane. Figure 2b is
a theoretical difference section through a similar linkage calculated for the hypothetical molecule Be(OH)$_4^{-2}$,[9] using *ab initio*
theory with a double zeta basis of (9s,5p) primitive Gaussian
functions contracted to a (4s,2p) set. The contour interval in
the theoretical map is 0.05 e/Å³. Although the theoretical map
mimics the bonding features observed for the OH bond, it is less

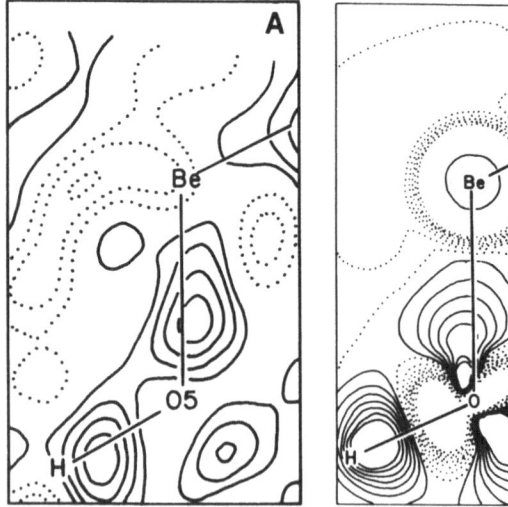

Figure 2.   Charge deformation maps through the BeOH linkage in
(a) euclase (anisotropic extinction) with contour interval
0.1 e/Å³ and (b) the molecule Be(OH)$_4^{-2}$. This theoretical map
was calculated using *ab initio* methods with a double zeta basis
of contracted Gaussian functions. Contour interval is 0.05 e/Å³.
Positive contours are solid, negative contours dotted, zero
contour is omitted.

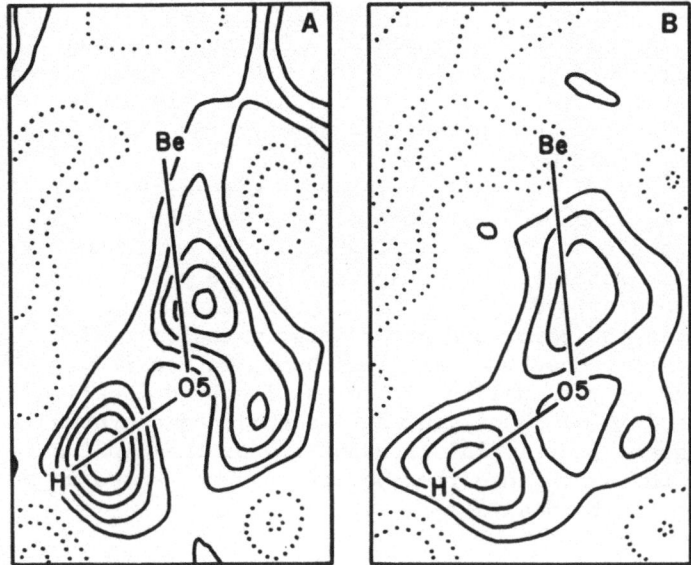

Figure 3. Charge deformation maps through the BeOH linkage in euclase. (a) $\Delta\rho_{X-X}$. (b) $\Delta\rho_{X-N}$ (isotropic extinction).

satisfactory in reproducing the peak in the Be-O bond. This may not be surprising as double zeta basis wave functions are generally believed to be of only qualitative significance. We have found several situations where polarization functions have a signifi-cant effect (Newton and Gibbs, in preparation).

It has been proposed that the position of the bond peaks in difference density maps may be related to the atomic radii of the bonded atoms.[9] Slater's atomic radius for Be is 1.05 $\mathring{A}$[10] which is in exact agreement with the separation between Be and the maxi-mum Be-O bond density in the experimental map.

Figure 3, a and b, are $\Delta\rho_{X-X}$ and $\Delta\rho_{X-N}$ (isotropic extinction) sections, respectively, through the same linkage as Figure 2a. The map calculated with F_{obs}^2 treated for isotropic extinction is very similar to the anisotropic case except that in the former the maximum bond densities are somewhat lower and the density is slightly more diffuse.

Figure 4a is a $\Delta\rho_{X-N}$ section through the Be-O2-Si linkage in euclase allowing comparison between the BeO and SiO bonds. The peak in the SiO bond profile is separated 0.97 Å from Si, compared to Slater's atomic radius for Si of 1.1 Å. Figure 4b is a theoretical deformation map[11] calculated through the SiOSi bond in disiloxane (H_6Si_2O), using *ab initio* SCF MO theory with a double zeta 66-31* basis (i.e., split valence sp with d-type polarization functions). Although for somewhat different linkages, the experimental BeOSi and theoretical SiOSi deformation maps show similar features. Most interesting perhaps is how the peaks in the bond lie off the bond axis displaced toward the other bond, with the displacement in the Be-O bond being greater than for Si-O. Similar features have been observed for SiOSi bonds in coesite.[12] Bent Si-O bonds may be rationalized in terms of hybrid localized molecular orbitals (LMO's) on oxygen where the SiO bond hybrids are at an angle of approximately 95° irrespective of the SiOSi angle. In SiOSi bonds, this may cause the hybrids to be directed as much as 30° away from the bond axis.[13]

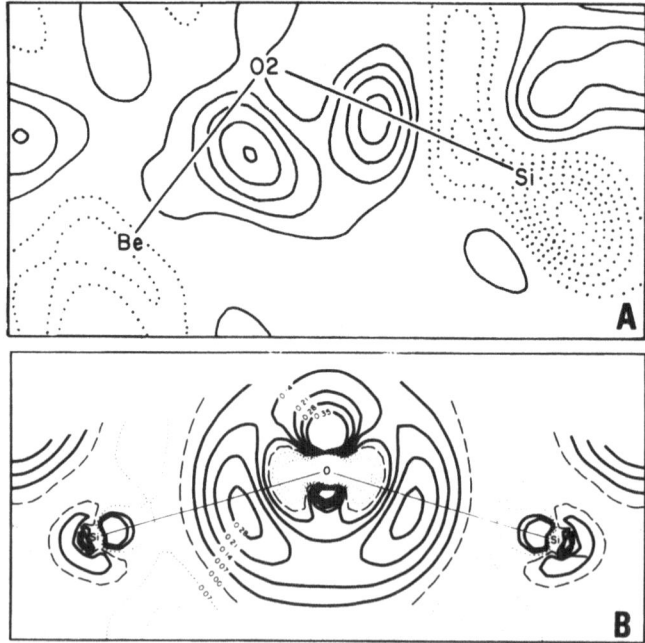

Figure 4.   (a) $\Delta\rho_{X-N}$ section through Be-O2-Si linkage in euclase. (b) Theoretical deformation map through SiOSi bond in disiloxane calculated using *ab initio* methods with a 66-31* basis.

Figure 5 is a $\Delta\rho_{X-N}$ map through the Be-02-Si bond after the kappa refinement. If the bonding were totally ionic, there would be an absence of bond density in the residual map after a kappa refinement. The residual peaks indicate that BeO and SiO bonds both have significant covalent character.

$\Delta\rho_{X-N}$ and $\Delta\rho_{X-X}$ sections through the Be-03-Be linkage in euclase are shown in Figure 6a and b, respectively. In addition 03 is bonded to a Si, located somewhat out of the plane. An obvious feature in the maps is the charge build-up in one BeO bond with almost a total lack in the other. The discrepancy in the two bond densities may be explained by referring to Figure 1. The large thermal ellipsoids represent hydrogen atoms which are each located 2.17 Å from 03 and 2.35 Å from the midpoint of the horizontal BeO3 bond (which lacks the bond density in Figure 6a). Figure 7 is a $\Delta\rho_{X-N}$ section (in the plane defined by H, 03 and Be) which shows the apparent BeO3 bond density bent out of the Be-03-Be plane toward the hydrogen, thus explaining the paucity of density in the Be-03-Be section. The distance of the hydrogen from 03 is sufficiently short to constitute hydrogen bond formation (within 2.4 Å). Interestingly, the 05-H···03 angle is 136° which compares favorably with angles from symmetrical, bifurcated hydrogen bonds in carbohydrates of 135°[14] and with similar angles of 132° optimized for carbohydrate molecules using *ab initio* HF calculations at the 4-31G level.[15] Although this is clearly not a conventional bifurcated type bond, it is similar in the sense that the H atom appears to be interacting with more than a single acceptor atom, i.e., as the density distortions of Figure 7 suggest. The deformation density visible in Figure 6a also seems to be elongated parallel to the H···H vector (refer to Fig. 2). The bond density

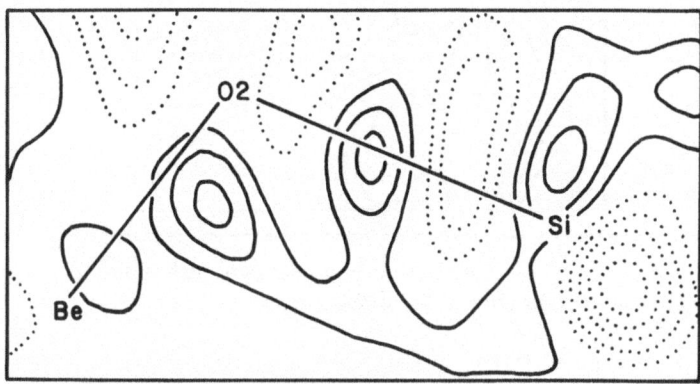

Figure 5. Residual map through the Be-02-Si bond in euclase after the monopole refinement.

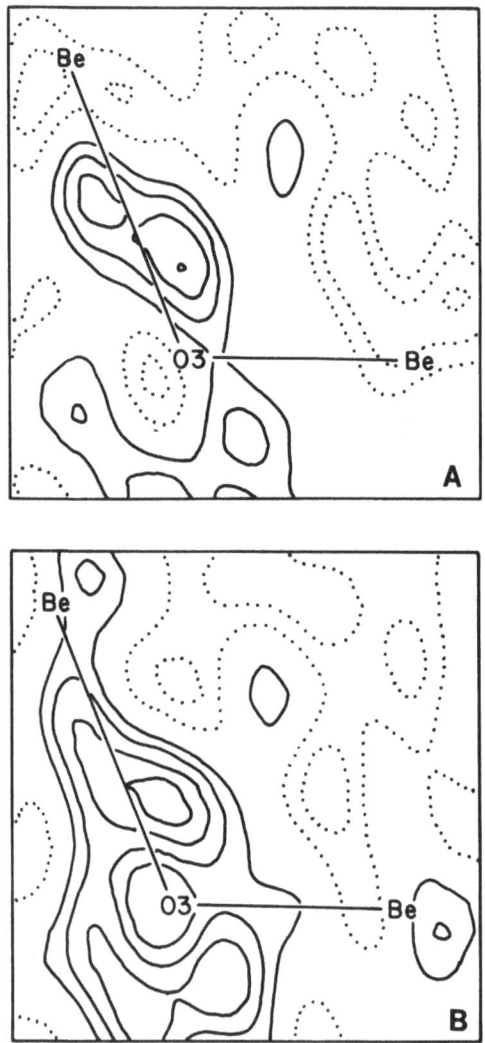

Figure 6.   (a) $\Delta\rho_{X-N}$ and (b) $\Delta\rho_{X-X}$ charge deformation maps
through the Be-03-Be bond in euclase.

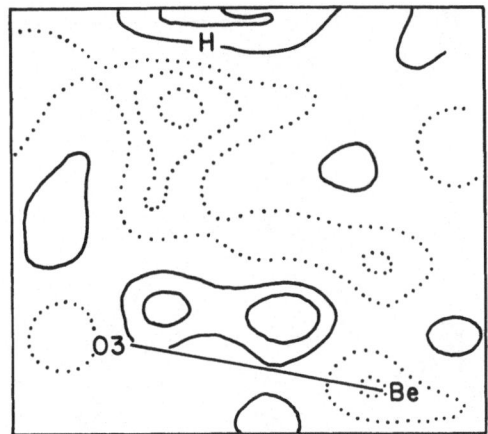

Figure 7. $\Delta\rho_{X-N}$ map in the H, 03, Be plane in euclase.

associated with the SiO bond visible in Figure 6b may also be distorted by the presence of the hydrogen atoms. The two BeO3 bond lengths are the longest in the BeO$_4$ tetrahedron; with the horizontal bond being 1.671 Å and the other being 1.666 Å. The relatively long Be-O bonds may result from the fact that the valence electrons on 03 are partitioned among three fairly covalent bonds (two Be and one Si). The difference in the two Be-O bond lengths (5σ) may be explained by weakening of the horizontal BeO3 bond due to interaction with the hydrogen atom.

STISHOVITE, SiO$_2$

Crystal Structure

Stishovite has the familiar rutile-type structure which consists of chains of edge-sharing SiO$_6$ octahedra oriented parallel to the z-axis (Fig. 8). Each chain is linked to four similar chains through three coordinate oxygen atoms. An ionic bonding model for stishovite would predict that due to cation-cation and anion-anion repulsion across the shared edge the SiO bonds in the shared edge would be longer than those not involved in edge sharing, and the O···O separations in the shared edge would tend to be shorter than normally observed. Interestingly, the SiO bonds in the shared edge are 0.05 Å (3 percent) *shorter* than those not involved in edge sharing. Furthermore, the O···O distance in the shared edge (2.29 Å) ranks among the shortest nonbonded O···O separations known.[3]

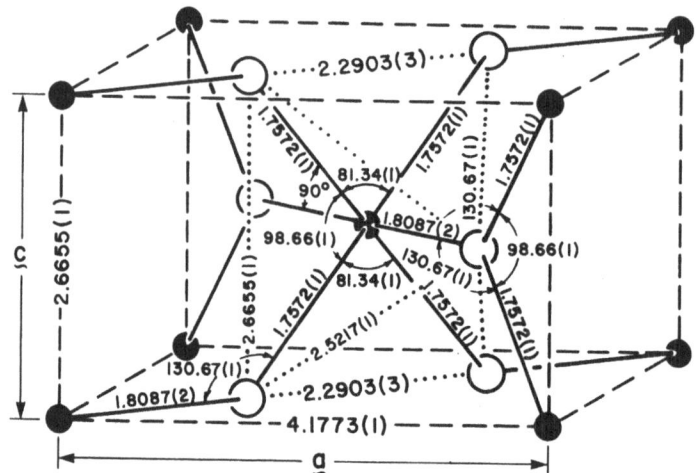

Figure 8.   The crystal structure of stishovite, SiO_2, including geometric data obtained from the monopole refinement.

Method

The positional, thermal, valence population, and radial parameters for stishovite were refined from a full set of X-ray data using the computer program RADIEL.[5] During refinement the observed data were treated for isotropic extinction (Type I, Lorentzian) and the total charge on the system was constrained to neutrality. Since a set of atomic parameters from an independent neutron refinement were not available for stishovite, the parameters from the full-data population/kappa refinement were though to be the least biased by bonding effects.[3] Deformation maps wer thus calculated using atomic parameters from the population/kappa refinement, including only the low order data $(\sin\theta/\lambda < 0.85 \text{ Å}^{-1})$. The average estimated error in the resulting $\Delta\rho_{X-X}$ maps is 0.031 $e/\text{Å}^3$ at a general position.

Net Atomic Charges and Kappas

The net atomic charges and kappas for stishovite from the population/kappa refinement are $q_{Si} = +1.70$, $\kappa_{Si} = 1.11(18)$, $q_O = -0.85(15)$, $\kappa_O = 0.961(8)$. As expected Si is contracted with a ne positive charge and oxygen is expanded with a negative net charge relative to the neutral free atoms.

Previous experimental estimations of the charge on Si have
ranged from near zero to +2.7 for various compounds. When those
compounds with constituents other than Si and O are excluded, as
in the silica polymorphs, the charge on Si is found to be near
unity when Si is in four coordination. The estimated net charge
on the six coordinate Si in stishovite of +1.70 is the largest
value obtained to date for any silica polymorph. This result suggests that the charge on Si increases with increasing coordination
number, CN(Si), as observed for Al when its coordination changes
from four in berlinite, $AlPO_4$,[16] to six in $CoAl_2O_4$[17] and diaspore,
$AlOOH$.[18]

Exploring this result further, the SiO bond lengths in the
molecules H_4SiO_4[11] and H_8SiO_6[3] were optimized with STO-3G molecular
orbital theory. The charges on Si, Q(Si), obtained in the calculations indicate that Q(Si) increases slightly from +1.34 for the
four-coordinate Si in H_4SiO_4 to +1.39 for the six-coordinate Si in
H_8SiO_6. However, because of the small increase in Q(Si), the SiO
bonds for three additional molecules (SiO, SiO_2 and planar H_2SiO_3)
were optimized. The resulting bond lengths obtained for these
three molecules and those obtained for H_4SiO_4 and H_8SiO_6 are
plotted against CN(Si) in Figure 9. In agreement with several
such trends reported by Shannon and Prewitt,[19] the correlation
between CN(Si) and SiO bond length is indicated to be highly
linear. In contrast, the trend between CN(Si) and Q(Si) is indicated to be nonlinear with the smallest increase in Q(Si) involving
a change in coordination number from four to six. Nevertheless,
the calculated results indicate, as suggested by the net charges
obtained in the least-squares refinement of the silica polymorphs,
that the ionicity of the SiO bond increases with coordination
number. (Note that H_8SiO_6 has four equatorial OH and 2 axial H_2O
groups, with equal Si-O distances).

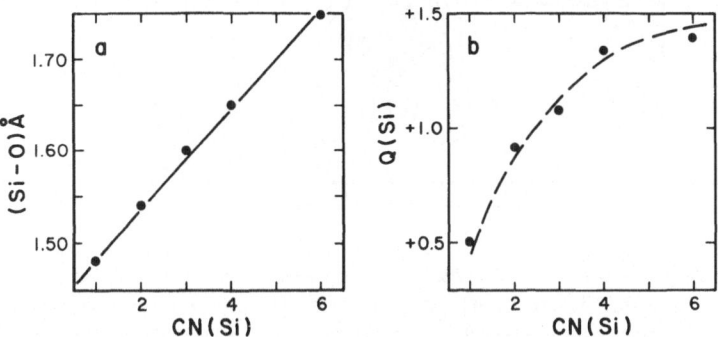

Figure 9.   STO-3G optimized SiO bond lengths and net atomic
charges on Si versus Si coordination for several molecules.

Electron Density Maps

      Figures 10 and 11 are $\Delta\rho_{X-X}$ sections through portions of the
stishovite structure which show significant peaks of electron
density in each of its bonds. The maximum bond densities are
0.47 and 0.30 e/Å^3 for bonds involved and not involved, respectively, in edge sharing. The peaks in the bonds are also observed
to be somewhat closer to oxygen than to Si. They occur approximately 1.1 Å away from Si in exact agreement with Slater's atomic
radius for Si. As expected, the shorter bond exhibits the larger
charge buildup. The similarity in the maximum density, position,
and shape of these chemically, but not crystallographically equivalent Si-O bonds lends credence to the accuracy of the results
and to the conclusion that these charge buildups represent an
accumulation of covalent bond density.

Figure 10. $\Delta\rho_{X-X}$ map including both non-equivalent SiO bonds in
stishovite. The silicon atom is located at the center of the
map. Positive contours solid, negative and zero contours dashed.
The maximum density in the bonds is 0.30 e/Å^3 for the horizontal,
and 0.47 e/Å^3 for the vertical SiO bonds. Contour interval is
.05 e/Å^3.

Figure 11. $\Delta\rho_{X-X}$ map through the octahedral shared edge in
stishovite.   Oxygen atoms are three-coordinate.   Contour interval
.05 e/Å^3.

Figure 11 is a $\Delta\rho_{X-X}$ section through the octahedral shared
edge in stishovite. Note that the bond maxima are displaced
slightly away from the shared edge suggesting strain in the four
membered ring system. Also, the four bond maxima are each connected by bridges of electron density of magnitude approximately
half that in the bonds themselves. These bridges may be expected
in that such density would seem to be energetically favorable due
to its shielding effect on the interacting nuclei across the
shared edge. Furthermore, their increase may also precipitate
the closer approach of Si to the bridging O atoms, in contraven-
tion of the predictions of an ionic model.

CONCLUSIONS

The deformation density of euclase suggests that the BeO and
SiO bonds are both significantly covalent. The BeO3 bond in
euclase appears to be significantly distorted by hydrogen bond
interactions which may bend, weaken, and lengthen the bond. The
charges obtained from kappa refinement suggest that the oxygen
atoms in euclase approach electric neutrality as the bonds reaching them become more covalent in nature.

Charge deformation maps for stishovite suggest that the Si-O bond is significantly covalent even when Si is in six coordination. Localized maxima of electron density on opposite sides of the shared edge appear to shield the Si atoms thereby reducing Si···Si repulsions. The favorable agreement between theoretical and experimental charge distributions for the BeO and SiO bonds lends credence to the notion that small molecules may serve as adequate models of the bonding in these materials.

ACKNOWLEDGMENTS

The authors gratefully acknowledge Dr. Mark Spackman for calculating the maps for stishovite, Ramonda Haycocks for typing the manuscript through several revisions and Sharon Chiang for drafting the figures. We thank the Research Division of VPI & SU for contributing large sums of money to help defray the computing costs incurred during this study. This study was supported by the National Science Foundation with grants EAR 77-23114 (GVG) and EAR 78-01780 (JAT) and by the Department of Energy (MDN).

REFERENCES

1. F. Ross, Trans. Am. Crystallogr. Assoc. 16:79 (1980).
2. J.W. Downs, M.S. Thesis, Virginia Polytechnic Institute and State University, Blacksburg, Virginia (1980).
3. R.J. Hill, M. Newton and G.V. Gibbs, in preparation.
4. F. Ross and D. Griffen, Neutron refinement of the crystal structure of euclase, unpublished.
5. P. Coppens, T. Guru Row, P. Leung, E. Stevens, P. Becker and Y. Yang, Acta Crystallogr. A35:63 (1979).
6. P. Coppens, Israel J. Chem. 16:159 (1977).
7. R. Stewart and M. Spackman, in: "Structure and Bonding in Crystals," M. O'Keeffe and A. Navrotsky, eds., Academic Press, New York (1981).
8. P. Coppens, Topics in Cur. Phys. 6:71 (1978).
9. A. Gupta, D. Swanson, J. Tossell and G.V. Gibbs, Am. Mineral. 66:601 (1981).
10. J. Slater, J. Chem. Phys. 41:3199 (1964).
11. M. Newton and G.V. Gibbs, in preparation.
12. G.V. Gibbs, R.J. Hill, F. Ross and P. Coppens, Geol. Soc. Am. Abstr. with Progr. 3:407 (1978).
13. M. Newton in: "Structure and Bonding in Crystals," M. O'Keeffe and A. Navrotsky, eds., Academic Press, New York (1981).
14. G. Jeffrey and S. Takagi, Acc. Chem. Res. 11:264 (1978).
15. M. Newton, G. Jeffrey and S. Takagi, J. Am. Chem. Soc. 101:1997 (1979).

16. N. Thong and D. Schwarzenbach, <u>Acta</u> <u>Crystallogr</u>. A35:658
     (1979).
17. K. Toriumi, M. Ozima, M. Akaogi and Y. Saito, <u>Acta</u> <u>Crystallogr</u>.
     B34:1093 (1978).
18. R.J. Hill, <u>Phys</u>. <u>Chem</u>. <u>Minerals</u> 5:179 (1979).
19. R. Shannon and C. Prewitt, <u>Acta</u> <u>Crystallogr</u>. B25:925 (1969).

ELECTRON DEFORMATION DENSITY IN CALCIUM BERYLLIDE

D.M. Collins and M.C. Mahar

Department of Chemistry
Texas A&M University
College Station, Texas   77843

ABSTRACT

Calcium beryllide serves as stereochemical exemplar for numerous intermetallic compounds of beryllium. An electron deformation density study based on X-ray diffraction data for $CaBe_{13}$ (Baker, T.W. Acta Crystallogr. 1962, 15, 175-9) gives clear evidence of multicenter bonding within the polyhedral array of beryllium atoms. There is in this study no evidence for the participation of calcium in a pattern of directed bonding and its primary role may be in charge stabilization of the electron deficient beryllium structure by which it is surrounded.

INTRODUCTION

Beryllium in its crystalline state has been the subject of many studies based on scattering experiments. Many of the recent studies have been focussed on bonding in explanation of the somewhat anomalous behavior of beryllium as a metal with covalent character. From three points of view[1-3] it has been established that a key feature of the bonding in beryllium is the sharing of electrons in multicenter interactions as might be expected for electron deficient compounds.

Intermetallic compounds of beryllium with high beryllium ratios are also electron deficient in that the numbers of near neighbors in known stereochemical patterns are far larger than the numbers of electron pairs formally available for bonding. Calcium beryllide[4] has each beryllium atom at characteristic bonding distances from 12 other atoms and it is clear that the ratio of near neighbors to available electron pairs is substantially larger than one, whatever construction is placed on the role of calcium; in similar types of boron polyhedra the smaller ratio is 20 polyhedral near neighbor interactions/13 available electron pairs.[5] In the present study we have sought experimental evidence for a pattern of multicenter bonding which could help explain the stability of $CaBe_{13}$ and related electron deficient compounds of beryllium.

CALCULATIONS

The present study employs data from the extraordinarily careful X-ray powder diffraction experiment of Baker.[4] Baker reported that $CaBe_{13}$ is cubic with $a = 10.312 \pm 0.001$ Å. The structure is of the $NaZn_{13}$ type[6] with space group O_h^6-Fm3c and eight Ca in $8(a)$: $\frac{1}{4},\frac{1}{4},\frac{1}{4}$;...; site symmetry 43; eight Be(I) in $8(b)$: $0,0,0$;...; site symmetry $m3$; and 96 Be(II) in 96 (i): $0,y,z$; site symmetry m.

The 42 independent intensity data,[4] corresponding to 57 structure factors for the three dimensional structure, were reported as reproducible to better than 2%. In the range of data, $0<\sin\theta/\lambda<0.62$, only intensities (200) and (664) were unreported. Baker's least squares refinement of a scale factor, an isotropic thermal parameter for Ca and one for both Be(I) and Be(II), and position parameters for Be(II) yielded a value for R based on intensities of 0.028^4 with an expectation of 0.014 for R based on structure factors[4]; the data/parameter ratio was 8.4.

Intensity data, corrected by Lorentz and polarization factors, from the powder diffraction experiment[4] were processed to simulate single crystal data. A majority of intensities correspond to re-solved reflection forms and were usable as reported. The remaining intensities correspond to exactly overlapping reflection-form pairs, except one triple in which the overlap is still exact. After accounting for the multiplicities appropriate to each form, the total observed intensity for each pair, and the one triple, was apportioned to the individuals using ratios of calculated structure factors. The resulting simulated single crystal data have for the customarily defined reliability index,[7] the value 0.014 in agreement with Baker's expectation. A least squares refinement of Baker's structure model was carried out using the simulated single crystal data with uniform weights and one additional isotropic thermal parameter so each atom type had its own. There was no significant change in the structure excepting the new thermal motion model. The parameters are given in Table 1; the corresponding calculated structure factors, the simulated single crystal observed structure factors and Baker's intensity data are given in Table 2.

Table 1. Atomic Position and Thermal Motion Coordinates in
          the Unit Cell

Atom Type	Fractional Coordinates[a]			Thermal Parameters,[a] B[b], in Å^2
	x	y	z	
Ca	¼	¼	¼	0.31 (2)
Be(I)	0	0	0	0.87 (9)
Be(II)	0	0.1770(1)	0.1121(1)	0.72 (3)

[a]Estimated standard deviations in the last figure follow in
parentheses.

[b]The thermal motion factor is of the form $\exp\{-B\ \sin^2\theta/\lambda^2\}$

The standard formulation of deformation density[8] was used and
a deformation density function was calculated by synthesis of
Fourier coefficients given by (F_0-F_c); F_0 and F_c are respectively
the observed and calculated structure factors, the latter based on
scattering factors for neutral isolated atoms.[9] In the absence of
an observed structure factor, the corresponding Fourier coefficient
was taken to be zero in the computation of deformation density.
Because there are no experimentally determined values for $\sigma(F)$, $\sigma(F)$
was approximated by $|F_0-F_c|$ and employed in the summation discussed
by Lehmann[10] to obtain $0.04e/\text{Å}^3$ for an estimated average esd in the
deformation density. This unusually small value can be attributed
to the extraordinarily well measured data as reflected in the
similarly small value, $R=0.014$. For a check a parallel calculation
was carried out for the most similar structure available, beryllium
metal. In this case our estimate produced an esd of $0.07e/\text{Å}^3$ which
is larger by a factor of 3.5 than the greatest value reported[3] in
the error distribution maps, excepting, of course, the unreliable
regions near atomic positions. Our estimate of average error in
deformation density is evidently conservative.

STRUCTURE AND RESULTS

Calcium beryllide is of structure type $D2_3$ discussed in detail
for sodium zinc,[6] $NaZn_{13}$. At the corners of a cube about each Ca
are eight nearly regular icosahedra having a Be(II) at each vertex
and a 13th atom, a Be(I), at the center of each icosahedra. Ca and

Table 2.   Intensity Data and Observed[a] and Calculated Structure
            Factors

hkl[b]	G^2(obs) (× 10^{-2})	F(obs)	F(calc)
200	–	–	60.0
220	1309	102.9	103.9
222	1227	122.0	123.9
400	851	117.3	117.1
420	4048	127.9	128.1
422	5251	145.7	143.9
440	417	58.1	58.4
531	2484	70.1	72.8
600 }	395	4.7	5.0
442 }		39.9	39.9
620	3143	114.9	113.2
622	845	58.5	58.2
444	1439	132.1	131.3
640	2763	105.7	105.4
642	4408	94.4	95.5
731	28	7.5	6.7
800	986	126.3	127.3
820 }	3734	88.1	87.6
644 }		85.7	84.9
822 }	1483	50.9	51.3
660 }		82.5	83.2
751	130	16.2	16.9
662	842	58.3	59.9
840	1091	66.4	65.1
664	–	–	29.9
753	571	34.0	34.4
842	1135	47.9	47.4
931	764	39.3	38.3
844	2576	105.6	105.7
10,0,0 }	170	14.7	15.6
860 }		25.2	25.1
10,2,0 }	5243	71.6	70.3
862 }		89.6	88.1
951	739	38.7	37.8
10,2,2 }	2100	71.9	70.8
666 }		99.9	98.9
953	1127	47.7	46.9
10,4,0 }	1530	62.6	62.3
864 }		33.7	33.4
10,4,2	1919	62.3	63.1
880	185	38.7	39.7
11,3,1 }	382	25.0	25.7
971 }		12.1	12.8
10,4,4 }	2747	51.0	50.5
882 }		92.2	91.0
10,6,0 }	3826	91.6	93.6
886 }		84.2	85.6
973	228	21.5	23.2
10,6,2	684	37.2	37.7
12,0,0 }	1476	97.5	96.7
884 }		59.9	59.1
12,2,2 }	2667	70.9	70.7
10,6,4 }		53.7	53.3
11,5,3 }	371	26.3	27.4
975 }		7.5	7.2
12,4,0	327	36.4	36.0
12,4,2 }		36.7	38.3
10,8,0 }	1291	39.6	41.1
886 }		31.0	32.1

[a] See the text for detalis concerning the derivation of F(obs) from G^2.

[b] Braces indicate reflection form groupings which occur at the same scattering angle.

Be(I) occupy positions of site symmetry 43 and *m*3 respectively, and
Be(II) lies in a mirror plane.

Ca is surrounded in symmetrically equivalent relationships by
24 Be(II) atoms at 3.039 Å. Be(I) is surrounded symmetrically by
12 Be(II) atoms at 2.161 Å, a distance shorter by more than 0.1 Å
than the hcp mean of 2.271 Å for the 12 near neighbor distances in
beryllium metal. Be(II) is also surrounded by 12 atoms: 9 other
Be(II) atoms, at an average distance of 2.262 Å, a Be(I) and two Ca.
If in an isolated unit of Be(II) and its 12 near neighbors the Ca
atoms were replaced by Be atoms at a distance from the center smaller
by 0.85 Å than the Ca-Be(II) distance, the result would be a polyhedron of Be atoms roughly approximating icosahedral symmetry around
the central Be(II); 0.85 Å is the difference between the metallic
radii of Ca and Be.[11] Moreover, the average distance between Be(II)
and its neighbors in such a unit would be 2.24 A which approximates
the hcp average Be-Be distance. There is, however, in spite of similarities the basic difference between the hcp $\overline{6}m2$ symmetry and the
quasi-icosahedral symmetry of the polyhedra within which each beryllium atom exists in $CaBe_{13}$. Interatomic distances and the frequency
of their occurrence is given in Table 3. Figure 1 shows a portion
of the structure which emphasizes the polyhedral environment of Be(I)
and Ca. Figure 2 portrays the environment of a Be(II) atom.

Table 3. Selected Bond Distances

Bond Type	Occurrence	Distance^a (Å)
Be(I) at (0,0,0) —		
Be(II) at (0,y,z)	12	2.161 (2)
Be(II) at (0,y,z) —		
Be(II) at (0,z,$\frac{1}{2}$-y)	2	2.275 (3)
Be(I) at (0,0,0)	1	2.161 (2)
Be(II) at (z,$\frac{1}{2}$-y,0)	2	2.222 (4)
Be(II) at (z,0,y)	4	2.262 (2)
Be(II) at (0,y,$\overline{z}$)	1	2.312 (4)
Ca at ($\frac{1}{4}$,$\frac{1}{4}$,$\frac{1}{4}$)	2	3.039 (2)
Ca at ($\frac{1}{4}$,$\frac{1}{4}$,$\frac{1}{4}$) —		
Be(II) at (0,y,z)	24	3.039 (2)

^{a}Estimated standard deviations are in parentheses.

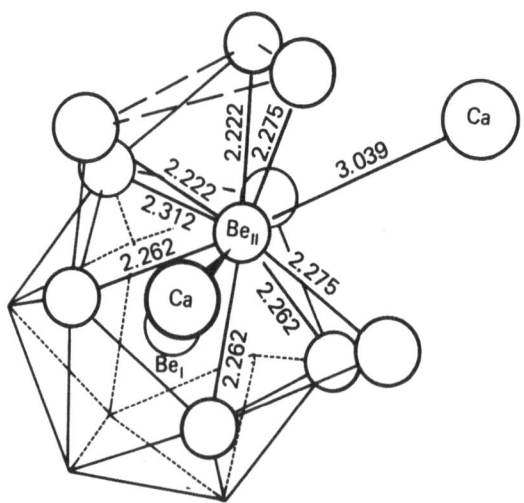

Figure 2.   The 12-neighbor environment of Be(II).

Figures 3 and 4 contain the significant sections of deformation
density contoured in approximate multiples of the mean esd with the
first solid contour at zero.   Figure 3 shows the zero level or
z = 0.0 section of the entire unit cell and some Be(I) and Be(II)
sites.   Figure 4 shows the ¼ level or z = 0.2500 section of the entire unit cell and some Ca sites.   It should be emphasized that a
true esd of deformation density greatly exceeds the mean esd near
atomic positions, and the density is so unreliable in those regions
as to preclude reasonable interpretation, even in qualitative terms.

The only feature significant by a 3σ criterion is that marked
by T in Figure 3.   This feature is part of a ridge of positive
density which runs through the tetrahedron based on a tetrahedral
face of an icosahedron.   The ridge extends at a significant level
about 0.5 Å beyond the tetrahedron's faces toward Ca atoms in both
directions.   The bimodal maximum of the ridge occurs within the
tetrahedron at a density of 0.14 $e/\text{Å}^3$ about 0.55 Å off the mirror
plane and it projects onto the point marked T.   The density is 0.11
$e/\text{Å}^3$ at T, its position is 0.31 Å from the center of the tetrahedron,
and in qualitative discussion it is taken to be at the center.

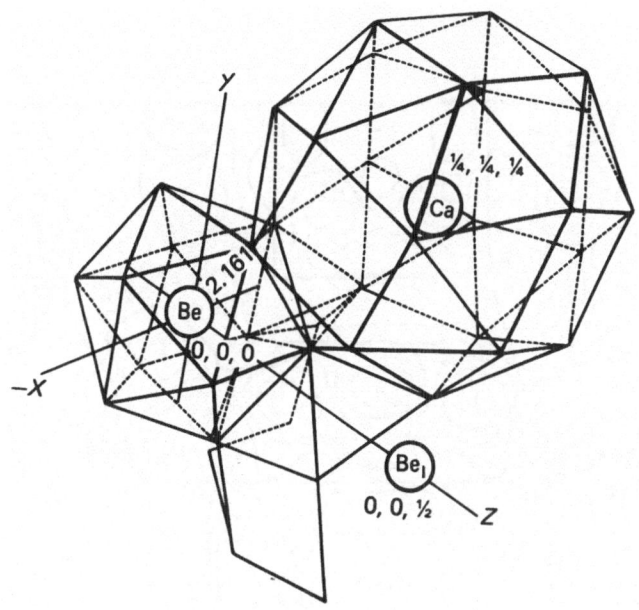

Figure 1.   Representative Be(II) polyhedra surrounding Be(I) and
            Ca in calcium beryllide.

        If, in reference to Figure 1, another icosahedron were placed
along the x axis, rotated 90° about the axis as required, with Be(I)
at $(-\frac{1}{2},0,0)$, there would be between the adjacent icosahedra a tetrahedron of Be(II) atoms sharing edges with the icosahedra.[6] For
present purposes it is more to the point that the two triangular
faces which in the first icosahedron are nearest the second, form
approximately regular external tetrahedra with the near vertices of
the second. If these faces are called "tetrahedral faces," then
reference to Figure 1 shows that 12 of the 20 icosahedral faces are
symmetrically equivalent tetrahedral faces. The remainining 8 lie
perpendicular to the diagonal 3-fold axes which pass through them
and may be distinguished by their required symmetry as "equilateral
faces."

Figure 3.   Deformation Density.   The section displayed is of the
            entire cell at z=0.0.   Contours are at multiples of
            0.033 e/Å³; the zero and positive contours are given by
            solid lines with T marking the one significant density
            feature.   The rectangles or portions thereof are super-
            imposed sections of the icosahedral units which show the
            sites of Be(II) at the apices and Be(I) at the centers.

Figure 4.   Deformation Density. The section displayed is of the
            entire unit cell at $z=\frac{1}{4}$. The contours are as in Fig. 3.
            The Ca sites are at $\pm\frac{1}{8}$, $\pm\frac{1}{8}$.

DISCUSSION

      The object of the present study was to determine what evidence
could be developed from X-ray diffraction data for directed, covalent
bonding in $CaBe_{13}$. This is only part of a larger problem relating
to the existence of many stable intermetallic compounds rich in
beryllium.   There is a rather common view that the stability and
stereochemistry of beryllium and its intermetallic compounds are
based upon packing effects to the effective exclusion of other considerations.   In the experimental studies of beryllium[1-3] it has
been shown conclusively that there is a significant pattern of directed, covalent, multicenter bonding which apparently stabilizes the
hcp structure of the metal.

$CaBe_{13}$ has the same formal valence electron makeup as Be metal since all atoms are from group IIA. It is clear of course that Ca differs from Be both in its size, which is greater by $\sim 0.85\text{Å}$ in metallic radius,[11] and in its chemistry which is predominately ionic in contrast to the covalent character of beryllium chemistry. In view of this difference in character and the electron deficient nature of the Be network, the presumption is that $CaBe_{13}$ may be usefully described as $(Ca^{2+})(Be_{13}{}^{2-})$. Deformation density maps provide no positive evidence for an ionic interaction of Ca with the Be network. There is, however, no significant localized deformation density within the snub cube of Be(II) atoms which surround Ca at 3.039Å, a distance which is formally short for a metallic interaction.[11] And the absence of such density is at least not in disagreement with the presumption that Ca contributes to stabilization of the $CaBe_{13}$ structure through an ionic type of interaction.

The one significant feature of deformation density lies within a nearly regular tetrahedron of Be(II) atoms. It will be clear that the $CaBe_{13}$ structure, and indeed every structure based on deltahedra, can be viewed as a collection of tetrahedra. But it seems useful to emphasize the icosahedral structure around Be(I) and keep the corresponding categorization of its faces as those nearest Ca, the equilateral faces, and those nearest the maxima of deformation density, tetrahedral faces. The deformation density feature can then be viewed both in association with the tetrahedron of Be(II) atoms in which it is centered and also with the infinite chains of icosahedra parallel to each axis. Along the axes the chains have icosahedra centered at multiples of 1/2 and the midpoints between are at odd multiples of 1/4 with site symmetry $\overline{4}2m$. Around these points four symmetrically equivalent deformation density features form a region of apparent multicenter bonding density which corresponds to each icosahedron being bonded to others through each of its tetrahedral faces. The four tetrahedra of Be(II) atoms which contain the symmetrically equivalent density features, themselves bound the density-free tetrahedron of Be(II) atoms formed by edge-sharing of adjacent icosahedra. Evidently Be(II) participates in an extended array of multicenter bonding interactions involving at least tetrahedral units constituted by interaction of adjacent icosahedra.

The role of Be(I) in stabilization of the $CaBe_{13}$ structure is, like that of Ca, revealed only through the absence of significant deformation density features anywhere inside the Be(II) polyhedron by which it is surrounded. It seems reasonable to expect that the occurrence of icosahedra in stable Be compounds is based on its provision of a means to share efficiently few electron pairs among many atoms. This is the accepted role for icosahedra in compounds of boron.[5] If the skeletal molecular orbitals are taken unchanged from the scheme developed for boron,[5] then the lowest lying orbital is totally symmetric and based on inward pointing atomic orbitals.

It would be readily populated by the two valence electrons of Be(I) which is in exactly the right location at the center of the icosahedron. If Be(I) is involved in the $CaBe_{13}$ structure through such a highly symmetric interaction, the corresponding deformation density would be symmetrically distributed through a rather large volume and probably not observable except possibly in experiments carefully designed to observe such a feature. It is in any case clear from the very short Be(I)-Be(II) distance of 2.161 Å that there is a strong interaction between Be(I) and its 12 neighbors. In addition, the absence of deformation density features requires that any associated electron distribution effect be of a diffuse nature.

CONCLUSION

CaBe$_{13}$ and other intermetallic compounds of Be are unusual materials with both metallic and covalent properties. In view of the electron deficient nature of Be compounds it is not surprising that there should be characteristic polyhedral subunits which provide a covalent vehicle for the sharing of electron pairs. The present work suggests that Be tends to form icosahedral subunits of 13 atoms, one at the center, which can interact in varied ways with neighboring structure elements. Of course, there are other kinds of polyhedral subunits, but, as in CaBe$_{13}$, it is probable they will all be found to support multicenter bonding as the primary basis for stability and stereochemical definition in the intermetallic compounds of beryllium.

This work was supported in part by the Robert A. Welch Foundation through grant A-742 and by the Research Corporation through a Cottrell Research Grant.

REFERENCES

1. Brown, P.J. *Phil. Mag.* 1972, 26, 1377-94.
2. Stewart, R.F. *Acta Crystallogr.*, Sect. A, 1977, A33, 33-8.
3. Yang, Y.W.; Coppens, P. *Acta Crystallogr.*, Sect. A, 1978, A34, 61-5.
4. Baker, T.W. *Acta Crystallogr.* 1962, 15, 175-9.
5. Wade, K. *Adv. Inorg. Chem. and Radiochem.* 1976, 18, 1-66.
6. Shoemaker, D.P.; Marsh, R.E.; Ewing, F.J.; Pauling, L. *Acta Crystallogr.* 1952, 5, 637-44.
7. Woolfson, M.M. "An Introduction to X-ray Crystallography"; Cambridge: Cambridge, 1970.
8. Coppens, P.; Stevens, E.D. *Adv. Quantum Chem.*, 1977, 10, 1-35.

9.   "International Tables for X-Ray Crystallography", Vol. IV; Ibers, J.A.; Hamilton, W.C., Eds.; Kynoch: Birmingham, England, 1974.

10. Lehmann, M.S. In "Electron and Magnetization Densities in Molecules and Crystals", Becker, P., Ed.; Plenum: New York, 1980.

11. Pauling, Linus, "The Nature of the Chemical Bond", 3rd ed.; Cornell: Ithaca, NY, 1960.

Section 4
MOLECULAR SOLIDS: THEORETICAL RESULTS

COMPUTATION AND INTERPRETATION OF ELECTRON DISTRIBUTIONS IN

INORGANIC MOLECULES

Michael B. Hall

Department of Chemistry
Texas A&M University
College Station, Texas   77843

INTRODUCTION

Although the electron density has always been a useful way of characterizing the electronic structure of a molecule, it attains special significance with the theorem of Hohenberg and Kohn[1], which provides a direct link between the electron density and the energy of the system.[2] The total electron density of a molecular system, like the total energy, differs little from that of its constituent atoms. Of course, it is just the small differences between the atomic density (energy) and the molecular density (energy) which are responsible for the chemical bond. Because of our interest in the chemical bond, we usually find the deformation density more useful than the total density. The deformation density, which is the total molecular density minus the density of spherical, ground-state atoms, can be interpreted as the changes in electron density which occur upon forming the molecule from its atoms.

Although this quantity is clearly defined and can be measured experimentally[3], at least within the limitations imposed by experimental resolution, by molecular and lattice vibrations[4], and by some poorly understood phenomena[5], there remain problems in the computation and interpretation of the deformation density. Despite considerable work on smaller molecules[6], it is not always obvious how accurate the wavefunction needs to be in order for qualitative or quantitative agreement with the experimental results. This problem manifests itself in the choice of basis set and in the problem of electron correlation. Even if we agree on the accuracy of the wavefunction, the interpretation of the deformation density in terms of our usual concepts of bonding is fraught with difficulties.

In this paper we report our investigations of these problems
with particular emphasis on the problems of interpretation. The
molecules discussed will include examples of electron-deficient
bonds, multiple metal-metal bonds, dative bonds, single metal-
metal bonds, and metal-cluster bonding. The theoretical approaches
to these problems will range from generalized molecular orbital
calculations[7] with large basis sets to approximate molecular orbital calculations[8] in near-minimal basis sets.

THEORETICAL

The generalized molecular orbital (GMO) approach consists of
a limited type of multiconfiguration self-consistent-field (MCSCF)
calculation, which is used to extract a valence orbital space from
the full orbital space. This calculation is followed by a configuration interaction (CI) calculation within this valence orbital
space. Previous work on small molecules has shown that this approach yields results in close agreement with more sophisticated
MCSCF-CI techniques.[7] Using this technique, we will examine the
deformation density of diborane, and of the molybdenum - molybdenum
triple bond.

The most important problems with respect to the interpretation
of the deformation density involve two questions. First, is the
accumulation of electron density between two atoms representative
of the bond between those atoms? Second, does the lack of accumulation between two atoms preclude the formation of a bond between
them? These questions arise because the deformation density corresponds to the total rearrangement in forming all the bonds in the
molecule. Thus, the value of the deformation density between two
particular atoms is influenced, not only by the bond between them,
but by bonds they make with other atoms in the molecule.

A particularly helpful addition to the interpretation of electron densities is the fragment deformation density, which is defined
as the density of the molecule minus the density of two or more
fragments of the molecule. By fragmenting the molecule in various
ways, an easy procedure for a theoretician, one can learn which
changes in the density correspond to which bonds. Use of this technique will be illustrated by comparing the covalent carbon-carbon
bond in ethane, C_2H_6, with the dative boron-nitrogen bond in the
borane-ammonia adduct, H_3BNH_3; and the Cr-CO bond in chromium hexacarbonyl.

For a number of years we have used approximate molecular orbital (MO) calculations[8] to help us understand the electronic structure of transition-metal complexes. They have been particularly
helpful in assigning and interpreting photoelectron spectra.[9]

These approximate MO calculations can not be expected to yield
quantitative results, and, as we shall see, will yield some qualitatively incorrect results. The problem is due, in part, to the
approximations and, in part, to the near-minimal basis set employed.
In spite of these errors the method can provide some valuable insight into the interpretation of the deformation density in large
systems. A particularly important part of their use in this capa-
city involves the fragment deformation density. Since both the
fragments and the molecule are calculated with the same approximations and in the same basis many of the errors will cancel. This
procedure will be applied to systems containing metal–metal bonds,
where experimentally one often finds little or no accumulation of
density between the two transition metals that are formally bonded
to each other.[10] With the fragment analysis we can concentrate our
attention on the formation of the metal–metal bond rather than the
total rearrangement, a great deal of which is induced by the ligand
environment of each metal. We will describe applications to
$Mn_2(CO)_{10}$ and to the cluster, $ClCCo_3(CO)_9$.

RESULTS AND DISCUSSION

Diborane

     Diborane, the simplest electron-deficient cluster, consists of
two BH_2 units bridged by two H atoms, 1. In our previous work

1

we calculated the dissociation energy of diborane in several basis
sets and at several levels of CI.[11] The best of these calculations
reproduced the dissociation energy ($B_2H_6 \rightarrow 2BH_3$) to within 3 kcal
mol^{-1} of the experimental value, 35 kcal mol^{-1}.

     Figure 1a, 1b and 1c show the deformation densities (molecule-
spherical atom in the same basis) in three different basis sets.
The terminal B–H plane is shown on the left and the bridging B–H–B
plane is shown on the right. Figure 1a shows the deformation densities for a minimal basis set, while Figure 1b shows those for a

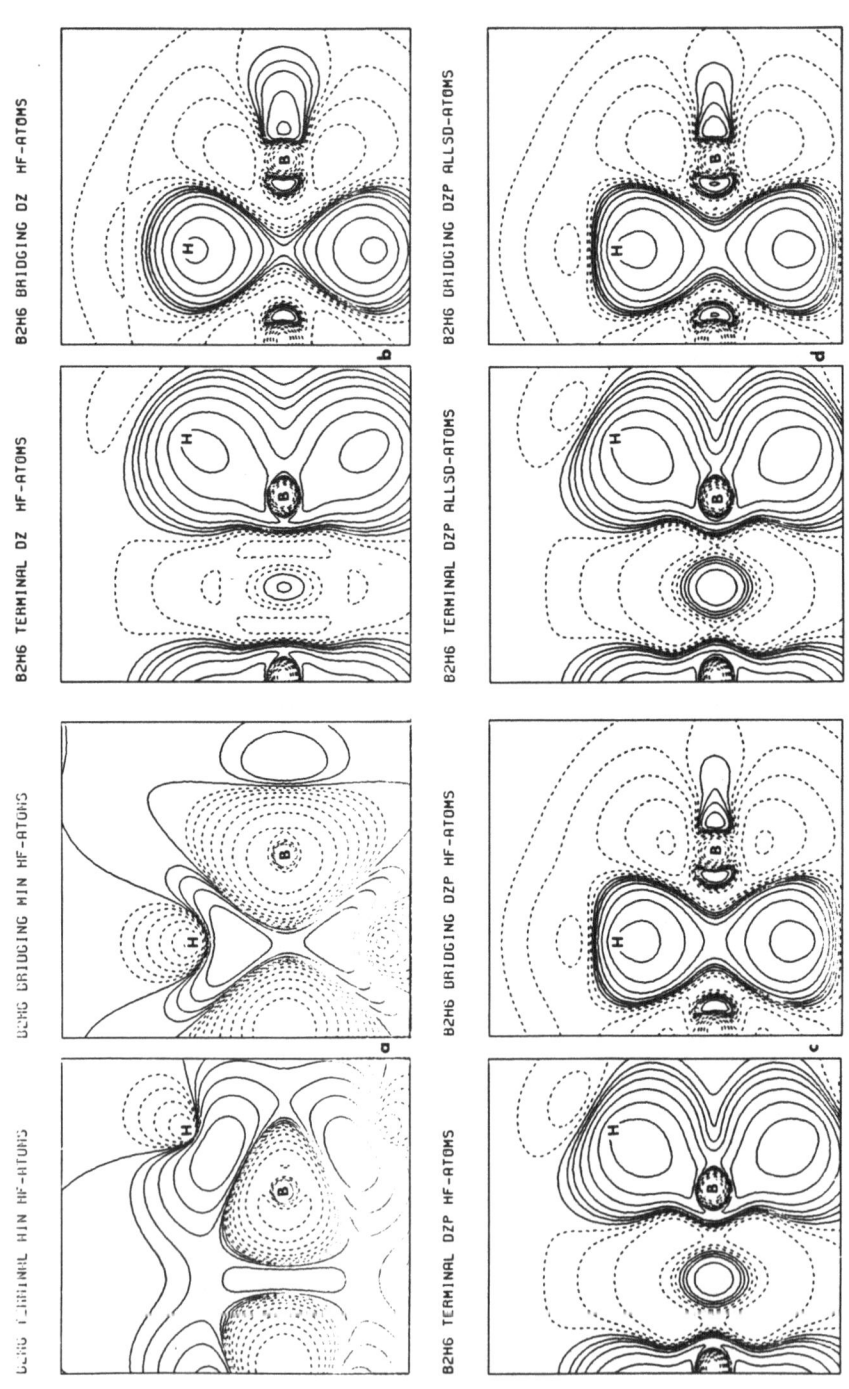

Fig. 1. Deformation Density of B_2H_6: a. Minimal basis, b. Double-zeta basis, c. Double-zeta basis plus polarization functions, d. Configuration interaction in basis set of c. The terminal plane is on the left and the bridging plane is on the right.

double zeta (DZ) basis set. The difference between these two calculations is substantial, and the minimal-basis-set results are not
even qualitatively close to those of the better basis set. Although
the addition of polarization functions on all atoms (d on B and p
on H) causes significant changes in the deformation density (Figure
1c), the results are qualitatively similar to the DZ basis set.
Polarization functions increase the electron density in the center
of the cluster and in the B-H bonds. The calculated dissociation
energy at the DZ level is only 13 kcal mol^{-1}. Addition of the
polarization functions increases the calculated value by 7 kcal
mol^{-1}.

The deformation density with extensive configuration interaction ($\sim$3000 configurations selected from 4956) is shown in Figure
1d. The change from the single-determinant HF result is almost im-
perceptible. A more quantitative examination[11] shows that the
largest change in the electron density when one adds CI is only $\frac{1}{4}$
of the change that one obtains when polarization functions are
added. However, CI accounts for over 12 kcal mol^{-1} of the dissociation energy.

Evidently, small changes in the electron density can be responsible for large changes in the dissociation energy. A similar
result is found for the total energy of diborane, which is lowered
by 1.1 eV on addition of polarization functions and by 6.0 eV on
addition of configuration interaction. Thus, polarization functions
are more important than configuration interaction for the electron
density but less important for the energy.

Although our results do not agree very well with previous
models for the experimental deformation density[12], they are in
agreement with previous theoretical work.[13] The terminal B-H bond
and the bridging B-H-B bond also look similar to those in 'a recent
experimental study of $B_{10}H_{14}$.[14]

Molybdenum to Molybdenum Triple Bond

Thermochemical results on $Mo_2(NMe_2)_6$, which formally possesses
a triple Mo-Mo bond, suggest a bond energy in the range 141 $\pm$ 47
kcal mol^{-1}.[15] Although our previous results[16] supported a bond

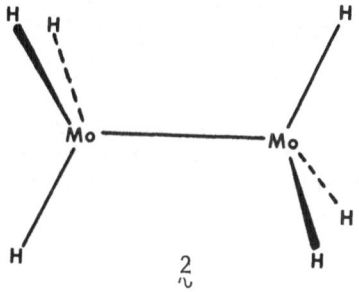

2

energy somewhat greater than 100 kcal mol^{-1}, our most recent calculations[17] suggest a bond energy of slightly less than 100 kcal
mol^{-1}. These theoretical results were obtained with the model
system Mo$_2$H$_6$, $\underset{\sim}{2}$. It may seem rather drastic to replace the NMe$_2^-$
ligand with H$^-$, but numerous structural determinations have shown
that the Mo≡Mo bond distance and, presumably, the bond energy are
relatively insensitive to the ligands.[18]

   Our theoretical approach begins with a generalized molecular
orbital (GMO) calculation, which is followed by a full configuration
interaction calculation within the 6 orbitals (bonding and antibonding) that define the triple bond. The basis set is triple zeta on

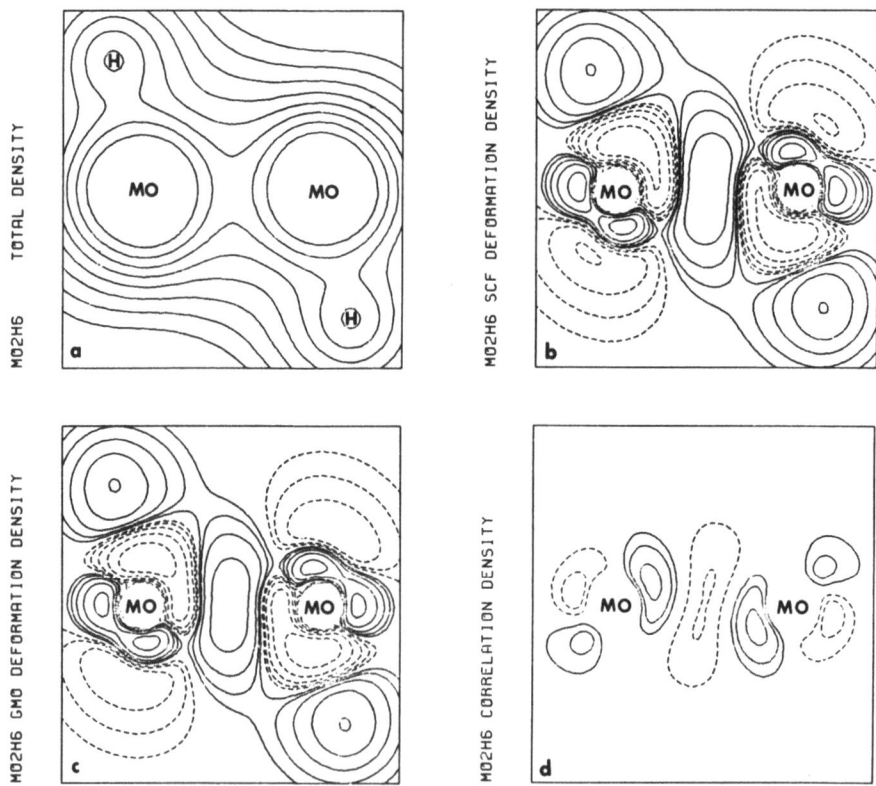

Fig. 2.   Mo$_2$H$_6$ Densities: a. Total density, b. Deformation density
          for single determinant, c. Deformation density after con-
          figuration interaction d. Correlation density, difference
          between b. and c.

Mo plus s and p bond-centered functions. However, the H is only represented in a minimal basis. Thus, we would not expect to obtain an accurate representation for the Mo-H bond, which is only hypothetical anyway, but we should obtain a fairly accurate representation of the Mo≡Mo bond.

The total density, the deformation density for the single determinant, and the deformation density at the GMO-CI level are shown in Figure 2a, 2b, and 2c, respectively. Because of the hypothetical nature of the H$^-$ ligand and because of the minimal H basis set, we will concentrate on the Mo-Mo region and ignore the Mo-H region. The deformation density clearly shows a build-up of density between the Mo atoms. This build-up is elongated perpendicular to the bond direction as one expects for multiple bonds. When we add CI, which is absolutely necessary for an accurate dissociation energy, we can see some changes in the density. The electron density is diminished in the bond region and increased close to the atoms (Figure 2d). Although the changes in electron density, when electron correlation is introduced, are larger in this system than in diborane, they remain a small fraction of the deformation density. Again, small changes in the deformation density are responsible for large changes in the bond energy.

Covalent vs. Dative Bonds

One of the features that distinguishes organic from inorganic chemistry is bond type. Organic chemistry is dominated by nearly homopolar covalent bonds, while inorganic chemistry, particularly transition metal chemistry, is dominated by heteropolar dative bonds. A good case can be made for the two bond types simply being a change of degree rather than being fundamentally different. To examine this question we will compare the deformation density of C_2H_6, which has a covalent C-C bond, with that of H_3BNH_3, which has a dative B ← N bond.

The calculations are at the single determinant SCF level in a standard double-zeta basis set.[19] Thus, the results will only be used for a qualitative comparison, and are not a quantitative representation of the "true" deformation density. In addition to the standard deformation density, for which one subtracts neutral atoms from the total density, we will introduce the fragment deformation density, for which one substracts preformed molecular fragments from the total density. For H_3BNH_3 we will subtract the two neutral molecules, BH_3 and NH_3. In both cases the fragments will remain distorted in the geometry that they have in the full molecule. By using molecular fragments to form the promolecule, we can concentrate our attention on the bond of interest rather than looking at all bonds forming simultaneously.

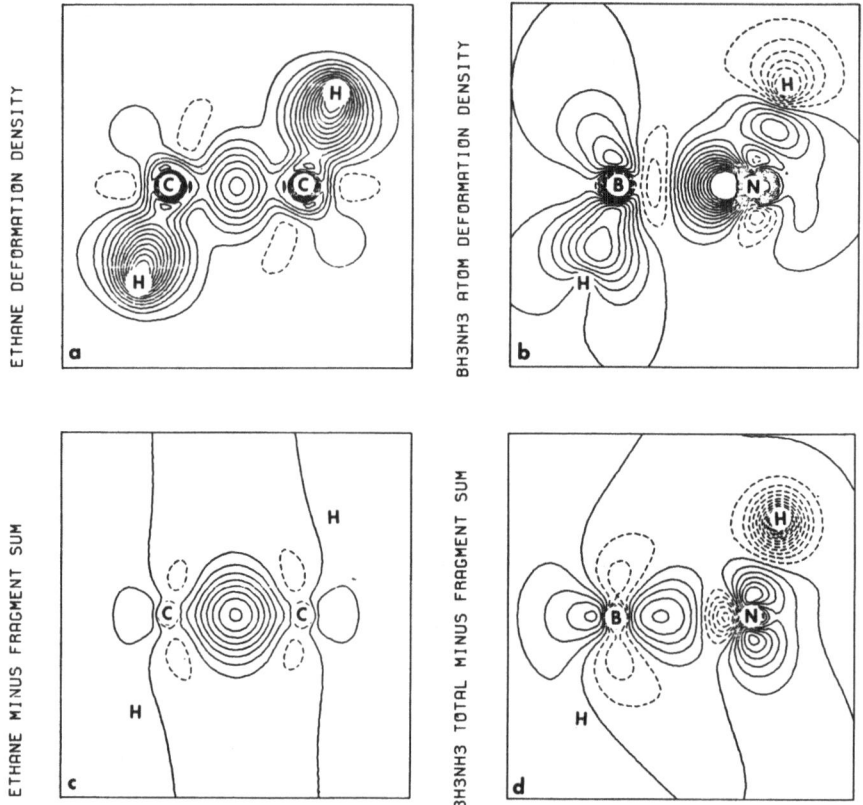

Fig. 3.   Comparison of C_2H_6 and H_3BNH_3: a. Atom deformation density
          of C_2H_6, b. Atom deformation density of H_3BNH_3, c. Frag-
          ment deformation density of C_2H_6, d. Fragment deformation
          density of H_3BNH_3.

The results are shown in Figure 3. The two maps on the top
(Figure 3a and 3b) are the standard deformation density for C_2H_6
and H_3BNH_3, respectively. One can see some change in the X–H bond
as X becomes progressively more electronegative along the series
B, C, and N. However, our main interest is in the C–C and B–N
bonds. For the former bond we see a symmetric, nearly spherical
build-up of density between the two C atoms. For the latter bond
we see a substantial build-up of density, but it is localized closer
to the N atom, and there is actually some loss of density near the
B atom. One recognizes the large build-up near N as lone pair
formation. At first, one is disturbed to see loss at B since one
expects bond formation to involve N to B donation. Clearly, we are
not looking only at B–N bond formation in this deformation density.

The gain in electron density at N is due to N–H bond formation, which creates a lone pair, and the loss at B is due to B–H bond formation, which creates a vacant orbital. This result for H_3BNH_3 raises a question about the deformation density of C_2H_6. How much of the feature along the C–C vector is just C–C bond formation and how much is due to C–H bond formation?

An examination of the fragment analysis, shown in Figure 3c and 3d, is extremely helpful in separating the features due to X–H bond formation from those due to X–X' bond formation. For C_2H_6 we have subtracted two CH_3 radicals from the molecular density. There is little change in the C–H bonds in forming the C–C bond and the build-up due to C–C bond formation alone is actually larger than the standard deformation density might imply. For H_3BNH_3 we have subtracted the neutral molecules from which the adduct is made. The most striking feature of this fragment deformation density is the clear representation of the donor–acceptor nature of the B–N bond. Along the bond axis one sees the BH_3 fragment gain electron density and the NH_3 fragment lose it. A substantial fraction of the density lost by NH_3 comes from the H, yet the H on the BH_3 is hardly affected by the gain. Notice that perpendicular to the bond axis there is a loss of density on BH_3 and a gain on NH_3. This π-type interaction is what inorganic chemists call backbonding and is usually invoked in low-oxidation-state transition metal systems. In all these systems backbonding reduces the net transfer of electrons and helps to preserve electroneutrality. Thus, as the NH_3 σ donates to the vacant BH_3 orbital, it also π accepts from the filled BH_3 π orbitals.

Although calculations in large basis sets may modify the actual magnitude of these density shifts, they should not alter the qualitative aspects of the results. The main purpose of comparing these two systems was to illustrate the interesting, and often quite different, perspective one obtains by examining the fragment deformation density.

Chromium Hexacarbonyl

Sometime ago Rees and Mitschler reported an experimental deformation density of chromium hexacarbonyl.[20] They interpreted their results within a framework of synergistic bonding i.e. the CO donates electrons from its 5σ orbital to the metal, while the metal donates electrons back to the CO's 2π orbital. Using results from scattered-wave $X\alpha$ calculations, Johnson and Klemperer have suggested that the only important interaction for the Cr–C bond is the σ donation to the metal.[21]

We have studied the dissociation of a single carbonyl group from chromium hexacarbonyl.[22] In this work we used an extended basis set on the chromium and on the dissociating carbonyl, while we represented the five remaining carbonyls in a minimal basis.

The overall accuracy of our results can be assessed by comparing our theoretical dissociation energy for the ion $Cr(CO)_6^+$ (31 kcal mol^{-1}) with that obtained from the mass spectrum (33 kcal mol^{-1}).[23] Partitioning the total energy into σ and π components, we found that the π orbitals contributed 25% of the Cr–C bond strength. Thus, the π bonding is important for the Cr–C bond. Only at very long Cr–C distances will CO function only as a σ donor. Since the 5σ donor orbital is antibonding between C and O, a molecule with a very long M–C bond should have a C–O bond shorter than that of free CO. This prediction was confirmed by our theoretical results and it has been observed in, at least, one case.[24]

Our theoretical deformation density[22] is similar to the experimental one.[20] A fragment analysis of the density shows that the CO's "lone pair" or 5σ orbital does not donate much density directly to the metal, but seems to rehybridize and to extend its "lone pair" density closer to the metal. The π interaction, however, does involve a net transfer of electron density from the metal to the ligand. Thus, in agreement with experiment[20] the metal possesses a net positive charge.

Single Metal–Metal Bonds: $Mn_2(CO)_{10}$

In several experimental studies of molecules which formally contain a metal–metal bond between two first-row transition metals one often finds that there is little or no accumulation of density between the two metal atoms.[10] This can be a little disconcerting since numerous studies on organic molecules usually show substantial accumulation near the line joining the nuclei. One begins to wonder if there is a metal-to-metal bond between the atoms in such a molecule. Dimanganese decacarbonyl is a key example of this problem.[25] In the solid state $Mn_2(CO)_{10}$ is found in the staggered[26] rather than in the eclipsed conformer shown in 3.

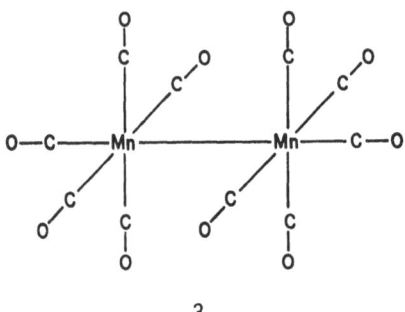

3

In either conformer two 17e$^-$ $Mn(CO)_5$ fragments are joined together without the support of any bridging ligands. Thus, there must be a direct metal–metal bond in this system.

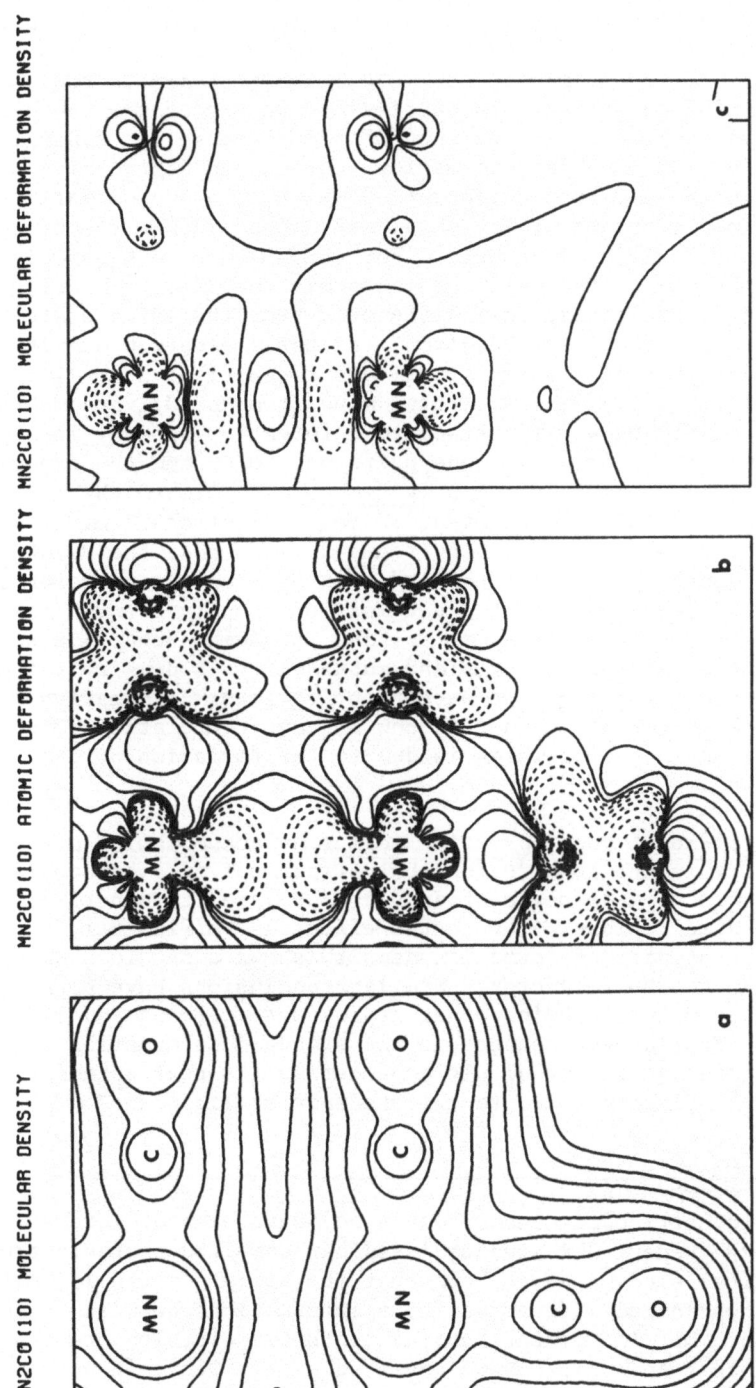

Fig. 4. $Mn_2(CO)_{10}$ Densities: a. Total density, b. Atom deformation density, c. Fragment deformation density.

We have examined the electron density of $Mn_2(CO)_{10}$ using the approximate MO method of Fenske and Hall.[8] Since the method results in an approximation to a Hartree-Fock wavefunction with a near minimal basis set, we can anticipate some significant errors in the absolute deformation densities. In particular, we expect the C-O bond in the carbonyl to be negative between C and O because the minimal basis overemphasizes lone pair formation, and causes destructive interference in the C-O bond region.[27] However, the approximate calculations should be better for metal-metal bonds since the overlap of the orbitals is smaller here. For convenience we have calculated the molecule in the eclipsed conformation, 3. The barrier to rotation is so small that this difference from the solid state structure will not effect our results. The total electron density, the standard deformation density (total minus spherical atoms), and the fragment deformation density (total minus two $Mn(CO)_5$ fragments) are displayed in Figure 4. The standard deformation density in Figure 4b shows the build-up of lone pairs on the CO and the expected loss of density between C and O (vide supra). Around the manganese, one can see the rearrangement of the d electrons from a spherical to a pseudo-octahedral environment, i.e. near the Mn atom there is loss along the Mn-C vector and gain between these vectors. The most remarkable feature is the net loss of density between the two manganese atoms where we expect to find the Mn-Mn bond. If we first prepare the $Mn(CO)_5$ radicals and subtract their density rather than that of spherical atoms, we obtain the fragment deformation density shown in Figure 4c. One can now see an accumulation of density along the Mn-Mn vector which is due to the formation of the metal-metal bond.

Thus, metal-metal bond formation causes an accumulation of electron density in the bond region, but attachment of 5 carbonyls to each metal causes a decrease of density (compared to spherical atoms) in the same regions. The latter effect is sufficiently strong that the standard deformation density shows a net loss of electrons along the Mn-Mn vector. Similar conclusions have been reached for this molecule using other approximate methods and larger basis sets.[28] The largest changes in the standard deformation density can be accounted for by simply considering the rearrangement of the Mn atom's electrons due to the presence of ligands.[29]

Clusters: $ClCCo_3(CO)_9$

The alkyldynetricobaltnonacarbonyls, 4, have been of interest for sometime.[30] Recent NQR studies[31] and our combined photoelectron, molecular-orbital study[32] have shown that the apical carbon is best viewed as sp hybridized with extensive π interactions between it and the Co_3 system. Details of the bonding in these molecules are discussed elsewhere[32] and will not be repeated here. Our main purpose in introducing the deformation density of these systems is to provide a second illustration of the power of the fragment approach in elucidating the bonding.

4

The calculations were done on $ClCCo_3(CO)_9$ by the approximate Fenske–Hall procedure with a minimal basis for all functions except the Co 3d, where Richardson's[33] double–zeta Slater basis was used. Again we should not expect quantitative accuracy in these maps. Especially poor will be the representation of the strong covalent bonds, such as C–O and C–Cl.

The results are shown in Figure 5. The two maps on the top show the plane through the three Co atoms. In the standard deformation density (Figure 5a) one clearly sees the spherical to pseudo-octahedral rearrangement of the electron density. The positive lobes outside the Co_3 triangle arise from the lone pairs of six carbonyl groups which are tilted slightly above this plane. As in $Mn_2(CO)_{10}$ there is no accumulation of charge near the line joining the Co atoms. However, if one calculates the fragment deformation density by subtracting the preformed fragments $Co(CO)_3$ and CCl (Figure 5b), one sees a significant accumulation of charge both in the center of the triangle and along the line joining the Co atoms. Again the combined rearrangement due to both Co–C and Co–CO bond formation results in a net loss between the Co atoms. The Co–Co bonds are revealed only if one removes the rearrangement of charge due to the Co–CO bonds as we have done in the fragment analysis.

Figure 5c and 5d show the corresponding deformation densities for the Cl–C–Co plane. The standard deformation density suggests a "bent" Co–C bond, which might be expected from the small Co–C–Co angle. The fragment deformation density clearly shows donation from σ lone pair on CCl to the Co_3 triangle and acceptance of π density from the Co_3 into the π orbitals on C and Cl. The involvement of Cl in this π system suggests that the degree of "bent" bond character may depend on the substituent on the apical carbon.

CONCLUSION

The interplay between theory and experiment is critically important in confirming and interpreting the deformation density.

CLCCO3(CO)9 MOLECULE MINUS NEUTL ATOMS
COBALT PLANE

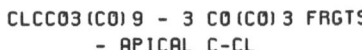

CLCCO3(CO)9 - 3 CO(CO)3 FRGTS
- APICAL C-CL

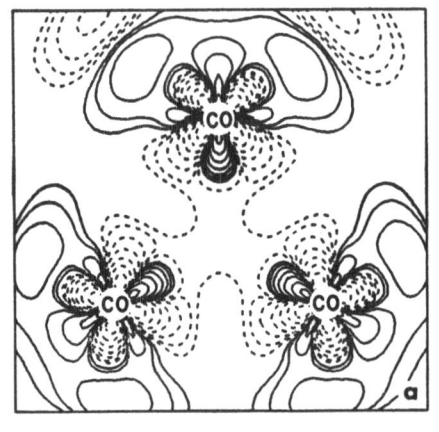

CLCCO3(CO)9 MINUS NEUTRAL ATOMS
COBALT-APICAL PLANE

CLCCO3(CO)9 - 3 CO(CO)3 FRGTS
- APICAL C-CL

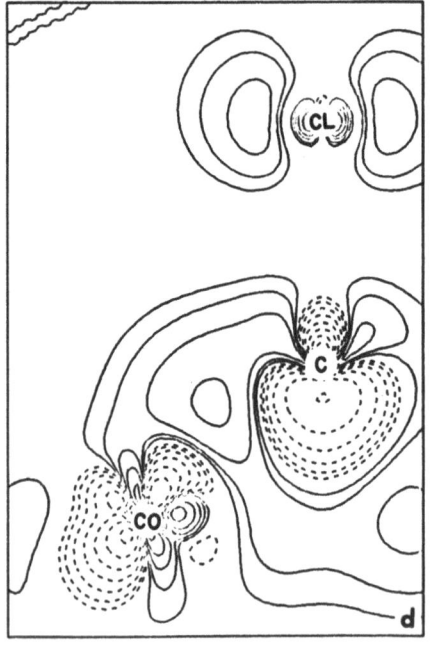

Fig. 5.   ClCCo$_3$(CO)$_9$ Densities: a. Atom deformation density in Co$_3$ plane, b. Fragment deformation density in Co$_3$ plane, c. Atom deformation density in Cl-C-Co plane, d. Fragment deformation density in Cl-C-Co plane.

Adequate basis sets appear to be more important for the deformation density than electron correlation. This even seems to be true for metal-metal bonds where correlation effects are large. The fragment analysis is especially helpful in separating effects occurring in different parts of the molecule. The fragment analysis revealed two quite different situations in which the standard deformation density might be misleading. In dative bonds, the net accumulation of charge between the atoms is greater than that due to bond formation. While in single metal-metal bonds, the net accumulation of charge between the metal atoms is less than that due to their bond formation. Thus, the amount of density accumulated in a bond is not proportional to its strength. Some systems may have an accumulation of density where there is no bond, while others may show a loss of density precisely where there is a bond.

ACKNOWLEDGEMENT

The author gratefully acknowledges the support of the National Science Foundation (CHE 79-20993), the Robert A. Welch Foundation (A-648), and of the National Resource for Computation in Chemistry under a grant from the National Science Foundation (CHE 77-21305) and the Basic Energy Sciences Division of the U.S. Department of Energy (W-7405-ENG-48). I would also like to thank Peter Chesky, Randall Kok, David Sherwood, and Trenton Taylor for preparing the density maps in this chapter.

REFERENCES

1. P. Hohenberg and W. Kohn, Phys. Rev. B. 136:864 (1964).
2. R. G. Parr, Chapter 2.1, in: "Electron Distributions and the
       Chemical Bond," P. Coppens and M. B. Hall, eds., Plenum, New
       York, (1982).
3. P. Coppens, Chapter 1.2 in: "Electron Distributions and the
       Chemical Bond," P. Coppens and M. B. Hall, eds., Plenum, New
       York (1982); P. Coppens and E. D. Stevens, Adv. Quantum Chem.
       10:1 (1977).
4. M. Breitenstein, H. Dannöhl, H. Meyer, A. Scheweig, and W.
       Zittlau, Chapter 4.3, in: "Electron Distributions and the
       Chemical Bond," P. Coppens and M. B. Hall, eds., Plenum,
       New York (1982).
5. J. M. Troup, M. W. Extine, and R. F. Ziolo, Chapter 5.1, in:
       "Electron Distributions and the Chemical Bond," P. Coppens
       and M. B. Hall, eds., Plenum, New York (1982).
6. V. H. Smith, Chapter 1.1, in: "Electron Distributions and the
       Chemical Bond," P. Coppens and M. B. Hall, eds., Plenum, New

York (1982); J. Bicerano, D. S. Marynick, and W. N. Lipscomb, J. Am. Chem. Soc. 100:732 (1978).

7. M. B. Hall, Chem. Phys. Letters 61:467 (1979);
    M. B. Hall, Int. J. Quantum Chem. 14:613 (1978);
    M. B. Hall, Int. J. Quantum Chem. Symp. 13:195 (1979).

8. M. B. Hall and R. F. Fenske, Inorg. Chem. 11:768 (1972).

9. T. F. Block and R. F. Fenske, J. Am. Chem. Soc. 99: 4321 (1977);
    D. L. Lichtenberger and R. F. Fenske, J. Am. Chem. Soc. 98:50
    (1976). B. J. Morris-Sherwood, B. W. S. Kolthammer, and M.
    B. Hall, Inorg. Chem. 20:2771 (1981).

10. R. Goddard and C. Kruger, Chapter 5.2 in: "Electron Distributions
    and the Chemical Bond," P. Coppens and M. B. Hall, eds.,
    Plenum, New York (1982); Y. Wang and P. Coppens, Inorg. Chem.
    15:1122 (1976); A. Mitschler, B. Rees, and M. S. Lehmann, J.
    Am. Chem. Soc. 100:3390 (1978).

11. T. E. Taylor and M. B. Hall, J. Am. Chem. Soc. 102:6136 (1980).

12. D. Mullen and E. Hellner, Acta Cryst. B33:3816 (1977); C.
    Scheringer, D. Mullen, and F. Hellner, Acta Cryst. B34:621
    (1978).

13. W. N. Lipscomb, Trans. Am. Cryst. Assoc. 8:79 (1972).

14. H. Dietrich and C. Scheringer, Acta Cryst. B34:54 (1978).

15. J. A. Connor, G. Pilcher, H. A. Skinner, M. H. Chisholm, and
    F. A. Cotton, J. Am. Chem. Soc. 100:7738 (1978).

16. M. B. Hall, J. Am. Chem. Soc. 102: 2104 (1980).

17. R. A. Kok and M. B. Hall, unpublished results.

18. F. A. Cotton, Acc. Chem. Res. 11:225 (1978).

19. T. H. Dunning, Jr., J. Chem. Phys. 53:2823 (1972).

20. B. Rees and A. Mitschler, J. Am. Chem. Soc. 98:7918 (1976).

21. J. B. Johnson and W. G. Klemperer, J. Am. Chem. Soc. 99:7132
    (1977).

22. D. E. Sherwood, Jr. and M. B. Hall, J. Am. Chem. Soc. submitted
    for publication.

23. G. D. Michels, G. D. Flesch, and H. J. Svec, Inorg. Chem. 19:479
    (1980).

24. Y. B. Koh and G. G. Christoph, Inorg. Chem. 18:1122 (1979).

25. B. Rees and A. Mitschler, to be published.

26. L. F. Dahl, E. Ishishi, and R. E. Rundle, J. Chem. Phys.
    26:1750 (1957).

27. W. Heijser, E. J. Baerends, and P. Ros, J. Mol. Struct. 63:109
    (1980).

28. W. Heijser, E. J. Baerends, and P. Ros, Farad. Disc. 14:211
    (1980).

29. M. Benard, Chapter 4.2, in: "Electron Distributions and the
    Chemical Bond," P. Coppens and M. B. Hall, eds., Plenum,
    New York, (1982).

30. D. Seyferth, Adv. Organomet. Chem. 14:97 (1976).

31. D. C. Miller and T. B. Brill, Inorg. Chem. 17:240 (1978).

32. P. T. Chesky and M. B. Hall, Inorg. Chem. in press.

33. J. W. Richardson, W. C. Nieuwpoort, R. R. Dowell, and W. F.
    Edgell, J. Chem. Phys. 36:1057 (1962).

ELECTRON DEFORMATION DENSITY DISTRIBUTIONS IN BINUCLEAR COMPLEXES OF TRANSITION METALS : COMPUTATION AND INTERPRETATION FROM AB INITIO MOLECULAR ORBITAL WAVEFUNCTIONS [1]

Marc Bénard

E.R. n° 139 du C.N.R.S.
Laboratoire de Chimie Quantique, Université L. Pasteur
4, rue Blaise Pascal, 67000 Strasbourg, France

ABSTRACT

The theoretical electron deformation density distributions obtained from ab initio MO wavefunctions for $\left[C_5H_5Fe(CO)_2\right]_2$, $(\eta^5-C_5H_5Ni)_2C_2H_2$, $Cr_2(O_2CH)_4$ and $Mo_2(O_2CH)_4$ are reviewed. Several sections of these distributions are displayed, together with their experimental counterpart. A systematic comparison is performed between the deformation density distributions and the results of the Mulliken population analysis obtained from the wavefunctions. This comparison provides a first rationalization of the electron deformation density pattern in the vicinity of a metal-metal axis. The strong dependence of this pattern upon the location of the metal in the periodic table, the specific character of d-orbitals relatively to the orbital expansion associated with covalent bonding, the influence of the depopulation of the external s orbital associated with transition metal complexation, are emphasized. It is shown that these factors, combined in some cases with the lack of a direct metal-metal bond, can account for the puzzling absence of charge accumulation detected in most deformation density distributions obtained for binuclear complexes.

INTRODUCTION

Recent advances in X-ray and neutron diffraction methods have made it possible to obtain accurate experimental information on charge distribution in crystals[2,3]. More specifically, several diffraction experiments performed in the last few years on binuclear complexes of transition metals have provided the first

insight into the electron distribution in the vicinity of a metal
atom [4-9]. Since the electron distribution is a sensitive property
straightforwardly related to the wavefunction of the system, the
experimental determinations can be considered as a challenge to
the quantum chemist [10]. The problem was especially exciting in the
case of binuclear complexes. Experimental determinations are still
scattered for these systems and apparently yield along the metal-
metal axis rather unexpected electron distribution patterns in
view of the previous experience mainly obtained from small organic
molecules.

A profitable collaboration established between our group and
several laboratories specialized in diffraction experiment (see
Acknowledgments) provides now a first rationalization of the
electron deformation density pattern in the vicinity of a metal-
metal axis. This rationalization is essentially based upon the
interpretation of ab initio MO calculations and upon the comparison
of computed and experimental deformation density distributions.
One of the purposes of this article is to collect the deformation
density maps recently computed in our group for several binuclear
complexes, namely the trans form of bis(dicarbonyl-π-cyclopenta-
dienyliron), $\left[C_5H_5Fe(CO)_2\right]_2$ (Fig. 1), μ-acetylene-bis(cyclopenta-
dienylnickel), $(\eta^5-C_5H_5Ni)_2C_2H_2$ (Fig. 6) and dichromium and
dimolybdenum tetraformates $Cr_2(O_2CH)_4$ and $Mo_2(O_2CH)_4$. These computed
distributions will be displayed side by side with the corresponding
experimental maps. The second object will be to present a general
discussion concerning these maps in order to define an expected
pattern for the density distribution between two metal atoms, as a
function of the valence orbital population of the metal in the
isolated atom and in the crystal.

Methodology and computational details

Even though the diffraction experiments can access the total
electron density ρ(observed) of the system, this observable is
usually not displayed directly. It is used to obtain a function
which is more sensitive to details of chemical bonding, the
deformation density distribution $\Delta\rho$, defined as

$$\Delta\rho = \rho_{(observed)} - \Sigma\rho_{(spherical\ atom)}$$

In this expression $\Sigma\rho_{(spherical\ atom)}$ corresponds to the density
distribution generated by the "promolecule" [11], that is the
assembly of the atoms of the system, each atom being considered
  - isolated
  - neutral
  - spherically averaged
  - in its ground state.
The theoretical deformation density maps were obtained from ab

initio LCAO-MO-SCF calculations carried out with the ASTERIX system
of programs [12]. The Gaussian basis sets were (13,9,7) for Mo,
(11,7,5) for Fe and Cr, (8,4) for the first row atoms and (4) for
hydrogen. For most calculations, the basis sets were contracted to
minimal basis sets for the inner shells and for the (n+1)s and
(n+1)p shells of the metal atoms, and to double-zeta basis sets
for the valence shells. The contribution of the promolecule to the
deformation density $\Delta\rho$ is obviously computed with the same basis
sets. Atomic wavefunctions are computed for the experimental ground
state and spherical averaging is obtained by equally distributing
the available electrons among the equivalent orbitals of the valence
shell.

Results and discussion

    1) Trans-$[(\pi-C_5H_5)Fe(CO)_2]_2$. This binuclear complex is charac-
terized by a symmetry point group C_{2h} for the isolated molecule,
yielding a planar cycle composed of the two metal atoms and of the
two carbons of the bridging carbonyl groups (Fig. 1). The electron

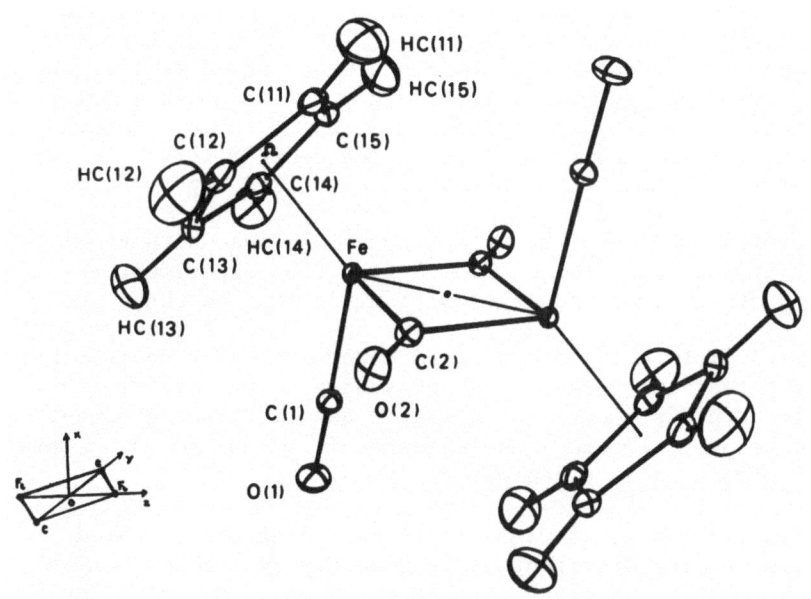

Fig. 1.   Molecular structure of trans-$[(\pi-C_5H_5)Fe(CO)_2]_2$, from
         Mitschler, Rees and Lehmann [8]. Reproduced with permission
         of the authors.

count on Fe is 17 electrons and, since the complex is diamagnetic,
a spin coupling is requested between the two metal atoms. Essentially from an analogy with the dibridged binuclear complexes
investigated by Dahl et al [13,17], it was generally thought that
this spin coupling was effective through a direct metal–metal bond
of the σ-type. However, both Extended Hückel and ab initio MO calculations have shown that the overall direct metal–metal interaction
was either nonbonding or slightly antibonding [14]. An analysis of
the HOMO has shown that the spin-coupling between the metal atoms
was induced by a stabilizing multicentric interaction involving
the d_{yz} orbital of the metal (z is the metal–metal axis and y the
direction of the bridging carbonyls) and the π antibonding orbitals
of the bridging ligands. This resulted in a strong backbonding
effect from the metals to the bridging carbonyls. This can be
detected from the relatively large values obtained for the bridging
C–Fe overlap populations, compared to the overlap populations
determined for a terminal C–Fe bond. The respective overlap populations for Fe–C (bridging) and Fe–C (terminal) are +0.152 and
+0.161. However one must keep in mind that each bridging carbon
contributes to two Fe–C bonds. Fig. 2 and 3 display the section
of the deformation density distribution by two perpendicular planes
containing a Fe atom and a terminal carbonyl. Section represented
in Fig. 2 contains the Fe–Fe line. Fig. 2a and 3a show the theoretical map obtained from the ab initio SCF wavefunction, and Fig. 2b
and 3b the experimental map determined by Mitschler, Rees and
Lehmann [8]. Fig. 4a and 4b display a section of the deformation
density distribution by the plane containing the Fe–Fe line and
the bridging carbonyls. Finally, Fig. 5a and 5b show a section of
the deformation density in a plane parallel to the cyclopentadienyl
ring and containing the iron nucleus. The contour intervals are
0.1 eA^{-3} for experimental maps and 0.2 eA^{-3} for computed maps.

The comparison with the experimental maps shows that all
significant features obtained from diffraction experiment are
reproduced by the computed wavefunction. As long as these most
prominent and significant features are concerned, the experimental
and computed distributions can be said topologically equivalent,
which means that the peaks, pits and saddle points are similarly
located in both distributions. A better quantitative agreement
could not be expected since the effects of thermal smearing have
not been introduced into the wavefunction.

However, a quantitative rationalization of the computed
deformation density distribution is directly available from the
Mulliken population analysis performed for both the promolecule
and the molecule. Effectively, in the case of transition metal
complexes, the deformation density maps are essentially a visualization of the interorbital charge transfers occuring when a superposition of isolated spherically averaged atoms is transformed

Fig. 2a. Computed density distribution of trans-
[(π–C₅H₅)Fe(CO)₂]₂ (ab initio SCF). Section containing the Fe–Fe line and the
terminal carbonyls. Contour interval
0.2 eA⁻³. Dashed contours for negative,
solid lines for 0 and positive deforma-
tion density. b. Experimental deformation density distribution
of trans–[(π–C₅H₅)Fe(CO)₂]₂ from 8. Section as
in a. Contour interval 0.1 eA⁻³. Reproduced
with permission.

Fig. 3a. Computed density distribution of trans–
$\left[(\pi-C_5H_5)Fe(CO)_2\right]_2$ (ab initio SCF). Sec–
tion containing one Fe atom and the ter–
minal carbonyl. Contours interval 0.2eA⁻³.

b. Experimental density distribution of trans–
$\left[(\pi-C_5H_5)Fe(CO)_2\right]_2$ from 8. Section as in a.
Contour interval 0.1 eA⁻³. Reproduced with
permission.

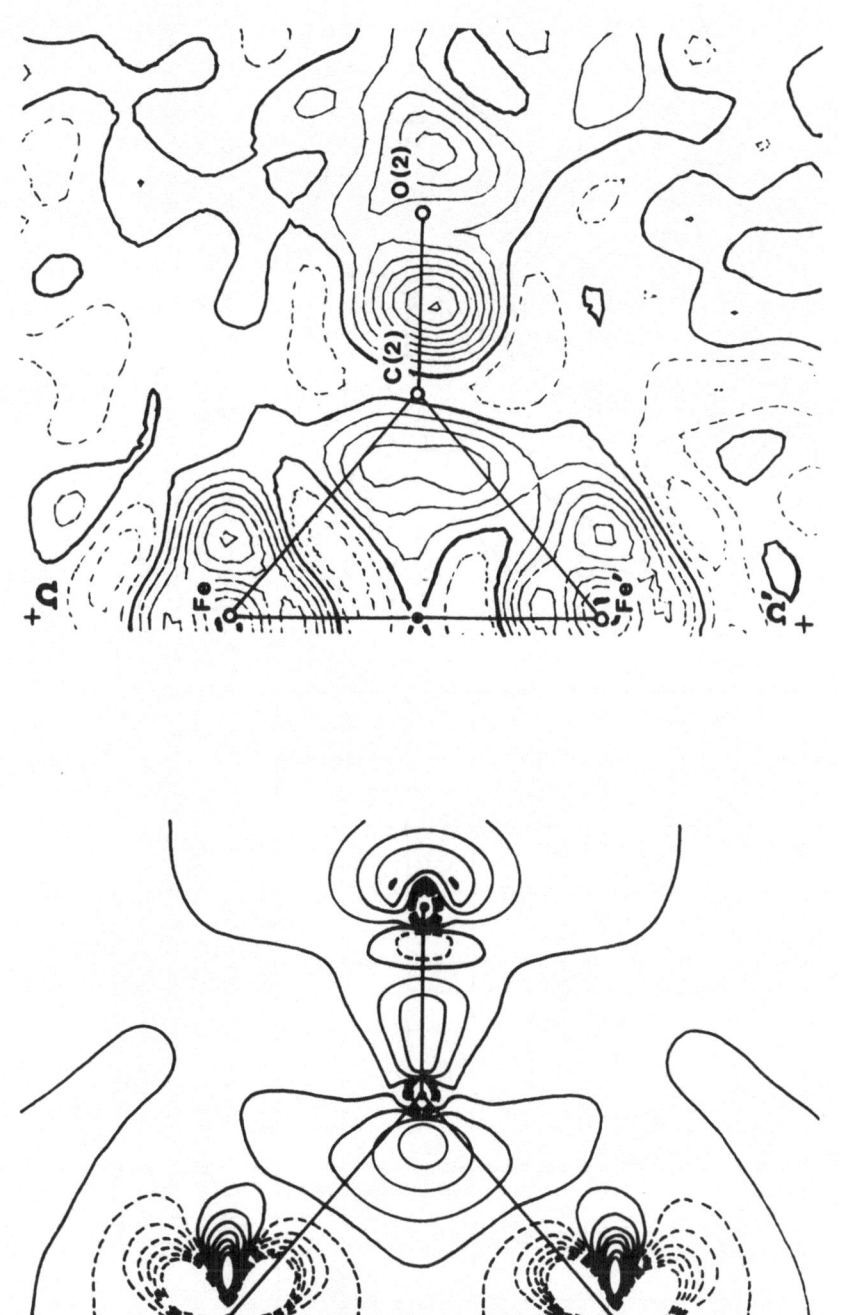

Fig. 4a. Computed density distribution of trans-$[(\pi-C_5H_5)Fe(CO)_2]_2$ (ab initio SCF). Section containing the Fe-Fe line and the bridging carbonyls. Contour interval 0.2 eA^{-3}.

b. Experimental density distribution of trans-$[(\pi-C_5H_5)Fe(CO)_2]_2$ from 8. Section as in a. Contour interval 0.1 eA^{-3}. Reproduced with permission.

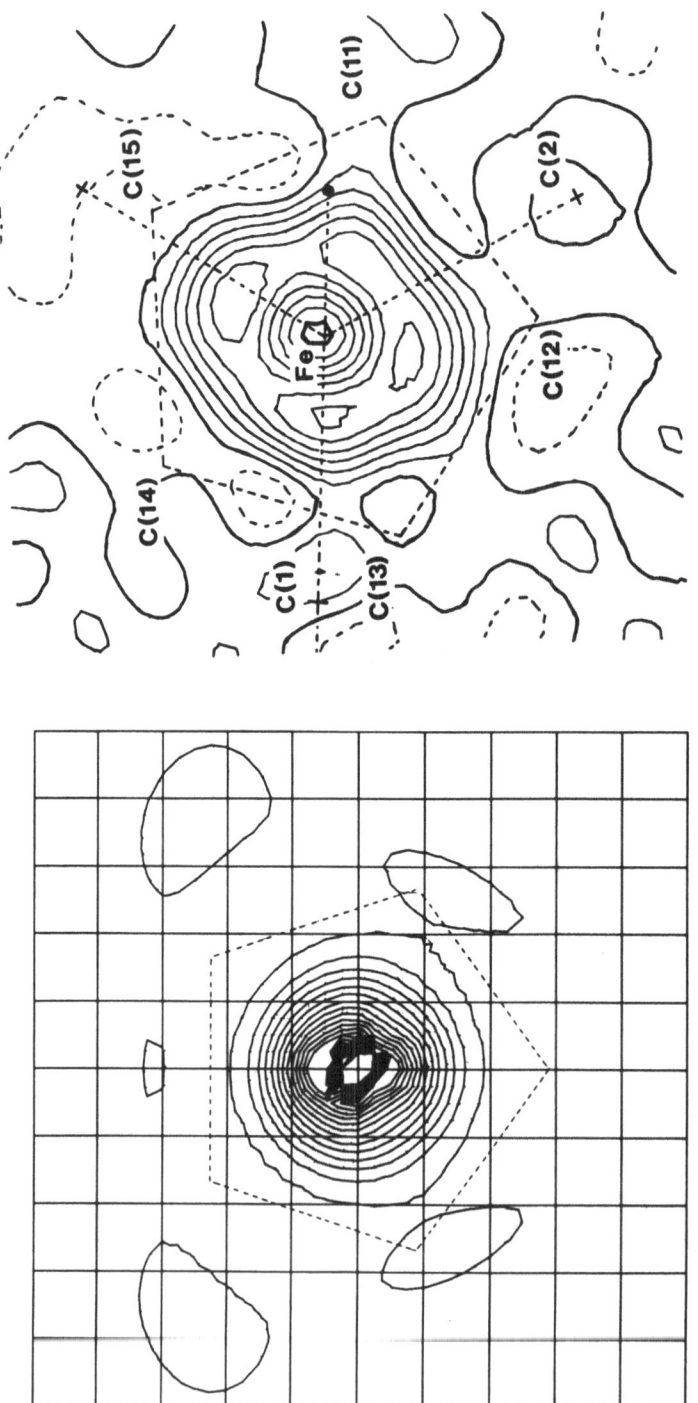

Fig. 5a. Computed density distribution of trans-
$\left[(\pi-C_5H_5)Fe(CO)_2\right]_2$ (ab initio SCF). Section parallel to the cyclopentadienyl
ring and containing the iron nucleus.
Contour interval 0.2 eA⁻³.

b. Experimental density distribution of trans-
$\left[(\pi-C_5H_5)Fe(CO)_2\right]_2$ from 8. Section as in a.
Contour interval 0.1 eA⁻³. Reproduced with
permission.

into a molecule. As it will be discussed later, intraorbital charge transfers (expansion or contraction of orbitals due to chemical bonding), contrary to the case of organic systems, are relatively fine effects which can be detected and analyzed in specific cases only.

The atomic ground state configuration of Fe is $3d^6 4s^2$ which yields an average population of 1.2 electron per d orbital. This will be the reference to which the orbital populations in the molecule will be compared in order to analyze the deformation density distribution around the Fe atoms. It will also be useful to decompose the e_g orbitals into their d_{x2}, d_{y2} and d_{z2} components, in order to compare the populations localized along each specific axis. Therefore, most of the orbital populations mentioned in the discussion will refer to the six d orbital system. For instance, the population of any of the d_{x2}, d_{y2} or d_{z2} orbitals for the spherically averaged Fe atom in the six d orbital system is 0.8 electron.*

These atomic and molecular populations are displayed in Table 1. There is a direct correlation between the population of

---

* The spherical averaging of the d shell corresponds to an even distribution of the charge among the five "physical" harmonics (xy), (xz), (yz), (x^2-y^2), $(2z^2-x^2-y^2)$. The orbital population of (xy), (xz) and (yz) is not affected by the six d↔five d transformation. The equipartition of the remaining charge among (x^2), (y^2) and (z^2) appears intuitively obvious from the equivalence of the three directions of space. It can be proved from the analytical expansion of the normalized harmonics

$$(x^2-y^2) = \sqrt{3}/2\left[(x^2)-(y^2)\right]$$
$$(2z^2-x^2-y^2) = \left[(z^2) - 0.5(x^2) - 0.5(y^2)\right]$$

and from the definition of the gross atomic orbital population given by Mulliken :

$$n(p) = \sum_i n(i)(C_{ip}^2 + \sum_{q \neq p} C_{ip}C_{iq}S_{pq})$$

In this definition, n(p) is the population of atomic orbital p ; n(i) the population of molecular orbital i and S_{pq} the overlap between p and q. Since $S_{(x2)(y2)} = S_{(x2)(z2)} = S_{(y2)(z2)} = 1/3$, it is clearly seen that

$$n(x^2) = n(y^2) = 1/2\, n(x^2-y^2) + 1/6\, n(2z^2-x^2-y^2)$$
$$n(z^2) = 2/3\, n(2z^2-x^2-y^2)$$

Table 1.   Spherically averaged atomic and molecular populations (in electrons) for metal d-orbitals (in terms of six d orbitals). The z axis is colinear to the metal-metal line.

	$[FeC_5H_5(CO)_2]_2$ [a]		$(NiC_5H_5)_2C_2H_2$ [b]			$Cr_2(O_2CH)_4$			$Mo_2(O_2CH)_4$	
	Atom	Mol. (SCF)	Atom	Mol. (SCF)	Mol. (SCF)[c]	Atom	Mol. (SCF)	Mol. (CI)	Atom[d]	Mol. (SCF)[d] [e]
d_{xy}	1.2	1.800	1.6	1.765	1.671	1.0	1.069	1.050	3.0	3.052
d_{xz}	1.2	1.912	1.6	1.863	1.984	1.0	1.010	1.040	3.0	2.999
d_{yz}	1.2	0.916	1.6	1.825	1.919	1.0	1.010	1.040	3.0	2.999
d_{x^2}	0.8	0.372	1.067	1.116	1.034	0.667	0.219	0.237	2.0	1.608
d_{y^2}	0.8	1.141	1.067	1.317	1.317	0.667	0.219	0.237	2.0	1.608
d_{z^2}	0.8	0.836	1.067	0.973	0.933	0.667	0.715	0.704	2.0	2.087
s(overall)	8.0	6.066	8.0	6.074	6.074	7.0	6.181	6.159	9.0	8.211

a) y axis colinear to the bridging carbonyls.
b) y axis colinear to the acetylene C-C bond.
c) After rotation of the reference system around the y axis by 17°.
d) Overall populations (3d+4d).
e) CI does not significantly modify orbital populations.

the d_{xz} orbital (1.2 electrons in the atom, 1.912 electrons in the
molecule) and the four peaks located around the Fe atom in Fig. 2a
and 2b (xz plane). The computed map (Fig. 2a) displays a large
and deep region of negative density along the x axis. This electron
deficient zone is also visible on the experimental map (Fig. 2b)
even though it is not apparent between the peaks related with the
d_{xz} orbital, probably because of thermal smearing. The existence
of this negative region is obviously correlated with the relative
electron deficiency of the d_{x^2} orbital (population in the spherical
atom : 0.8 el., in the molecule 0.372 el. (Table 1)). The lack of
significant features along the metal–metal axis has to be put in
parallel with the very small change in the population of the d_{z^2}
orbital due to the molecular environment (0.836 el. instead of
0.8 el. in the atom). However, it has to be noticed that there is a
slight increase in population in the molecular orbital whereas the
corresponding region in both the theoretical and experimental maps
is very slightly negative. The cumulative contribution of other
orbitals and more specifically the quasi complete depopulation of
the 4s orbital of iron (Table 1) has to be taken into account for
explaining this effect. The influence on the deformation density of
the depopulation of the external s orbital of metal atoms, which
is a fairly general phenomenon in transition metal complexes, has
never been analyzed yet. Since this orbital is very diffuse, the
spherical electron deficient region generated by its depopulation
is not deep enough to be immediately visible. However, it might be
responsible for some general features displayed by deformation
density maps of transition metal complexes. For instance, the
slightly negative environment frequently observed around metal
atoms between 0.5 and 1.0 Å and the puzzling problem of the lack
of significant electron accumulation at the center of a metal–metal
bond [9,15] could be related with the removal of 1 or 2 electrons
from the (n+1)s orbital (see Table 1). Experimental distributions
obtained by Hino and Saïto for $Mo_2(O_2CCH_3)_4$ are fully consistent
with this interpretation (see discussion section 4).

In trans−$\left[(\pi-C_5H_5)Fe(CO)_2\right]_2$, the relative depopulation of the
d_{yz} orbital (0.916 el. instead of 1.2 el. in the atom) is visible
in the computed map (Fig. 4a) but partially smeared on the experimental map (Fig. 4b). However, the accumulation regions related
with the d_{y^2} orbital (1.141 el. instead of 0.8 el.) are well defined
on both Fig. 4a and 4b. The peaks due to the same d_{y^2} orbital are
smeared with the neighbouring accumulation zones generated by the
d_{xy} orbital (1.800 el. instead of 1.2 el). This yields the two
large accumulation regions displayed in Fig. 3a and 3b symmetrically
with regard to the Fe–Co axis.

A similar analysis is also of interest when the reference
system is rotated around the y axis so that the xy plane becomes
parallel to the cyclopentadienyl rings. Sections of both the computed

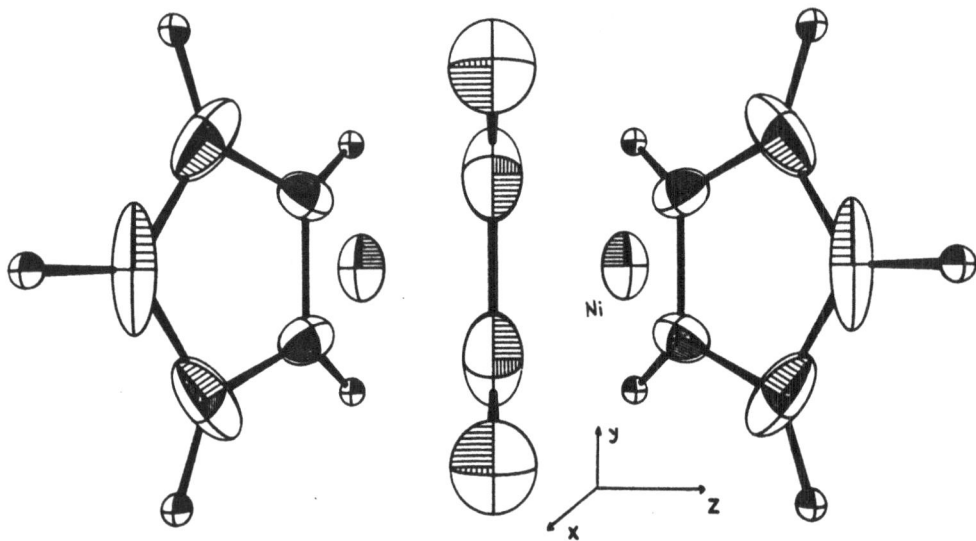

Fig. 6. Molecular structure of $(\eta^5 C_5 H_5 Ni)_2 C_2 H_2$, from Wang and
         Coppens [4]. Reproduced with permission of the authors.

The main feature of the experimental distribution is a continuous region of positive electron density along the metal-metal
axis with two maxima at about 0.5 Å from the adjacent nucleus.
Interpreted in terms of d-orbital population analysis, such an
accumulation of electron density just means that the population of
the metal d-orbital oriented along the z axis is higher in the
molecule than in the spherically averaged atom. However, the
electron configuration of the Ni atom in its ground state is $d^8 s^2$,
which yields a spherically averaged population as big as 1.6 electrons per d-orbital. Consequently, according to the experimental
distribution, the population of the $d_{2z^2-x^2-y^2}$ nickel orbital in
the molecule might reach a very high value, probably not too far
from 2 electrons [24]. Such an occupancy is more compatible with
the presence of a bonding-antibonding orbital pair than with the
existence of a unique σ bonding orbital along the z axis.

Effectively, the ideal population of an individual atomic
orbital engaged in covalent bonding is one electron, whatever may
be the orbital type and the population of this orbital from the
spherically averaged atom. Since the average population of the
valence orbitals increases when going to the right side of the
periodic table, the visualization of a covalent bond in the deformation density distribution will be strongly dependent upon the

and the experimental deformation density distributions in a plane
parallel to the C_5H_5 rings and containing one Fe atom are almost
perfectly centrosymmetric around the metal (Fig. 5a and 5b). In the
new reference system, the respective populations of d_{x^2}, d_{y^2} and
d_{xy} orbitals are 1.141, 1.137 and 1.600 el., so that the difference
with regard to the promolecule appears positive and approximately
constant all around the metal (Table 1).

If we consider the orbital populations of the bridging CO
carbon atoms, the important back-bonding effect detected from the
analysis of the wavefunction is corroborated by the increase in
charge of the p_z orbital (0.898 el. instead of 0.667 el. in the
spherically averaged C atom). For the sake of comparison, the population of the p_x orbital, which is not affected by backbonding, is
0.621 el., and the populations of similar orbitals in the terminal
CO carbons are 0.589 and 0.621 el. The comparison, on experimental
maps, of the accumulation regions centered on the carbon lone pair
for a terminal (Fig. 2b and 3b) and for a bridging carbonyl (Fig. 4b)
is significant. For a terminal carbonyl, this accumulation region
consists in a sharp peak centered on the lone pair, but for a
bridging carbonyl, the electron-rich zone extends over the whole
σ and π regions.

2) $(\eta^5\text{-}C_5H_5Ni)_2(C_2H_2)$ (Fig. 6). In agreement with electron counting and with the very short Ni-Ni distance of 2.345 Å [4], an analysis
of the ab initio wavefunction obtained for this system, and more
specifically of the HOMO, concludes to the presence of a direct
metal-metal bond of the σ type [16]. This bond is bent away from the
acetylene as had been found already for several dimers of the type
$Fe_2(CO)_6X_2$ [17].

However, the theoretical deformation density maps obtained
for this complex are in serious disagreement with the distribution
obtained from X-ray diffraction experiment by Wang and Coppens [4],
especially in the vicinity of the nickel atoms and along the metal-
metal bond. If z is colinear to the metal-metal axis and y to the
acetylene C-C bond, Fig. 7 displays a cut of the deformation density
distribution by the xz plane, containing the two Nickel atoms and
the midpoint of the C-C bond of C_2H_2. Fig. 7a shows the theoretical
map deduced from our ab initio SCF wavefunction. Fig. 7b represents
a computed map obtained by Ros [18] from LCAO Hartree-Fock-Slater
calculations [19]. Fig. 7c reproduces the experimental map obtained
by Wang and Coppens. Fig. 8a, b, c represent a section of the same
distributions by the plane xy containing the acetylene molecule
and the midpoint of the Ni-Ni axis. Finally, Fig. 9a and 9b, respectively, display a section of the ab initio SCF and experimental
distributions by a plane containing the two Ni atoms and parallel
to the acetylene C-C bond.

location in the periodic table of the atoms involved in bonding.
For first-row atoms however, this effect is partially masked by an
important electron shift inside the orbitals engaged in bonding.
Electrons are moving along the bond axis, expanding the orbital in
order to maximize the overlap, thus yielding the well-known charge
accumulation region at the center of the bond. However, in transition metal complexes, the metal-metal bonding distances are usually
very large and the overlap between two valence-shell orbitals just
involve the "tail" of the d orbitals.* Even though the orbital
expansion is still detectable for d-orbitals involved in bonding,
it is not predominant anymore in the deformation density distributions [25]. Consequently, the overlap of d-orbitals will generally
yield at the center of the bond a large and diffuse accumulation
region of low density gradient, sometimes at the limit of significancy, rather than a sharp peak in a limited region. In the other
regions, that is, along the metal-metal axis and in the vicinity
of the metal atoms, the deformation density distribution will be
essentially correlated with the difference between the population
of the d-orbitals in the molecule and in the spherically averaged
atom. Along a metal-metal bond, this difference will decrease from
a positive value to zero and then to a negative value when the
ground state configuration of the considered atom varies for instance
from d^1s^2 to d^5s^1 or d^5s^2 and then to d^8s^2. Consequently, in
contradiction with the intuitive feeling, a σ bond between two
nickel atoms (and, more generally, between two metal atoms with
high d population) cannot be visualized with a significant charge
accumulation at the center, but rather with an <u>electron deficient</u>
<u>region</u> along the bond axis.

   This scheme is illustrated by the theoretical deformation
density distribution obtained from the ab-initio wavefunction. In
the xy plane (Fig. 7a) each nickel atom is surrounded by four sharp
charge accumulation regions in an electron deficient environment [26].
More specifically, two electron deficient regions, each of them
centered on one Ni atom, are pointing toward each other. Their
axis is not colinear with the Ni-Ni line, but generates an angle
of approximately 17°. It is important to notice that these electron
deficient regions <u>are colinear with the metal-metal bent bond.</u> This

---

* See for instance the sketch of d-d overlaps given by W.C. Trogler
in Ref. [48]. The fact that the maximum of the d-d overlap lies at
distances much shorter than any known metal-metal bond length is
in direct relation with the compactness of 3d orbitals. Another
equivalent formulation is to say that the 3d-3d overlap in a metal-
metal bond is weak because it just involves the diffuse part
(the "tail") of the d orbitals.

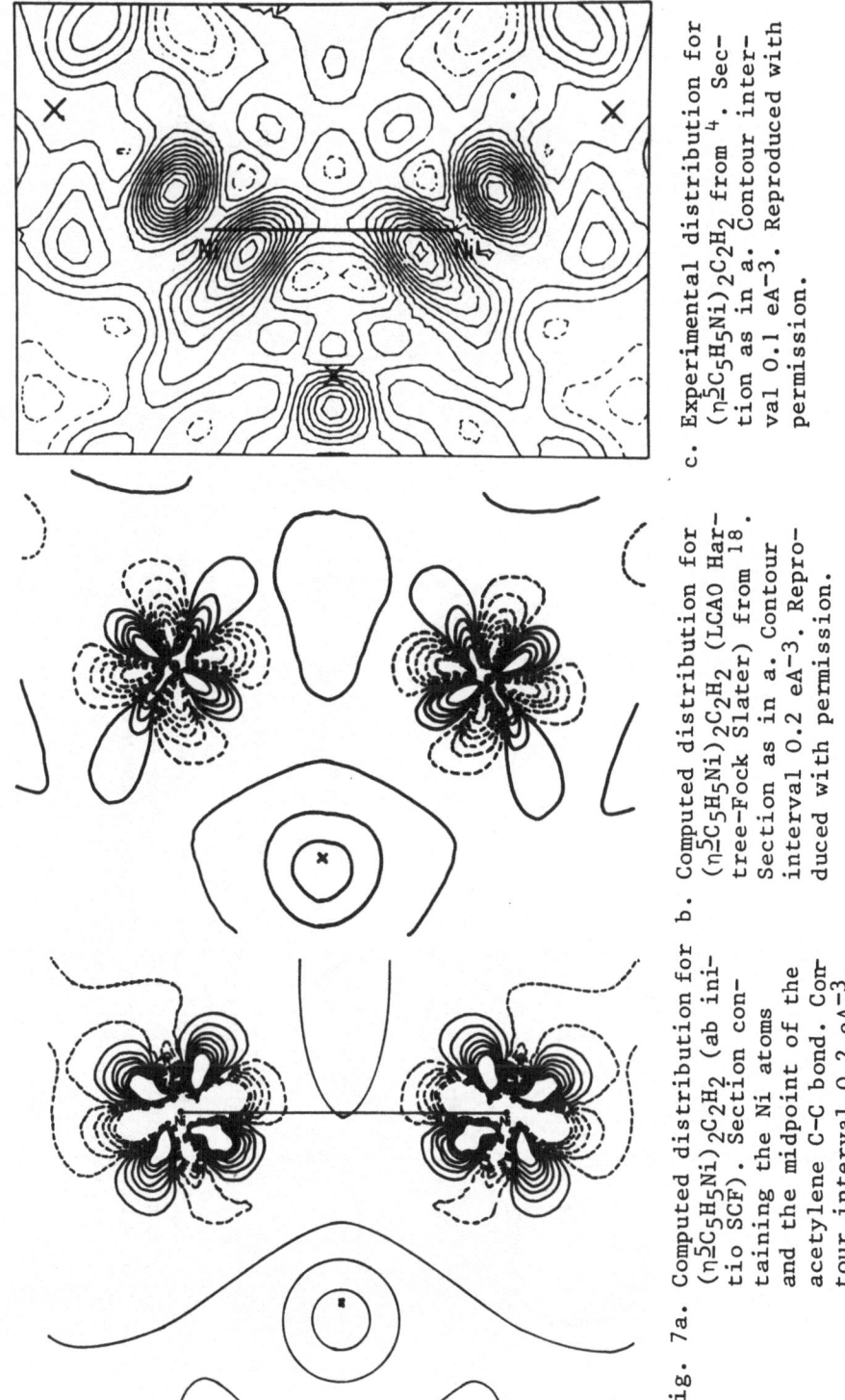

Fig. 7a. Computed distribution for $(\eta^5 C_5H_5Ni)_2C_2H_2$ (ab initio SCF). Section containing the Ni atoms and the midpoint of the acetylene C-C bond. Contour interval 0.2 eA^{-3}.

b. Computed distribution for $(\eta^5 C_5H_5Ni)_2C_2H_2$ (LCAO Hartree-Fock Slater) from [18]. Section as in a. Contour interval 0.2 eA^{-3}. Reproduced with permission.

c. Experimental distribution for $(\eta^5 C_5H_5Ni)_2C_2H_2$ from [4]. Section as in a. Contour interval 0.1 eA^{-3}. Reproduced with permission.

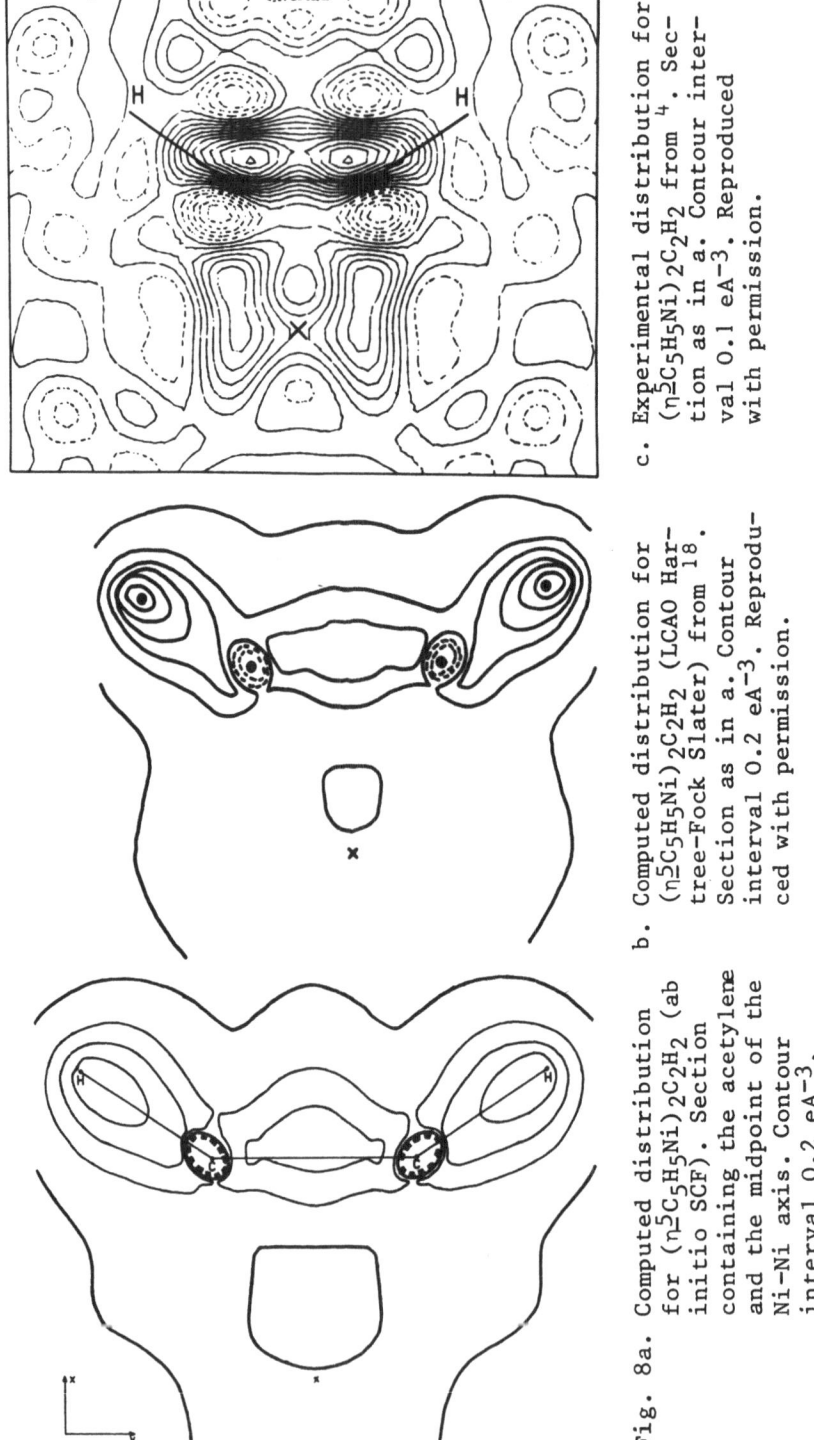

Fig. 8a. Computed distribution for ($\eta^5C_5H_5Ni$)$_2C_2H_2$ (ab initio SCF). Section containing the acetylene and the midpoint of the Ni–Ni axis. Contour interval 0.2 eA^{-3}.

b. Computed distribution for ($\eta^5C_5H_5Ni$)$_2C_2H_2$ (LCAO Hartree-Fock Slater) from 18. Section as in a. Contour interval 0.2 eA^{-3}. Reproduced with permission.

c. Experimental distribution for ($\eta^5C_5H_5Ni$)$_2C_2H_2$ from 4. Section as in a. Contour interval 0.1 eA^{-3}. Reproduced with permission.

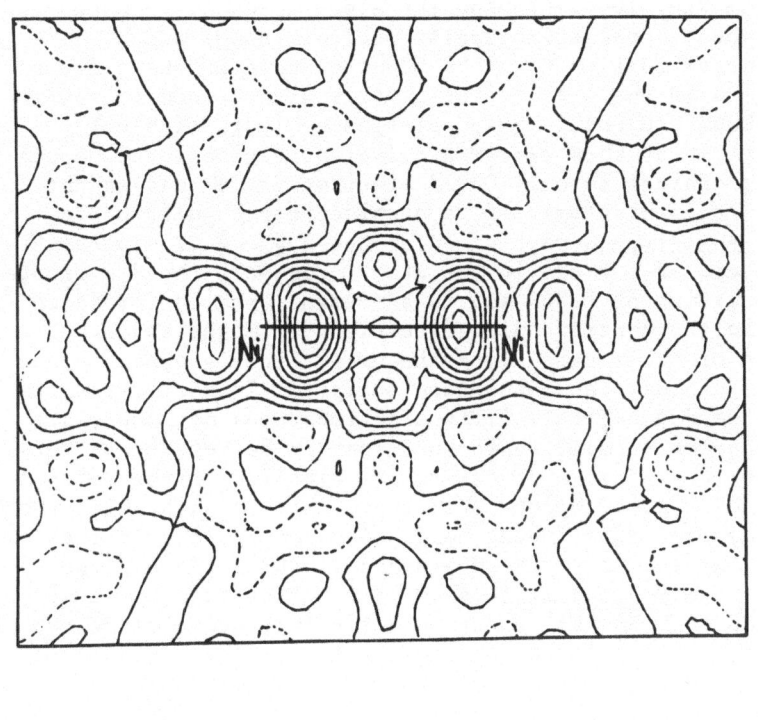

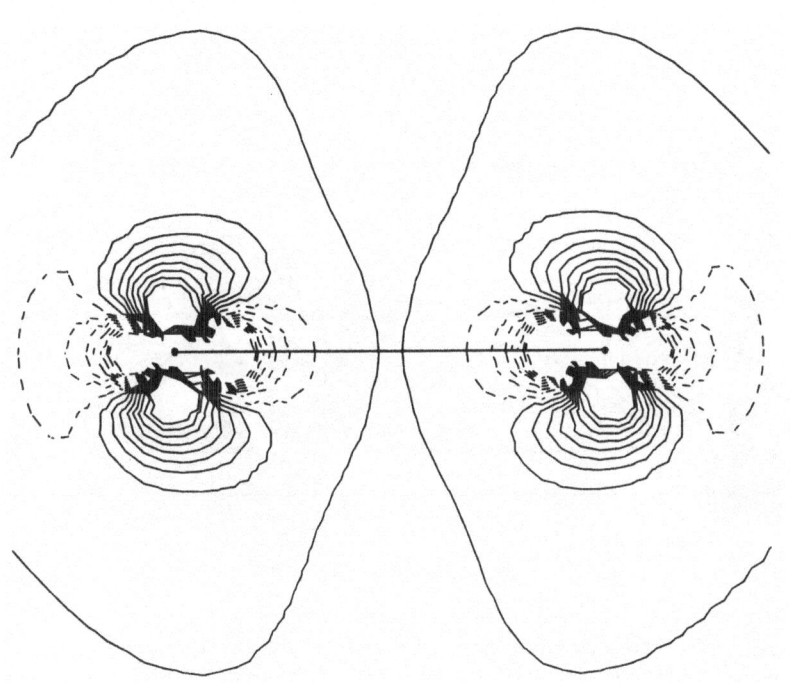

b. Experimental distribution for $(\eta_{\underline{-}}^{5}C_5H_5Ni)_2C_2H_2$ from [4]. Section as in a. Contour interval 0.1 eÅ^{-3}. Reproduced with permission

Fig. 9a. Computed distribution for $(\eta_{\underline{-}}^{5}C_5H_5Ni)_2C_2H_2$ (ab initio SCF). Section containing the Ni atoms and parallel to the acetylene C—C bond. Contour interval 0.2 eÅ^{-3}.

can be seen from Fig. 10 displaying the electron density contour
map for the highest occupied molecular orbital (HOMO) which
describes the metal-metal bond : the bending angle of the contour
lines with regard to the metal-metal axis is also around 15° away
from the acetylene C-C bond. The population analysis (Table 1)
confirms the origin of this electron deficient region : the population of the d_{z^2} orbital is only 0.973 electron (1.067 el. in the
spherically averaged Ni atom). If we consider the population in
terms of five d orbitals, the population of the $d_{2z^2-x^2-y^2}$ orbital is
1.485 el. (instead of 1.6 el. in the atom). However, the d_{xz} orbital
has a high population of 1.865 el. in agreement with the presence
of four peaks in the corresponding direction. It might appear that
the orientation of the z axis along the Ni-Ni line is probably not
the best one to discuss the population analysis. In particular, the
population of the d_{x^2} orbital, 1.116 electrons does not correlate
with the presence of a small region of electron deficiency in this
region of space (Fig. 7a). A direct rotation of 17° of the reference
system around the y axis, which makes the z axis colinear with the
direction of the bent metal-metal bond, allows a better correlation
(Table 1). The population of the d_{z^2} and d_{x^2} orbitals are respectively decreased to 0.933 and 1.034 electrons which is now consistent
with the existence of a small electron deficient region along the
x axis. However, the population of the d_{xz} orbital has been increased

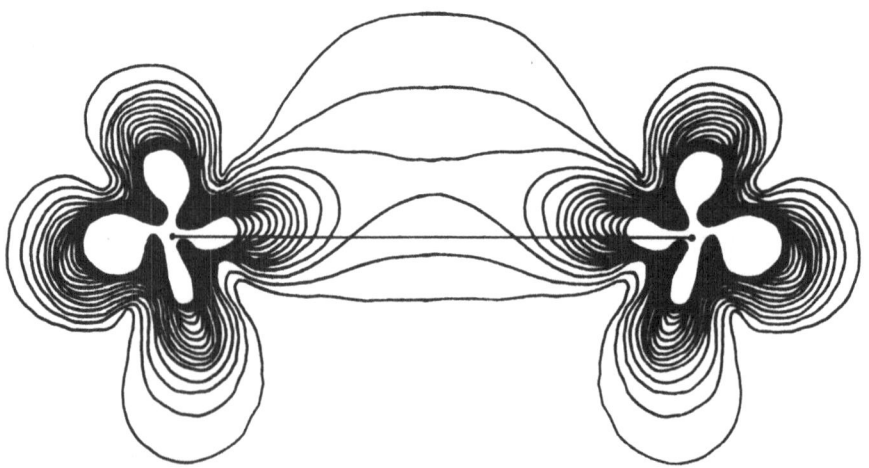

Fig. 10. Electron density contour map for the highest occupied
         molecular orbital of $(\eta^5C_5H_5Ni)_2C_2H_2$ (ab initio SCF
         wavefunction). Section as in Fig. 7a. Contour interval
         0.01 e(a.u.)$^{-3}$.

to 1.984 el. With the exception of d_{z^2} and d_{x^2}, the population of every d-orbital is bigger than in the spherically averaged atom.

The computed map displays another important feature which had not received previously sufficient attention, since it is close to the limit of significancy of both calculations and experiment. A very smooth and diffuse accumulation region is observed above the center of the Ni-Ni line in the direction opposite to the acetylene C-C bond. This accumulation region culminates (at less than 0.1 eA^{-3}) slightly above the center of the Ni-Ni bent bond. This diffuse accumulation region might visualize the "overlap" between the two metal d-orbitals involved in bonding, or, in other words, the orbital expansion associated with the formation of the bond. It is highly probable that this positive region would be significantly enhanced if the spherically averaged nickel atoms were already accounting for the quasi-complete depopulation of the 4s orbital (Table 1).

The possible origin of the disagreement between the experimental and the theoretical deformation density distribution has been discussed previously [14] and is still an open question. The acentric crystal structure had been a priori considered as a possible source of complications in the calculation of difference densities from X-ray data [4]. However, the effect of basis set truncation had to be considered as a possible source of error for a property which is known to be very sensitive to the level of accuracy of the wave-function. In that sense, the theoretical deformation density maps computed by Ros [18] using the LCAO-Hartree-Fock Slater method [19,20] are of great interest (Fig. 7b and 8b). They first show that a completely different way of constructing and solving the Hartree-Fock equations [22] yields a deformation density distribution very similar to what is obtained from the ab initio wavefunction. Furthermore, the double-zeta STO basis set used in this calculation is close enough to the Hartree-Fock limit [27] to cancel the objection concerning the basis set function [28].

It seems that the only possibility for the computed distribution to be erroneous would be to consider a large inter-orbital charge transfer in the metal induced by configuration interaction. Even though the results obtained by Shim, Dahl and Johansen (1979) on the Ni_2 system indicate that such a possibility cannot be excluded, it seems rather improbable in view of the correct description obtained at the SCF level for $[FeC_5H_5(CO)_2]_2$.

3) $\underline{Cr_2(O_2CH)_4}$. The cage structure of dimetal tetracarboxylates $M_2(O_2CR)_4$ is known with various transition metals. According to a classical scheme [30], when the metal is a d^4 species such as Cr(II), Mo(II) or Re(III), this structure gives rise to a metal-metal quadruple bond : one σ bond, two equivalent π bonds and one weak δ

bond obtained from lateral overlap of the d_{xy} orbitals (z is oriented along the M-M axis). In agreement with the electron population of M-M bonding MO's, the corresponding electronic configuration is called $\sigma^2\pi^4\delta^2$. However, in the case of dichromium tetracarboxylates, the ground state found from ab initio calculations at the SCF level and for the experimental separation of 2.36 Å [31] is not the expected $\sigma^2\pi^4\delta^2$ configuration. A formally nonbonding, apparently dissociative configuration, $\sigma^2\delta^2\delta^{*2}\sigma^{*2}$ (where stars designate an antibonding orbital),has been assumed to be the SCF ground state [32,33]. A detailed analysis of this inversion of states has shown that this unexpected result was due to an incorrect description of the multiple Cr-Cr bond at the SCF level [34]. A correct, non dissociative ground state is restored when configuration interaction is introduced. However, in this CI expansion, the weight of the strongly bonded $\sigma^2\pi^4\delta^2$ configuration is 18 % only at the equilibrium geometry. Consequently, the large contributions of metal-metal nonbonding and antibonding configurations tend to annihilate the strength of the quadruple bond, so that a formal bond order considerably less than 4 could be derived from this wavefunction [35].

Sections of the deformation density distributions by a plane containing the metal atoms and two carboxylate ligands and by a plane bisecting the two ligand planes are respectively displayed in Fig. 11 and 12. Subscripts respectively correspond : a) to the distribution computed from the quadruply bonding configuration at the SCF level ; b) to the distribution computed from the CI expansion ; c) to the experimental distribution obtained by Coppens et al [9] from X-ray diffraction data. The computed maps obtained from SCF and CI wavefunctions are nearly identical in the region of the Cr-O and O-C-O bonds, as expected since orbitals involved in the CI expansion have almost pure metal character. The localization of the density peaks along these bonds and in the region of the oxygen lone pair is in agreement with the picture obtained from diffraction techniques. As usually observed, both electron accumulation and deficiencies are often found larger by the calculation than by the experiment, as thermal motion and limited resolution reduce the sharpest features in the experimental maps [36]. Quantitative discrepancies are also expected to arise from the lack of flexibility of the double-zeta basis set used for valence shells in the theoretical calculation. However, a calculation recently performed at the SCF level (Fig. 13) with a more extended basis set (double-zeta for first-row atoms and internal shells of the metal atoms, triple-zeta for the d-shell) [37] did not modify significantly the computed map even at the quantitative level.

Around the center of the metal-metal axis, the SCF wavefunction exhibits a large, diffuse and continuous accumulation zone over the σ, π (Fig. 11a) and δ (Fig. 12a) bonding regions, culminating at the center of symmetry of the system with a peak of 0.2 eA^{-3}.

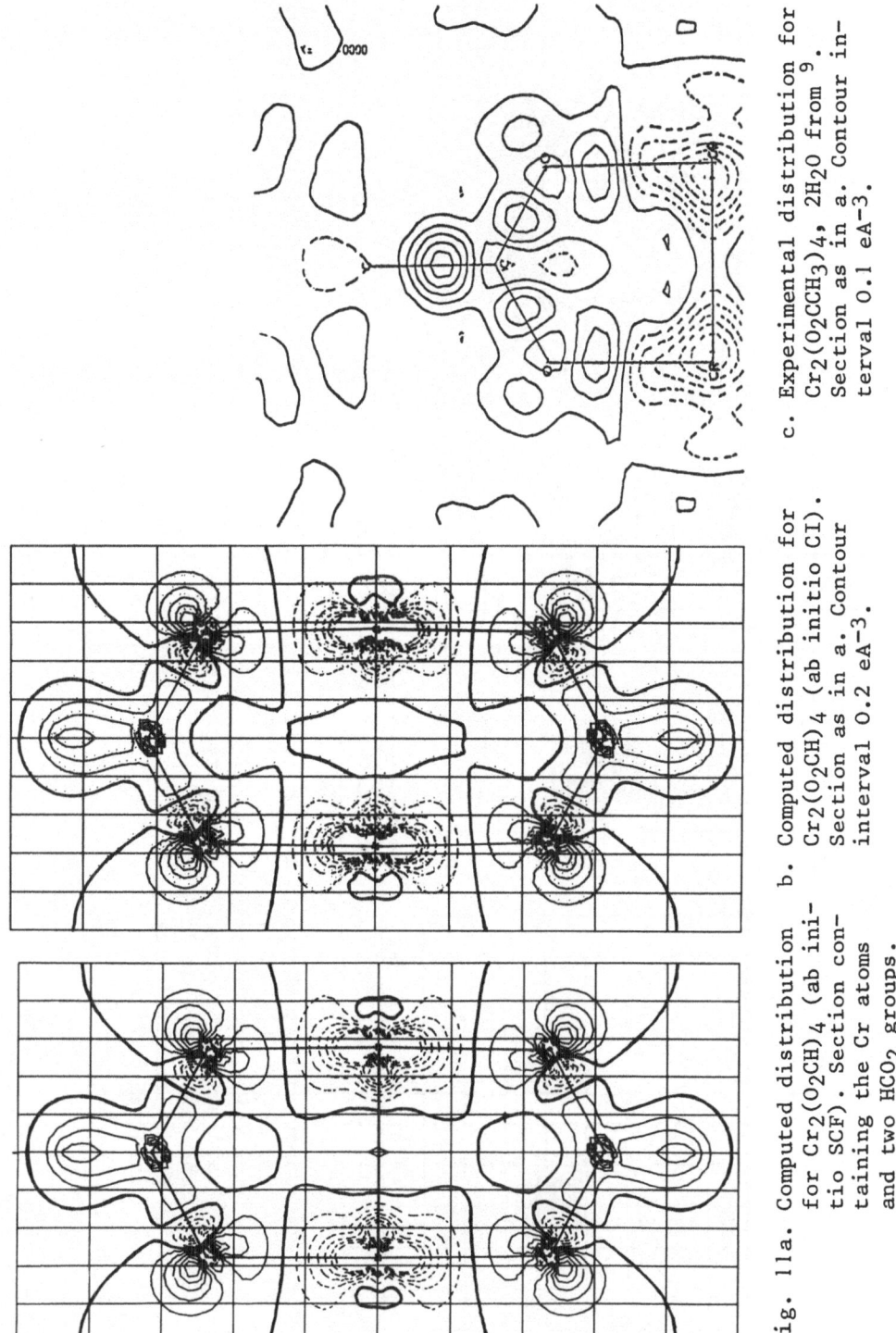

Fig. 11a. Computed distribution for $Cr_2(O_2CH)_4$ (ab initio SCF). Section containing the Cr atoms and two HCO_2 groups.

b. Computed distribution for $Cr_2(O_2CH)_4$ (ab initio CI). Section as in a. Contour interval 0.2 eA^{-3}.

c. Experimental distribution for $Cr_2(O_2CCH_3)_4$, $2H_2O$ from 9. Section as in a. Contour interval 0.1 eA^{-3}.

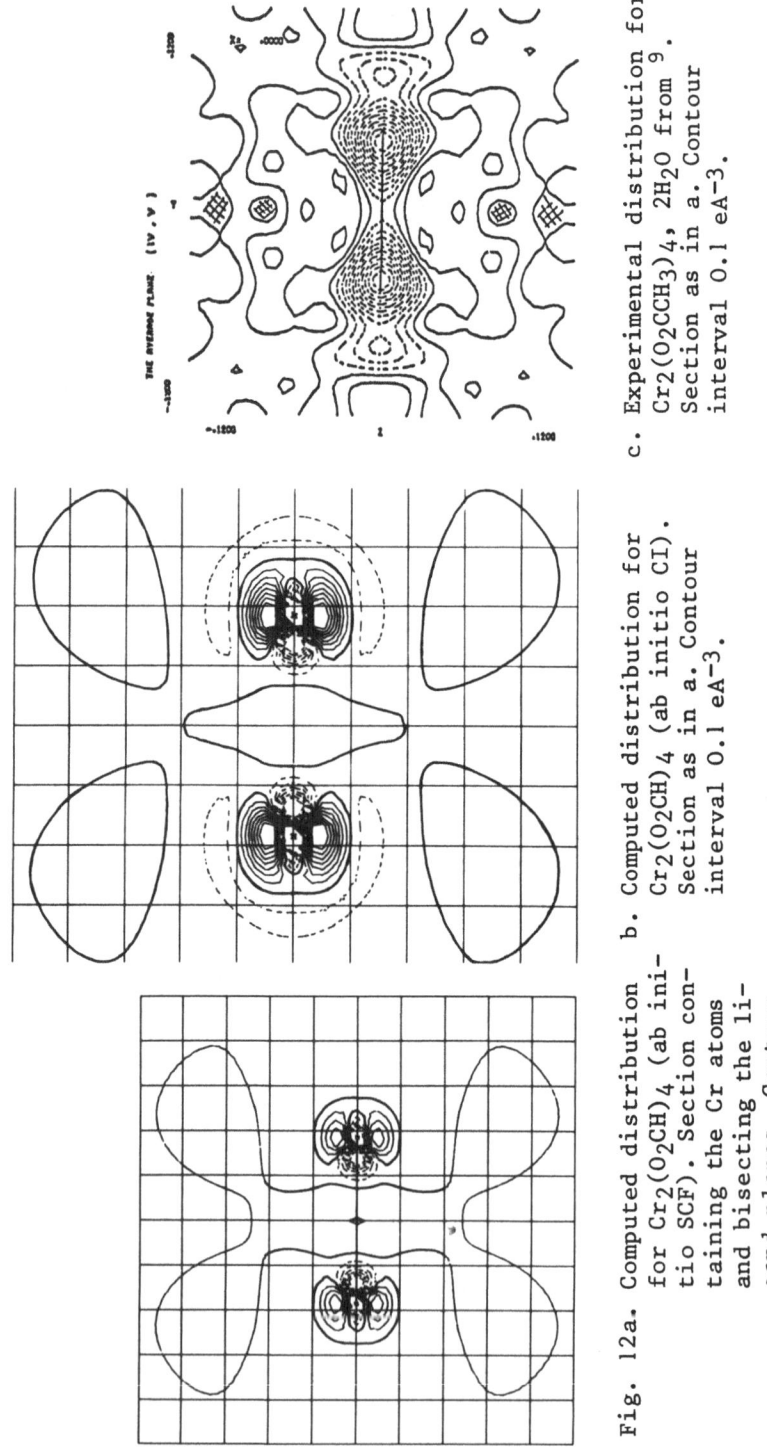

Fig. 12a. Computed distribution for Cr₂(O₂CH)₄ (ab initio SCF). Section containing the Cr atoms and bisecting the ligand planes. Contour interval 0.2 eA⁻³.

b. Computed distribution for Cr₂(O₂CH)₄ (ab initio CI). Section as in a. Contour interval 0.1 eA⁻³.

c. Experimental distribution for Cr₂(O₂CCH₃)₄, 2H₂O from 9. Section as in a. Contour interval 0.1 eA⁻³.

Fig. 13. Computed distribution for $Cr_2(O_2CH)_4$ (ab initio SCF, triple-zeta basis set for metal d shell)[37]. Section as in Fig. 11a. Contour interval 0.1 eA^{-3}.

Since the ground state configuration of the Cr atom is d^5s^1, the average population of each d-orbital is one electron in the five d-orbital system. Each d-orbital, except the $d_{x^2-y^2}$, is involved in a metal–metal bond and consequently their population in the molecule is not expected to deviate significantly from 1 electron. Populations listed in Table 1 confirm this prediction : therefore, contrary to the case of systems discussed above, the residues observed along the metal–metal bonds will be a consequence of intra-orbital rather than interatomic or interorbital charge transfer. This is especially obvious along the z axis, where an electron deficient region is observed in the vicinity of the metal, and an accumulation zone at the center of symmetry of the system. This behaviour is qualitatively similar to what is observed along a σ bond between first-row atoms. However, due to the weak overlap of d-orbitals, the accumulation region is exceedingly smooth with very low peak height and density gradient. The map obtained form the CI wavefunction (Fig. 11b) still displays excess density in the bonding region, but, consistent with the introduction of antibonding contributions, the value of ρ_{max} has still been reduced to 0.1 eA^{-3} and the area of the positive region has become significantly smaller. In agreement with the decrease of electron shielding due to the expansion of the d_{z^2} orbital, a significant contraction of the d_{xy} orbital is visible on computed distributions obtained at both SCF and CI levels (Fig. 12a and 12b).

Some disagreement between the computed and the experimental distributions occurs at the center of symmetry of the system. In both SCF and CI distributions, this point corresponds to the maximum of a large and continuous accumulation region. However, the experimental distribution (Fig. 11c and 12c) depicts this region as slightly electron-deficient and corresponding to a saddle point between two distinct accumulation regions, above and below the Cr-Cr axis. This discrepancy was tentatively attributed to an effect of the axial H_2O ligands which were not introduced into the computed wavefunction [9]. However, a similar discrepancy occurs for $Mo_2(O_2CCH_3)_4$, even though tetracarboxylates of molybdenum are not axially coordinated [38].

4) $\underline{Mo_2(O_2CH)_4}$. The system $Mo_2(O_2CH)_4$ is isoelectronic to $Cr_2(O_2CH)_4$. However, structural and chemical properties (metal-metal distance [39], sensitivity to axial coordination [38,46]) are noticeably different. Consistently, ab initio molecular orbital calculations on $Mo_2(O_2CH)_4$ and $Cr_2(O_2CH)_4$ give a rather different picture of metal–metal bonding. As a matter of fact, the SCF approximation does not fail, as in $Cr_2(O_2CH)_4$ to give the correct ground state corresponding to the quadruply bonding $\sigma^2\pi^4\delta^2$ configuration. Even though the effect of CI expansion is still important from the energy viewpoint, the mixing with nonbonding and antibonding configurations is far from being as crucial as in $Cr_2(O_2CH)_4$.

Therefore, in dimetal tetracarboxylates, the Mo-Mo bond is much
closer than the Cr-Cr bond to the classical model of the quadruple
bond. In a parallel direction, the "non-analiticity effects" [41]
introduced at the SCF level by symmetry constraints and an erroneous
description of the metal-metal bond dissociation are much less
important in this range of M-M distances for the $Mo \equiv Mo$ than for
the $Cr \equiv Cr$ system [34]. Chemically speaking, one could say that,
around 2.10 A, the more diffuse d-orbitals of molybdenum allow a
better overlap than the d-orbitals of chromium. Recent synthesis
of the complexes displaying a "supershort" $Cr \equiv Cr$ bond indicate
that for this type of bonding the region of most efficient overlap between two chromium atoms corresponds to the range
1.84 - 1.90 A [47].

Figures 14 and 15, respectively, display the section of the
density distribution in two planes intersecting along the Mo-Mo
axis, one containing two R-COO groups and the other bisecting the
two ligand planes. Subscript a) shows the theoretical maps obtained
from the ab initio CI wavefunction. Subscript b) corresponds to
sections of the experimental distribution, D_0, obtained by Hino and
Saïto with reference to neutral spherical atoms [15]. Fig. 14c and
15c present another distribution, D_1, obtained by the same authors
from the X-ray determined electron density by substracting the
density given by a superposition of spherical atoms with effective
charges refined by electron population analysis. It must be noticed
than only D_0 can be directly compared to the computed distribution.

The experimental deformation density distribution D_0 obtained
at room temperature [15] displays around the metal atoms big spherical
accumulation regions with steep density gradient (Fig. 14b and 15b).
The origin of these sharp peaks appears controversial since they
could be attributed either to a contraction of the valence electron
density associated with bond formation or with an artefact in
possible relation with the scale factor [42].

The other features of the experimental D_0 distribution can be
easily correlated with the computed maps and with similar characters
observed for $Cr_2(O_2CCH_3)_4$, $2H_2O$. This is the case for the region of
electron deficiency observed around Mo atoms. These negative density
regions are approximately spherical, but much deeper in the direction of the x and y axes. Another, less important minimum is observed along the z axis (z colinear to the metal-metal axis). The
region of electron deficiency along the x and y axes are obviously
associated with the important depopulation of the $d_{x^2-y^2}$ orbital
in the molecule with respect to the spherically averaged atomic
orbital. The Mulliken population analysis of the computed wavefunction indicates that in terms of six d orbitals, the overall population
of the d_{x^2} orbital (3d+4d), identical to that of d_{y^2}, is 1.608
electrons, instead of 2 electrons in the spherical atom. Neverthe-

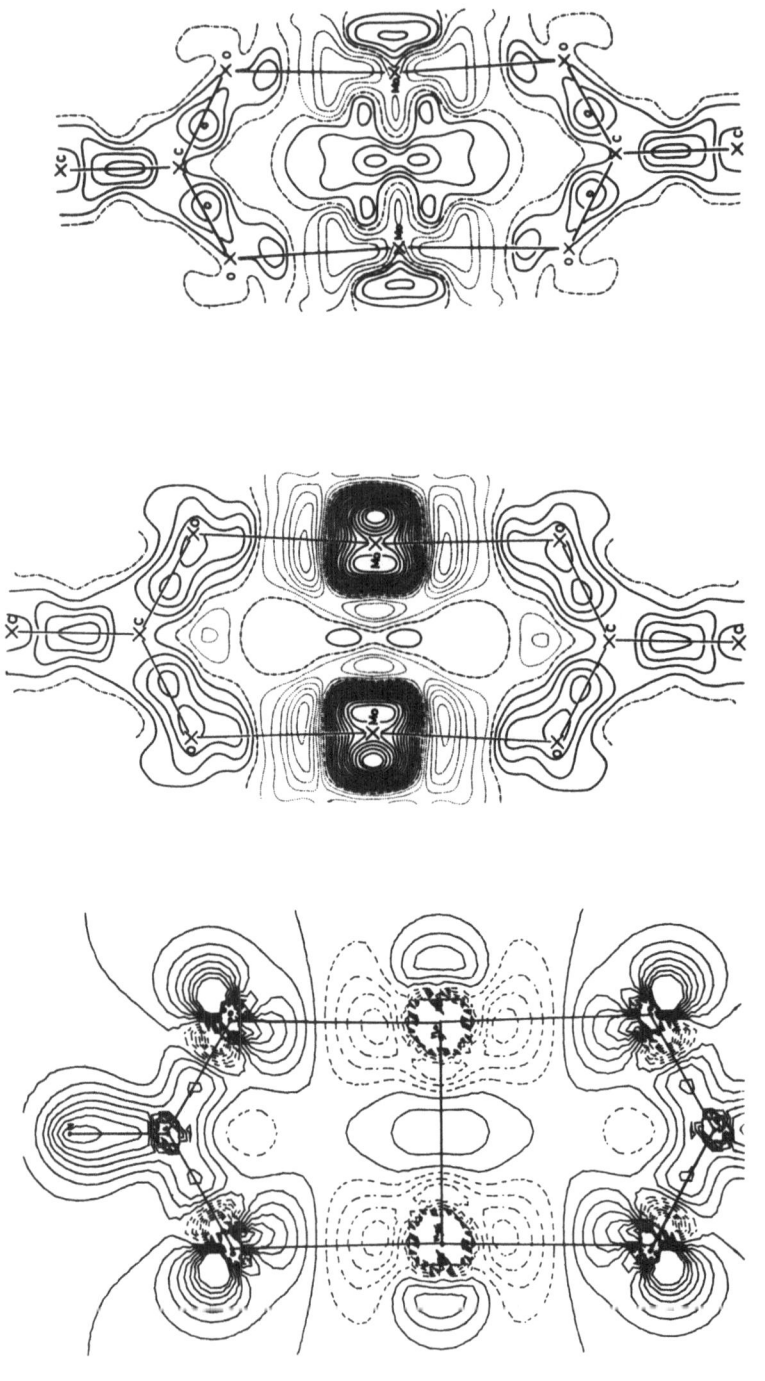

Fig. 14a. Computed distribution for Mo2(O2CH)4 (ab initio CI). Section containing the Mo atoms and two HCO2 groups. Contour interval 0.1 eA^{-3}.

b. Experimental distribution for Mo2(O2CCH3)4 with reference to neutral atoms (D0) [15]. Section as in a. Contour interval 0.1 eA^{-3}.

c. Experimental distribution for Mo2(O2CCH3)4 with reference to charged atoms (D1) [15]. Section as in a. Contour interval 0.1 eA^{-3}.

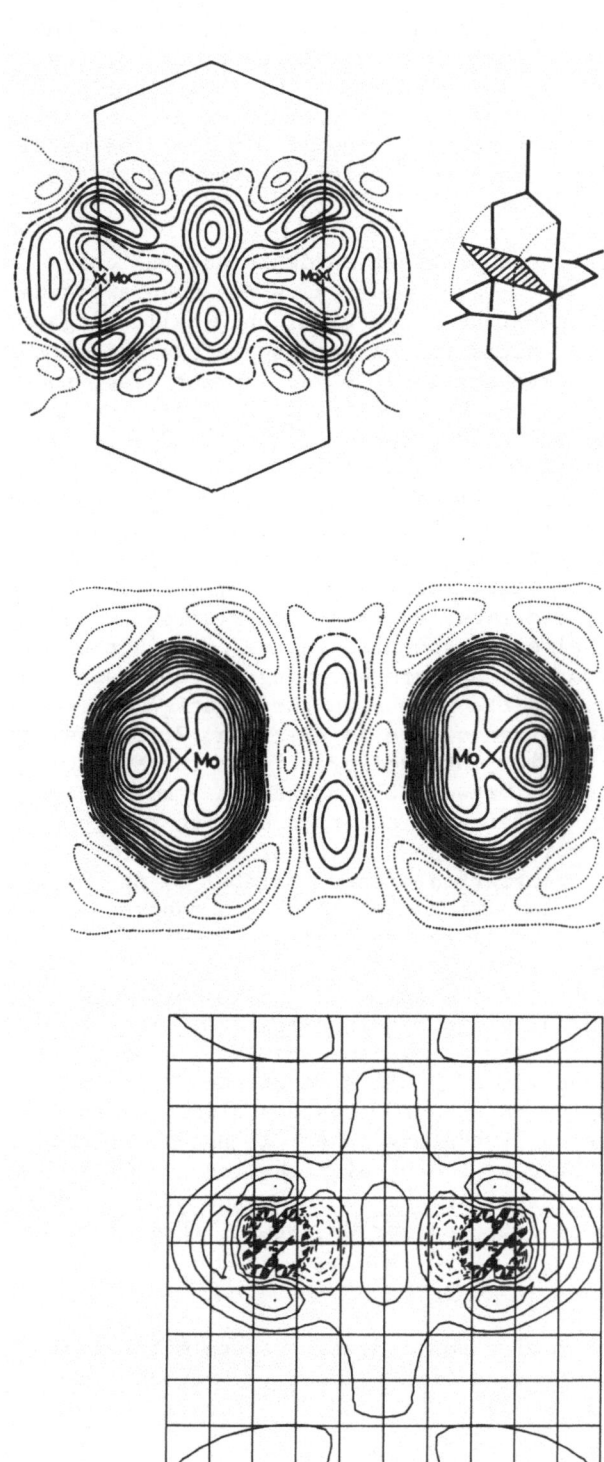

Fig. 15a. Computed distribution for $Mo_2(O_2CH)_4$ (ab initio CI). Section containing the Mo atoms and bisecting the ligand planes. Contour interval 0.1 $eÅ^{-3}$.

b. Experimental distribution for $Mo_2(O_2CCH_3)_4$ with reference to neutral atoms $(D_0)^{15}$. Section as in a. Contour interval 0.1 $eÅ^{-3}$.

c. Experimental distribution of $Mo_2(O_2CCH_3)_4$ with reference to charged atoms $(D_1)^{15}$. Section as in a. Contour interval 0.1 $eÅ^{-3}$.

less, there is a small increase of the overall population of the
d_{z^2} orbital (2.087 electrons instead of 2). However, as for the
tetracarboxylate of chromium, the electron population clearly
moves toward the external part of the d_{z^2} orbital, thus generating
along the z axis an electron deficient zone at 0.5 Å of the metal
and then a positive region around the center of symmetry of the
system.

Even though the occurrence of electron deficient regions
specifically centered along the x, y and z axis has been explained
above, the existence of a continuous spherical and shallow negative
region joining these deeper sites raises once again the question
of the external s orbital depopulation. In the present case, some
insight can be obtained from the D_1 experimental deformation density
maps computed by Hino and Saïto [15] from the molecular density by
subtracting the contribution of <u>charged</u> spherical atoms (Fig. 14c
and 15c). Comparing to the experimental D_0 maps (Fig. 14b and 15b),
the spherical negative environment has vanished, except along the
axes. Furthermore, the height of the accumulation zone located in
the bonding region has been significantly enhanced, from 0.2 to
0.4 eA^{-3}. Since in the D_1 distribution, the depopulation of the
molybdenum 4s orbital is already accounted in the contribution of
the charged spherical atoms, it appears enticing to correlate the
rise of the central peaks and the breaks in the negative environment to the previous elimination of the 4s electron. Similarly,
in the theoretical distribution obtained by Heijser for $Mn_2(CO)_{10}$[43],
the small accumulation region between the manganese atoms increases
somewhat when the density difference is performed with respect to
two $Mn(CO)_5$ fragments in which the population of the external s
orbital has already been transferred either to metal d- or to ligand
orbitals. A similar effect is noticed when isolated manganese atoms
are introduced with a $d^7 s^0$ configuration instead of $d^5 s^2$ [43].*

Another characteristic of the computed density map might be
of interest from the chemical viewpoint in comparison with chromium
tetracarboxylates. A significant charge accumulation (0.25 eA^{-3}) is
visible along the z axis on the external side of each Mo atom
(Fig. 14a and 15a). This region is unfortunately hidden on the
experimental D_0 distribution by the big spherical accumulation
region discussed above. However, the D_1 distribution displays a
similar positive residues (Fig. 14c and 15c). On the other hand,
no similar feature had been detected for $Cr_2(O_2CH)_4$ (Fig. 11b and

---

* A significant build-up of electron density between metal atoms
when electron density is computed with respect to fragments has
also been found by M. Hall for $[Co(CO)_3]_3CH$ (see the present
proceedings).

12b). This could be correlated with the difference in sensitivity to axial coordination of chromium and molybdenum tetracarboxylates, extensively documented by Cotton, et al [38]. These authors have demonstrated that tetracarboxylates of chromium were extremely sensitive to axial coordination. Most of these complexes are either actually associated with axial ligands or displaying strong inter-molecular interactions in axial position. On the other hand, tetra-carboxylates of molybdenum are almost completely insensitive to axial coordination. This could be correlated with a shielding effect due to the electron-rich region close to the molybdenum atom and protecting the metal from the approach of a ligand in axial position.

Finally, it can be pointed out that, even though the metal-metal bond is stronger and the distance shorter in $Mo_2(O_2CH)_4$ than in $Cr_2(O_2CH)_4$ [44], the peak height at the center of symmetry is approximately the same in both computed distributions, at the CI level. If the external s orbital depopulation effect is supposed to be similar in both cases (Table 1), it appears then plausible that the bigger strength in Mo-Mo bonding is balanced by the more diffuse character of the overlapping d orbitals, thus yielding a similar pattern for the deformation density distribution.

CONCLUSION

The early studies performed on deformation density distribu-tion of binuclear complexes have raised rather puzzling questions, so unexpected were the results. Most of these questions were re-lated with the lack of significant accumulation along metal-metal bonds. Such a visualization of the orbital overlap had been expected essentially from an analogy with the well documented carbon-carbon bond, and, more generally, with covalent bonds involving first-row atoms. Even though diffraction experiments on binuclear complexes are still scattered, the lack of significant charge accumulation at the center now appears to be a specific pattern of the metal-metal bond. For complexes displaying a double carbonyl bridge between the metal atoms, the lack of charge accumulation can be correlated with the absence of a direct metal-metal bond detected in this type of complexes. However, the examples of $Mn_2(CO)_{10}$, $Cr_2(O_2CH)_4$ and $Mo_2(O_2CH)_4$ have shown that the phenomenon was not restricted to carbonyl-bridged complexes and required a more general investigation. The systematic comparative analysis of experimental and computed deformation density distributions has provided a rationalization based on the following points :
– metal-metal distances are usually very large and d-d overlaps very weak compared to first-row atom valence orbitals. This will favor large and diffuse rather than steep accumulation regions.
– even though intra-orbital charge transfers (orbital expansion) can still be detected in d-orbitals in relation with bonding, this

effect is in most cases partially or totally hidden by <u>inter-orbital</u> charge transfers occuring when the neutral, spherically averaged metal atom transforms into a charged, polarized metal atom engaged in a complex. This phenomenon makes the deformation density distribution extremely dependent upon the electron configuration of the neutral metal atom in its ground state. - Among these inter-orbital charge transfer processes, the quasi complete depopulation of the external s orbital of the metal occurs in most transition metal complexes. Since this orbital is exceedingly diffuse, its depopulation tends to induce a shallow, spherical negative environment between 0.5 and 1 Å from each metal atom and contributes to annihilate the charge accumulation at the center of metal-metal bond.

It can be finally examined whether and in which conditions it might be possible to find a real and significant charge accumulatinn along a metal-metal bond. A first possibility would be to induce from the spherically averaged atom a net increase of electron population into the d orbital involved in metal-metal bonding. This would be realised with an atom displaying a ground state configuration $d^n s^2$ with n < 5, thus yielding an average population per atomic orbital less than 1 electron. Therefore, a single σ bond between two titanium or two vanadium atoms would probably display a significant charge accumulation along the bond, probably with a double maximum. *

Another possibility, more similar to the classical scheme of a covalent bond between first row atoms would be to detect a particularly enhanced orbital expansion, yielding a significant peak at the center of the metal-metal bond. According to the preliminary X-ray diffraction experiment results obtained by Rees and Mitschler [45], it seems that the "supershort" Cr $\equiv$ Cr bonds recently discovered by Cotton et al. [47] are good candidates for displaying such a high central charge accumulation.

ACKNOWLEDGMENTS

It is a pleasure for me to thank the various laboratories which collected and kindly provided to me the experimental results

---

* The position of each maximum would correspond to the point of highest electron density in the atomic orbitals. An experimental study of Ti_2O_3 for which a single metal-metal bond is assumed shows the occurrence of such a double maximum between the titanium atoms. However, the authors do not interpret this feature in terms of orbital populations [49].

referred to in this article, namely P. Coppens and coworkers (State University of New-York at Buffalo, USA), Y. Saïto and K. Hino (Institute for Solid State Physics at the University of Tokyo, Japan), B. Rees and coworkers (Laboratoire de Cristallochimie et Chimie Structurale, Université Louis Pasteur, Strasbourg, France).

All the calculations have been carried out at the Centre de Calcul de Strasbourg-Cronenbourg. I thank the staff for their cooperation.

REFERENCES

1. Presented at the Symposium "Electron Distributions and the Chemical Bond" 181st ACS National Meeting, Atlanta (1981).
2. P. Coppens, MTP International Review of Science : Physical Chemistry, Ser. 2, Vol. 11 (1975).
3. P. Coppens, E.D. Stevens, Adv. Quantum Chem. Vol. 10 (Academic Press), pp. 1-35 (1977).
4. Y. Wang, P. Coppens, Inorg. Chem. 15:1122 (1976).
5. B. Rees, Acta Crystallogr. A32:483 (1976).
6. J.L. Staudenmann, P. Coppens, J. Muller, J. Solid State Comm. 19:29 (1976).
7. W. Geibel, G. Wilke, R. Goddard, C. Krüger, R. Mynott, J. Organometal. Chem. 160:139 (1978).
8. A. Mitschler, B. Rees, M.S. Lehmann, J. Am. Chem. Soc. 100:3390 (1978).
9. M. Bénard, P. Coppens, M.L. De Lucia, E.D. Stevens, Inorg. Chem. 19:1924 (1980).
10. E.D. Stevens, J. Rys, P. Coppens, J. Am. Chem. Soc. 100:2324 (1978).
11. F.L. Hirshfield, S. Rzotkiewicz, Molec. Phys. 27:1319 (1974).
12. M. Bénard, A. Dedieu, J. Demuynck, M.-M. Rohmer, A. Strich, A. Veillard, ASTERIX, a system of programs for the Univac 1110, unpublished.
    M. Bénard, M. Barry, Comp. Chem. 3:121 (1979).
13. J.M. Coleman, L.F. Dahl, J. Am. Chem. Soc. 89:542 (1967). L.F. Dahl, E.R. De Gil, R.D. Feltham, J. Am. Chem. Soc. 91:1653 (1969).
14. M. Bénard, Inorg. Chem. 18:2782 (1979).
15. K. Hino, Y. Saïto, M. Bénard, Acta Crystallogr., in press.
16. M. Bénard, J. Am. Chem. Soc. 100:7740 (1978).
17. B.K. Teo, M.B. Hall, R.F. Fenske, L.F. Dahl, Inorg. Chem. 14:3103 (1975).
18. P. Ros, unpublished (1979).
19. The LCAO-Hartree-Fock Slater method [20] replaces the nonlocal exchange potential occuring in Hartree-Fock equations by the Xα exchange potential. However, the HFS equations are not solved by introducing the muffin-tin approximation (scattered wave Xα method [21]), which is known to give a

rather poor representation of the true molecular potential[22].
These equations are expanded in linear combinations of
STO's and solved through the discrete variation method [23].

20. E.J. Baerends, P. Ros, Chem. Phys. 2:52 (1973).
    Int. J. Quantum Chem. (Symp.) 12:169 (1978).

21. K.H. Johnson, J. Chem. Phys. 45:3085 (1966).

22. E.J. Baerends, P. Ros, Chem. Phys. 8:412 (1975).

23. D.E. Ellis, G.S. Painter, Phys. Rev. B2:2887 (1970).
    Computational Methods in Band Theory (Plenum, New-York),
    pp. 271-277 (1971).

24. The population of this orbital also depends to some extent
    on the population of d_{x2} and d_{y2} which are more difficult
    to estimate from the experimental maps. However, the expe-
    rimental distribution does not display any significant
    electron deficient region susceptible to decrease the popu-
    lation of the $d_{2z2-x2-y2}$ orbital [4].

25. Except in specific cases, such as complexes of chromium, molyb-
    denum, manganese or any metal displaying in its ground
    state configuration an average population of 1 electron per
    d-orbital. In such cases the orbital expansion phenomena
    along the bond axis are not masked anymore by differences
    in orbital population between the molecule and the spheri-
    cally averaged atom (see for instance the discussion concer-
    ning $Cr_2(O_2CH)_4$ and $Mo_2(O_2CH)_4$).

26. The overall d-electron population of nickel is increased from
    8 el. in the atom to 8.86 el. in the molecule. The electron
    deficient environment external to the peaks related with
    d-orbitals is obviously due to the decrease in population,
    of the diffuse 4s orbital.

27. E. Clementi, IBM J. Res. Devel. Suppl. 9:2 (1965).

28. It has been shown [29] that addition of polarization functions
    to the double-zeta STO basis set did not modify signifi-
    cantly the electron density distribution.

29. W. Heijser, E.J. Baerends, P. Ros, J. Molec. Struct. 63:109
    (1980).

30. F.A. Cotton, Inorg. Chem. 4:334 (1965).

31. F.A. Cotton, B.G. De Boer, M.D. La Prade, J.R. Pipal,
    D.A. Ucko, J. Am. Chem. Soc. 92:2926 (1970).
    Acta Crystallogr. B27:1664 (1971).

32. M. Bénard, J. Am. Chem. Soc. 100:2354 (1978).

33. C.D. Garner, I.H. Hillier, M.F. Guest, J.C. Green, A.W.
    Coleman, Chem. Phys. Letters 41:91 (1976).

34. M. Bénard, J. Chem. Phys. 71:2546 (1979).

35. P.J. Hay, J. Am. Chem. Soc. 100:2897 (1978).

36. B. Rees, A. Mitschler, J. Am. Chem. Soc. 98:7918 (1976).
    P. Coppens, E.D. Stevens, Israël J. Chem. 16:175 (1977)
    E.D. Stevens, J. Rys, P. Coppens, Acta Crystallogr.
    A33:333 (1977).

37. R. Wiest, M. Bénard, unpublished (1980).

38. F.A. Cotton, M. Extine, L.D. Gage, Inorg. Chem. 17:172, 17;176 (1978).

39. 2.36 Å for $Cr_2(O_2CCH_3)_4$, $2H_2O$ [31], 2.09 Å for $Mo_2(O_2CCH_3)_4$ [40].

40. F.A. Cotton, Z.C. Mester, T.R. Webb, Acta Crystallogr. B30:2768 (1974).

41. A.R. Gregory, Chem. Phys. Letters 11:271 (1971). G. Berthier, B. Levy, L. Praud, Gazzetta Chim. Ital. 108:377 (1978).

42. B. Rees, Acta Crystallogr. A32:483 (1976) ; A34:254 (1978) ; Israël J. Chem. 16:180 (1977).

43. W. Heijser, Ph. D. Thesis, Amsterdam (1979). W. Heijser, E.J. Baerends, P. Ros, Proceedings of the Faraday Symposium 14, Manchester (UK), in press (1980).

44. In fact, contrary to the case of $Cr_2(O_2CH)_4$, the CI expansion did not yield any significant modification of the deformation density distribution, relatively to the quadruply bonding configuration obtained at the SCF level. This confirms that the electronic structure of $Mo_2(O_2CH)_4$ resembles more the classical scheme of the quadruple bond than that of $Cr_2(O_2CH)_4$.

45. B. Rees, A. Mitschler, unpublished (1980).

46. F.A. Cotton, W.H. Ilsley, W. Kaim, J. Am. Chem. Soc. 102:3464 (1980).

47. F.A. Cotton, S. Koch, M. Millar, J. Am. Chem. Soc. 99:7372 (1977). A. Bino, F.A. Cotton, W. Kaim, J. Am. Chem. Soc. 101:2506 (1979).

48. W.C. Trogler, J. Chem. Education 57:425 (1980).

49. M.G. Vincent, K. Yvon, A. Grüttner, J. Ashkenazi, Acta Cryst. A36:803 (1980).

EXPERIMENTAL VERSUS THEORETICAL ELECTRON DENSITIES:

METHODS AND ERRORS [+]

Martin Breitenstein, Helmut Dannöhl, Hermann Meyer,
Armin Schweig, and Werner Zittlau

Fachbereich Physikalische Chemie, Universität Marburg
D-3550 Marburg, W.-Germany

INTRODUCTION

The experimental and theoretical determination of electron
densities of atoms and molecules has met with an ever increasing
interest over the last twenty years. Many review articles dealing
with theoretical and/or experimental aspects are available [2-17].
Since 1975 we have been working on the problem of calculating electron densities (in the form of deformation densities) that agree
with experimental densities within the experimental accuracy. Steps
to be taken on this way were: (1) Finding some insight into the
accuracy of theoretical and experimental densities, (2) developing
a simple method for thermal smearing of the calculated static densities, (3) estimating the influence of electron correlation on
electron densities, (4) finding and testing a suitable economical
basis set (i.e. the 4-31G basis set augmented by bond functions,
BF's, or bond polarization functions, BP's) for approximate Hartree-
Fock (AHF) calculations of electron densities, and (5) estimating
the influence of crystal forces on electron densities in salts.
Below we present some more general aspects of this work whereas
calculational details will be given elsewhere [18].

[+] Part 14 of "Comparison of Observed and Calculated Electron
Densities". Part 13: see ref.1.

METHODS

Density distributions (A) to (E) are of importance in electron density work [+]:

(A) The one-electron probability density distribution or simply
the electron density $\rho(\vec{r})$ is defined as dp/dv where dp is the probability of finding an electron in a volume element dv at a point
$\vec{r}$ in space. $\rho(\vec{r})$ is given by (using $\vec{r} \equiv \vec{r}_1$)

$$\rho(\vec{r}) = N\!\int |\Psi(\vec{r}_1,s_1,\vec{r}_2,s_2,\ldots,\vec{r}_N,s_N)|^2$$
$$\cdot\, ds_1 ds_2 \ldots ds_N d\vec{r}_2 d\vec{r}_3 \ldots d\vec{r}_N \tag{1}$$

$\Psi(\vec{r}_1,s_1,\vec{r}_2,s_2\ldots\vec{r}_N,s_N)$ is the many electron (N electron) wavefunction of the system. $\vec{r}_1,\vec{r}_2,\ldots,\vec{r}_N$ and $s_1,s_2,\ldots,s_N$ are the space
and spin coordinates of the electrons, respectively.

(B) The electron radial distribution function $D(r)$ is by definition equal to dp/dr where dp is the probability of finding an electron at a distance between r and $r + dr$ from the origin of the coordinate system. $D(r)$ is calculated from $\rho(\vec{r})$ through

$$D(r) = r^2 \!\int \rho(\vec{r}) d\Omega \tag{2}$$

with $d\Omega = \sin\theta\, d\theta\, d\phi$. θ and ϕ are the polar and azimuthal angles of $\vec{r}$
in a spherical coordinate system.

(C) The two-electron probability density distribution or simply
the two-electron density $\Gamma(\vec{r}_1,\vec{r}_2)$ is defined as $dp/dv_1 dv_2$ where dp
is the probability of finding an electron in a volume element dv_1
at point $\vec{r}_1$ and simultaneously another electron in a volume element
dv_2 at point $\vec{r}_2$. $\Gamma(\vec{r}_1,\vec{r}_2)$ is given by

$$\Gamma(\vec{r}_1,\vec{r}_2) = \frac{N(N-1)}{2} \int |\Psi(\vec{r}_1,s_1,\vec{r}_2,s_2,\ldots,\vec{r}_N,s_N)|^2$$
$$\cdot\, ds_1 ds_2 \ldots ds_N d\vec{r}_3 d\vec{r}_4 \ldots d\vec{r}_N \tag{3}$$

(D) The electron pair probability density $\rho_c(\vec{r})$ is equal to dp/dv
where dp is the probability of finding an electron in a volume element dv at a position $\vec{r}$ relative to another electron that can be
found anywhere in space. $\rho_c(\vec{r})$ is available from $\Gamma(\vec{r}_1,\vec{r}_2)$ through

---

[+] See, e.g., ref.3.

$$\rho_c(\vec{r}) = \int \Gamma(\vec{r},\vec{R}) d\vec{R} \tag{4}$$

where $\vec{r} = \vec{r}_1 - \vec{r}_2$ and $\vec{R} = (\vec{r}_1 + \vec{r}_2)/2$.

(E)   The electron pair correlation function P(r) is defined as dp/dr where dp is the probability of finding two electrons separated by a distance lying between r and r + dr. P(r) is available from $\rho_c(\vec{r})$ through

$$P(r) = r^2 \int \rho_c(\vec{r}) d\Omega \tag{5}$$

Note that $\rho(\vec{r})$ and D(r) are normalized to the number of electrons in the system and $\Gamma(\vec{r}_1,\vec{r}_2)$, $\rho_c(\vec{r})$ and P(r) to the number of electron pairs (ij and ji taken as one electron pair).

Calculations of $\rho(\vec{r})$ are usually based on eq.(1). Ψ can be obtained from Hartree-Fock (HF as e.g. for atoms)-, approximate Hartree-Fock (AHF by using Roothaans' method and limited Slater-type orbital [STO] or uncontracted Gaussian-type orbital [GTO] or contracted Gaussian-type orbital [CGTO] basis sets)- or AHF-calculations with subsequent configuration interaction (CI) calculations [+]. In these cases we refer to the corresponding densities as $\rho_{HF}(\vec{r})$-, $\rho_{AHF}(\vec{r})$-, and $\rho_{AHF\ CI}(\vec{r})$ densities.

Experimental determinations of one-electron densities are chiefly based on intensity measurements of X-ray photons elastically scattered by a crystal. The basic relationships are [*]:

$$I_{elast}(\vec{H}) = |F(\vec{H})|^2$$

$$= |\int_{\substack{unit \\ cell}} \rho(\vec{r}) \exp(2\pi i \vec{H}\cdot\vec{r}) d\vec{r}|^2 \tag{6}$$

$$\rho(\vec{r}) = V^{-1} \sum_H F(\vec{H}) \exp(-2\pi i \vec{H}\cdot\vec{r}) \tag{7}$$

The elastically scattered photon intensity I_{elast} is expressed in units of the scattering intensity of a classical electron. $\vec{H}$ is the reciprocal-lattice vector. $F(\vec{H})$ is the structure factor which is according to eq.(6) generally complex. V is the volume of the unit cell.

[+]   See textbooks of quantum chemistry.

[*]   See, e.g., ref.12.

The sum in eq.(7) runs over all vectors $\vec{H}$. Note from eqs.(6) and (7) that the scattering intensity only provides the amplitudes of the structure factors whereas, for the computation of the electron density, the phases are needed, too (phase problem). Furthermore, the Fourier series of eq.(7) must be truncated in practical cases (truncation- or termination-error) and the amplitudes of $F(\vec{H})$ must be put on an absolute scale (scaling). Finally it should be noted that $\rho(\vec{r})$ derived from eq.(7) is the thermally smeared or dynamic (i.e. vibrationally averaged) density which is often written as $<\rho(\vec{r})>$.

Thus in principle (note, however, the phase problem, the truncation and scaling problems mentioned above as well as further problems connected with the measurement of I_{elast}) $<\rho(\vec{r})>$ can be obtained from an X-ray scattering experiment, on the base of eqs. (6) and (7)[19-21]. Within this method, that might be referred to as the X-method, the phases are estimated from the simplest electron density model available, i.e., the free-, isolated-, or independent atom model (IAM) [+].

In this model the density is the superimposition of the densities of the constituent atoms centered at the nuclear positions of the system. Commonly spherical averaged densities of the individual atoms are used.

More appropriate model densities (in the sense that $<\rho(\vec{r})>$ is more closely described than in the IAM model) have been proposed and are now extensively used. These are the aspherical (or pseudo) atom density models (with refinements based thereupon) on the one hand[22-42] and the bond density models (with corresponding refinements) on the other[43-52]. We refer to the methods working with these models as M-methods (M from molecule) because molecular densities (in the sense that effects of chemical bonding are included in the model densities) are refined in a least square procedure, thus leading to more appropriate "molecular" phases.

One-electron densities based on eqs.(6) and (7) are "molecular" densities in a crystal, i.e. subject to effects of intermolecular forces, and are thus a priori not the densities of an isolated molecule. For that reason, the recent attempts to derive one-electron densities from high energy electron diffraction measurements in the gas phase or to provide relationships between the intensities of the scattered electrons and one- and two-electron densities of the free molecules merit special attention [*], [53-69].

[+]   See, e.g., ref.8.

[*]   See, e.g., refs. 4, 15, and 17.

The basic relationships are, for the total intensity of scattered electrons $I_{tot}(s)$:

$$I_{tot}(s) = \sum_{A,B} Z_A Z_B j_o(sr_{AB}) - 2 \sum_A Z_A \int D_A(r) j_o(sr) dr$$

$$+ 2 \int P(r) j_o(sr) dr + N \tag{8}$$

$$P(r) - \sum_A Z_A D_A(r) =$$

$$\frac{1}{\pi} \int sr \left[I_{tot}(s) - \sum_{A,B} Z_A Z_B j_o(sr_{AB}) - N \right] \sin srds \tag{9}$$

and for the intensity of elastically scattered electrons $I_{elast}(s)$:

$$I_{elast}(s) = \sum_{A,B} Z_A Z_B j_o(sr_{AB}) - 2 \sum_A Z_A \int D_A(r) j_o(sr) dr$$

$$+ \frac{2N}{N-1} \int P_\rho(r) j_o(sr) dr \tag{10}$$

$$\frac{N}{N-1} P_\rho(r) - \sum_A Z_A D_A(r) =$$

$$\frac{1}{\pi} \int sr \left[I_{elast}(s) - \sum_{A,B} Z_A Z_B j_o(sr_{AB}) \right] \sin srds \tag{11}$$

The intensities $I_{tot}(s)$ and $I_{elast}(s)$ are expressed in units of the scattering intensity of the classical electron. s is the scattering variable, $(4\pi/\lambda)\sin\theta/2$, θ the scattering angle, λ the electron wavelength, Z_A the nuclear charge of nucleus A, $j_o(x)$ the zeroth-order spherical Bessel function and r_{AB} the distance between nuclei A and B. $D_A(r)$ is the electron radial distribution function referred to nucleus A as origin. The summations are over all nuclei of the molecule and the integrations are to be carried out from $r = 0$ to $r = \infty$ and from $s = 0$ to $s = \infty$, respectively. $P_\rho(r)$ can be deduced from eqs. (12) to (14)

$$P_\rho(r) \qquad = r^2 \int \rho_{cp}(\vec{r}) d\Omega \tag{12}$$

$$\rho_{cp}(\vec{r}) \qquad = \int \Gamma_\rho(\vec{r},\vec{R}) d\vec{R} \tag{13}$$

$$\Gamma_\rho(\vec{r}_1,\vec{r}_2) \qquad = \frac{N-1}{2N} \rho(\vec{r}_1)\rho(\vec{r}_2) \tag{14}$$

Eqs.(9) and (11) show that only radial distribution functions of electrons relative to nuclei [D(r)] and relative to other electrons [P(r) or $P_0(r)$] - and not $\rho(\vec{r})$ or $\Gamma(\vec{r}_1,\vec{r}_2)$ - could be deduced from electron scattering experiments. Moreover, only the sum of both sorts of distributions would be available from an experiment. Finally, if P(r) would be derived from an X-ray scattering experiment in the gas-phase, D(r) could be deduced only for homonuclear diatomic molecules. In addition, the deduced distributions would be vibrational averages: <P(r)> and <D(r)>. It should be further mentioned that, in deriving the radial distribution functions from gas phase scattering experiments there are truncation and scaling problems similar to those in the crystal case.

In spite of these principal difficulties insight into one-electron densities of molecules (N_2 and O_2) was sought by electron diffraction in the following way[63,69]. A model density (of the aspherical or pseudoatom variety) was assumed and $I_{tot}(s)$ was written as a function of the parameters of the model density according to eq.(8). The parameters were then determined by a least-squares fit of $I_{tot}(s)$ to the measured intensity distribution. We refer to this procedure as the ED-method.

It must be emphasized that the determination of electron densities from both X-ray diffraction from crystals as well as from electron diffraction from gases has to deal with model densities so that the results might depend upon the models assumed. Hitherto some experience on this problem has been gained in X-ray diffraction work but not in electron diffraction work where only two diatomics (N_2 and O_2) have been considered so far. The main merits of the electron diffraction work appears to lie in the precise intensity measurement so that the quality of wavefunctions (theoretical densities) might be tested against the observed intensity distributions.

Usually, instead of the molecular density $\rho(\vec{r})$ itself one considers the function

$$\Delta\rho(\vec{r}) = \rho(\vec{r}) - \rho(\vec{r})_{IAM} \tag{15}$$

where $\rho(\vec{r})_{IAM}$ is the density distribution of the independent atom model. $\Delta\rho(\vec{r})$ is known as the electron density difference[70], Roux function, bond density[71] or deformation density[35]. Accordingly, the experimentalists calculate $\Delta\rho(\vec{r})_{X-X \text{ or } X-N \text{ or } M-A}$ densities from eq.(16) or $\Delta\rho(\vec{r})_{ED}$ densities from eq.(17) instead of $\rho(\vec{r})_{X \text{ or } M}$ densities from eq.(7) or $\rho(\vec{r})_{ED}$ densities from eq.(8), respectively:

$$\Delta\rho(\vec{r})_{\substack{X-X \text{ or} \\ X-N \text{ or} \\ M-A}} = V^{-1} \sum_H \Delta F(\vec{H}) \exp(-2\pi i\vec{H}\cdot\vec{r}) \tag{16}$$

$$\Delta I_{tot}(s) = -2 \sum_A Z_A \int \Delta D_A(r) j_o(sr) dr + 2\int \Delta P(r) j_o(sr) dr \tag{17}$$

where $\Delta F(\vec{H})$, $\Delta I_{tot}(s)$, $\Delta D_A(r)$, and $\Delta P(r)$ are the differences of the respective molecular and IAM quantities.

If the data needed to calculate the structure factors of the IAM (i.e. positional and thermal parameters) are taken from an independent neutron diffraction study the resulting density is called an X-N density, if the data are taken from a high-order X-ray refinement the final density is termed an X-X density. The pseudoatom or bond density model deformation densities are referred to as M-A densities (A stands for atom).

To summarize, the following density difference functions are presently used in the analyses of electron densities:

(A) The AHF deformation density

$$\Delta\rho_{AHF}(\vec{r}) = \rho_{AHF}(\vec{r}) - \rho_{AHF}(\vec{r})_{IAM} \tag{18}$$

(B) The AHF CI deformation density

$$\Delta\rho_{AHF\ CI}(\vec{r}) = \rho_{AHF\ CI}(\vec{r}) - \rho_{AHF}(\vec{r})_{IAM} \tag{19}$$

(C) The correlation density

This is the density difference between the one-electron densities with and without taking into account electron correlation:

$$\Delta\rho_{corr}(\vec{r}) = \rho_{AHF\ CI}(\vec{r}) - \rho_{AHF}(\vec{r}) \tag{20}$$

(D) The X-N deformation density

(E) The X-X deformation density

(F) The M-A deformation density

(G) The ED deformation density

For the definition and calculation of the density differences (D) to (G), see above.

ERRORS

Much work [+, 72-79] has been devoted to the expected accuracy
of ab initio one-electron properties including one-electron densities and deformation densities[80-89] of an isolated molecule. The
sources of error in such calculations have been divided into the
following categories[79]:

(A)   Born-Oppenheimer approximation

This approximation assumes that the nuclear and electronic motions
are separable, i.e. the total wavefunction is written as a product
of a nuclear and an electronic wavefunction the latter depending
parametrically on the nuclear coordinates. No estimate of the error
introduced into deformation densities by this approximation has
been made so far.

(B)   Relativistic effects

Relativistic effects in the Hamiltonian are normally neglected.
However recently[90] the relativistic effects in the deformation densities of $ZnCu_2$, $CdCu_2$, and $HgCu_2$ have been calculated to be at
most approximately 0.02 e/A^3, 0.03 e/A^3, and 0.14 e/A^3, respectively.

(C)   Variation with internuclear distances

Since (in the Born-Oppenheimer approximation) the electronic wavefunction depends parametrically on the nuclear coordinates, the
deformation density also depends on the internuclear distances.
The effect of nuclear motion on $\Delta\rho(\vec{r})$ is then found by averaging
$\Delta\rho(\vec{r})$ with respect to the nuclear wavefunction. Vibrational averaging has been performed for the diatomics H_2^+ [91], N_2 [92], BeH [*], and
CO [*]. Considerable effects arise only near the nuclei whereas the
bonding regions are little affected. Results on larger molecules
on the same level of approximation are not available.

Note, however, that vibrational averaging over lattice vibrations in crystals is crucial because of the large effects involved
(changes of about 50% in bonding and lone pair regions are quite
common).

(D)   Finite expansion basis set

(E)   Correlation effect

We refer to the difference between the AHF (finite expansion basis
set) and the HF densities of a system in a point of space as the

---

[+]   See ref.4, p.1.

[*]   See ref.10, p.119.

basis set error and to the difference between the exact and the HF
densities as the correlation error (in all cases non-relativistic,
Born-Oppenheimer densities without vibrational averaging).

It had been argued that on the HF level one-electron properties can be calculated for closed-shell molecules with no low-
lying excited singlet states with an accuracy of about 1%[75]. AHF
densities were expected to be accessible with an accuracy of about
2-5% for atoms, diatomics, and small polyatomic molecules [+]. Actual
calculations of densities[18,93] suggest that these values are quite
reasonable in discussing errors of (theoretical) deformation densities.

Table I summarizes for some small molecules the values of the
AHF density $\rho_{AHF}(\vec{r})$, the deformation density $\Delta\rho_{AHF}(\vec{r})$ and the correlation density $\Delta\rho_{corr}(\vec{r})$ in the mid-point of the respective bonds
(or in the peaks of $\Delta\rho_{AHF}(\vec{r})$ for O-H of H_2O and H_2O_2) [18,93]. Details of the AHF calculations including some sets of polarization
functions and the CI calculations leading to about 60-70% of the
correlation energy will be given elsewhere[18]. $\Delta\rho_{corr}(\vec{r})$ may be
considered as a good approximation[18] to the exact correlation error. As it can be seen from Table I $\Delta\rho_{corr}(\vec{r})$ fulfills the 1% rule
of thumb remarkably well, with one notable exception, however, the
F_2 molecule.

For N_2 and CO, HF one-electron densities are available [*,94].
A comparison with our AHF densities shows that the AHF densities
deviate from the exact HF ones by about 0.1 e/A^3 or 2% in the mid-
point of the respective bonds. In these cases, the basis set and
correlation errors partially compensate each other, but in general
one has to take into consideration additivity of these errors,
thus arriving at a total error of about 3%. This nicely fits the
aforementioned 2-5% accuracy anticipated for AHF densities. Since
for most molecules the densities in bonding regions are considerably smaller than for N_2 and CO, the absolute total error will
in general not essentially exceed ±0.1 e/A^3, in the mid-point of
a bond.

The error in the deformation density in the mid-point of a
bond will be nearly the same as in the molecular density itself.
Since in crystallography usually (relativistic) HF atoms are used
the correlation error needs not to be taken into account in the
IAM. Furthermore the AHF density differs only insignificantly (by
about 0.01 e/A^3 for the nitrogen atom, for example) from the (relativistic) HF density for the first-row atoms, at the appropriate
distance from the nucleus.

[+] See ref.4, p.1.

[*] Estimated from orbital density maps.

Table I. One-electron density data in the mid-point of the respective bonds (e/A^3) [a]

Molecule	Bond	$\rho_{AHF}(\vec{r})$ [b]	$\Delta\rho_{AHF}(\vec{r})$ [c]	$\Delta\rho_{corr}(\vec{r})$ [d]	$\dfrac{\Delta\rho_{corr}}{\rho_{AHF}}$ (%)
N_2	N≡N	4.85	1.34	-0.06	-1.2
O_2	O=O	3.70	0.32	-0.09	-2.4
F_2	F–F	1.50	-0.05	-0.10	-6.7
CO	C≡O	4.28	1.12	-0.06	-1.4
H_2O	O–H [e]	2.50	0.71	-0.02	-0.8
HC≡CH	C≡C	2.85	0.86	-0.02	-0.7
	C–H	2.17	0.66	-0.02	-0.9
H_2O_2	O–O	1.88	0.09	-0.04	-2.1
	O–H [e]	2.46	0.71	-0.04	-1.6

a) Based on calculations specified elsewhere [18]

b) AHF one-electron density

c) AHF deformation density from eq.(18) of text

d) Correlation density from eq.(20) of text

e) Density data in the deformation density maximum of the bond (not in the mid-point)

To summarize, for bonding regions of small molecules, the dominating error in the deformation densities is (in spite of large
CGTO basis sets) the basis set error (about 2% of the molecular
electron density). The correlation error (about 1% of the molecular electron density) can be nearly quantitatively corrected by
performing large-scale CI calculations (i.e. taking account of all
singly and doubly excited configurations as, e.g., by using the
PERTCI method[95]). Errors due to the other approximations (Born-Oppenheimer approximation, neglect of relativistic effects and of
vibrational averaging over the internal modes) appear to be negligible for the small molecules considered. However as mentioned
above, vibrational averaging over the lattice vibrations is un-
avoidable. This effect reduces bond densities near the mid-point of
a bond by roughly 50%. Hence we can state that we have presently to
deal with errors of about ± 0.1 e/A^3 in the bonding regions of a molecule at rest and of about ± 0.05 e/A^3 in the same regions for a
molecule moving in a crystal, in the best theoretical AHF CI deformation densities available.

Error analyses of the experimental crystal diffraction density
determinations (in conjunction with finding out optimum experimental conditions or suitable theoretical models for applying corrections) are available [+,96-113].

Errors are connected with the various stages of the experiment
and the calculation of $\Delta\rho(\vec{r})$ from the experimental data, namely
with:

(A) The measurement of intensities

These errors may arise from the experimental set-up, the sample,
and the measurement conditions and can be statistical and/or systematic.

(B) Corrections applied to intensities

These errors are systematic and are associated with the correction
of side effects as absorption, extinction, thermal diffuse scattering (TDS), multiple scattering and anomalous scattering.

(C) Scaling of intensities

(D) Phasing of structure amplitudes

(E) Series termination (in eq.(7))

These errors known as scalefactor-, phase-, and series termination
(or truncation) error are due to experimental limitations and are
systematical errors.

+ See ref.10, p.119; ref.11, pp.103, 149, 154, and 180; ref.17,
  pp.287 and 355.

It has been shown [+] that statistical (random) errors including systematic contributions that influence the error distributions are typically ± 0.05 e/A^3 in the bonding regions of deformation densities, as measured by the standard deviation.

Much less is known about the systematic errors in densities that shift the error distribution but do not change the distribution itself.

The presence of such uncorrected or not appropriately corrected systematic errors can be best detected from comparing all the deformation densities of a system that have been obtained by various research groups using various methods. Two such examples will be presented below.

Figure 1 displays the experimental deformation densities of cyanuric acid that are presently available. Table II summarizes the peak heights in these densities. The cyanuric acid case is particularly informative since all determinations are based on the same experimental data. Nonetheless peak heights generally vary by about 0.1-0.3 e/A^3.

Table III presents the peak heights for tetracyanoethylene. As it is obvious from these data the variation in the peak heights (especially in the C≡N bond) is even more pronounced in this case.

These results must be taken with certain reservations. Systematic errors in a deformation density may arise from errors in $\rho(\vec{r})$ and errors in $\rho(\vec{r})_{IAM}$. Taking the same X-ray data there are two ways to settle the IAM reference density in an M-A density determination. One can fix $\rho(\vec{r})_{IAM}$ to the N (neutron) IAM density or one can use the appropriate M (molecule) IAM density. In the first case the deviation of $\Delta\rho(\vec{r})_{M-A}$ from $\Delta\rho(\vec{r})_{X-N}$ gives an indication of the error in the M density as compared to the X density but no indication of the error in the $\Delta\rho(\vec{r})_{X-N}$ density. The latter might be excellent or drastically wrong. In the second case where a reference state is used that is specific to a certain method the resulting $\Delta\rho(\vec{r})_{M-A}$ can principally be better or worse than the corresponding $\Delta\rho(\vec{r})_{X-N}$ counterpart since there is no guarantee that a neutron reference state is the best choice to be made. Thus a comparison between $\Delta\rho(\vec{r})_{X-N}$ and $\Delta\rho(\vec{r})_{M-A}$ under these conditions is indicative of systematic errors.

Amongst the five deformation density maps of Fig.1 there are three groups of maps: (i) maps (a) and (c), (ii) maps (b) and (e), and (iii) map (d). Each group has its own reference state. Thus strictly only maps (a), (b), and (d) can be used to get an estima-

---

[+] See ref.10, p.119; ref.17, pp.287 and 355

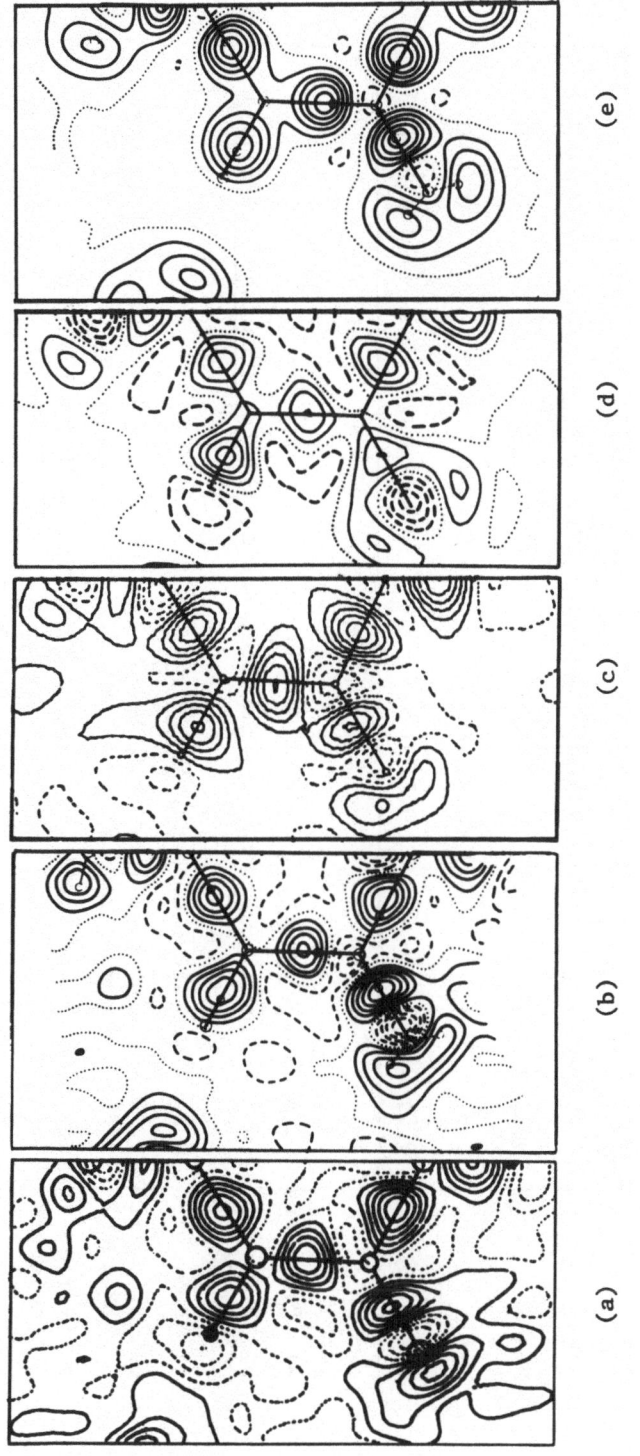

Fig.1.   Experimental low-temperature deformation densities of cyanuric acid in the molecular plane. Only one half of the molecule is shown with one of the carbonyl groups directed upwards. (a), (b) X-N maps taken from refs.114 and 50, respectively. (c), (d), (e) M-A maps taken from refs.115, 47, and 50, respectively. Contour interval 0.1 e/A^3; positive contours are full lines, negative contours are dashed.

(a)   (b)   (c)   (d)   (e)

Table II. Peak heights in $\Delta\rho_{\text{X-N}}$ or M-A (e/A^3) for cyanuric acid (low-temperature data)

Method	C=O [e]	C-N [e]	N-H [e]	lone pair [e]
X-N [a]	0.5, 0.3	0.5, 0.4, 0.5	0.5, 0.4	0.5, 0.3
X-N [b]	0.5, 0.3	0.4, 0.4, 0.4	0.4, 0.4	0.4, 0.3
M-A [c]	0.4, 0.2	0.5, 0.5, 0.4	0.5, 0.5	0.3, 0.2
M-A [d]	0.3, 0.3	0.3, 0.3, 0.3	0.4, 0.4	0.2, 0.2
M-A [b]	0.5, 0.5	0.5, 0.5, 0.5	0.4, 0.4	0.3, 0.3

a) P.Coppens and A.Vos, Acta Cryst. B27:146 (1971).

b) H.Dietrich, C.Scheringer, H.Meyer, K.-W.Schulte, and A.Schweig, Acta Cryst. B35:1191 (1979).

c) D.S.Jones, D.Pautler, and P.Coppens, Acta Cryst. A28:635 (1972).

d) C.Scheringer, A.Kutoglu, E.Hellner, H.L.Hase, K.-W.Schulte, and A.Schweig, Acta Cryst. B34:2162 (1978). C.Scheringer, A.Kutoglu, and E.Hellner, Acta Cryst. B34:2670 (1978).

e) Various values refer to crystallographically different bonds or lone pairs in the molecule.

Table III. Peak heights in $\Delta\rho_{X-N}$ or X-X or M-A (e/A^3) for tetracyanoethylene (room temperature data)

Method	C≡N	C=C	C-C	lone pair
X-N [a]	0.9	0.4	0.6	0.4
X-N [b,c]	0.8	0.3	0.5	0.4
X-N [d,e]	0.6; 0.4[g]	0.3	0.3; 0.2[g]	0.2; 0.2[g]
X-X [d,e]	0.5; 0.2[g]	0.3	0.2; 0.1	0.1
M-A [b,f]	0.8	0.4	0.4	0.4

a) P.Becker, P.Coppens, and F.K.Ross, J.Am.Chem.Soc. 95:7604 (1973).

b) N.K.Hansen and P.Coppens, Acta Cryst. A34:909 (1978).

c) As a) but an improved extinction model used.

d) V.Drück and H.Guth, personal communication.

e) Study on a monoclinic crystal, T = 295 K, $\sin\theta/\lambda = 0.90$ A^{-1}, $R_i(I) = 3.25\%$, $R(F_o) = 3.64\%$.

f) Multipole deformation density from refinement with positional and thermal parameters fixed at neutron values.

g) Values for crystallographically different bonds or lone pairs.

tion of possible systematic errors in the experimental deformation
densities of cyanuric acid.

The X-N [a], X-N [b,c], and M-A [b,f] data of Table III are based on
deformation densities that are referenced to the same state. The
X-N [d,e] and the X-X [d,e] results were calculated from independent
measurements and again both use differing reference states. Thus
the X-N [b,c], X-N [d,e], and the X-X [d,e] results are useful in getting
some insight into possible systematic errors in the experimental
deformation densities of tetracyanoethylene.

One must conclude from these results that the error in experimental deformation densities (mainly due to systematic errors) can
be as large as about $\pm$0.3 e/A^3 in bonding regions. Thus the accuracy of experimental deformation densities can be appreciably lower
than the often quoted precision of about $\pm$0.05 e/A^3.

It is well known that the error from the neglect of TDS, the
scalefactor as well as truncation errors are most serious within
0.3 A from the atomic centers. In order to assess the truncation
error for itself we calculated the static and dynamic deformation
densities of formamide as a function of the experimental resolution
$\sin\theta/\lambda$ with θ being the maximum Bragg angle and λ the X-ray wavelength. The results are shown in Figures 2 and 3 below.

The dynamic density maps (which are the maps to be compared
with experimental maps) of Figure 2 suggest that for $\sin\theta/\lambda = 1.0$
A^{-1} the final deformation density has already been reached in all
regions (including those near the nuclei). Even a resolution of
$\sin\theta/\lambda = 0.8$ A^{-1} leads to deformation density maxima differing by
less than 0.1 e/A^3 from the most exact ones (here $\sin\theta/\lambda = 1.4$ A^{-1}).
Similar results were obtained for other cases[116]. Thus in view of
the other errors made in experimental density determinations the
truncation error appears to be negligible when working with a resolution of at least 0.8 to 1.0 A^{-1}.

Note, however, that the theoretical model is based on numerically exact high-order structure factors and the same temperature
factors for the "molecular" and the IAM structure factors. The experimental situation is less ideal here because it is difficult to
measure high-order reflexes with the same accuracy as the low-order
ones and the determination of the "correct" thermal parameters may
be a problem. Thus the theoretical result has, in practice, to be
taken with a grain of salt.

The convergence of the static deformation densities (Figure 3)
with increasing resolution is much slower than in the dynamic case.
The rapid changes in the deformation density in the nuclear regions
are inadequately described even in the best finite resolution used

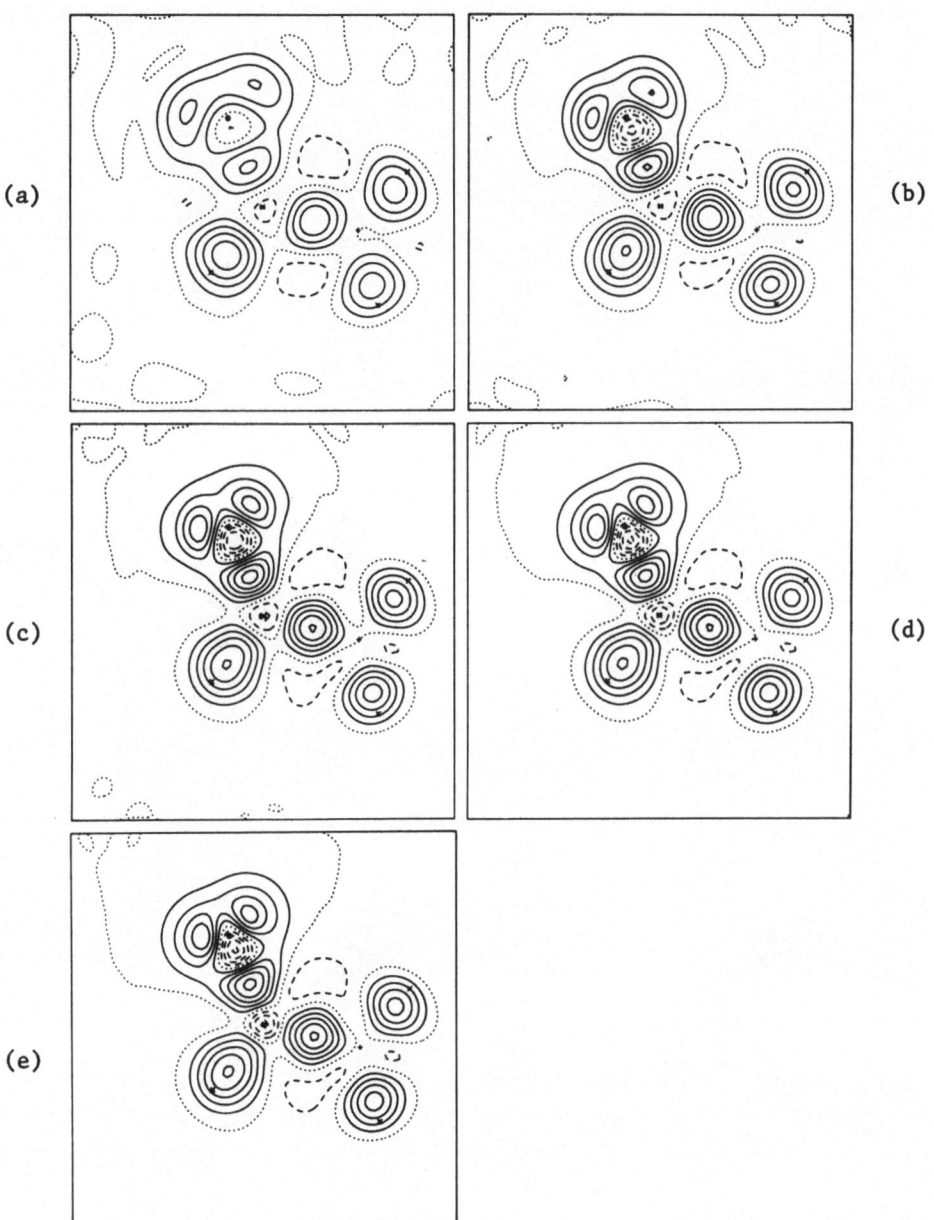

Fig.2.   Series termination effect on the (theoretical) dynamic de-
         formation density of formamide. The $\sin\theta/\lambda$ values used
         are: (a) 0.6 A^{-1}, (b) 0.8 A^{-1}, (c) 1.0 A^{-1}, (d) 1.2 A^{-1},
         (e) 1.4 A^{-1}. Contours as in Fig.1.

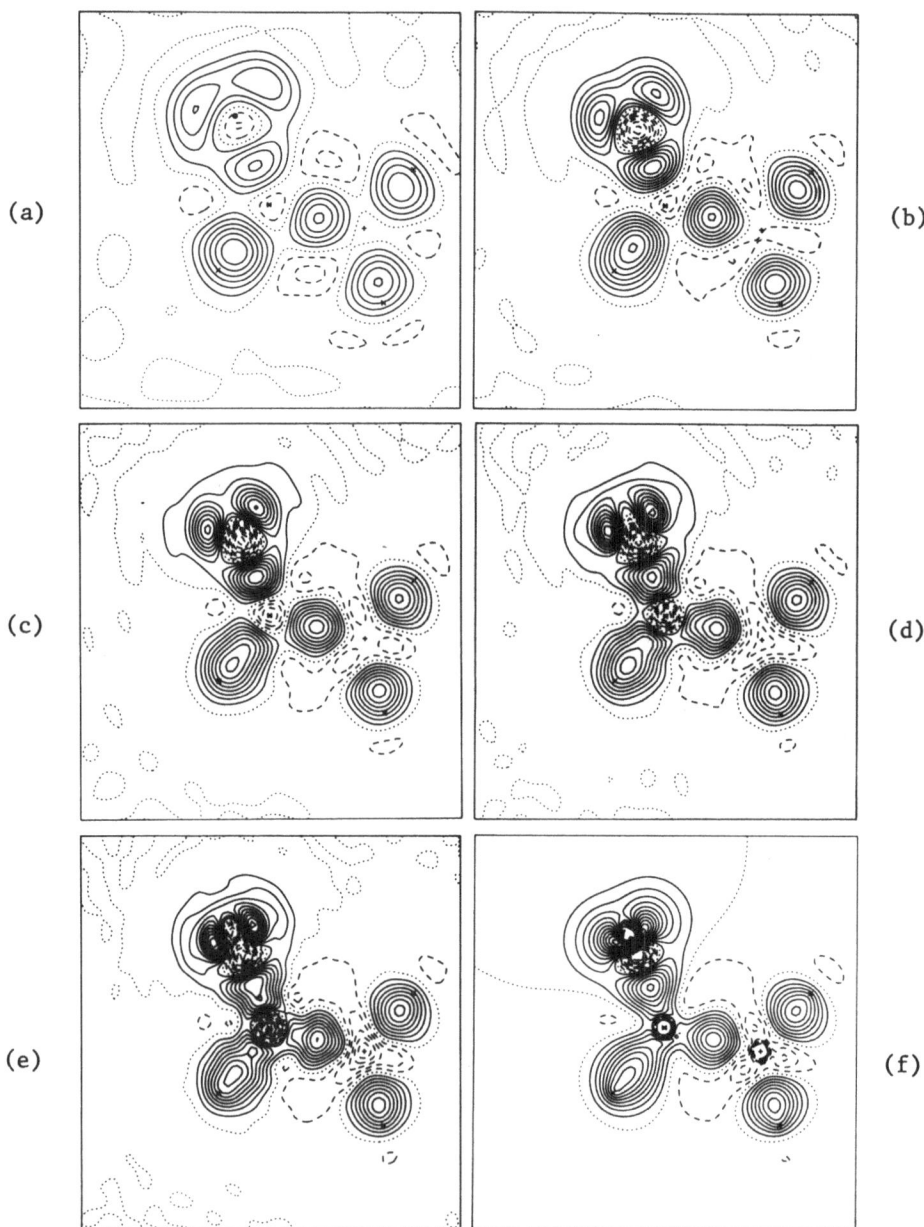

Fig.3.   Series termination effect on the (theoretical) static de-
         formation density of formamide. (a) to (e): $\sin\theta/\lambda$ values
         as in Fig.2.(a)-(e); (f) infinite resolution. Contours as
         in Fig.1.

$(\sin\theta/\lambda = 1.4 \text{ A}^{-1})$. Nevertheless the maxima in the bonding regions are already well reproduced with $\sin\theta/\lambda = 1.0 \text{ A}^{-1}$.

The accuracy of electron diffraction deformation densities has not been assessed so far.

VIBRATIONAL AVERAGING

The basic relationships for vibrational averaging over the internal vibrations (i.v.) and the lattice vibrations (l.v.) are [+]:

$$<\rho(\vec{r})>_{i.v.} = \int \rho(\vec{r};\vec{R}_1,\vec{R}_2,\ldots\vec{R}_M) \, P(\vec{R}_1,\vec{R}_2,\ldots\vec{R}_M)$$

$$\cdot \, d\vec{R}_1 d\vec{R}_2 \ldots d\vec{R}_M \tag{21}$$

$$<\rho(\vec{r})>_{\substack{i.v. \\ l.v.}} = \int <\rho(\vec{r} - \vec{u})>_{i.v.} P(\vec{t},\vec{\omega}) d\vec{t} d\vec{\omega} \tag{22}$$

with

$$\vec{u} = \vec{t} + \vec{\omega} \times \vec{r} \quad \text{(for small } \vec{\omega}) \tag{23}$$

$P(\vec{R}_1,\vec{R}_2,\ldots\vec{R}_M)$ is the probability distribution function of the nuclei $dp/d\vec{R}_1 d\vec{R}_2 \ldots d\vec{R}_M$ where dp is the probability of finding simultaneously nucleus 1 within $d\vec{R}_1$, nucleus 2 within $d\vec{R}_2$, and so on, and nucleus M within $d\vec{R}_M$. $P(\vec{t},\vec{\omega})$ is the rigid body[117,118] probability distribution function $dp/d\vec{t} d\vec{\omega}$ where dp is the probability of finding the rigid body molecule translated by $\vec{t}$ and rotated by $\vec{\omega}$. $\vec{t}$ is the instantaneous translation of the molecular mass center and $\vec{\omega}$ is a vector directed parallel to the axis that describes the instantaneous rotation of the molecule from its equilibrium position. The length of $\vec{\omega}$ is equal to the rotational angle. $\vec{u}$ is the instantaneous displacement of a point $\vec{r}$ in the molecule due to the rigid body translation $\vec{t}$ and the rotation $\vec{\omega}$.

No strictly theoretical calculations based on eqs.(21) and/or (22) have been performed so far because $P(\vec{R}_1,\vec{R}_2,\ldots\vec{R}_M)$ for polyatomic molecules and $P(\vec{t},\vec{\omega})$, which would imply crystal calculations as well as $\rho(\vec{r};\vec{R}_1,\vec{R}_2,\ldots\vec{R}_M)$ as a function of $\vec{R}_1,\vec{R}_2,\ldots\vec{R}_M$, are not generally accessible, for computational reasons. Therefore all attempts made so far [119-128] to achieve vibrational averaging rely

---

[+] See, e.g., ref.11.

on experimental methods as spectroscopic and diffraction methods
for finding the probability distribution functions needed in eqs.
(21) and (22), respectively. Unfortunately, diffraction methods
cannot differentiate between internal and lattice vibrations.

Approximate procedures for the evaluation of eq.(22) along
with eq.(21) are available[101,120,121,123,124,128]. The basic relationships and approximations of the method used in the present authors' group[120, 121] are as follows:

$$\rho(\vec{r}) = \sum_{\mu,\nu} P_{\mu\nu} \phi_\mu \phi_\nu \tag{24}$$

$$F(\vec{H}) = \sum_{\mu,\nu} P_{\mu\nu} f_{\mu\nu} T_{\mu\nu} \exp(2\pi i \vec{H} \cdot \vec{r}_{\mu\nu}) \tag{25}$$

$$T_{\mu\nu} = \exp(-2\pi^2 \vec{H}^T \underline{U}_k \vec{H});$$

$$\phi_\mu \text{ and } \phi_\nu \text{ on atom k} \tag{26}$$

$$T_{\mu\nu} = \exp(-2\pi^2 \vec{H}^T \underline{U}_{k1} \vec{H});$$

$$\phi_\mu \text{ on atom k, } \phi_\nu \text{ on atom 1} \tag{27}$$

$$\underline{U}_{k1} = \underline{T} + \underline{V}_{k1} \underline{L} \underline{V}_{k1}^T - \underline{V}_{k1} \underline{S} - (\underline{V}_{k1} \underline{S})^T \tag{28}$$

$$\underline{V}_{k1} = \begin{pmatrix} 0 & -z_{k1} & y_{k1} \\ z_{k1} & 0 & -x_{k1} \\ -y_{k1} & x_{k1} & 0 \end{pmatrix} \tag{29}$$

$$\underline{U}_{k1} = (\underline{U}_k + \underline{U}_1)/2 \tag{30}$$

$\rho(\vec{r})$ in eq.(24) is the electron density in terms of the den-
sity units $\phi_\mu \phi_\nu$ and the population matrix $P_{\mu\nu}$ where ϕ_μ and ϕ_ν are
members of a GTO basis set. Eq.(25) describes the evaluation of
appropriate structure factors and is based on various model assumptions. It makes use of the convolution approximation[129]. Accordingly, one temperature factor $T_{\mu\nu}$ is assigned to each density
unit $\phi_\mu \phi_\nu$ centered at $\vec{r}_{\mu\nu}$. Formulas for the evaluation of the form
factors $f_{\mu\nu}$ of these density units $\phi_\mu \phi_\nu$ are available in literature [33,130,131]. The $T_{\mu\nu}$ are computed from the experimental high

order X-ray or neutron atomic vibration tensors U_k according to eqs.
(26) to (30). Eqs.(28) to (30) need a further comment. T, L, and S
in eq.(28) are the translation tensor, the libration tensor and a
screw tensor that describes the coupling between translation and
libration, respectively. The tensors are determined from U_k tensors
by a least squares fitting procedure. V_{kl} is an antisymmetric tensor where x_{kl}, y_{kl}, and z_{kl} are the components of the position vector of the   two-centre density $\phi_\mu \phi_\nu$ (ϕ_μ on atom k, ϕ_ν on atom 1).
Alternatively, U_{kl} can be calculated from eq.(30)[132].   In an ac-
tual calculation it is always tried at first to evaluate the U_{kl}
from eqs.(28) and (29), and only if the $T\,L\,S$ analysis is not
satisfactory, eq.(30) is used. Finally, an approximate $\langle\rho(\vec{r})\rangle_{\substack{i.v.\\l.v.}}$

is gained from eq.(7) using the $F(\vec{H})$ of eq.(25).

A few notes are to be added. Since the approximate evaluation
of $\langle\rho(\vec{r})\rangle_{\substack{i.v.\\l.v.}}$ is made in reciprocal space the experimental X-ray

resolution can be taken account of by using the experimental
$\sin\theta/\lambda$ in the averaged density [eq.(7)] calculations. For simplicity, the $F(\vec{H})$ are evaluated in an orthorhombic pseudo-crystal.
One molecule is enclosed in the pseudo-unit cell with lattice parameters chosen large enough as to make overlap of electron density
from neighbouring cells negligible.

Experience has shown that $\langle\Delta\rho(\vec{r})\rangle_{\substack{i.v.\\l.v.}}$ based on $\langle\rho(\vec{r})\rangle_{\substack{i.v.\\l.v.}}$
calculated in this way is rather insensitive to the particular mo-
del chosen for the evaluation of the U_{kl} and, within the $T\,L\,S$ model, to the positioning of the two-centre densities. Therefore,
the U_{kl} computed for the mid-point of the bonds are used throughout.

$\langle\Delta\rho(\vec{r})\rangle_{\substack{i.v.\\l.v.}}$ has been compared to experimental deformation

densities in quite a few cases; the agreement is satisfactory within the experimental error if appropriate GTO basis sets are used [+].

Till now, no comparisons between the various theoretical procedures[101,120,121,123,124,128] are available in literature. For
that reason such a first comparison, however limited to the methods of refs.120, 121, and 123, is presented in Figure 4 for the
4-31G density of oxalic acid. As it can be seen both vibrationally
averaged densities [*] are practically the same.

[+] See, e.g., ref.18.

[*] Fig.4.(a) and (b) are taken from refs.133 and 134, respectively.

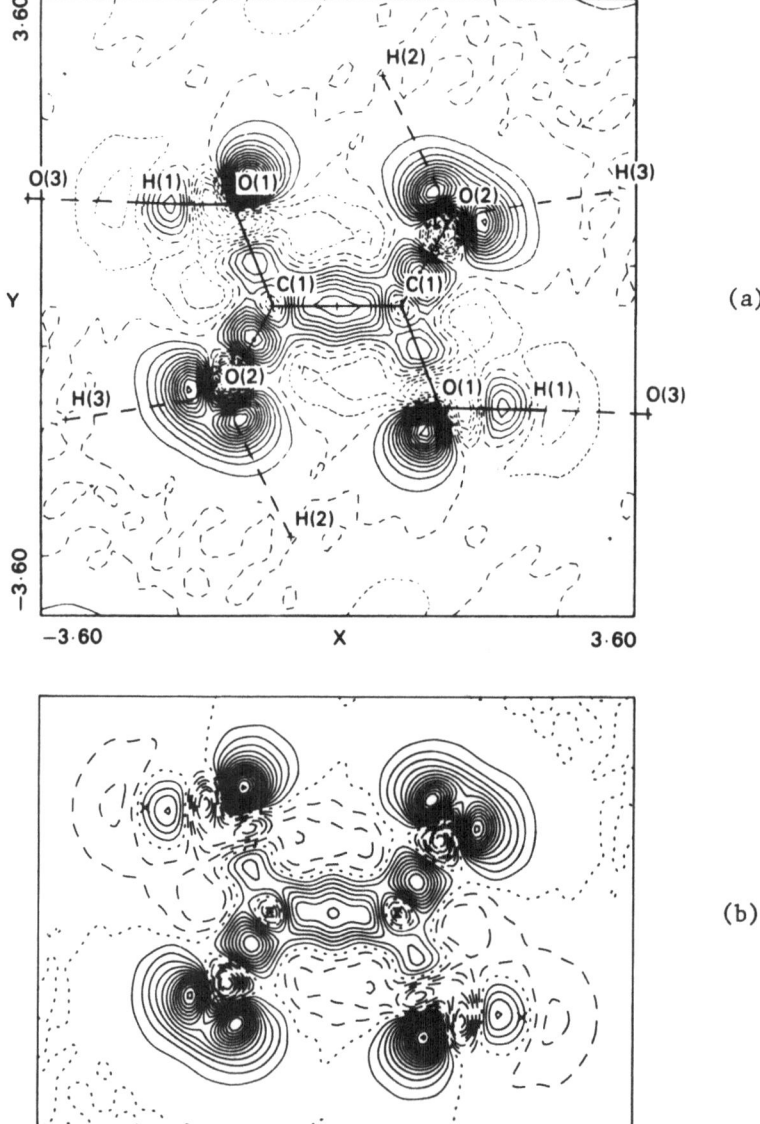

Fig.4.   Vibrationally averaged (dynamic) 4-31G deformation den-
         sity of oxalic acid. (a) method of ref.123, (b) method of
         refs.120 and 121. Contour interval 0.05 e/A³. Positive
         contours are full lines, zero and negative contours are
         dashed.

The deformation density of Fig.4.(b) is based on the following data: $\sin\theta/\lambda = 1.2 \text{ A}^{-1}$; pseudo-unit cell dimensions: $|\vec{a}| = 8$ A, $|\vec{b}| = 6$ A, and $|\vec{c}| = 4$ A; the $\underline{U}_k$ from ref.135, refinement II; the $\underline{U}_{k1}$ from eq.(28); number of structure factors calculated: 5571; computation time: 5 1/2 h for the structure factors and 2 1/2 h for the deformation density using 12012 data points in the plane considered, on a Telefunken TR 440 computer.

In general the computation time for the structure factor calculations is approximately proportional to the volume of the pseudo-unit cell V, to $(\sin\theta/\lambda)^3$ and to N^2, where N is the number of basis set orbitals.

CONCLUSION

Now it is possible to calculate electron densities (in the form of deformation densities) that agree with experimental densities within the experimental accuracy. The experimental accuracy can be appreciably lower than the often quoted precision of about $\pm 0.05 \text{ e/A}^3$, say, for bonding regions. It is just this accuracy that presently is attainable in these regions for the "best" correlated GTO deformation densities of small molecules. In the present authors' opinion experimental determinations of deformation densities of small molecules have no chance to outperform corresponding theoretical calculations, at present; for "larger" molecules both approaches are considered to be at level[18]. There is no indication, at the present time, that thermal smearing presents a problem in the theoretical calculations. In future, however, more comparisons between the various thermal smearing procedures are to be made. Unless the experimental error is brought down to ± 0.05 e/A^3 or $\pm 0.1 \text{ e/A}^3$ with certainty, for small or large molecules, respectively a most direct interplay between theory and experiment will continue being helpful, for both sides.

ACKNOWLEDGEMENT

This work was supported by the Deutsche Forschungsgemeinschaft, SFB 127, Project H2, and the Fonds der Chemischen Industrie. The quantum-chemical calculations were carried out on the TR 440 computer of the Rechenzentrum der Universität Marburg and the Cyber 174 of the Rechenzentrum der Universität Gießen.

REFERENCES

1. A.Kutoglu, C.Scheringer, H.Meyer, and A.Schweig, Acta Cryst. 1981, in press.

2. R.Brill, Solid State Physics 20:1 (1967).
3. R.A.Bonham, Record of Chemical Progress 30:185 (1969).
4. Several articles in: "Transaction of the American Crystallo-
    graphic Association", Proceedings of the Symposium on "Ex-
    perimental and Theoretical Studies of Electron Densities",
    Albuquerque, New Mexico (1972).
5. R.F.Stewart, "Critical Evaluation of Chemical and Physical
    Structural Information", National Academy of Sciences,
    Washington, D.C. (1974), p.540.
6. B.Dawson, "Advances in Structure Research by Diffraction Me-
    thods", Pergamon Press, Oxford and Friedr.Vieweg & Sohn,
    Braunschweig (1975).
7. R.F.W.Bader, Int.Rev.Sciences, Theoretical Chemistry, Physi-
    cal Chemistry, Series 2, Vol.1., Butterworths, London (1975)
    p.43.
8. P.Coppens, Int.Rev.Science, Chemical Crystallography, Physi-
    cal Chemistry, Series 2, Vol.11, Butterworths, London (1975)
    p.21.
9. A.Hinchliffe and J.C.Dobson, Chem.Soc.Rev. 5:79 (1976).
10. Several articles in: Physica Scripta 15:66-157 (1977).
11. Several articles in: Isr.J.Chem. 16:87-229 (1977).
12. P.Coppens and E.D.Stevens, Adv.Quant.Chem. 10:1 (1977).
13. P.Coppens, Angew.Chem. 89:33 (1977).
14. I.Absar and J.R.Van Wazer, Angew.Chem. 90:86 (1978).
15. R.A.Bonham, in: Electron Spectroscopy Vol.3, Academic Press,
    London (1979), pp.127-187.
16. H.Fuess, in: Modern Physics in Chemistry Vol.2, Academic
    Press, London (1979), pp.1-193.
17. Several articles in: "Electron and Magnetization Densities in
    Molecules", Plenum Press, New York (1980).
18. M.Breitenstein, H.Dannöhl, H.Meyer, A.Schweig, R.Seeger, U.
    Seeger, and W.Zittlau, to be published.
19. H.Witte and E.Wölfel, Z.Phys.Chem.Neue Folge 3:296 (1955).
20. A.Weiss, H.Witte, and E.Wölfel, Z.Phys.Chem.Neue Folge 10:98
    (1957).
21. S.Göttlicher and E.Wölfel, Z.Elektrochemie 63:891 (1959).
22. R.McWeeny, Acta Cryst. 4:513 (1951).
23. R.McWeeny, Acta Cryst. 5:463 (1952).
24. R.McWeeny, Acta Cryst. 7:180 (1954).
25. B.Dawson, Acta Cryst. 17:990 (1964).
26. R.F.Stewart, E.R.Davidson, and W.T.Simpson, J.Chem.Phys.
    42:3175 (1965).
27. J.J.DeMarco and R.J.Weiss, Phys.Rev.A137 : 1869 (1965).
28. B.Dawson, Proc.Roy.Soc. A298:255 (1967).
29. B.Dawson, Proc.Roy.Soc. A298:264 (1967).
30. B.Dawson, Proc.Roy.Soc. A298:379 (1967).
31. K.Kurki-Suonio, Ann.Sci.Fenn.Ser.AVI 241:3 (1967).
32. K.Kurki-Suonio, Acta Cryst. A24:379 (1968).
33. R.F.Stewart, J.Chem.Phys. 51:4569 (1969).

34. R.F.Stewart, J.Chem.Phys. 53:205 (1970).
35. F.L.Hirshfeld, Acta Cryst. B27:769 (1971).
36. M.Harel and F.L.Hirshfeld, Acta Cryst. B31:162 (1975).
37. R.F.Stewart, Acta Cryst. A32:565 (1976).
38. N.K.Hansen and P.Coppens, Acta Cryst. A34:909 (1978).
39. P.F.Price, J.N.Varghese, and E.N.Maslen, Acta Cryst. A34:203
       (1978).
40. G.Vidal-Valat, J.P.Vidal, and K.Kurki-Suonio, Acta Cryst.
       A34:594 (1978).
41. E.D.Stevens, Mol.Phys. 37:27 (1979).
42. G.S.Chandler, M.A.Spackman, and J.N.Varghese, Acta Cryst.
       A36:657 (1980).
43. R.Brill, H.Dietrich, and H.Dierks, Acta Cryst. B27:2003 (1971).
44. E.Hellner, Acta Cryst. B33:3813 (1977).
45. D.Mullen and E.Hellner, Acta Cryst. B33:3816 (1977).
46. H.Dietrich and C.Scheringer, Acta Cryst. B34:54 (1978).
47. C.Scheringer, A.Kutoglu, E.Hellner, H.L.Hase, K.-W.Schulte,
       and A.Schweig, Acta Cryst. B34:2162 (1978). C.Scheringer,
       A.Kutoglu, and E.Hellner, Acta Cryst. B34:2670 (1978).
48. D.Mullen and E.Hellner, Acta Cryst. B34:1624 (1978).
49. C.Scheringer, D.Mullen, E.Hellner, H.L.Hase, K.-W.Schulte,
       and A.Schweig, Acta Cryst. B34:2241 (1978).
50. H.Dietrich, C.Scheringer, H.Meyer, K.-W.Schulte, and A.Schweig,
       Acta Cryst. B35:1191 (1979).
51. D.Mullen and E.Hellner, Acta Cryst. B34:2789 (1978).
52. A.Kutoglu, C.Scheringer, H.Meyer, and A.Schweig, Acta Cryst.
       1981, in press.
53. J.Geiger, Z.Phys. 175:530 (1963).
54. R.A.Bonham and T.Iijima, J.Chem.Phys. 42:2612 (1965).
55. M.Fink and J.Kessler, J.Chem.Phys. 47:1780 (1967).
56. D.A.Kohl and L.S.Bartell, J.Chem.Phys. 51:2891 (1969).
57. D.A.Kohl and L.S.Bartell, J.Chem.Phys. 51:2896 (1969).
58. M.Fink and R.A.Bonham, Rev.Scient.Instr. 41:389 (1970).
59. J.W.Liu, J.Chem.Phys. 59:1988 (1973).
60. J.W.Liu and R.A.Bonham, Chem.Phys.Lett. 14:346 (1972).
61. A.Haberl and J.Haase, Z.Naturforsch. 29a:1023 (1974).
62. R.C.Ulsh, H.F.Wellenstein, and R.A.Bonham, J.Chem.Phys.
       60:103 (1974).
63. M.Fink, D.Gregory, and P.G.Moore, Phys.Rev.Lett. 37:15 (1976).
64. J.W.Liu and V.H.Smith Jr., Chem.Phys.Lett. 45:59 (1977).
65. J.Epstein and R.F.Stewart, J.Chem.Phys. 67:4238 (1977).
66. M.Fink, P.G.Moore, and D.Gregory, J.Chem.Phys. 71:5227 (1979).
67. A.Lahmann-Bennani, A.Duguet, and H.F.Wellenstein, J.Phys.
       B12:461 (1979).
68. S.Shibata, F.Hirota, N.Kakuta, and T.Muramatsu, Int.J.Quant.
       Chem. 18:281 (1980).
69. M.Fink, personal communication 1980.
70. M.Roux, S.Besnainou, and R.Daudel, J.Chim.Phys. 53:218 (1956).
71. R.F.W.Bader, W.H.Henneker, and P.E.Cade, J.Chem.Phys. 46:3341
       (1967).

72. J.Goodisman, J.Chem.Phys. 38:304 (1963).

73. J.Goodisman and W.Klemperer, J.Chem.Phys. 38:721 (1963).

74. G.G.Hall, Adv.Quant.Chem. 1:241 (1964).

75. J.Gerratt, Ann.Rep. A65:3 (1968).

76. R.G.Clark and E.T.Stewart, Quart.Rev. 24:95 (1970).

77. E.Steiner, Ann.Rep. A67:5 (1970).

78. E.A.Laws, R.M.Stevens, and W.N.Lipscomb, J.Chem.Phys. 56:2029
     (1972).

79. S.Green, Adv.Chem.Phys. 25:180 (1974).

80. M.Roux and M.Cornette, J.Chem.Phys. 37:933 (1962).

81. J.L.J.Rosenfeld, Acta Chem.Scand. 18:1719 (1964).

82. C.W.Kern and M.Karplus, J.Chem.Phys. 40:1374 (1964).

83. P.R.Smith and J.W.Richardson, J.Phys.Chem. 69:3346 (1965).

84. P.R.Smith and J.W.Richardson, J.Phys.Chem. 71:924 (1967).

85. M.J.Hazelrigg Jr. and P.Politzer, J.Phys.Chem. 73:1008 (1969).

86. D.B.Boyd, J.Chem.Phys. 52:4846 (1970).

87. R.M.Stevens, E.Switkes, E.A.Laws, and W.N.Lipscomb, J.Am.Chem.
     Soc. 93:2603 (1971).

88. E.A.Laws and W.N.Lipscomb, Isr.J.Chem. 10:77 (1972).

89. E.A.Laws, R.M.Stevens, and W.N.Lipscomb, J.Am.Chem.Soc.
     94:4461 (1972).

90. P.Ros, J.G.Snijders, and T.Ziegler, Chem.Phys.Lett. 69:297
     (1980).

91. M.W.Thomas, Chem.Phys.Lett. 20:303 (1973).

92. H.Meyer and A.Schweig, unpublished results.

93. G.Lauer, H.Meyer, K.-W.Schulte, A.Schweig, and H.L.Hase,
     Chem.Phys.Lett. 67:503 (1979).

94. P.A.Christiansen and E.A.McCullough Jr., J.Chem.Phys. 67:1877
     (1977).

95. H.L.Hase, G.Lauer, K.-W.Schulte, and A.Schweig, Theoret.Chim.
     Acta 48:47 (1978).

96. P.Coppens and W.C.Hamilton, Acta Cryst. B24:925 (1968).

97. E.N.Maslen, Acta Cryst. B24:1172 (1968).

98. R.F.Stewart, Acta Cryst. A24:497 (1968).

99. P.Coppens, Acta Cryst. A25:180 (1969).

100. P.Coppens, Acta Cryst. B30:255 (1974).

101. A.F.J.Ruysink and A.Vos, Acta Cryst. A30:497 (1974).

102. A.F.J.Ruysink and A.Vos, Acta Cryst. A30:503 (1974).

103. P.J.Becker and P.Coppens, Acta Cryst. A31:417 (1975).

104. E.D.Stevens and P.Coppens, Acta Cryst. A31:612 (1975).

105. P.Coppens, Acta Cryst. A31:S218 (1975).

106. B.Rees, Acta Cryst. A32:483 (1976).

107. Y.Wang, R.H.Blessing, F.K.Ross, and P.Coppens, Acta Cryst.
     B32:572 (1976).

108. E.D.Stevens and P.Coppens, Acta Cryst. A32:915 (1976).

109. G.deWith, S.Harkema, and D.Feil, Acta Cryst. B32:3178 (1976).

110. R.B.Helmholdt and A.Vos, Acta Cryst. A33:38 (1977).

111. B.Rees, Acta Cryst. A34:254 (1978).

112. C.Scheringer, A.Kutoglu, and D.Mullen, Acta Cryst. A34:481
        (1978).
113. A.J.C.Wilson, Acta Cryst. A35:122 (1979).
114. P.Coppens and A.Vos, Acta Cryst. B27:146 (1971).
115. D.S.Jones, D.Pautler, and P.Coppens, Acta Cryst. A28:635
        (1972).
116. H.Dannöhl, H.Meyer, and A.Schweig, unpublished results.
117. D.W.J.Cruickshank, Acta Cryst. 9:747,754 (1956).
118. V.Schomaker and K.N.Trueblood, Acta Cryst. B24:63 (1968).
119. J.W.McIver Jr., P.Coppens, and D.Nowak, Chem.Phys.Lett.
        11:82 (1971).
120. H.Reitz, Thesis 1975, Marburg.
121. H.-L.Hase, H.Reitz, and A.Schweig, Chem.Phys.Lett. 39:157
        (1976).
122. C.Scheringer and H.Reitz, Acta Cryst. A32:271 (1976).
123. E.D.Stevens, J.Rys, and P.Coppens, Acta Cryst. A33:333 (1977).
124. J.W.Bats and D.Feil, Chem.Phys. 22:175 (1977).
125. R.Bianchi, P.Cremaschi, G.Morosi, and M.Simonetta, Chem.Phys.
        22:267 (1977).
126. C.Scheringer, Acta Cryst. A33:426 (1977).
127. C.Scheringer, Acta Cryst. A33:430 (1977).
128. G.DeWith, Chem.Phys. 32:11 (1978).
129. C.A.Coulson and M.W.Thomas, Acta Cryst. B27:1354 (1971).
130. R.McWeeny, Acta Cryst. 6:631 (1953).
131. K.J.Miller and M.Krauss, J.Chem.Phys. 47:3754 (1967).
132. P.Coppens, T.V.Willoughby, and L.N.Csonka, Acta Cryst.
        A27:248 (1971).
133. E.D.Stevens, Acta Cryst. B36:1876 (1980).
134. M.Breitenstein, H.Dannöhl, and A.Schweig, unpublished results.
135. E.D.Stevens and P.Coppens, Acta Cryst. B36:1864 (1980).

Section 5
MOLECULAR SOLIDS: EXPERIMENTAL RESULTS

EXPERIMENTAL OBSERVATION OF TELLURIUM LONE-PAIR AND MOLYBDENUM-MOLYBDENUM TRIPLE AND QUADRUPLE BOND DENSITIES

J.M. Troup, M.W. Extine   R.F. Ziolo

Molecular Structure Corporation   Xerox Corporation
3304 Longmire Drive   Webster Research Ctr.
College Station, Texas   77840   Webster, N.Y.   14580

ABSTRACT

Electron density distribution maps have been calculated for $(CH_3)_2TeCl_2$, $(CH_3)_3TeCl$ and several multiple metal-metal bonded compounds. Previously for atoms as heavy as tellurium one might have expected such a tremendous accumulation of errors that experimental electron densities would be meaningless; however, an approach has been developed for the extension of (X-X) deformation density calculations to compounds of this type. The lone-pair density in $(CH_3)_2TeCl_2$ occupies an equatorial position in the trigonal bipyramidal coordination of tellurium and appears to be directed toward chlorine atoms which are involved in secondary bonding interactions. The first conclusive experimental observation of metal-metal bonding electrons in M≡M and M≣M compounds was observed as significant electron density build-up between molybdenum centers in $Mo_2(N(CH_3)_2)_4(CH_3)_2$ and $Mo_2(O_2CPh)_4$ respectively. An explanation for "ghost" peaks observed in previous studies on transition metal and other heavy atom compounds is postulated.

INTRODUCTION

Experimental electron density studies have been performed for the first time on compounds containing atoms as heavy as tellurium and molybdenum. Previously, only a few compounds containing atoms as heavy as first row transition elements have been investigated. Existing studies on Cr, Mn, Fe, and Ni[1a,b,c,d] complexes have led to a great deal of controversy with regard to metal-metal and metal-ligand bonding since many chemically unreasonable electron density features are observed that do not agree well with existing theoretical calculations.[1a,d] Our work in this area was prompted by this controversy especially with regard to metal-metal bonding.

STUDIES ON $Te(CH_3)_3Cl$ AND $Te(CH_3)_2Cl_2$

We first observed[2] during a routine x-ray structural investigation of $Te(CH_3)_3Cl$ that electron density is present around the tellurium atom in a tetrahedral position with respect to the methyl groups; where one might expect to see tellurium IV lone-pair density (Fig. 1). We observed that these electron density features could be enhanced relative to random noise by using a $\sin \theta/\lambda$ cutoff of 0.40 Å^{-1} for $Te(CH_3)_3Cl$. This technique is similar to the method used to enhance hydrogen atom density since the scattering factor for bonding and lone pair (b-lp) electron density is similar to that of hydrogen atoms.[3] The possible observation of tellurium IV lone-pair density led to a detailed electron density analysis of the compounds $Te(CH_3)_3Cl$, $Te(CH_3)_3Br$ and $Te(CH_3)_2Cl_2$[4] at low temperatures where redundant data were collected and averaged. The investigation of these tellurium compounds as well as several molybdenum and chromium compounds has allowed us to establish some ground rules for examining the experimental electron density of compounds containing heavy atoms.

The electron density study of $Te(CH_3)_2Cl_2$ was performed. The high angle refinement method with x-ray data collected at low temperatures as discussed by P. Coppens and others was used to obtain an approximation of spherical Hartee-Fock type atoms, termed the "promolecule". Promolecule structure factors were substracted from the observed data x-ray structure factor which contains the electron density information. The result is a residual deformation density map, where core and some valence electron density have been subtracted leaving only b-lp electrons. This method is abbreviated as the X-X method as opposed to X-N where neutron diffraction is used to produce the "promolecule". We find that for compounds with atoms as heavy as molybdenum and tellurium full data maps calculated as above contain a considerable amount of random error that has accumulated due to the higher angle data where little electron density information is present. By omitting the high angle data at a suitable $\sin \theta/\lambda$ level a good compromise between resolution and clarity can be obtained. For $Te(CH_3)_2Cl_2$ the maps were calculated with $\sin \theta/\lambda < 0.40 \text{Å}^{-1}$ data and show carbon-tellurium as well as carbon-hydrogen bond density. Carbon-hydrogen bond density is surprising in the presence of a tellurium atom where one might expect that it would be hard to, first of all locate hydrogen atoms, not to mention see carbon-hydrogen bonding density. For the first time tellurium IV lone-pair density was clearly observed.

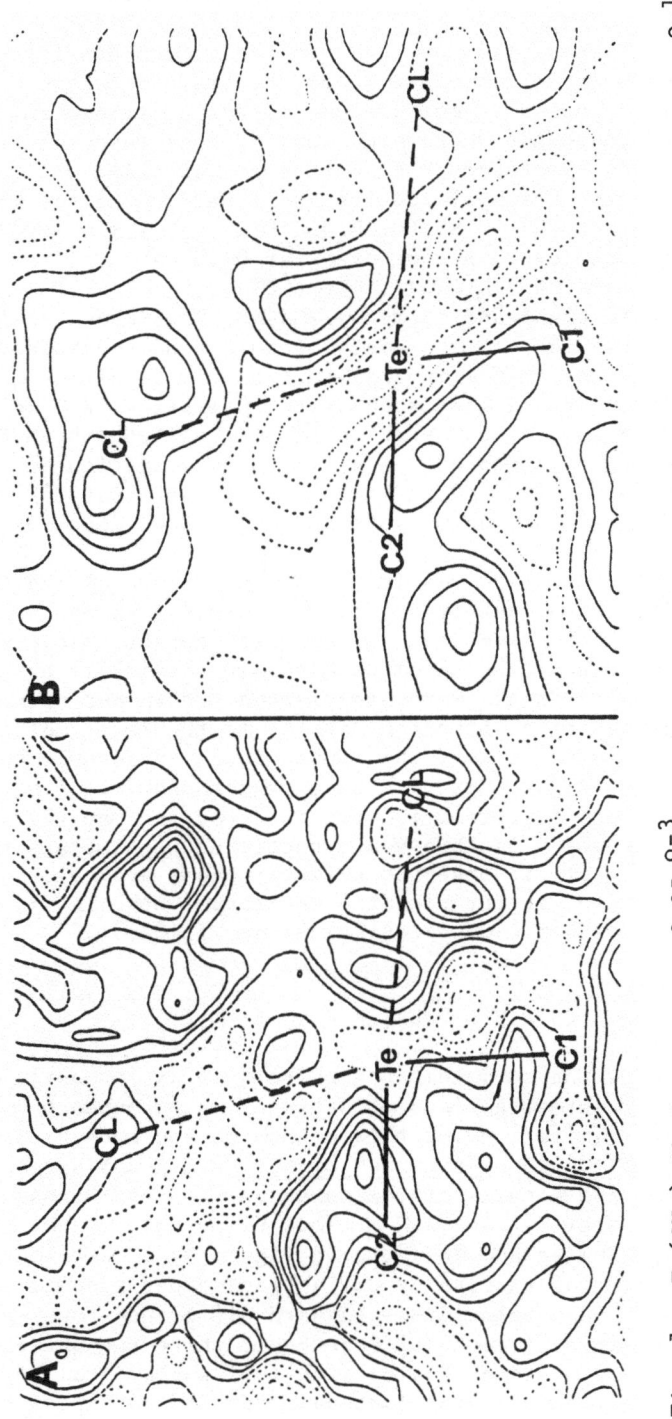

Fig. 1. Te$(CH_3)_3$Cl, contours at 0.05e$Å^{-3}$ starting at zero; A. Full data map; B. sin $\theta/\lambda \leq$ 0.40$Å^{-1}$.

ROOM TEMPERATURE TRIPLE AND QUADRUPLE METAL-METAL BOND SYSTEMS

Our orignial observation of reasonable electron density in
$Te(CH_3)_3Cl$ using data collected at   room temperature triggered an
investigation of a series of molybdenum-molybdenum triple and
quadrulpy bonded compounds where high quality room-temperature data
had been collected. Professor F.A. Cotton of Texas A&M University
allowed us to use the orignial routine x-ray data collected in his
laboratory for $Mo_2(O_2CPh)_4$,[6] $Mo_2(N(CH_3)_2)_4(CH_3)_2$,[7] and $Cr_2(TMP)_4$[8] for
several test calculations. A contoured (at $0.05e$ Å^{-3}) difference   .
Fourier map for $Mo_2(N(CH_3)_2)_4(CH_3)_2$ (Fig. 2) in the
dimethylimido-Mo-Mo plane and in methyl-Mo-Mo plane show the first
experimental observation of metal-metal triple bond densities.
Molybdenum-molybdenum, -nitrogen and -carbon bond densities as
well as other features are consistent with theoretical Xα calculations by B. Bursten et al[9] and ab initio calculations by M.B.
Hall[10] on this type of compound. A similar investigation of room
temperature data on $Mo_2(O_2CPh)_4$ (Fig. 3) shows the first molybdenum-
molybdenum quadruple bond and molybdenum-oxygen bond densities.

The $Cr_2(TMP)_4$ structure was investigated because of the
similarities with the structure of $Cr_2(OAc)_4$ investigated by
Bernard et al[1a]. In the electron density maps for $Cr_2(OAc)_4$ large
peaks were observed near the chromium atoms in chemically unreasonable positions. These peaks were unexplained and in our opinion
made the interpretation of electron density in the metal-metal bond
difficult. $Cr_2(TMP)_4$ is one of the "super-short" quadruple bond
compounds which one might expect to have significant metal-metal
density. In the room temperature difference Fourier map (Fig. 4)
weak build-up $(0.05e$Å$^{-3})$ is observed directly between metal atoms
however, higher density is observed slightly above and below the
metal-metal vector and in the dihedral section. These results
compare favorably with SCF-CI calculations on $Cr_2(OAc)_4$.[1a]

LIMITATIONS OF THE DIFFERENCE FOURIER METHOD

The success of the room temperature relatively low angle
difference Fourier investigations of heavy atom compounds prompted
us to examine the limitations of this method of approach by looking at a well studied system examined by both X-X and X-N techniques. The electron density of the furanose ring of sucrose was
previously examined by X-X and X-N methods using molybdenum
radiation to high angles and is shown in Figure 5.[11] We examined
the electron density of sucrose by collecting copper radiation
data to 150° in 2θ and calculated standard difference Fourier maps
(i.e., and algebraic difference of Fobs-Fcalc where the phase of
Fobs is equal to the phase of Fcalc) with full angle data. All
furanose ring densities and even lone-pair density on the oxygen

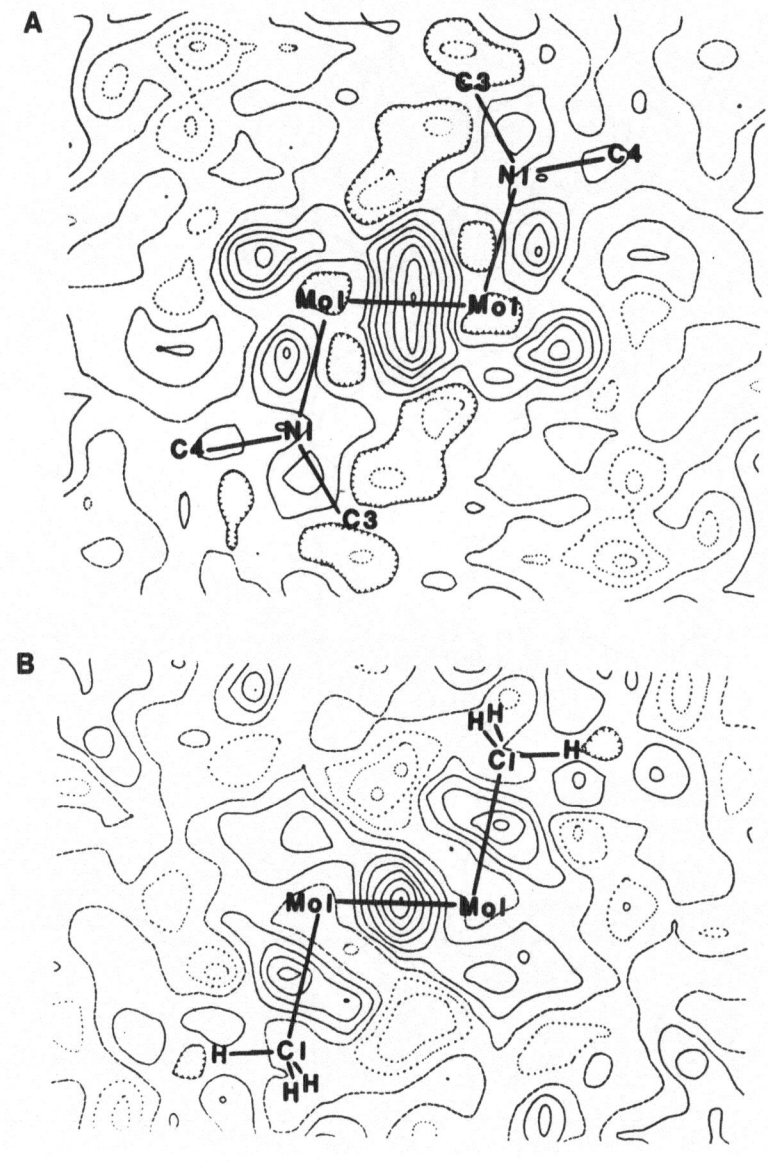

Fig. 2. $Mo_2(N(CH_3)_2)_4(CH_3)_2$ contours at $0.05e\text{\AA}^{-3}$ starting at zero; A. Mo_2-amide; B. Mo_2-methyl plane.

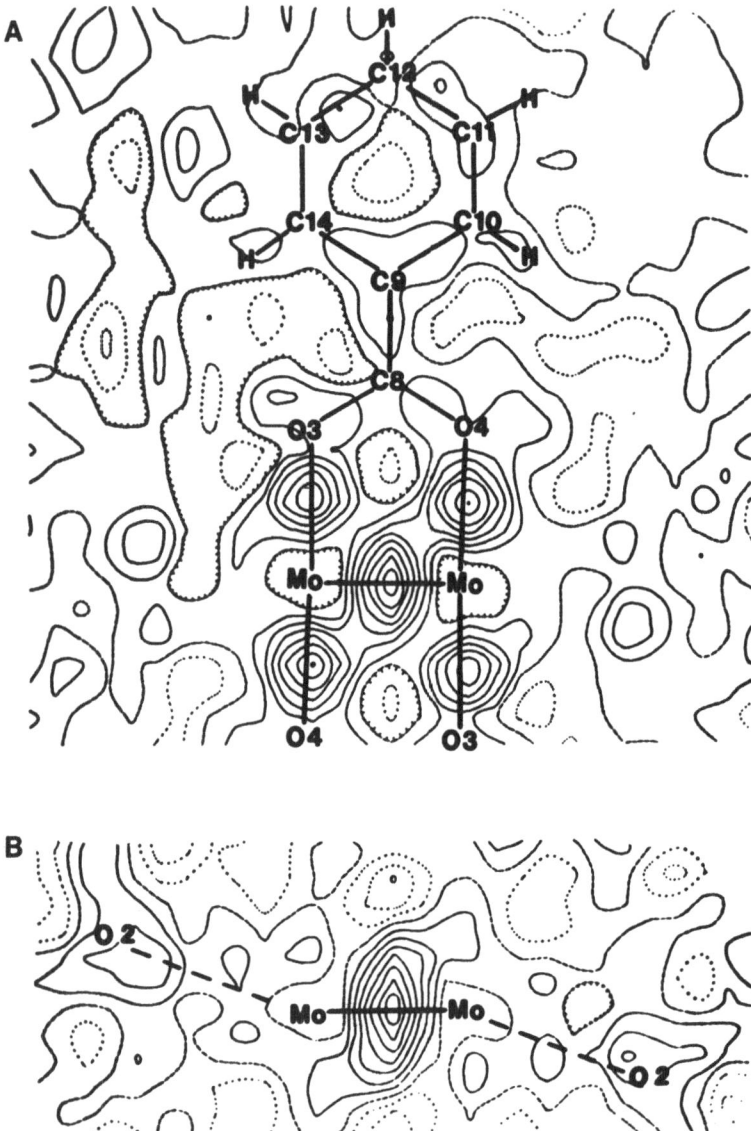

Fig. 3. $Mo_2(O_2CPh)_4$ contours at $0.05e\text{Å}^{-3}$ starting at zero: A. Mo_2O_2C plane; B. dihedral plane.

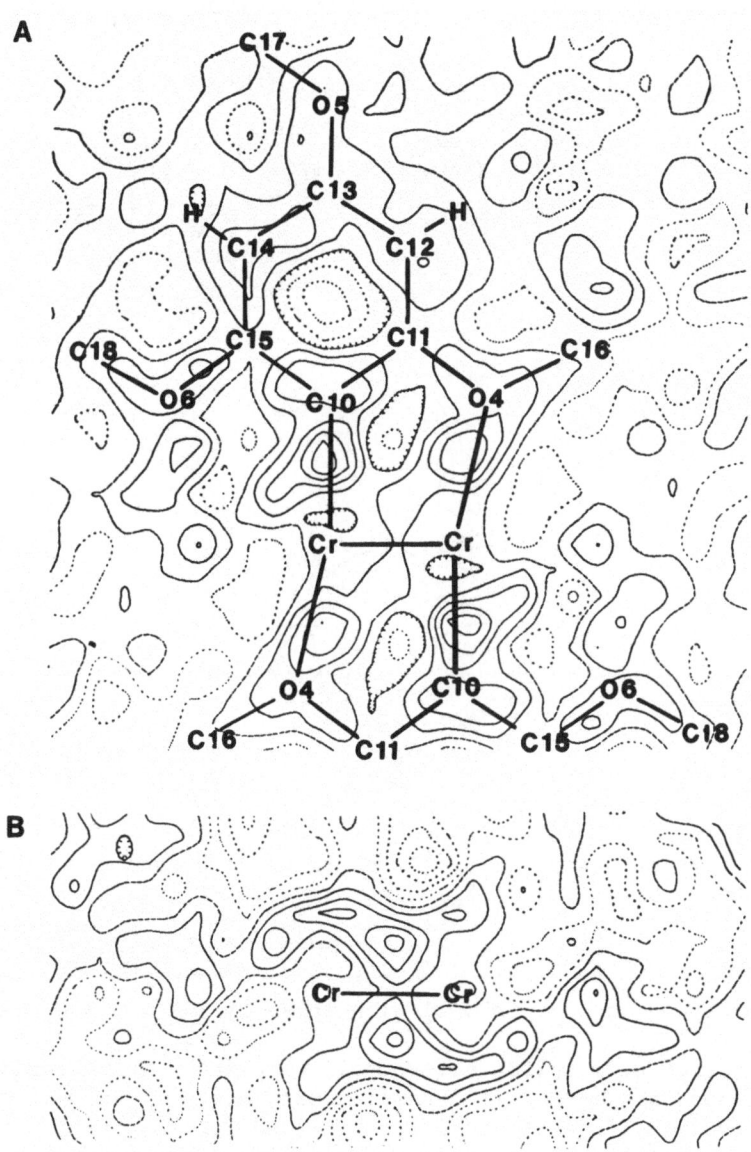

Fig. 4. $Cr_2(TMP)_4$ contours at $0.05e\text{Å}^{-3}$ starting at zero; A. Cr_2-ligand plane; B. dihedral plane.

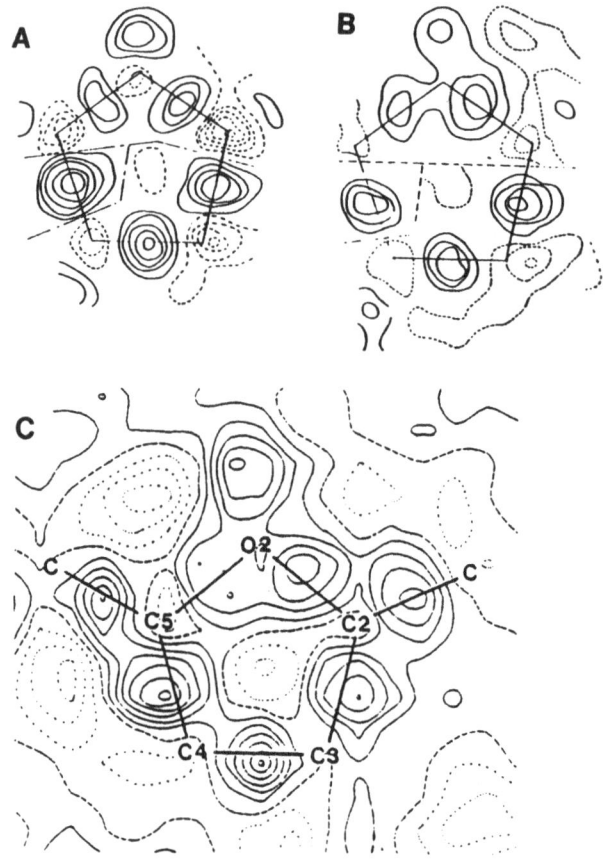

Fig. 5   Sucrose-Furanose ring all contours at $0.05e\text{Å}^{-3}$; A. X-N map,
MoK$_\alpha$ data[11], B. X-X$_{HO}$ map, MoK$_\alpha$ data[11], C. Difference
Fourier map, CuK$_\alpha$ data.

atom were found to compare well with those by X-X and X-N methods.

A heavier atom sulfur heterocycle compound[12] which contains
multiple bond and lone-pair densities was chosen to examine these
features using different refinement methods of electron density
analysis. In Figure 6a, a standard difference Fourier map was
calculated and in Figure 6b the map was calculated using a very
limited high angle refinement $(0.6 < \sin\theta/\lambda < 0.7\text{Å}^{-1})$ to help
maximize the multiple bonded regions.

We find as have others that multiply bonded first row element
densities suffer from an asphericity shift of the atoms toward the
b-lp density which tends to reduce the observable b-lp features,
however, for singly bonded first row elements (especially in a

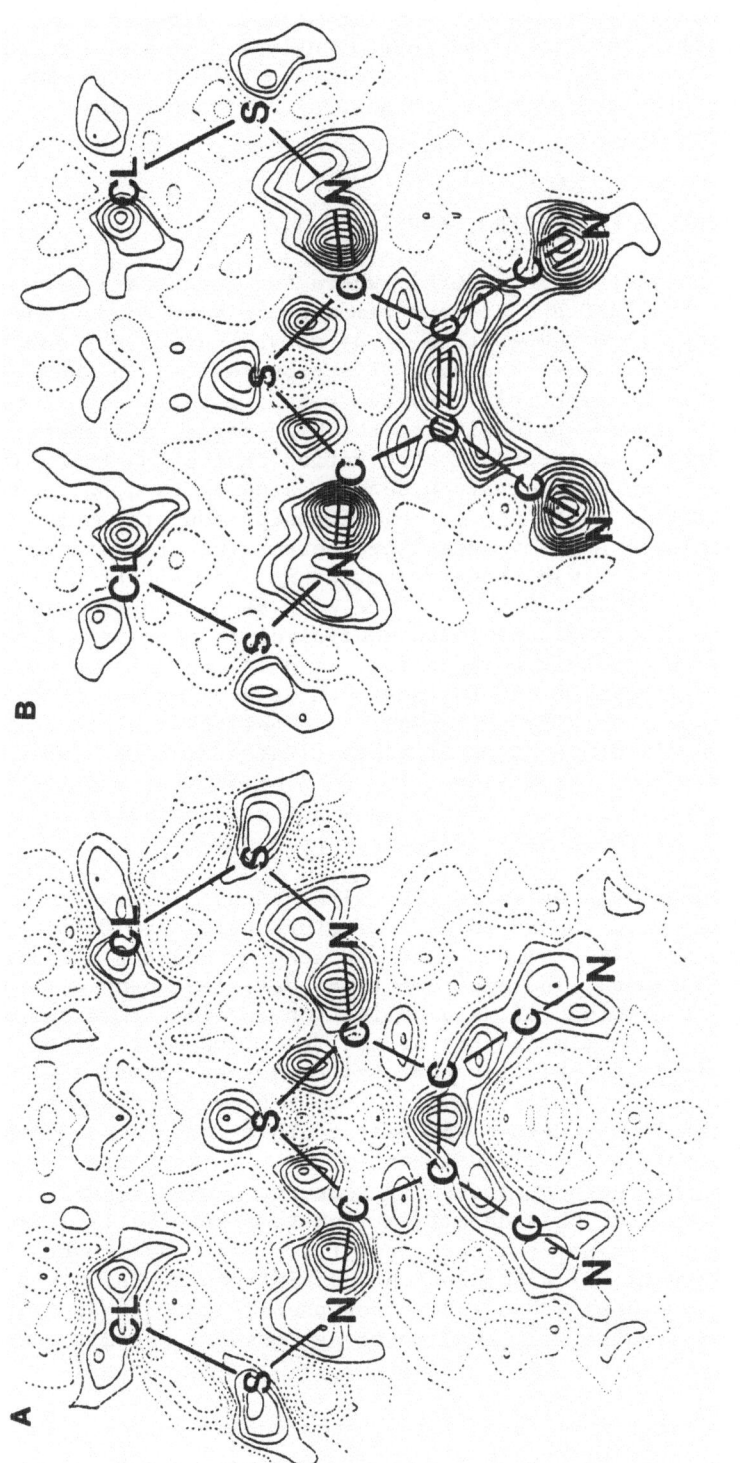

Fig. 6. $C_6Cl_2N_4S_3$ contours at $0.05e\text{Å}^{-3}$; A. Full data map; B. Full data map, high angle refinement.

symmetrically bonded environment) and for heavier atoms little
asphericity shift is observed and consequently a high angle refinement has little or no detectable effect on the b-lp density maps.
This effect was observed as well for several molybdenum and
tellurium compounds.

DETAILED ANALYSIS OF $Mo_2(O_2CPh)_4$ AND $Te(CH_3)_3Br$

Detailed analyses on $Mo_2(O_2CPh)_4$ where redundant data were
collected at both -112°C and at room temperature were conducted.
As in the previous room temperature study molybdenum-molybdenum
quadruple bond density was observed. Electron density in the
organic region was considerably enhanced relative to the earlier
low resolution study due primarily to the high angle refinement
$(0.6Å^{-1} < \sin \theta/\lambda)$. An important observation in the low temperature
study and other heavy atom electron density studies is a build-up
of "ghost" density in a 180 degree relationship about the heavy
atoms with associated negative densities perpendicular to these
densities. This effect was observed but unexplained in studies
on $Cr_2(OAc)_4$[1a] and $Cr(CO)_3(C_6H_6)$[13]. Our work shows the effect is
considerably reduced or absent in an equivalent room-temperature
study. We believe that since the effect is only present around
the heavy atoms it must be due to the "core" electrons and not a
bonding or lone-pair density phenomenon. The effect could be
explained as a small dislocation of atoms (static disorder) which
is reflected as a peak proportional to the Z of an atom and too
weak to be observed for light atoms. A study was conducted on the
compound $Te(CH_3)_3Br$ where this effect was pronounced.

Extensive data sets on $Te(CH_3)_3Br$ with redundant data to
high angles were collected at room temperature using the $\theta-2\theta$
scan technique, at -112°C with the $\theta-2\theta$ scan method and at -112°C
with ω scan techniques. The experiment was designed to only vary
the temperature and method of data collection in order to examine
the enhancement at low temperature of the "ghost" densities, as
well as the possibility that truncation of peaks by normal $\theta-2\theta$
scan methods might be responsible if a true ω broading disorder
effect is present in the data. We observed that some "ghost" peak
build-up is present to a limited extent at room temperature
superimposed on the reasonable density features, however, a
definite enhancement of this density at low temperatures is
observed. A greater "ghost" peak effect from $\theta-2\theta$ data over
that from ω scan data was also apparent. This result would tend
to support the hypothesis that a disorder phenomenon is responsible;
however, similar studies will need to be performed to verify these
findings.

Our observations suggest that the "ghost" peak build-up and
hole generation could be due to a thermal stress, forcing a

dislocation of the structure which cannot be accommodated by
anisotropic thermal parameters. We have observed the build-up
phenomenon in different crystals of the same compound in several
different directions and also in vastly different size crystals
thereby reducing the possibility of absorption and extinction errors
as explanations. It also appears that the direction of greatest
"ghost" peak build-up is the same as the longest physical length of
the crystal, however, this correlation will have to be confirmed
in future studies. The remedy at present appears to be the use of
room-temperature data in some cases for experimental electron
density studies of heavy atom compounds (even though the data
contains greater thermal diffuse scattering).

STUDIES ON $Fe(CO)_5$ AND $Fe_2(CO)_9$

Two first row transition element experimental electron density
studies were also performed and were found to support the earlier
finding for heavier atoms. A study of iron pentacarbonyl[14] was
found to show reasonable Fe-C and C-O bonding density in the
equatorial plane of the molecule. The lone-pair density on the
carbon of the carbonyl groups was rather broad extending into the
carbonyl antibonding orbital region. This may be evidence for
considerable π backdonating character as well as sigma character
in the equatorial iron-carbon bonds. The axial carbonyl section
shows an intense peak between the iron and carbon atom and is
believed due to the "ghost" peak effect enhanced by low temperatures
rather than true bonding or lone-pair electron density.

A related electron density study on diiron nonacarbonyl[15] also
performed at low temperature (with redundant data collected and
averaged), shows no "ghost" build-up effects around the heavy atom.
Reasonable carbon-oxygen and bridging carbonyl carbon lone-pair
densities were observed; however, the terminal carbonyl carbon
lone-pair densities are reduced from what one might have expected.
This is the first observation of positive density ($0.20 e\text{\AA}^{-3}$) in a
first row transition element metal-metal bond as contrasted to the
study of $Fe_2(C_5H_5)_2(CO)_4$ where no metal-metal density was found.
This structure as well as $Te(CH_3)_2Cl_2$ are cases where low temperatures studies do not show "ghost" peak build-ups of density but in
fact benefit from the reduced thermal parameters.

CONCLUSIONS

The use of routine room temperature data for a preliminary
examination of experimental electron density is recommended
whenever the electron density is expected to be of interest. Low
temperatures may enhance the formation of "ghost" peaks around

heavier atoms proportional to the Z of the atoms that are not b-lp electron density features. High angle refinements for the purpose of determining a "promolecule" appear to be necessary for multiply bonded first row atoms, less important for singly bonded first row atoms, and much less important for configurations of heavier atom types.

The above methods rapidly led to the conclusive experimental observation of metal-metal triple and quadruple bonding electrons and tellurium IV lone-pair electron density. The application of experimental electron density determinations to transition metal and other heavy atom compounds creates vast possibilities with respect to the understanding of chemical bonding and to the enhancement of theoretical chemistry.

ACKNOWLEDGEMENT

We are grateful to Professor F.A. Cotton for kindly allowing us to use the original reflection data sets for our calculations and for his support and to Professor P. Coppens for helpful discussions at the onset of this work.

REFERENCES

1a. Benard, M.; Coppens, P.; DeLucia, M.L.; and Stevens, E.D. Inorg. Chem., 19:1924 (1980).
 b. Mitschler, A.; Rees, B., unpublished results.
 c. Mitschler, A.; Rees, B.; Lehmann, M.S. J. Am. Chem. Soc., 100:3390 (1978).
 d. Wang, Y.; Coppens, P. Inorg. Chem. 15:1122 (1976).
 2. Troup, J.M.; Ziolo, R.F., work in progress.
 3. Coppens, P. Acta Crystallogr. B30:255 (1974).
 4. Ziolo, R.F.; Troup, J.M., submitted for publication.
 5. Extine, M.W.; Troup, J.M., work in progress.
 6. Cotton, F.A.; Extine, M.W.; Gage, L.D. Inorg. Chem. 17:172 (1978).
 7. Chisholm, M.H.; Cotton, F.A.; Extine, M.W.; Murillo, C.A. Inorg. Chem. 17:2338 (1978).
 8. Cotton, F.A.; Koch, S.A.; Millar, M. Inorg. Chem. 17:2087 (1978)
 9. Bursten, B.E.; Cotton, F.A.; Green, J.C.; Seddon, E.A.; Stanley, G.G. J. Am. Chem. Soc. 102:4579 (1980).
10. Hall, M.B. Prodeedings of the Am. Chem. Soc. symposium on "Electron Distributions and the Chemical Bond", Spring 1981, Atlanta, Ga.
11. Hanson, J.C.; Sieker, L.C.; Jensen, L.H.: Transactions of the Amercian Crystallographic Association , 133-148 (1972).
12. Wudl, F.; Zellers, E.T. J. Am. Chem. Soc. 102:4283 (1980).
13. Rees, B.; Coppens, P. Acta Crystallogr. B29:2516 (1973).
14. Troup, J.M., preliminary unpublished results.

DEFORMATION DENSITY DETERMINATION (X-X, X-N) OF ORGANOMETALLIC

COMPOUNDS

Richard Goddard and Carl Krüger

Max-Planck-Institut fürKohlensorschung, Lembkestr. 5
D-4330 Mülheim a.d. Ruhr
West Germany

The electronic distribution and symmetry properties of
starting or intermediate organometallic complexes may be important
factors in determining the pathways of chemical reactions at
metallic centres and the course of metal-induced catalytic reactions.

As recent results show the feasibility of deducing electronic
density distributions in first-row transition metal complexes using
the X-X and X-N methods[1], detailed structural investigations of
this type have been undertaken for several key organometallic
complexes. On the basis of these experiments we are confident that
application of these methods to organometallic complexes of the
first-row transition metals can lead to chemically interesting
results.

We have used oxalic acid dihydrate to check our experimental
and computational procedures. The results of these studies (X-X,
X-N)[2] are shown in Figs. 1 and 2 and the experimental X-ray data
are given in Tab. 1. Features of the electron distribution are
very similar to those obtained by other experimentalists[3] and the
electronic deformation density generally matches the theoretically
calculated one.[4]

Our first electronic deformation density determination was
on a product of the reaction of bis(cyclopentadienyl)chromium
with cyclooctatetraene-disodium, $\lfloor (\eta^5 - C_5H_5Cr)_2 C_8H_8 \rfloor$ (Fig. 3),(I).
This compound has been studied by neutron diffraction to supplement
an earlier (X-X) charge density determination[5] and to give a more
accurate description of the electron density in the compound (X-N).

297

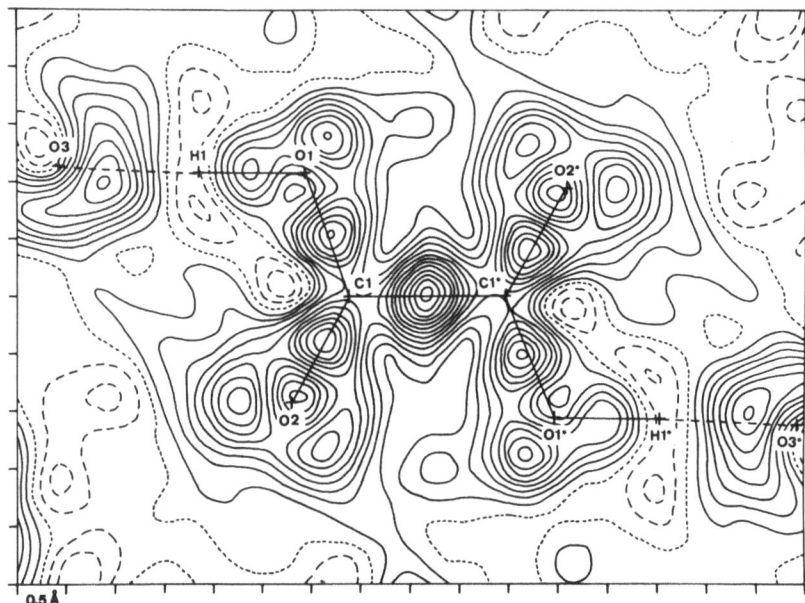

Fig. 1: The electronic deformation density in the plane of the oxalic
acid molecule (X-X). Contours are at 0.05 eÅ^{-3} with negative
contours dashed and zero contours dotted. $S_{max} = 0.7$ Å^{-1}.

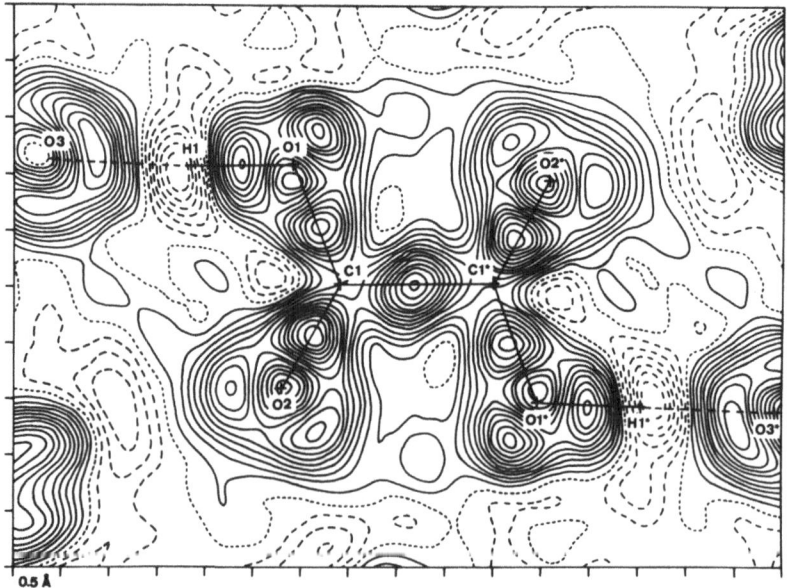

Fig. 2: Deformation density in the plane of the oxalic acid
molecule (X-N). Contours as in Fig. 1.

Table 1. Experimental X-ray data for α-oxalic acid dihydrate

Oxalic Acid

α -$(COOH)_2 \cdot 2H_2O$
a = 6.1108(9) Å
b = 3.5109(4) Å
c = 11.9780(8) Å
β = 105.767(8)°
$P2_1/n$,   Z = 2

T = -170°C
5799 reflections measured (sin θ/λ) max = 1.4 Å^{-1}
$R_{average}$ = 0.012

$1.7 < 2\theta < 150.8°$
Mo-Kα X-radiation (graphite monochromated) $\bar{\lambda}$ = 0.71069 Å
All Data refinement: 3091 reflections, I > 2.0 σ(I)
R = 0.043, R' = 0.054 (49 parameters)
High-angle refinement: 1400 reflections ($\frac{\sin \theta}{\lambda} > 1.0$ Å^{-1})
R = 0.052, R' = 0.047, Figure of merit = 1.26

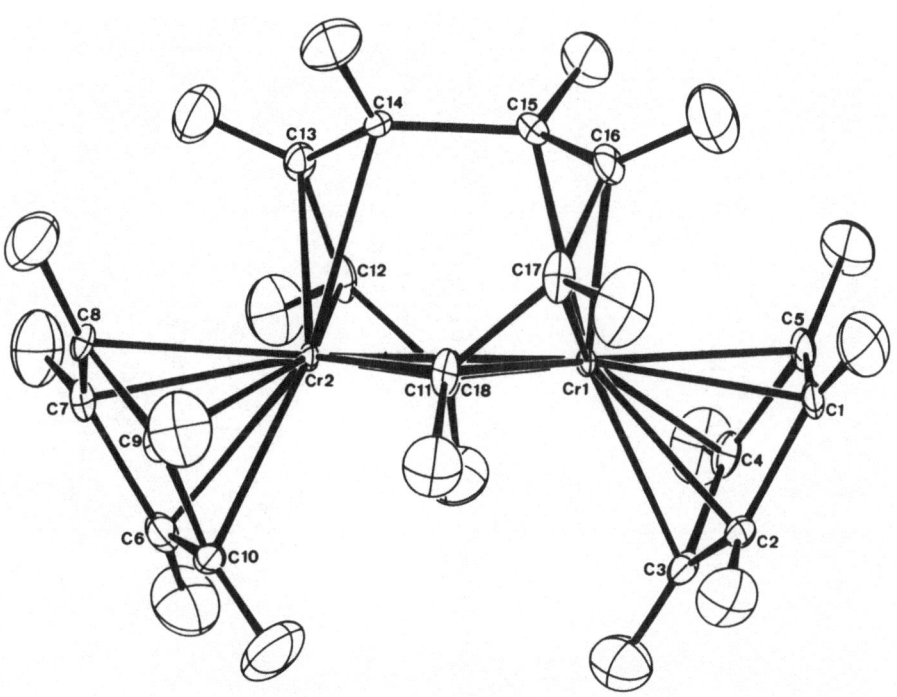

Fig. 3: The molecular structure of $(\eta^5\text{-}C_5H_5Cr)_2C_8H_8$.

Table 2. Experimental data for $(\eta^5-C_5H_5Cr)_2C_8H_8$, (I)

	X-Ray Data Collection Enraf-Nonius-CAD-4, Graphite-Monochromator			Neutron Data Collection ILL-Grenoble
		Crystal A	Crystal B	Crystal C
Wavelength		$Mo(K\alpha)\lambda = 0.71069$Å		$\lambda = 1.1486(2)$Å
Temperature (K)	291°	103° ±0.2	103° ±0.2	103° ±0.5
Crystal size (mm)		0.70×0.49×0.23	0.19×0.25×0.26	0.61×0.74×1.10
Cell data (max. 88 Reflexions)				
a =	9.6173(6)Å	9.565(2)Å	9.545(2)Å	9.537(9)Å
b =	12.1567(9)	12.094(3)	12.038(3)	12.034(9)
c =	12.6647(6)	12.556(3)	12.550(3)	12.568(13)
β =	107.991(6)°	108.40(2)°	108.40(2)°	108.34(7)°
μ (cm^{-1})		15.36		2.14
Transmission			0.6301-0.8072	0.8172-0.9016
$P2_1/n$; Z = 4				
R_{Fin} (full data set)	0.042	0.04	0.056	0.06
Reflexions measured	$[\pm h,k,l]$	$[\pm h,\pm k,l]$	1.6-27.4° $[\pm h,\pm k,\pm l]$; 27.4-50° $[\pm h,k,l]$	$[\pm h,k,l]$
θ-Range	1°<θ<28.4°	1.5°<θ<40°	1.60<θ<50°	3.6°<θ<49.5°
averaged to	3865	9034	18217	2793
obs.	1998	3349	7425	1817
unobs. (2σ)	1713	1168	6790	592
R_M	0.022	0.02	0.025	0.03

We chose to investigate this compound not only because of its interesting bonding features (C_8H_8-chain-metal interaction, metal-metal interaction), but also because of its helpful symmetry properties. The compound crystallizes in a centric space group with a high point group symmetry (monoclinic) but with no symmetry restraints on the individual molecules. The non-crystallographic twofold axis through the molecule thus serves as an internal standard for our results.

The electron density determination of (I) is the result of several diffraction experiments (Tab. 2). The X-ray intensities were collected on a Nonius CAD-4 four circle diffractometer once from each of two different crystals in different mounted orientations. The crystals were mounted in Lindemann glass capillaries under argon gas and cooled by a stream of nitrogen gas to a temperature of -170°C. Neutron diffraction data were collected from a larger crystal of the same sample on the D8 four-circle diffractometer at the Institut Laue-Langevin high flux beam reactor in Grenoble (France). Low temperature, in this case, was maintained with an 'Air Products' one-stage closed-loop displex refrigerator using helium as the coolant, and the sample was kept at the same temperature as for the X-ray experiment. Intensity data from both experiments were corrected for Lorentz, polarisation and absorption effects (extinction effects were corrected for during refinement using an isotropic extinction model).

Table 3. The Refinement of $(\eta^5\text{-}C_5H_5Cr)_2C_8H_8$, (I).

	X-ray full set	X-ray low angle	X-ray high angle	Neutron isotropic	Neutron anisotropic
Number of independent reflections (m)	7425	3042	3325	1817	1817
Minimum $\sin\Theta/\lambda$	0.04	0.04	0.80	0.05	0.05
Maximum $\sin\Theta/\lambda$	1.08	0.70	1.08	0.66	0.66
Intensity requirement	$I>2.0\sigma(I)$	$I>2.0\sigma(I)$	$I>2.0\sigma(I)$	$I>2.0\sigma(I)$	$I>2.0\sigma(I)$
Number of variables (n)	253	1	181	153	344
$R(F)=\Sigma\|\|F_0\|-k\|F_c\|\|/\Sigma\|F_0\|$	0.056	0.043	0.072	0.085	0.062
$R_w(F)=\{\Sigma w(\|F_0\|-k\|F_c\|)^2/\Sigma w(F_0)^2\}^{\frac{1}{2}}$	0.048	0.038	0.068	0.054	0.037
$S(F)=\Sigma w(\|F_0\|-k\|F_c\|)^2/(m-n)$	1.643		1.196	3.12	2.29
Extinction parameter $(\times10^{-4})$	0.02 (1)	0.02			0.44 (1)
Average shift to error ratio in the final cycle	0.0001	0.0	0.0001	0.011	0.008

Five types of refinement were considered (Tab.3). In the first,
atomic parameters obtained from an earlier X-ray study were refined
with isotropic thermal parameters against the neutron intensity data,
and anisotropic thermal motion was allowed in the second. The
remaining three refinement procedures were carried out using the
X-ray data. The results of conventional X-ray refinement are given
under X-ray "full-set". "High-angle" refinement was done with those
3325 intensities with $I > 2.0\,\sigma(I)$ and $\sin\theta/\lambda > 0.8\ \text{\AA}^{-1}$. Hydrogen
atom positional and anisotropic thermal parameters were fixed at
values obtained from neutron diffraction, and chromium atom and
carbon atom parameters, a scale factor and an extinction parameter
were allowed to refine.

The deformation density maps show the results of the calcula-
tion of the deformation density $\Delta\rho = \rho_{cryst} - \sum\rho_{free\ atoms}$, which
was computed by Fourier transformation of $F_{obs,X-ray} / k\text{-}F_{calc}$
for all X-ray data with $\sin\theta/\lambda < 0.7\ \text{\AA}^{-1}$.
The positional parameters and the thermal parameters of the
hydrogen atoms as well as positional parameters for the carbon
atoms were considered to be better determined in the neutron study
and these together with the remaining parameters determined by the
high-angle X-ray refinement, were used to calculate F_{calc} using
form factors of spherical atoms in their ground state.   The
symmetry of the deformation density map in the vicinity of the
molecule so calculated closely resembles the <u>chemically</u> imposed
2-fold axis of symmetry. A definition of some planes in the molecule
is given in Fig. 4.

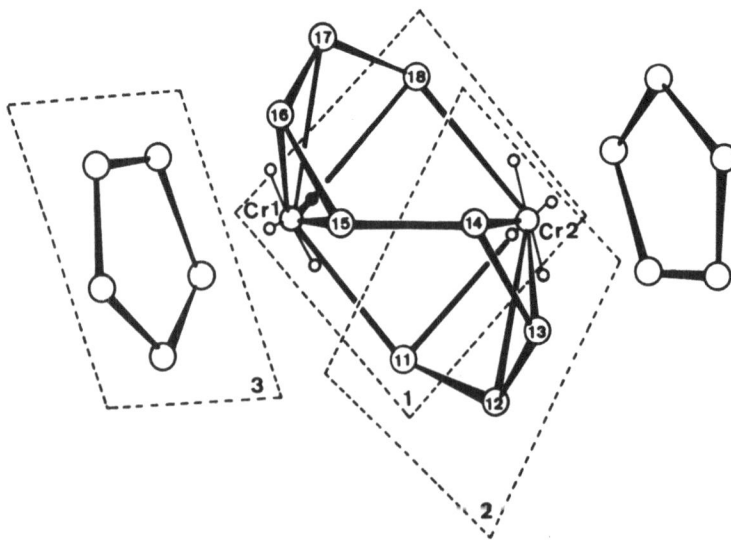

Fig.4: Definition of various planes in the $(\eta^5\text{-}C_5H_5Cr)_2C_8H_8$ molecule

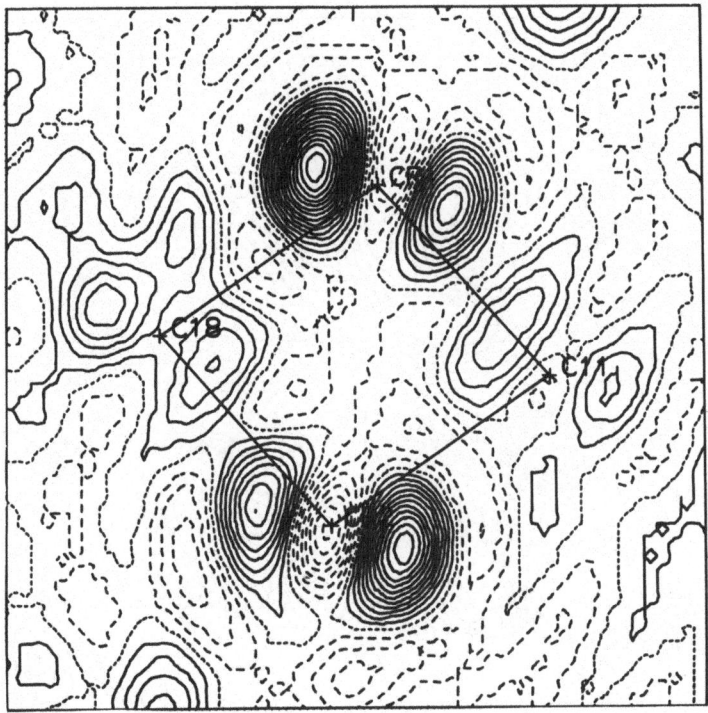

Fig. 5: Deformation density map of (1) in a mean plane defined by
        Cr1, Cr2, C11 and C18, perpendicular to the approximate
        2-fold axis of symmetry. (Plane 1, Fig. 4)

        Fig. 5 gives a view of the deformation density map in a mean
plane defined by Cr1, Cr2, C11 and C18, perpendicular to the approximate 2-fold axis of symmetry. Contours are at 0.1 eÅ^{-3} (zero contours
are dotted, and negative contours dashed).

        Fig. 6 is a slice through the deformation density perpendicular
to that mentioned above and passing through the two chromium atoms,
C14 and C15. The deformation density in the plane of the butadiene
moiety η^4-bonded to Cr2 is shown in Fig. 7 and its equivalent
counterpart appears in Fig. 8. Figs. 9 and 10 give the deformation
density distribution in the plane of the two crystallographically
independent C_5H_5 rings. Accumulation of electron density is observed
in all C-C bonds, but there is remarkably little electron density
along the metal-metal vector, Figs. 5 and 6. Regions of high
electron density within the Cr-C-Cr-C ring (Fig. 5) are consistent
with the formulation that the bridging carbon atoms are each
σ-bonded to one chromium atom and π-bonded to the other. The

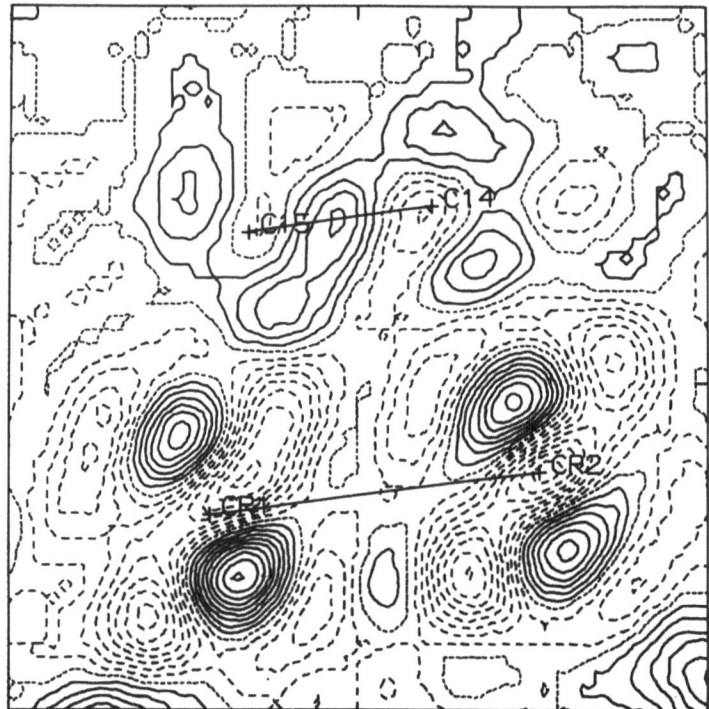

Fig. 6: A slice through the deformation density of (I), perpendicular to that in Fig. 5, and passing through the two chromium
atoms, C14 and C15.

positions of the hydrogen atoms on each of the bridging carbon
atoms and the bridging carbon to chromium atom distances (mean
2.023(2) and mean 2.101(2) Å) are in accord with this picture.
In addition, four relatively strong peaks (1.1-1.4 e$Å^{-3}$), lying
approximately 0.8 Å from the metal, surround each chromium atom.
Each metal and its attendant peaks are coplanar and the planes so
formed are parallel to one another.

It seems worthwhile mentioning that the derived results from
the different data sets are similar in a qualitative way. There
are, however, slight differences on a quantitative level. Based on
the experimentally determined geometry for (I) molecular orbital
calculations of the extended Hückel type were undertaken. The
molecular orbital picture of a M_2Cp_2 unit[6] is well known, as are
those of polyenes, and the bonding of the two metal atoms is
characterised by a stabilisation of metal-metal antibonding
orbitals by the bridging polyene.

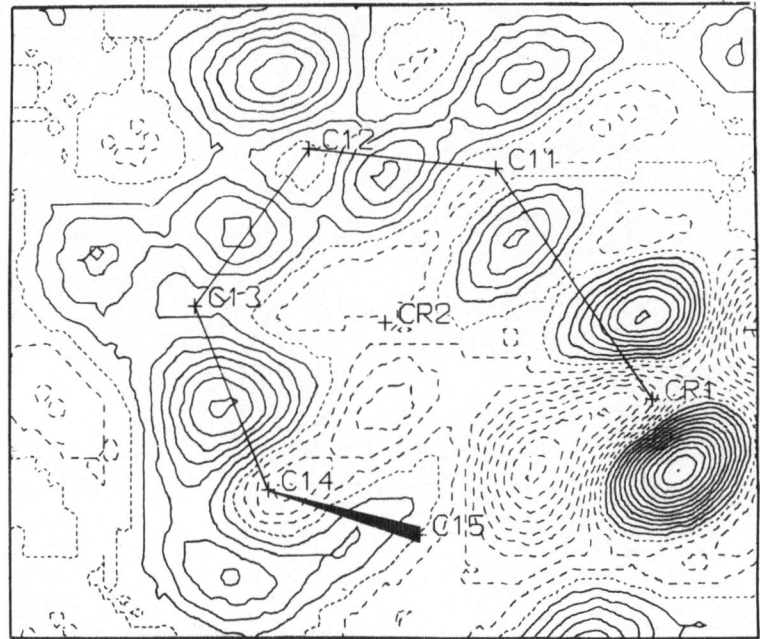

Fig. 7: Deformation density of (I) in the plane of the butadiene
moiety η^4-bonded to Cr2. (Plane 2, Fig. 4.)

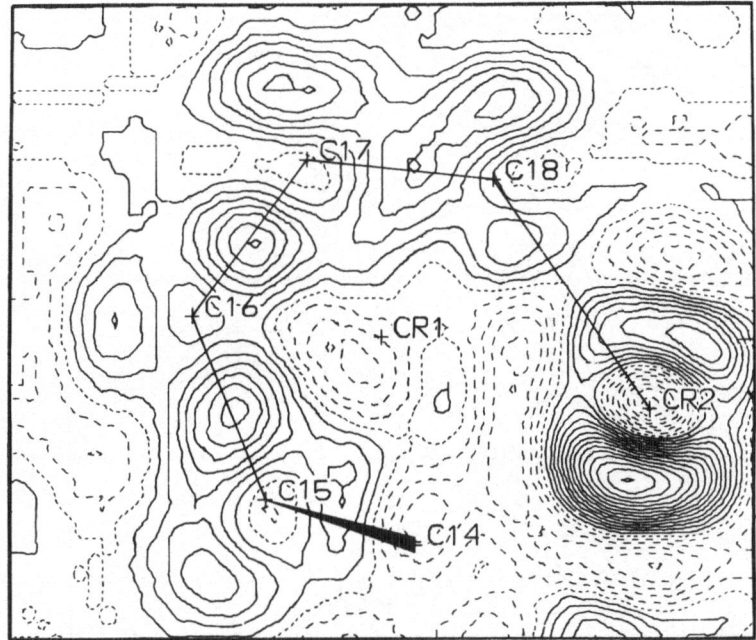

Fig. 8: Deformation density of (I) in the plane of the butadiene
moiety η^4-bonded to Cr1.

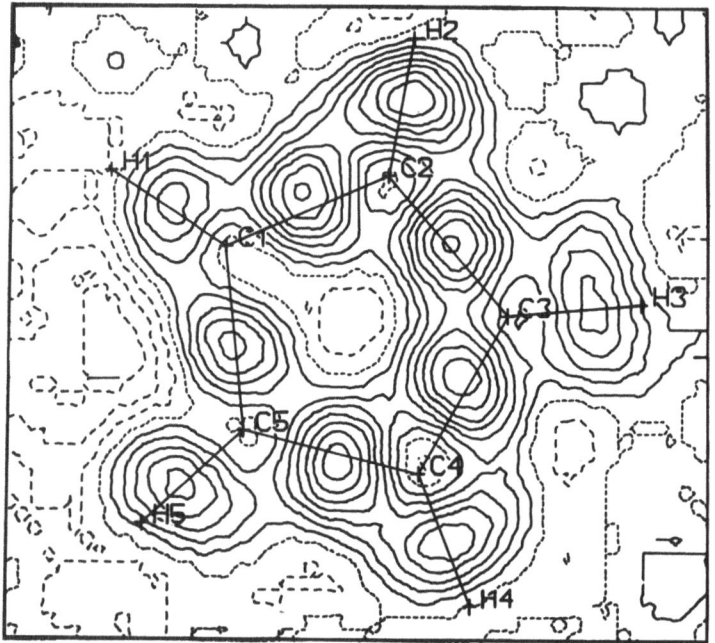

Fig. 9: Deformation density in the plane of the C_5H_5-ring bonded
to Cr1 in (I). (Plane 3, Fig. 4.)

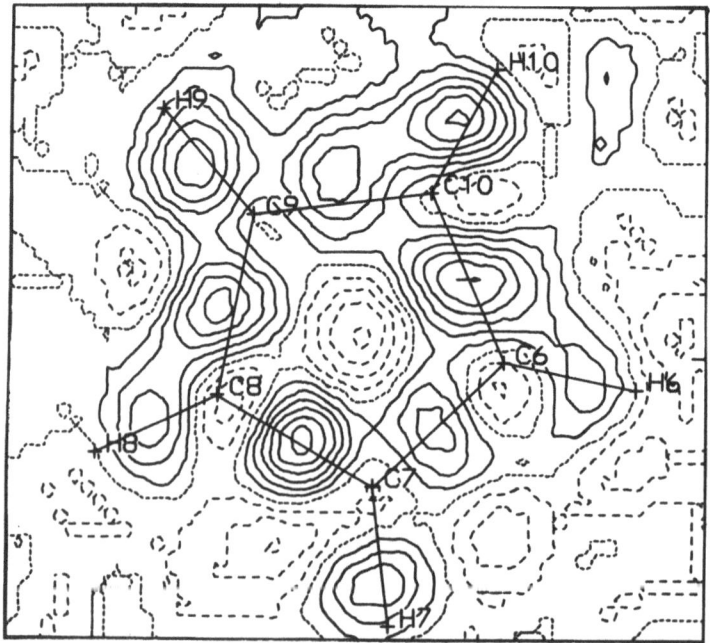

Fig. 10: Deformation density in the plane of the C_5H_5-ring bonded
to Cr2 in (I).

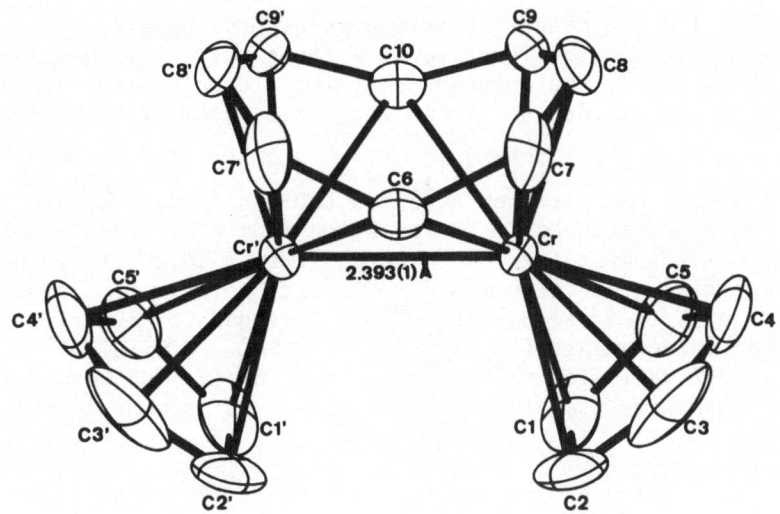

Fig. 11: The molecular structure of (II).

Until now we have been unable to obtain suitable crystals of the ring-closed isomer (II) (Fig. 11) for a similar study. Formally, (II) has a Cr-Cr triple bond.

We then extended our studies to mixed metal compounds, which are interesting in their own right since each metal possesses its own specific ligand-bonding ability. As an example we chose η^4, η^4-cyclooctatetraene-tricarbonylchromium(0)η^5-cyclopentadienylcobalt(1), (III)[7]. This compound exhibits yet another feature:

$$OC-Cr \leftarrow Co$$

(III)

namely, that the two metals in this complex have apparently different electronic configurations. Without the dative metal-metal bond shown in (III), the compound contains a cobalt atom with 18 electrons in the valence shell and a chromium atom with 16 electrons.

17,267 reflections ($\left[\pm h \pm k l\right]$, θ = 0-19.5°; $\left[\pm h k l\right]$, θ = 19.5-39.1°) were collected on a Nonius CAD-4 diffractometer using graphite monochromated Mo-Kα X-radiation from a crystal of (III) maintained at a temperature of -170 $\pm$ 0.5°C by a cold stream of N$_2$ gas. Intensities were corrected for Lorentz, polarisation and absorption effects (μ = 19.43 cm^{-1}) and averaged to give 11,964 independent measurements (R$_{av}$ = $\sum |I - <I>|$ / $\sum <I>$ = 0.013). Of these, 7842 satisfied the criterion I > 2.0 σ(I) and were deemed "observed" and were used in the solution and refinement of the structure.

The deformation density was calculated as above. Non-Hydrogen atom positional and thermal parameters were obtained from high-angle-data least-squares refinement (sin θ/λ > 0.55 Å^{-1}, 4523 obs. refns., R = 0.042; R$_w$ = 0.038, w = 1/σ (F)^2). Hydrogen atom parameters were taken from a previous full-data refinement (7842 refns., R = 0.032, R$_w$ = 0.034, goodness of fit 1.50). The scale factor, k, was determined by least-squares refinement using all the data with sin θ/λ < 0.55 Å^{-1} (3319 refns.) and fixed atom parameters.

There are two independent molecules in the unit cell which, within experimental error, are identical in their molecular parameters. The structure of one molecule of (III) is shown in Fig. 12. A C$_8$H$_8$ ring is attached "diene" fashion to each a chromium and a cobalt atom. Carbon-carbon bond lengths within the ring suggest extensive delocalisation in the two butadiene moieties (C-C$_{mean}$ 1.421 Å) which does not extend throughout the whole ring. There are significantly longer C-C distances in the ring at C11-C18 and C14-C15 (mean 1.457 Å). The carbonyl Cr-C distance trans to the Cr-Co bond at 1.806 Å is appreciably shorter than the equivalent Cr-C distances of the cis carbonyl groups (mean 1.863 Å) and probably reflects the relatively weak Cr-Co interaction, indicated by the long metal-metal bond length (2.923 Å). The molecule as a whole has an idealised mirror plane passing through Cr, Co, Cp2 and the midpoints of the bonds C12-C13 and C16-C17.

Fig. 13 shows the deformation density of the molecule in this plane (contours are at 0.1 eÅ^{-3} with negative contours dashed and zero contours dotted, S$_{max}$ = 0.55 Å^{-1}). The two large peaks above the metal-metal bond can be attributed to electron density in the C12-C13 and C16-C17 bonds. The peak of higher electron density corresponds to the bond closest to the metal with the higher electron count. The elongation of these peaks is consistent with

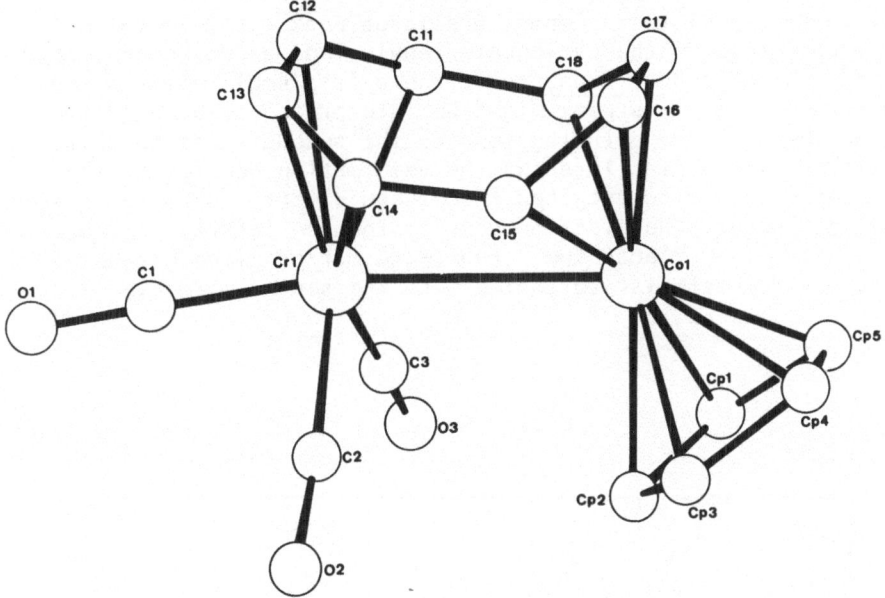

Fig. 12: The molecular structure of (III).

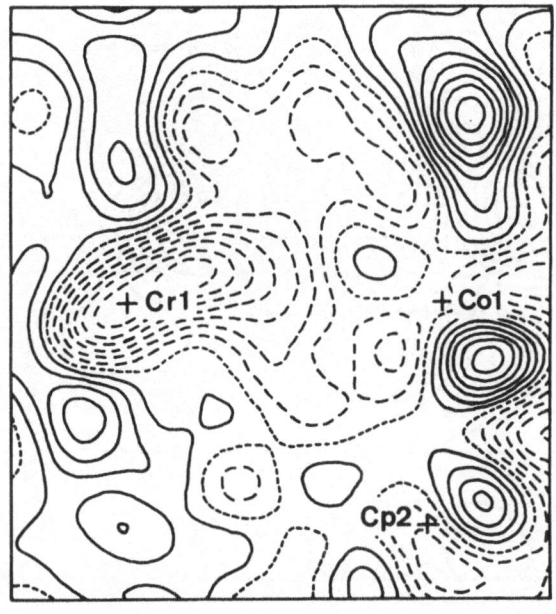

Fig. 13: Deformation density in the mirror plane of one of the
two crystallographically independent molecules of (III).

the π-character of these bonds. One large peak close to Co1 points
in the direction of the cyclopentadienyl ring, as has been observed
in other Cp-compounds. <u>Trans</u> to it there is a small peak in the
direction of Cr1. This peak might be interpreted as being a lone
pair on the Co atom, pointing towards the second electron deficient
transition metal. Fig. 14 shows the deformation density in the
cyclopentadienyl ring attached to Co1 and electron density is seen
in all C-C bonds. The electron density in the vicinity of the
complex closely reflects the approximate mirror plane of the
molecule and is similar to that around the second molecule.

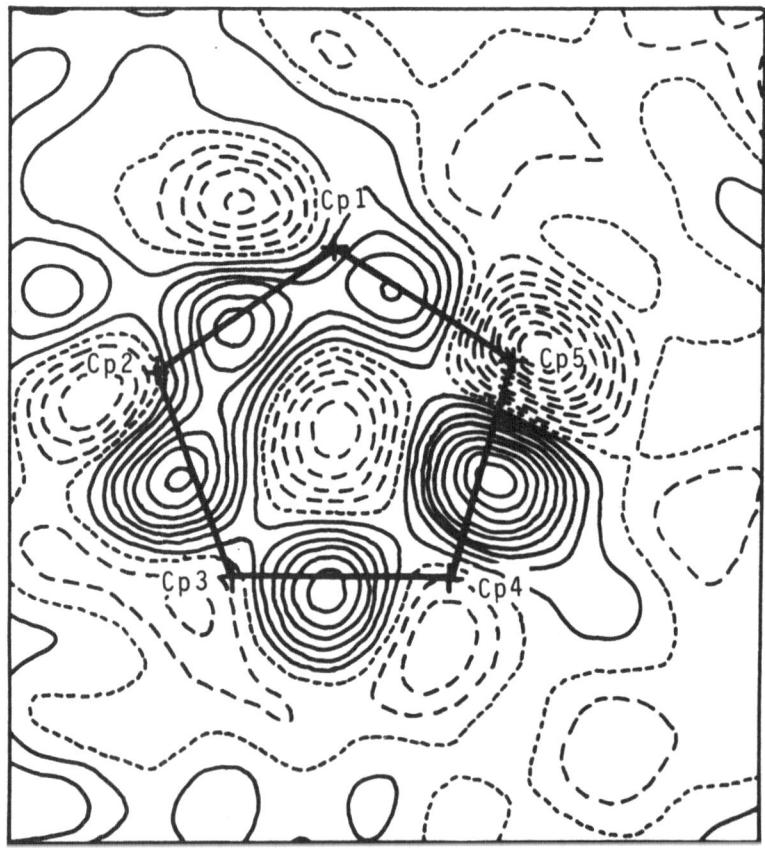

Fig.   14: The deformation density in the mean plane through the
           cyclopentadienyl ring in (η^4: η^4-cyclooctatetraene)-
           tricarbonylchromiumcyclopentadienylcobalt, (III).

The examples given so far prompted us to extend our investigations to other ligand systems. Thus we started a comparative study of η^2-bonded olefin and of η^3-allylic compounds as well as carbenes and carbynes.

Allylnickel complexes are well known as catalysts for the oligomerisation of olefins. It has been recently shown that allylnickelalkoxy complexes catalyse the linear trimerisation of isoprene, whereas the corresponding thio-alkoxy complexes do not [8]. The alkoxy and thio-alkoxy compounds have different structures in the solid state [9]. The η^3-2-methyl-allylnickel-methanolate has a planar Ni-X-Ni-X ring, where X = OMe, while the similar η^3-allyl-nickelthiomethanolate has a puckered one (X = SMe), with a dihedral angle between the coordination planes of the metal atoms of 117° (Figs. 15 and 16, respectively).

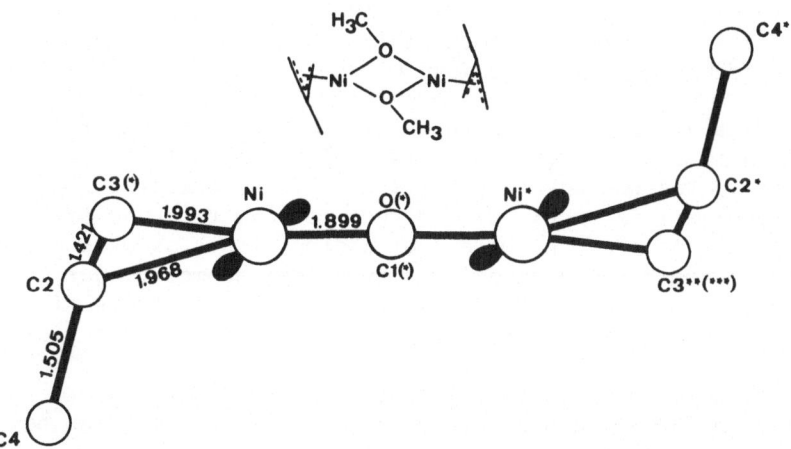

Fig. 15: The molecular structure and schematic formula of bis(2-methyl-π-allyl)nickelmethanolate, (IV).

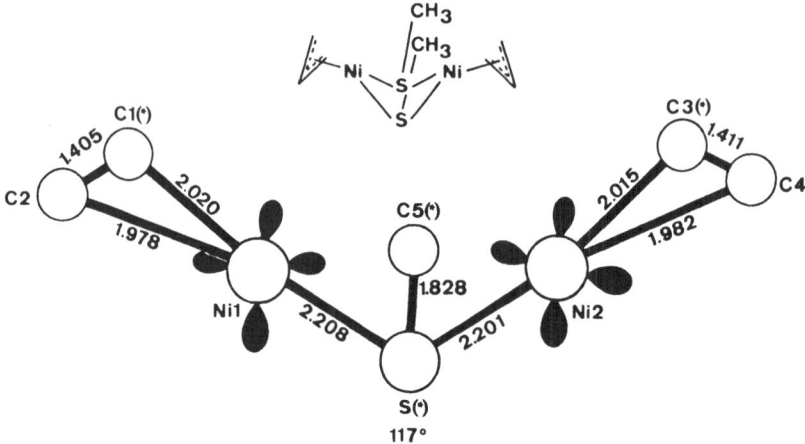

Fig. 16: The molecular structure and schematic formula of
bis(π -allyl)nickelthiomethanolate, (V).

A similar puckering of the Ni-X-Ni-X ring has been observed
in bis $\left[- \mu \text{-methyl-1,3-dimethyl-}\eta^3\text{-allylnickel}\right]$,VI , where the
corresponding dihedral angle is the same[10], and we have also
investigated this compound by (X-X) methods. Molecular orbital
calculations on bis $\left[-\eta^3\text{-allyl-PdCl}\right]$ shows the potenial energy
surface for puckering of the central $(Pd-Cl)_2$ ring to be relatively
flat[11]. N.M.R. evidence indicates this is not the case for the
sulphur-bridged complex (Fig. 16). [9]

The metal-metal distance in these complexes is thus reduced
from 2.91 Å to 2.75 Å when oxygen is replaced by sulphur. We were
interested to see whether the shorter Ni-Ni distance for the
sulphur complex is a result of a direct metal-metal bond stabilised
by the better donor at the bridge or whether it is as a result of
the sulphur atom's preference for a pyramidal geometry. Our
preliminary results of (X-X) deformation density analysis of
(IV) and (V) with low temperature data support the former view.

Experimental details for these three investigations are
summarised in Table 4 and described for the sulphur complex below.

Intensity measurements using graphite monochromated Mo-Kα
X-radiation were made in the range $1.8 < \theta < 59.7°$ on a crystal of

Table 4: Experimental data for IV, V and VI.

	$[(\pi-CH_3C_3H_4)NiOCH_3]_2$	$[(\pi-C_3H_5)NiSCH_3]_2$	$[(\pi-C_5H_9)NiCH_3]_2$
	IV	V	VI
Space group	Pnnm	Pnma	P21/n
a	11.973(1) Å	9.870(1) Å	9.541(2) Å
b	4.586(1) Å	11.575(1) Å	9.307(1) Å
c	10.984(1) Å	9.927(1) Å	14.921(2) Å
β	90.0°	90.0°	94.43°
Temperature	-170°	-170°	-170°
θ-range	1.0 - 50.0°	1.0 - 33.5°	1.0 - 60.0°
Reflections measured	$[\pm h, \pm k, \bar{1}]$	$[\pm h, k, \bar{1}]$	$[\pm h, k, \bar{1}]$
No. of reflections	10475	6196	24095
θ-range	49.95 - 75.0°	33.45 - 60.0°	
Reflections measured	$[h, k, \bar{1}]$	$[h, k, \bar{1}]$	
No. of reflections	3512	7050	
$\sin\theta/\lambda$ max	1.36 Å^{-1}	1.22 Å^{-1}	1.22 Å^{-1}
Averaged to	6446	8715	22934
R_{av}	0.025	0.014	0.026
Scan width	1°(2θ/Ω), 60 sec. max. measurement time, multiscan until intensity differences less than 2σ. Data corrected for absorption and isotropic extinction.		

the sulphur complex maintained at a temperature of 103°K. The 13,246
reflections collected were corrected for Lorentz,polarisation and
absorption effects, and they were averaged to 8,715 (R_{av} = 0.014),
of which 4,798 had I > 2.0 σ (I). Refinement with all the data gave
R = 0.04, R_w = 0.034, and included an isotropic extinction parameter.

The deformation density map was calculated as usual from
$\Delta\rho$ = $\rho_{crystal}$ - $\sum\rho_{free\ atoms}$. Non-hydrogen atom positions were
obtained from high-angle-data least-squares refinement (sin θ/λ >
0.65 $Å^{-1}$, 3538 refns., R_w = 0.04) and hydrogen atoms were located
from the full-data refinement. The scale factor, k, was determined
by least-squares refinement using all the data with sin θ/λ < 0.75$Å^{-1}$
and fixed atom parameters.

Crystal quality and thermal instability of VI, however, was
such that derived results suggest a reinvestigation for confirmation.
Essential features of the deformation density are nevertheless
similar to those observed for (IV).

Comparable cuts through sections of all three molecules are
given in Figs. 17 to 22. Figs. 17, 18 and 19 show the resulting

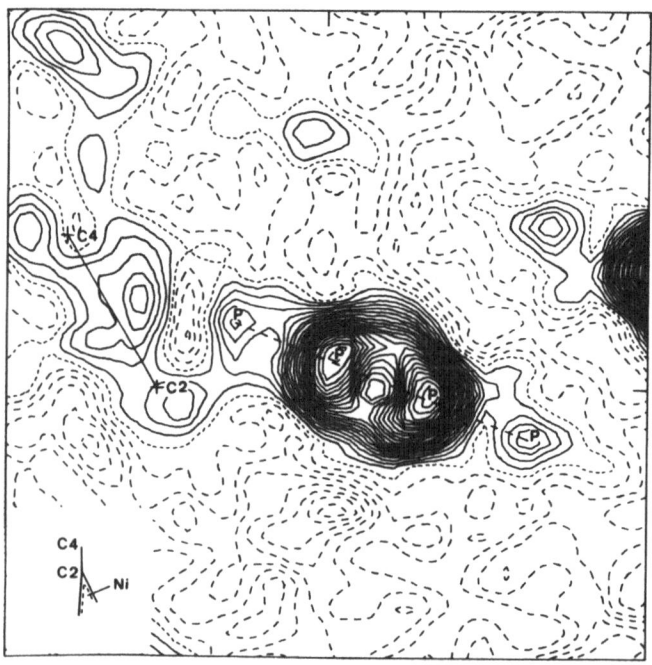

Fig. 17: Deformation density of (IV) through nickel, C2 and C4
perpendicular to the plane of the Ni-0-Ni-0 ring.

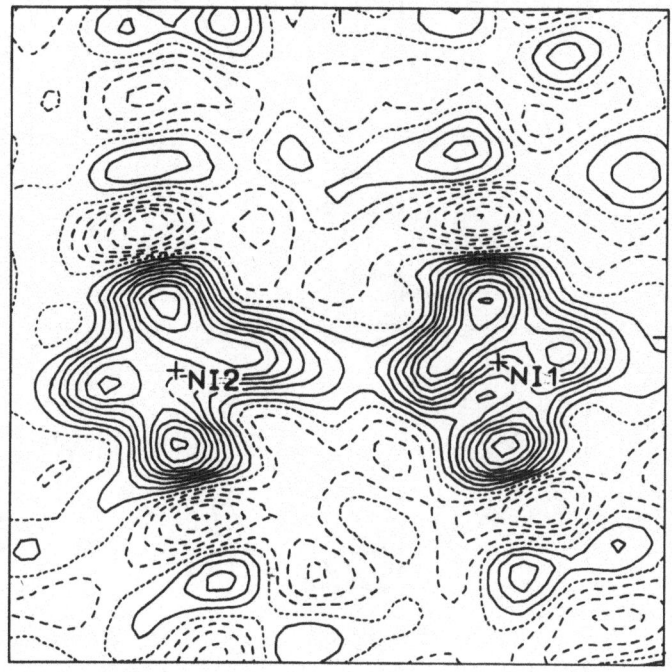

Fig. 18: Deformation density of (V) through the two nickel atoms
perpendicular to the S----S vector.

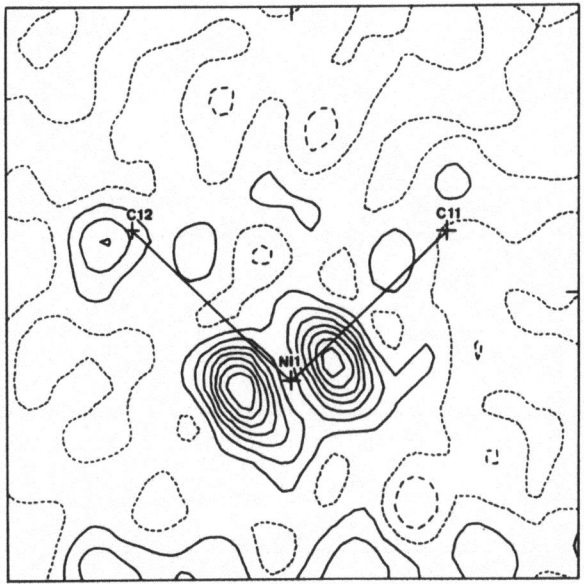

Fig. 19: Deformation density of (VI). A slice through the C-Ni-C
portion of the Ni-C-Ni-C ring.

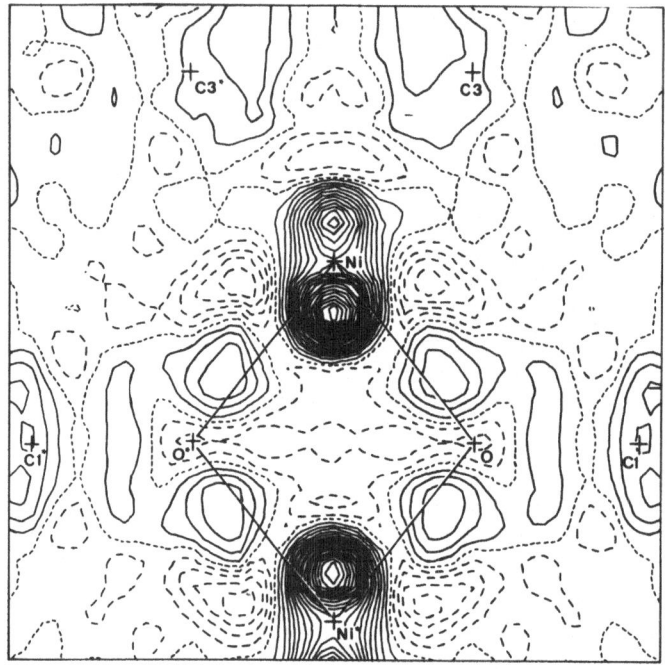

Fig. 20: Deformation density of (IV) through the Ni-O-Ni-O ring.

Fig. 21: Deformation density in the plane of the Ni1-S-Ni2 triangle
          of (V).

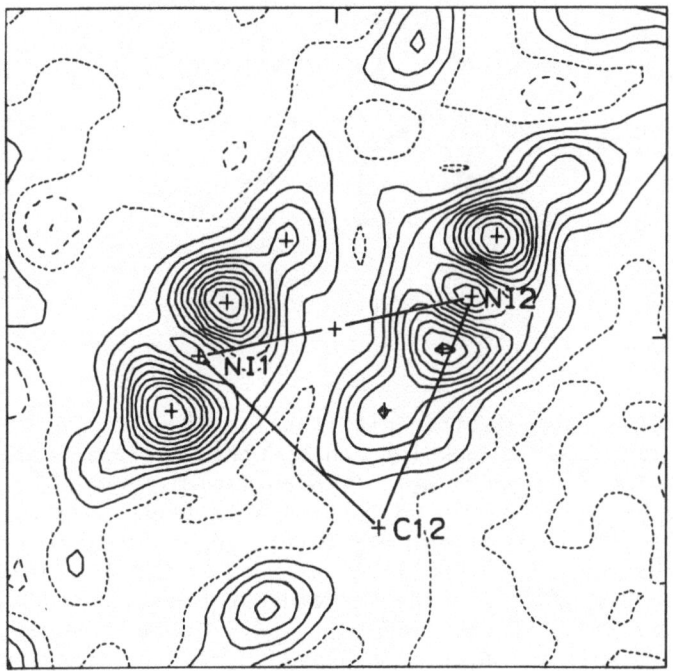

Fig. 22: Deformation density through Ni-C-Ni in the Ni-C-Ni-C ring of (VI).

maps drawn in planes containing the two nickel atoms and perpendicular to the vector joining the bridging atoms (0, S, C) (contours are at 0.1 eÅ^{-3}; continuous, dotted, and dashed lines correspond respectively to positive, zero, and negative contours).

Whereas in (IV) and (VI) (0, C) two intense lobes surround each nickel atom, in (V) (S), four such peaks are arranged so that one lobe on each metal atom lies along the metal-metal axis, indicating a nickel-nickel interaction. Other peaks on the deformation maps lie between the carbon atoms of the η^3-allyl ligands, in the C-S bond and around the sulphur atom itself. Fig. 21 shows the deformation density in the plane of the two metals and one of the bridging sulphur atoms.

This apparent metal-metal interaction prompted us to extend our studies into related classes of compounds. Thus sets of neutron and X-ray data of (VII) are presently in the stage of data processing. This research on biologically relevant compounds is performed in cooperation with Prof. D. Seyferth, MIT.

(VII)

Knowledge of the electron distribution about the η^3-allyl groups encouraged us to look into an acentric problem containing allyl groups. An X-ray data set of tetraallyldichromium[12], (VIII), (space group $P\bar{4}2_1c$), a compound that is a suitable catalyst for the polymerisation of olefins, was collected at low temperature ($a = b = 18.050(2)$, $c = 7.406(1)$ Å, $Z = 8$; $[\pm h \pm k \pm l]$ 24,192 refl., $\sin \theta/\lambda_{max} = 0.85$ Å^{-1}) and the structure refined as described above ($R = 0.039$, 5,560 obs. refns., $I > 2.0\sigma(I)$). However, using those techniques which have proved useful for centric problems have not led to sensible results thus far case. At present, more exact phases are being sought by applying multipole refinement and similar strategies.

As various combinations of transition metal-organic compounds and main group metal-organic compounds form valuable catalytic systems for polymerisation of olefins etc., a study of the nature of the interaction of these different metals within model compounds should prove to be a very interesting problem.

The structures of transition metal organometallic complexes containing lithium are usually ionic. They comprise an organo-metallic anion and lithium cation separated by distances greater than the sums of their respective van der Waals radii. However, there is increasing structural evidence that in certain organo-metallic complexes, notably those of low valent transition metals, the lithium is directly bound to the complex[13]. Fig. 23 (IX) shows one such example, there are others[13]. Some insight into the nature of this metal-metal interaction is the object of our investigation.

By abstracting the lithium atom from (IX) with tritylchloride it has been possible to obtain crystals of the paramagnetic organometallic complex alone, (X) (Fig. 24). We have measured X-ray and neutron diffraction data from this compound at -170 °C.

If the interaction of this complex with lithium were solely

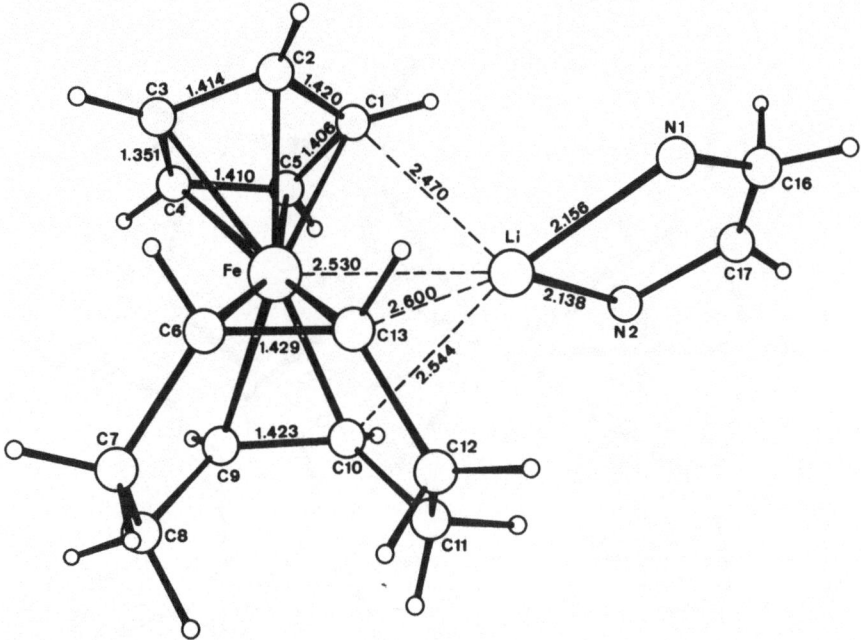

Fig. 23: The molecular structure of (π-cyclopentadienyl)-
(η^{4}-1,5-cyclooctadiene)iron-lithium(tetramethylethylene-
diamine), (IX). (The methyl groups on the two nitrogen
atoms have been omitted for clarity.)

ionic, then the lithium would be expected to position itself
closest to the maximum electron density. The more covalent the
interaction the more the position of the lithium will be dictated
by the best orbital interaction. The lithium atom has an sp^3-hybrid
pointing away from the solvating ligands and a p-orbital perpendicular to this. Fig. 25 shows the schematic representation of the
orbitals of (C_5H_5) $(C_2H_4)_2Fe$ built up from those of two fragments,
one of $(C_5H_5)Fe$ and the other of two ethylene molecules[14].
In a covalent interaction the sp^3-hybrid on lithium would be
ideally suited to interact with the HOMO in (X), the seat of the
paramagnetism (Xa). The determination of the spin density in
$(C_5H_5)(C_8H_{10})Fe$ will be the subject of a future investigation
using polarised neutrons.

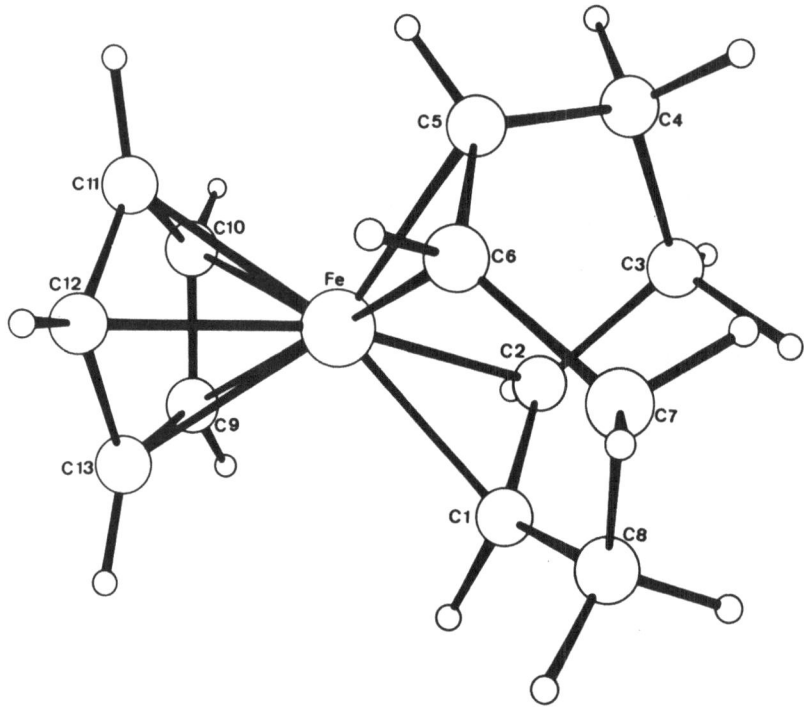

Fig. 24: The molecular structure of the (π-cyclopentadienyl)
(η^4-1,5-cyclooctadiene)iron radical, (X).

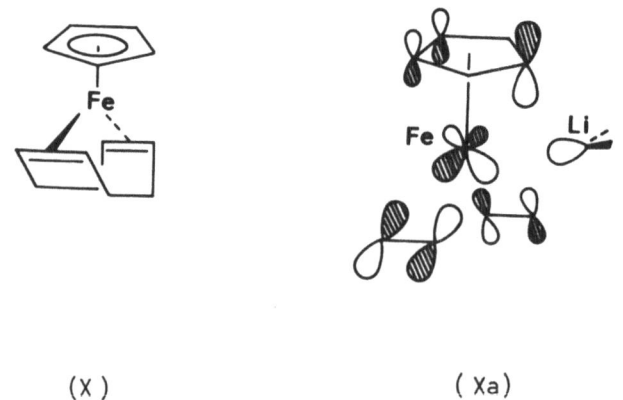

(X)   (Xa)

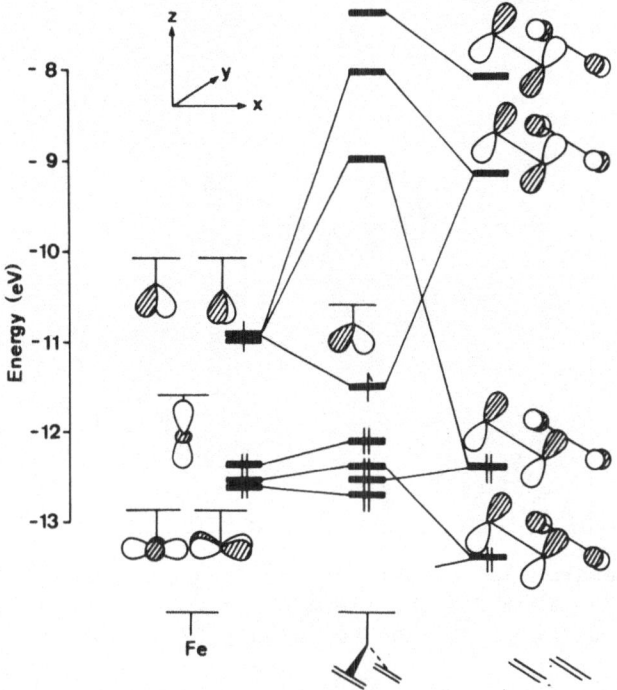

Fig. 25: The orbitals of $(C_5H_5)(C_2H_4)_2Fe$ built up from those of two fragments, one of (C_5H_5) Fe and the other of two ethylene molecules.

So far, 18,637 diffracted intensities have been measured on this compound at -170°C in the range $1.8 < \theta < 70.0°$ on a Nonius CAD-4 diffractometer using graphite monochromated Mo-K_α X-radiation. Nonhydrogen atom positions were obtained from high-angle-data least-squares refinement, $\sin \theta / \lambda > 0.8$ Å^{-1} (3156 obs. refns., $R_w = 0.050$), hydrogen atoms being located from full data refinement (6845 obs. refns., R = 0.048) and included to give a better model fit. The scale factor, k, was determined by least-squares refinement using all the data with $\sin \theta/\lambda < 0.75$ Å^{-1} and fixed atom parameters.

Fig. 26 shows the electronic deformation density of (X) in a plane defined by the iron atom and the midpoint between C2 and C5 (C2,5) and between C1 and C6 (C1,6) (contours are at 0.1 eÅ^{-3} with negative contours dashed and zero contours dotted, $S_{max} = 0.75$ Å^{-1}). Electron density is clearly seen in the bonds C3, C4 and C8, C7, which cut the plane at their midpoints. Electron density around the iron atom is divided into three lobes, two pointing towards the cyclopentadienyl ring and one towards the centre of the cyclooctadienyl ligand. It is interesting to note

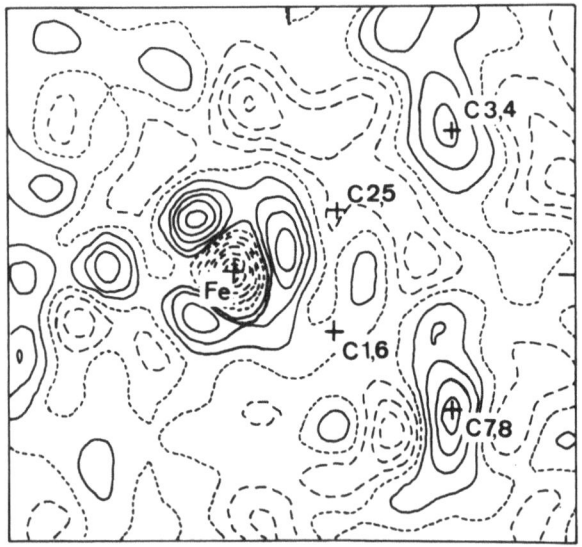

Fig. 26: Deformation density in (X) in a plane defined by the
         iron atom and the midpoints between C2 and C5 (C2,5)
         and between C1 and C6 (C1,6), (X-X).

that the direction of approach of the lithium atom in (IX) is
along one of the regions of high electron density on the metal in
(X). Again, the symmetry of the deformation density map in the
vicinity of the molecule closely reflects the local chemical
pseudo - symmetry (2).

    Finally, we have extended our studies to transition metal-
carbon multiple bonds. As an initial investigation we have chosen,
in collaboration with Prof. E. 0. Fischer, Munich, one each of
the simplest carbyne (XI) and carbene (XII) complexes.

                    (XI)   (XII)

It was only in 1973 that the first organometallic complexes containing a metal-carbon triple bond were prepared[15]. In the meantime numerous structures have appeared in the literature[16]. These compounds are characterised by a linearly coordinated carbyne carbon atom and by short metal-carbon distances. The highly unsaturated nature of the metal-carbon bond makes these compounds particulary important chemically. An electron density study of one of the simplest carbyne complexes, (XI), has been undertaken to investigate the nature of this metal-carbon triple bond by the (X-X) method[1].

Conventional full data refinement of the structure using the low temperature X-ray data revealed a rather unusual distortion. Instead of the $Cl(CO)_4CrC$ part of the molecule assuming its ideal C_{4v} symmetry it is reduced to C_{2v}. Significant differences in the Cr-C and to a lesser extent in the C-O distances in the four carbonyl groups are observed. These are given in Fig. 27, which shows the molecular structure. Although the hydrogen atoms are poorly located in the X-ray experiment it is interesting to note that one hydrogen atom, H3, is found to lie in the plane of the two carbonyl groups which have the shortest Cr-C distances.

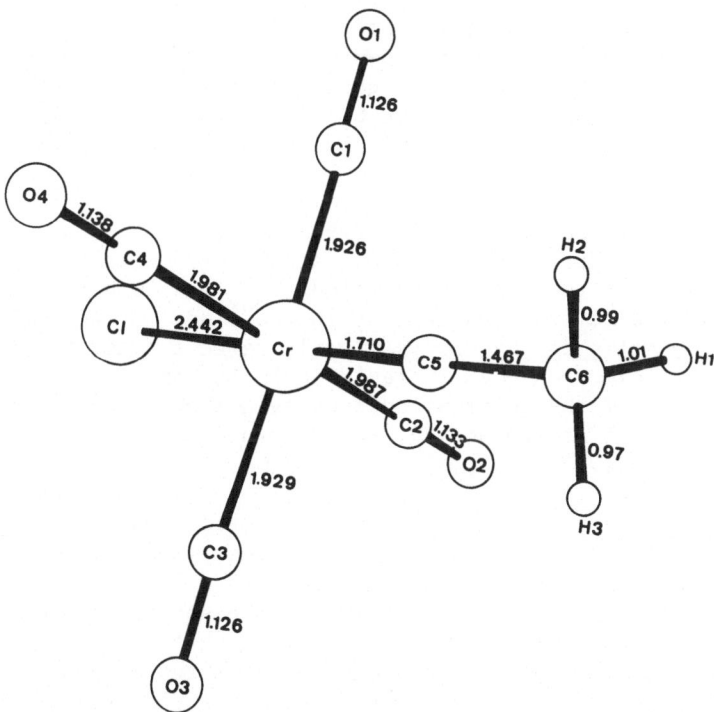

Fig. 27: The molecular structure of the carbyne complex tetracarbonylchloroethynechromium, (XI).

Is it possible that there is an interaction between the ligands on the metal and the hydrogen atoms on the methyl group through the $M \equiv C$ bond as is shown below?

$$Cl(CO)_4Cr \equiv C-C\begin{array}{c} H \\ \diagdown H \\ H \end{array} \longleftrightarrow Cl(CO)_4Cr = C = C\begin{array}{c} H \\ \diagup \\ H \end{array} \quad H \oplus$$

X-ray reflections were collected ($[\pm h \pm kl]$, $\theta = 1.0$-$29.0°$; $[\pm hkl]$, $\theta = 29.0$-$45.9°$) using graphite-monochromated Mo-Kα X-radiation from a yellow crystal of (XI)[17] mounted on a Nonius CAD-4 diffractometer and cooled to -170 ± 0.2 °C with a cool stream of N_2 gas. The 4742 measured intensities were corrected for Lorentz, polarisation and absorption effects ($\mu = 15.2$ cm^{-1}; t_{max} 0.890, t_{min} 0.843) and were averaged to give 1836 independent observed ($I > 2.0 \sigma(I)$) reflections $[R_{average} = \sum ||I - <I>| / \sum <I> = 0.013]$. Least-squares refinement with all the data converged at $R = 0.025$; $R_w = 0.026$, $w = 1/\sigma(F)^2$.

The deformation density map was calculated as described above. Atomic parameters of the promolecule[18] were obtained from high-angle-data least-squares refinement (sin $\theta/\lambda > 0.55$ Å^{-1}, 775 refns., $R = 0.030$, $R_w = 0.025$; goodness-of-fit 1.10). The scale factor, k, was determined by least-squares refinement using all the data with sin $\theta/\lambda < 0.56$ Å^{-1} and fixed atom parameters (1120 refns.).

Resulting sections through the (X-X) deformation density are shown in Figs. 28-30 (contours are at 0.1 eÅ^{-3} with negative contours dashed and zero contours dotted, $S_{max} = 0.56$ Å^{-1}). Fig. 28 shows the deformation density in the plane of the $Cr(CO)_4$ moiety. The symmetry of the map is clearly reduced from C_4 to C_2 with the higher electron density residing in the shorter Cr-C bonds. Two views of the deformation density along the Cl-$Cr \equiv C$-C axis are shown in Figs. 29 and 30, each is in a plane containing two carbonyl groups.

Particularly noteworthy is the high electron density in the C5-C6 bond and the extended nature of the electron density in the $Cr \equiv C$ bond. One explanation of this feature is a hyperconjugative type of interaction of the methyl group at the sp-carbon atom with the transition metal as indicated above. This is in accord with spectroscopic evidence on nitrogen substituted carbyne complexes[19]. Such an interaction could lead to a distortion within the $Cr(CO)_4$ group, so that alternate Cr-C distances significantly differ from one another. Apart from the possible effect of crystal packing, the influence of the C_{3v}-symmetry of the CH_3 group appears

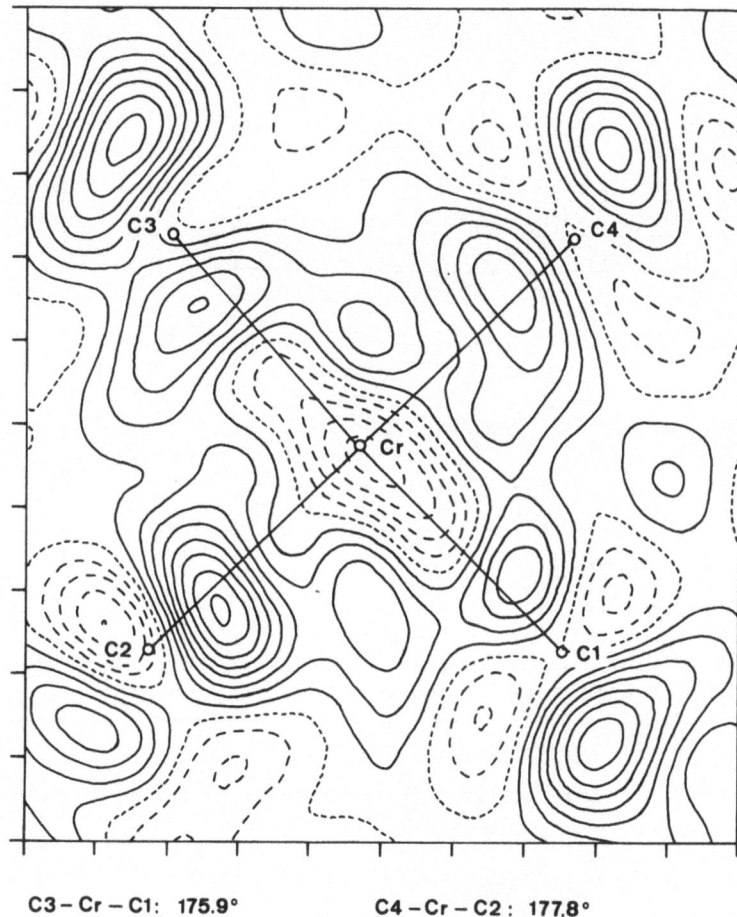

C3 – Cr – C1: 175.9°   C4 – Cr – C2: 177.8°

Fig. 28: Deformation density in the plane of the Cr(CO)$_4$
group in (XI).

to be the only perturbation which could cause the reduction of the local symmetry of the Cl(CO)$_4$Cr group from C_{4v} to C_{2v}. The symmetry of the deformation density reflects the approximate mirror symmetry of the molecule. A simple molecular orbital description of this type of complex, without taking hyperconjugative effects into account, is given in Fig. 31. All subsequent theoretical calculations on ClCr(CO)$_4$CCH$_3$[20] and (CO)$_5$CrCNEt$_2^{\oplus}$ agree with these results[21].

X-ray and neutron data are now available for pentacarbonyl-chromiummethoxymethylcarbene and its deformation density is presently being investigated.

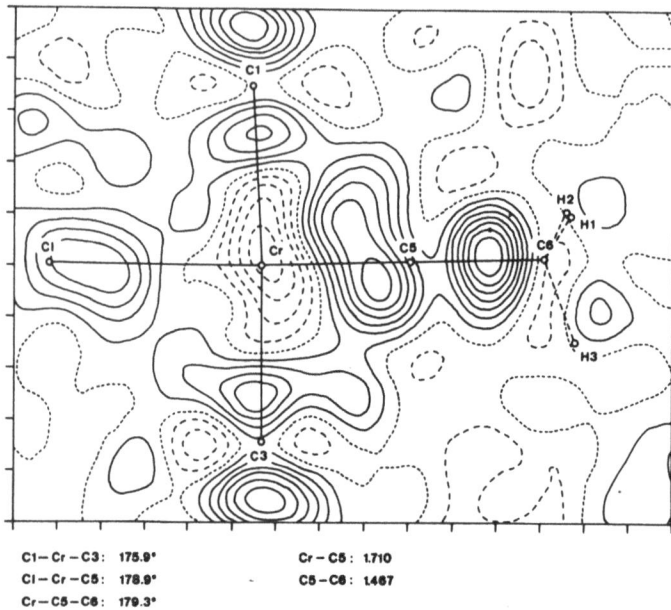

C1−Cr−C3: 175.9°   Cr−C5: 1.710
Cl−Cr−C5: 178.9°   C5−C6: 1.487
Cr−C5−C6: 179.3°

Fig. 29: A view of the deformation density in (XI) along the
         Cl-Cr≡C-Me axis in the plane of two carbonyl groups.

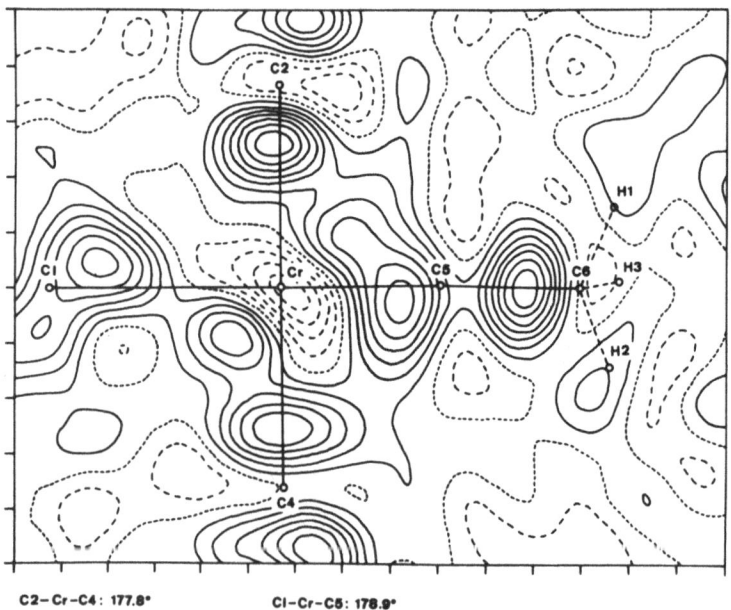

C2−Cr−C4: 177.8°   Cl−Cr−C5: 178.9°

Fig. 30: The deformation density of (XI) along the Cl-Cr≡C-Me
         axis perpendicular to that in Fig. 29.

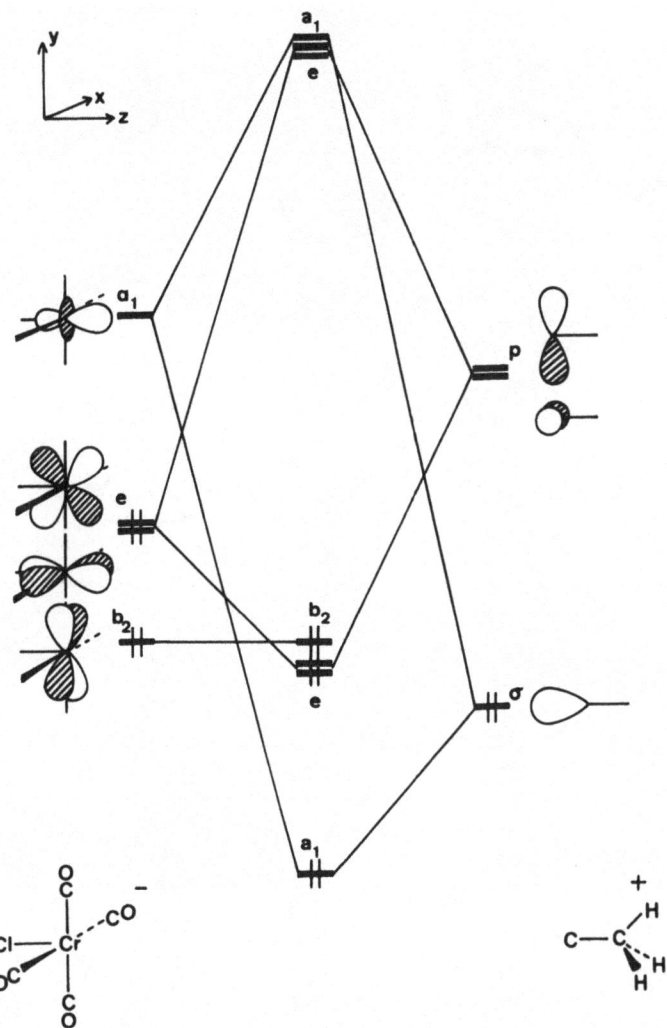

Fig. 31: The orbitals of the methyne complex showing the interaction
of C-Me with a Cl(CO)$_4$Cr unit. A distortion such
as the one observed leads to a splitting of the e levels.

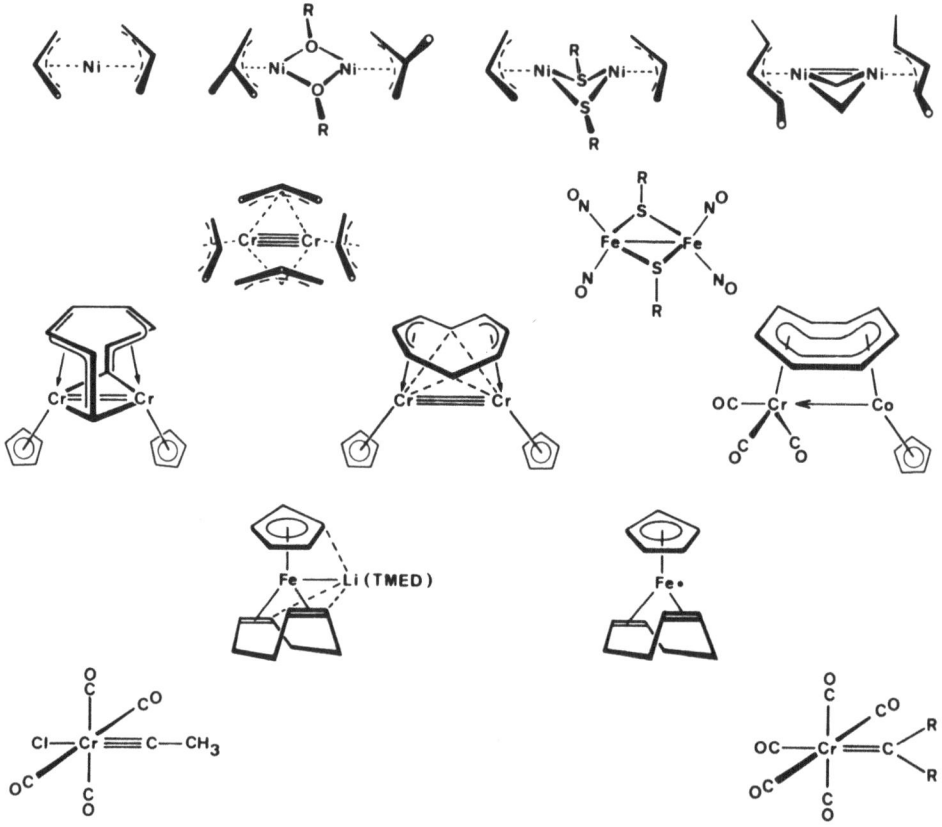

Fig. 32

Summary

Figure 32 shows the organometallic compounds that we are
currently studying. We have consistently found peaks in the
deformation densities of these compounds in all carbon-carbon
bonds, in the metal-metal-ligand bonding region and around
the transition metals themselves. As simple theoretical considerations provide only a partial explanation for what is observed
around the metal, improvements in both experimental and theoretical
methods are being sought.

Acknowledgements

This research was sponsored by the Deutsche Forschungsgemein-schaft and a research grant of the Max-Planck-Gesellschaft to R. G., which is greatfully acknowledged. In addition, we are indebted to the local staff at the I.L.L., Grenoble, for their valuable help.

References

1. Bat-Sheva Seminar on electron density mapping in molecules and crystals, April 1977; Israel. J. Chem., 16:Nos 2-3 (1977).
    P. Coppens and E. D. Stevens, Adv. Quantum Chem., 10:1 (1977).
    P. Coppens, Angew. Chem., 89:33 (1977).
    Electron and Magnetization Densities in Molecules and Crystals, NATO ASI Series B, Physics Vol. 48, Plenum Press (1979).
    P. Coppens, Trans. A.C.A., 16:59 (1980).
2. Neutron data by M. Lehmann and R. Feld as part of the oxalic acid project.
3. E. D. Stevens and P. Coppens, Acta Cryst., B36:1864 (1980).
    R. Feld, Dissertation Univ. Marburg (1980).
4. H. Johannson, Acta Cryst., A35:319 (1979).
    E. D. Stevens, Acta Cryst., B36:1876 (1980).
5. W. Geibel, G. Wilke, R. Goddard, C. Krüger and R. Mynott, J. Organomet. Chem., 160:139 (1978).
6. D. L. Thorn and R. Hoffmann, Inorg. Chem., 17:126 (1978).
7. The compound was kindly supplied by Dr. A. Salzer, Univ. Zürich.
8. S. Akutagawa, T. Taketomi, H. Kumobayashi, K. Takayama, T. Someya and S. Otsuka, Bull. Chem. Soc. Japan, 51:1158 (1978).
9. B. Gerding, Dissertation, Univ.-GHS-Essen (1979).
10. C. Krüger, J. C. Sekutowski, H. Berke and R. Hoffmann, Z. Naturforsch., 33b:1110 (1978).
11. R. H. Summerville and R. Hoffmann, J. Amer. Chem. Soc., 98:7240 (1976).
12. W.Oberkirch, Dissertation, TH-Aachen (1963).
    T. Aoki, A. Furusaki, Y. Tomiie, K. Ono and K. Tanaka, Bull. Soc. Chim. Japan, 42:545 (1969).
13. K. Jonas and C. Krüger, Angew. Chem., 92:513 (1980).
    Angew. Chem. Int. Ed. Engl., 19:520 (1980).
    K. Jonas, R. Mynott, C. Krüger, J. C. Sekutowski and Y.-H. Tsay, Angew. Chem., 88:682 (1976).
14. For a description of the method see: M. Elian, R. Hoffmann, Inorg. Chem., 14:1058 (1975).
15. E. O. Fischer, G. Kreis, C. G. Kreiter, J. Müller, G. Huttner and H. Lorenz, Angew. Chem. Int. Ed. Engl., 12:564 (1973).
16. G. Huttner, A. Frank and E. O. Fischer, Israel J. Chem., 15:133 (1977).
17. The crystalline sample was kindly supplied by Prof. E. O. Fischer, Univ. Munich.

18. F. L. Hirshfeld, Theoret. Chim. Acta, 44:129 (1977).
19. U. Schubert, D. Neugebauer, H. Fischer, P. Hofmann and
    E. A. Schilling, personal communication.
20. D. Saddei, H. J. Freund and G. Hohnleicher, personal communica-
    tion.
21. N. M. Kostic and R. F. Fenske, personal communication.

ANALYSIS OF ELECTRONIC STRUCTURE FROM ELECTRON DENSITY

DISTRIBUTIONS OF TRANSITION METAL COMPLEXES

Edwin D. Stevens*

Department of Chemistry
State University of New York at Buffalo
Buffalo, New York 14214

INTRODUCTION

Accurate experimental determination of the electron density distribution in molecular solids has been realized in the past decade. For molecules containing only light atoms, careful experiments are found to yield charge densities in excellent agreement with the most sophisticated theoretical calculations.[1,2] In the majority of cases, however, the observed electron distributions of light atom systems could have been predicted, at least qualitatively, from simple valence bond models. Experimentally determined electron distributions are therefore potentially more useful in understanding the electronic structures of transition metal complexes were the mechanisms of metal-metal and metal-ligand bonding are often far less obvious.

Few accurate studies of systems containing transition metal atoms have been reported. The quantity of interest, the valence electron distribution, contributes a smaller fraction to the total x-ray scattering. Also, experimental problems such as errors in the corrections for absorption and extinction become more significant when transition metal or other heavy atoms are present. However, the accuracy of the experimental methods may be expected to continually improve. In addition, the prominent role of transition metal complexes in many aspects of chemistry insure that a

----

*Present address: Department of Chemistry
                    University of New Orleans
                    New Orleans, Louisiana 70122

great many studies of such complexes will be reported in the
future.

No attempt will be made to review all the reported results
on systems containing transition metals. This contribution will
be concerned with techniques for the analysis and interpretation
of experimental charge density maps of molecules containing
transition metal atoms.

DEFORMATION MAPS

Let us assume that an accurate set of x-ray intensity measurements have been collected to high resolution, and reduced to a
set of experimental structure factors, F_{obs}, with phases obtained
from a model structure. An experimental electron distribution
can then be calculated by a Fourier summation over the observed
structure factors. The rearrangement of the electron distribution
as a result of chemical bonding is more clearly revealed in the
deformation density,

$$\Delta\rho = \frac{1}{V} \Sigma (F_{obs} - F_{ref}) e^{2\pi i \underset{\sim}{S} \cdot \underset{\sim}{r}} \qquad (1)$$

defined as the difference between the observed density and the
density calculated for a superposition of neutral, spherical
atoms.

Since considerable time and effort must be invested in
obtaining the experimental electron distribution, it is obviously
desirable to extract as much information as possible from each
study. In general, there are three approaches to interpretation
of the experimental maps:

1. Direct inspection,
2. Modeling with deformation functions,
3. Comparison with theoretical calculations.

Although this paper is concerned with the interpretation of the
deformation density, it should be noted that even if the ultimate
interest is in some other one-electron property which may be
calculated from the experimental density, such as the distribution
of net atomic charges, dipole moments, electric field gradients,
or electrostatic potentials, the same approaches to the levels of
interpretations apply.

Much information may be obtained by simple examination of
the experimental map. The deformation density in the region of
organic ligand molecules shows peaks associated with covalent
bonding and non-bonding ("lone pair") electron distributions. In

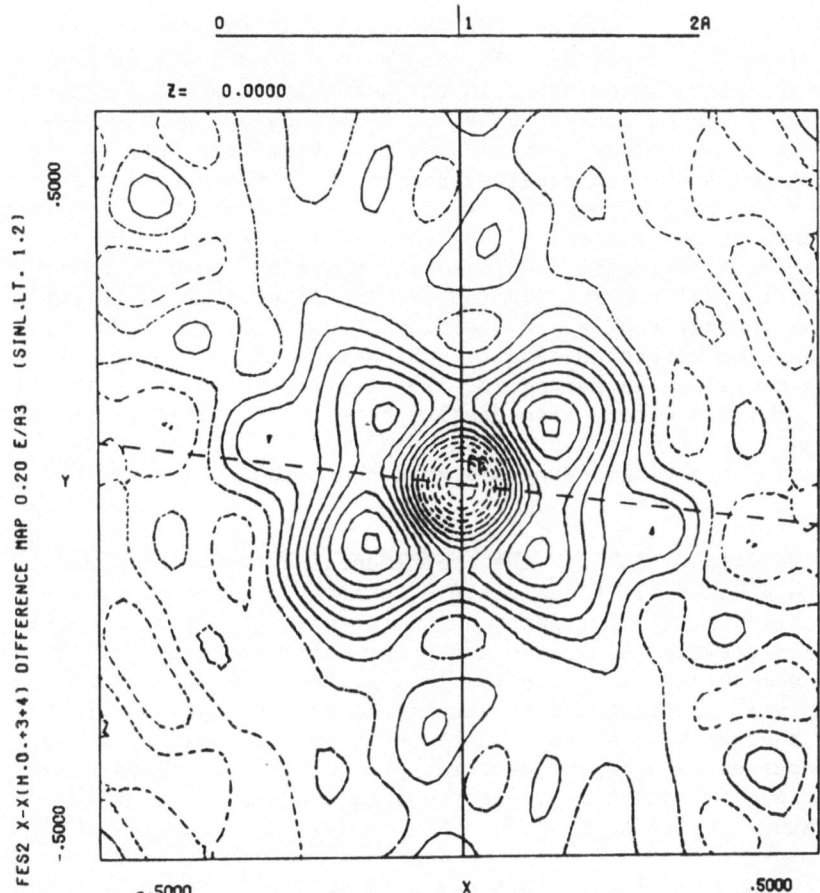

Figure 1. Experimental deformation density around the iron atom
in FeS_2. The plane is defined by a line joining the
iron atom with one of the sulfur atoms and a line
(dashed) bisecting two other sulfur atoms. Contours at
0.2 e/$Å^3$ intervals with zero and negative contours
broken.

addition, sharp peaks are frequently observed near transition atoms with partially filled d-shells (Fig. 1). In many cases, this asphericity near the metal is attributable to crystal field splitting of the d-orbitals and preferential occupancy of the lower energy levels by the d-electrons.[3-6]

In some cases, however, the appearance of the density map may be misleading. For example, an absence of density accumulation is occasionally observed in the deformation density between two formally bonded atoms. However, this does not necessarily indicate a weak bond or lack of bonding interaction between the atoms. A small or even negative bond peak in the deformation density often occurs when the reference density which is subtracted corresponds to an atom with a nearly filled valence shell. Consider the F_2 molecule, for example, where the bonding interaction will involve primarily the overlap of p-orbitals oriented along the molecular axis and occupied by one electron from each atom. For the reference density subtracted, the occupancy of the p-orbitals will be an average of 5/3 electrons each, resulting in a large negative deformation density between the atoms. The same effect will be seen in transition complexes containing metal-metal bonds involving d-orbitals when the d-shell is more than half full.

In addition, although the experimental deformation density conveys a great deal of qualitative information, the extraction of reliable quantitative information is considerably more difficult. For example, since the x-ray intensity measurements extend only to some finite limit in resolution, features in the deformation map will be truncated to varying degrees by series termination. Truncation of sharp features such as lone pair peaks is often observed, and can be especially severe for non-bonding d-electron peaks,[7] which may contribute to the x-ray scattering beyond $\sin\theta/\lambda = 1.4 \ \text{Å}^{-1}$.

A further difficulty in extracting quantitative information is the fact that the experimental deformation density corresponds to an average over all the vibrational motions in the crystal. The effect of thermal smearing must be considered in any quantitative discussion of the deformation density, such as a comparison of peak heights, or of properties of the deformation density, such as dipole moments.

At the other extreme, in terms of complexity, is analysis of the experimental deformation density by comparison with theoretical calculations. First, a molecular wavefunction is calculated using the observed atomic coordinates, and the corresponding theoretical deformation density calculated. In order to compare the theoretical result with experiment, the theoretical density

should be convoluted with the atomic or molecular thermal motions
as observed in the crystal, and series terminated to the same
extent as the experiment. By examining discrepancies between
theory and experiment, improvements may be made in the theoretical model, and the process repeated until satisfactory agreement
is obtained throughout the molecule, taking into consideration
the magnitude and distribution of errors in the experimental map.
The three-dimensional nature of the information in the electron
density distribution and the slow rate of convergence of the
deformation density with improvements in the theoretical basis
insure a rather high quality in a theoretical result which is
capable of reproducing an accurate experimental deformation
density within typical experimental uncertainties.

Once agreement between theory and experiment has been achieved, the theoretical wavefunction may be used for a complete
analysis of the electronic structure. Such a procedure is obviously difficult and time consuming, and to date has only been approach-
ed with light atom structures.

Intermediate in complexity between simple inspection of the
deformation density and rigorous comparison with theoretical
calculations are approaches involving modeling of the observed
density. Parameters describing the distortions of atoms due to
the chemical environment are added to the usual structural model
used in least-squares refinement of the x-ray data. Thus, by a
straightforward extension of standard refinement techniques,
quantitative information is obtained without resort to theoretical
calculations. Problems associated with finite experimental
resolution and thermal smearing are, at least formally, avoided.
In addition, it has recently been shown that density models
currently in common use are capable of providing, within the
limits of the model, quantitative information on the distribution
of electrons among d-orbitals in transition metal complexes.[4,8]
It appears this approach will be most useful for studies of
transition metal complexes in the near future, and it is discussed
in detail here.

DENSITY MODELING

Density models seek to represent the (static) electron
density by a finite number of analytical functions, $g_i(r)$, with
corresponding populations, C_i, such that

$$\rho_{\text{model(static)}} = \sum_i C_i g_i(\underset{\sim}{r}) \tag{2}$$

where the populations are to be determined by least-squares
refinement of the x-ray data, along with the usual crystallo-

graphic parameters (scale factor, positions, thermal parameters, extinction, etc.). It is assumed that the functions are not deformed during vibrational motion. The dynamic density is then a convolution over the thermal motion probability distributions, $P_i(r)$, of each fragment,

$$\rho_{model(dynamic)} = \sum_i C_i \int g_i \ (\underset{\sim}{r} - \underset{\sim}{u}) \ P_i(\underset{\sim}{u}) \ d\underset{\sim}{u} \tag{3}$$

Several different density models have been developed for the analysis of charge density measurements. One of these, a multipole expansion about atomic centers with spherical harmonic angular functions,[9,10] is found to be especially useful for the analysis of transition metal complexes. In this model, the electron density is represented by a sum of "pseudoatom" densities,

$$\rho_{model} = \sum_i \rho_{i,pseudoatom,} \tag{4}$$

with each pseudoatom given as a finite multipole expansion of the form

$$\rho_{i,pseudoatom} = \rho_{core} + \sum_{\ell m} C_{\ell m} \ R_\ell \ (r) \ y_{\ell m}(\theta,\phi) \tag{5}$$

The $R_\ell(r)$ are radial functions $y_{\ell m}(\theta,\phi)$ are spherical harmonic angular functions, and the $C_{\ell m}$ are populations of each of the multipole function to be determined. The radial functions are Slater-type

$$R_\ell(r) = N \ r^{n_\ell} \exp(-\xi r) \tag{6}$$

where N is a normalization constant, and the expansion is usually carried to the "hexadecapole" level ($\ell = 4$).

This model is found to be quite efficient in the analysis of light atom structures. It is generally possible to fit experimental densities within the accuracy of the experiment, while maintaining an adequate ratio of observation to parameters. This approach to obtaining the electron distribution from diffraction data offers several advantages over Fourier methods:

1. By direct summation over the density functions, a static electron distribution is obtained. This assumes a proper model for thermal motion is included in the formalism, as well as an adequate model for the electron distribution, and that sufficient high resolution data to distinguish between thermal motion and density deformations are available.

2. The electron distribution is obtained at infinite resolution. Note that since the experiment extends only to some

finite resolution, features of the model are substituted where
experimental information is lacking.

3.   Since the set of deformations has limited flexibility,
much of the random noise in the experimental data will be filtered
out. The effects of filtering and resolution are evident in the
appearance of model density of FeS_2 (Fig. 2) compared with the
observed density (Fig. 1).

4.   Each observation is properly weighted during refinement
according to its statistical uncertainty, whereas each observation
receives equal weight in a calculation of the density by Fourier
summation.

5.   Additional constraints on the electron density distribution may be easily imposed during refinement. These often include
electroneutrality of the crystal and symmetry constraints on
chemically equivalent parts of the molecule when they are not
incorporated into the crystallographic symmetry.

6.   For non-centrosymmetric crystal structures, the model
provides a much better estimate of the phase of the observed
structure factor. The influence of the phases on the appearance
of the deformation map is often dramatic.

7.   The results of the experiment, the $C_{\ell m}$'s, are in a more
compact and convenient form, facilitating communication of the
results and calculation of properties of the density.

8.   Finally, for transition metal complexes, a correspondence exists between the populations of multipole deformation
functions on the metal atom and the d-orbital occupancies.

d-ORBITAL OCCUPANCIES

In many transition metal complexes, other than those containing metal-metal bonds, the d-electrons are essentially non-
bonding or are involved only to a slight extent by ligand σ-
donation and π-backbonding. In such cases, the electronic
structure may be analyzed, to a first approximation, using crystal
field theory, which neglects the covalent interactions between
the metal atom and ligands.

Let us further assume that the atomic wavefunction for the
metal atom may be expressed as a Slater determinant with an
atomic basis containing a single set of d-orbitals. Each d-
orbital may be occupied by 0, 1, or 2 electrons, and the distribution of electrons among the orbitals will be determined by the

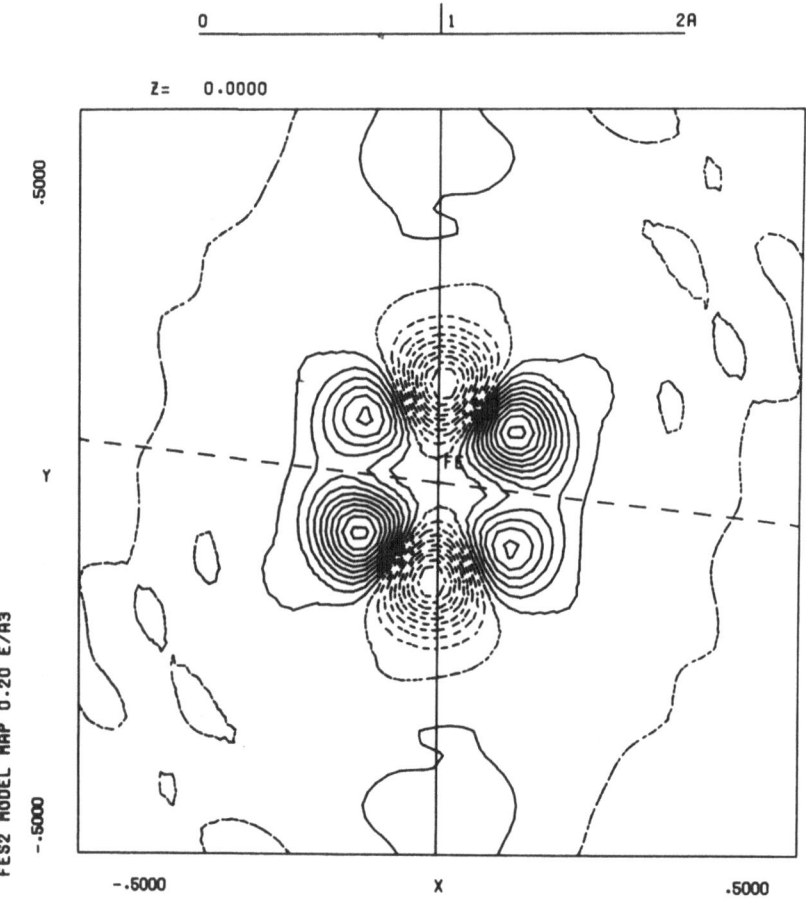

Figure 2. Dynamic multipole model density around the iron atom in
FeS$_2$, plotted in the same plane as Figure 1. Contours
at 0.2 e/A^3.

strength of the crystal field, the pairing energy, and the symmetry of the site. If each orbital has an occupancy n_m, then the density corresponding to the d-electrons will be given by

$$\rho_{d-electrons} = (R_2(r))^2 \sum_m n_m (y_{2m})^2, \tag{7}$$

where $R_2(r)$ is the common radial part of the d-orbital functions.

The products of spherical harmonic functions may be expressed as a linear combination of spherical harmonic functions.[11,12]

$$(y_{\ell m})^2 = \sum_{\ell'm'} J_{\ell'm'} y_{\ell'm'}, \tag{8}$$

where the summation over ℓ' is from 0 to 2ℓ. Thus the electron density distribution corresponding to the d-electrons will be given by a summation over spherical harmonic functions to the hexadecapole level,

$$\rho_{d-electrons} = (R_2(r))^2 \sum_m n_m (\sum_{\ell'm'} J_{\ell'm'} y_{\ell'm'}). \tag{9}$$

A typical multipole refinement model for a first row transition metal used in the refinement of x-ray data would have the following form,

$$\rho_{metal} = \rho_{core} + C_{4s}\rho_{4s} + R_4(r) \sum_{\ell m} C_{\ell m} y_{\ell m} \tag{10}$$

where ρ_{core} is the density of the K, L, and M shell core, C_{4s} is the population of the 4s orbital, and the last term represents the 3d density. Scattering factors corresponding to ρ_{4s} are taken from Hartree-Fock atomic wavefunctions. It is straight-forward to equate the coefficients in eq. 9 with those in eq. 10 corresponding to the same angular function $y_{\ell m}$. This yields a set of linear equations for $C_{\ell m}$ in terms of the orbital occu-pancies, n_m. Once the $C_{\ell m}$ have been obtained from refinement of the x-ray data, the n_m may be obtained by solving the set of linear equations.

The number of d-orbital occupancies to be determined will be 5 or less depending on the local symmetry of the crystal field, and the number of equations (equal to the number of allowed multipole functions) will depend on the crystallographic site symmetry. In cases of low site symmetry, the number of allowed multipoles may be larger than the number of orbital occupancies to be determined, but many of these terms cannot arise as products of d-orbitals. In cases of low symmetry, the orientation of the

set of d-orbitals with respect to the ligands may become a
variable as well.

The occupancies of the d-orbitals on the metal may therefore
be obtained directly from a multipole refinement of the x-ray
data. This is demonstrated below using several examples with
metal atoms at sites with various symmetries. Note that this
procedure is equally applicable to the analysis of spin densities
obtained from polarized neutron diffraction measurements.[13,14]

IRON PYRITE (FeS$_2$)

In pyrite,[4] each iron atom is surrounded by six disulfide
ions. The crystallographic site symmetry is $\bar{3}$, but if only the
nearest neighbor sulfur atoms are considered, the local symmetry
of the ligands is a slight trigonal distortion ($\bar{3}m$) of an octahedral field. In an octahedral field, the 5 d-orbitals will be
split into a set of 3 t_{2g} orbitals of lower energy and a set of 2
e_g orbitals of higher energy. The slight trigonal distortion
further splits the t_{2g} orbitals into a single a_g along the 3-fold
axis and a set of 2 e_g orbitals. Since the iron has a formal
charge of +2, there are 6 d-electrons which are expected to
occupy the a_g and lower set of e_g orbitals in a strong crystal
field.

If n_1, n_2, and n_3 are the occupancies of the a_g, lower e_g,
and upper e_g orbitals, respectively, then the following relations
to the multipole density functions, C_{00}, C_{20}, C_{40}, and C_{43+}, may
be obtained.

$$C_{00} = (n_1 + n_2 + n_3) \, (1/4\pi)^{1/2}$$

$$C_{20} = (2n_1 - n_2 - 2^{1/2}n_4) \, (5/19\pi)^{1/2}$$

$$C_{40} = (6n_1 - (2/3)n_2 - (7/3)n_3$$
$$+ (50^{1/2}/3)n_4) \, (1/196\pi)^{1/2}$$

$$C_{43+} = (-n_2 + n_3 + (1/3)n_4) \, (5/56\pi)^{1/2} \qquad\qquad (11)$$

where an additional term n_4 is present to account for possible
mixing of the upper and lower e_g orbitals. This set of linear
equations can be solved to yield expressions for the orbital
occupancies as functions of the multipole population parameters.

An additional multipole, y_{43-}, is allowed by the 3 site symmetry. If this parameter is included, an additional result can be obtained, the orientation angle about the 3-fold axis between the sulfur polyhedron and the d-orbital coordinate frame.

For iron pyrite, multipole populations have been obtained from refinement of high-resolution ($\sin\theta_{max}/\lambda = 1.46$ Å^{-1}) single-crystal x-ray intensity measurements. The d-orbital occupancies obtained from the multipole populations are compared with ideal high- and low-spin configurations in Table 1. The total occupancy of the d-shell was constrained to 6 electrons during refinement. The population of the y_{43-} multipole was found to be insignificant confirming the validity of treating the local crystal field as $\bar{3}$m aligned with the nearest neighbor sulfur atoms.

TABLE 1. Apparent d-Orbital Occupancies of Iron in FeS_2.

		High Spin	Low Spin	Observed
n_1	(a_o)	1.20	2.00	2.00(13)
n_2	$(1e_g)$	2.40	4.00	3.24(16)
n_3	$(2e_g)$	2.40	0.00	0.76(15)
n_4		0.00	0.00	-0.11(26)

POTASSIUM IRON DISULFIDE (KFeS$_2$)

In contrast to FeS_2, the d-electrons of the iron atom in KFeS$_2$ are in a high-spin configuration. The structure consists of infinite chains of edge sharing iron-sulfur tetrahedra. The energy levels of the d-electrons are split into a lower e_g set ($d_{z^2}, d_{x^2-y^2}$) and a higher t_{2g} set (d_{xy}, d_{xz}, d_{yz}) by the tetrahedral field of the S^{2-} ions. The z axis is chosen along the chain axis, with x and y axes perpendicular and bisecting the S-Fe-S angles. The energy levels are further split by a short iron-iron distance (2.701(1)Å) along the chain direction leaving only the d_{xz} and d_{yz} orbitals degenerate.

Using n_1, n_2, n_3, and n_4 for the occupancies of the $d_{x^2-y^2}$, d_{z^2}, d_{xy}, and (d_{xz}, d_{yz}) orbitals, the following relationships to the multipole populations can be derived,[15]

$$C_{00} = (n_1+n_2+n_3+n_4)\ (1/4\pi)^{1/2}$$

$$C_{20} = (-2n_1+2n_2-2n_3+n_4)\ (5/196\pi)^{1/2}$$

$$C_{40} = (n_1+6n_2+n_3-4n_4)\ (1/196\pi)^{1/2}$$

$$C_{44+} = (-n_1+n_3)\ (5/28\pi)^{1/2}. \tag{12}$$

COBALT(II)TETRAPHENYLPORPHYRIN

In cobalt tetraphenylporphyrin,[17] the metal atom is located in a 4-coordinate square planar environment at a crystallographic $\bar{4}$ site. The crystal field may be considered as a distorted octahedron which further splits the t_{2g} and e_g orbitals leaving only the d_{xz} and d_{yz} orbitals degenerate. The equations relating the d-orbital occupancies to the multipole populations are identical to those for the distorted tetrahedral field given above. Occupancies resulting from refinement of low temperature (100K) x-ray data are compared with idealized high and low-spin configurations in Table 3. The assignment of the unpaired spin to the d_{z^2} orbital for the low-spin configuration is based on epr results.

TABLE 2. Apparent d-Orbital Occupancies of Iron in $KFeS_2$.

	High-Spin	Low-Spin	Observed
$n_1\ (d_{x^2-y^2})$	1.00	2.00	1.11(5)
$n_2\ (d_{z^2})$	1.00	2.00	0.81(8)
$n_3\ (d_{xy})$	1.00	0.33	0.99(5)
$n_4\ (d_{xz},d_{yz})$	2.00	0.67	2.09(11)

TABLE 3. Apparent d-Orbital Occupancies of Cobalt in CoTPP.

		High-Spin	Low-Spin	Observed
n_1	$(d_{x^2-y^2})$	1.4	0.0	1.0(2)
n_2	(d_{z^2})	1.4	1.0	1.0(2)
n_3	(d_{xy})	1.4	2.0	1.3(2)
n_4	(d_{xz},d_{yz})	2.8	4.0	3.7(3)

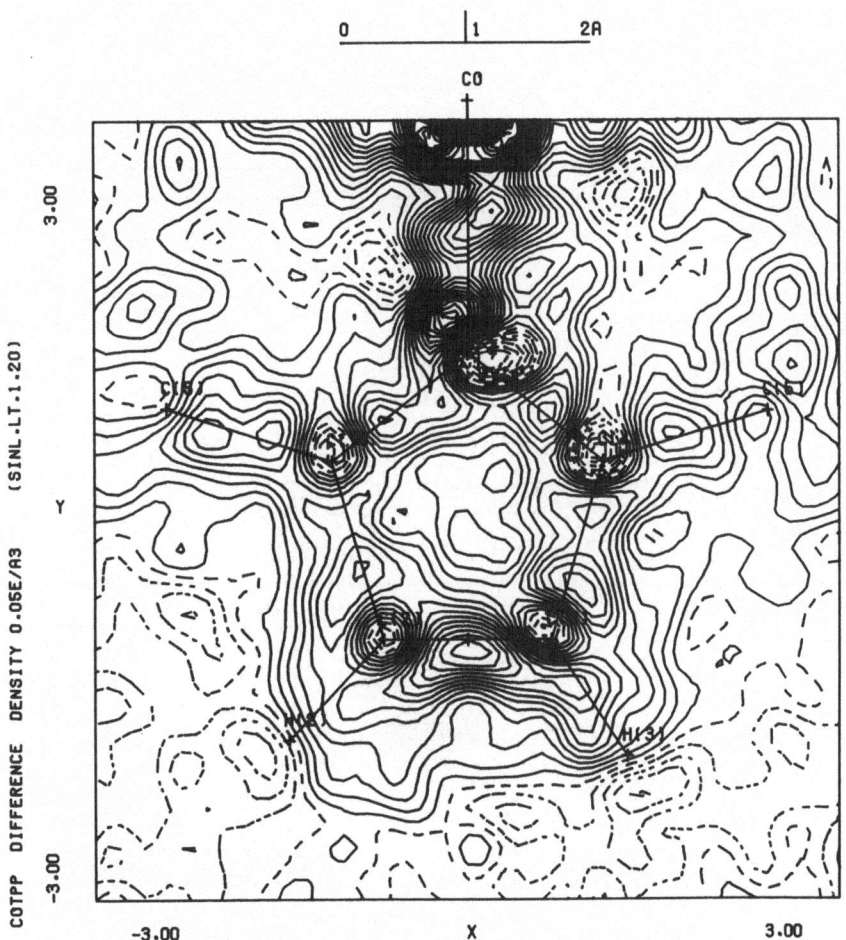

Figure 3. Experimental deformation density of cobalt(II) tetraphenylporphyrin (CoTPP) at 100K plotted in the plane of
the pyrole ring. Contours are at 0.5 e/Å³ with zero
and negative contours broken.

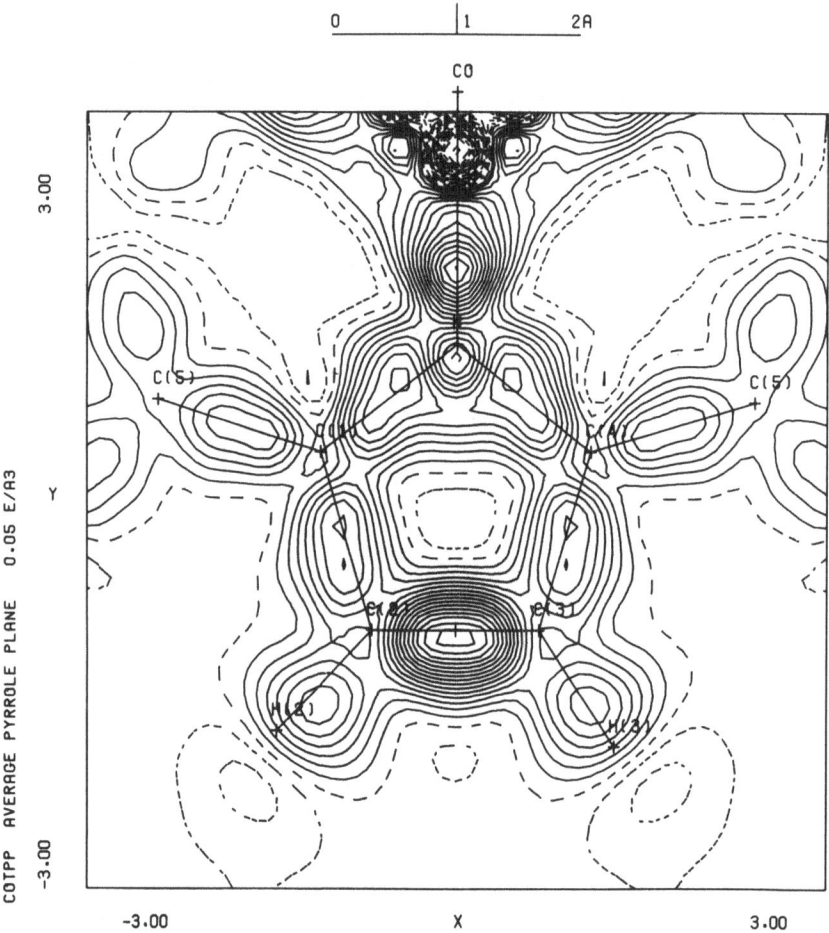

Figure 4. Dynamic multipole model density of CoTPP plotted in
the plane of the pyrole ring. Contours as in Figure 3.

Several of the advantages of density modeling are evident in the deformation densities calculated for CoTPP. Fig. 3 shows the experimental deformation density in the plane of one of the pyrole rings of the porphyrin molecule. Fig. 4 shows the corresponding dynamic model density. The filtering of the experimental noise by density modeling is quite evident. Also, chemical (left-right) symmetry has been imposed on the model during refinement. In addition, better resolution of the density features, apparent in the model density, is due to the improved estimates of the structure factor phases for this non-centrosymmetric crystal structure. Since proper treatment of thermal smearing and series termination would be impractical for theoretical calculations on a large system such as this, a less satisfactory comparison can be made between the static model density and static theoretical calculations. Static model densities of CoTPP in two sections of the molecule are plotted in Figures 5 and 6.

LIMITATIONS

In many cases, the type of analysis described above should only be considered as a first approximation to a description of the electronic structure. The limitations of the method are due to the initial assumptions. The most serious of these is likely to be the neglect of covalent overlap. This is especially true for the study of metal-metal bonds, for example, where the covalent interaction is often of primary interest.

In the case where covalent bonding is significant between the metal atom and ligands, the orbital occupancies derived from the multipole functions will recover only that part of the charge with the shape and symmetry of the density functions. If the covalent interaction between the metal and ligand orbitals is represented by a molecular wavefunction,

$$\psi = C_M \phi_M + C_L \phi_L, \tag{13}$$

where C_M and C_L are coefficients of the molecular orbital in terms of atomic basis orbitals ϕ_M and ϕ_L on the metal and ligand atoms, then the density will be given by

$$\rho = n(C_M^2 \phi_M^2 + 2C_M C_L \phi_M \phi_L + \phi_L^2). \tag{14}$$

The apparent occupation number obtained from an atom centered density model will be nC_M^2, plus perhaps some contribution from the overlap term included by adjustment of the radial function during refinement. Since $C_M^2 \leq 1$, it is not surprising that non-integer values are observed for the apparent orbital occupancies.

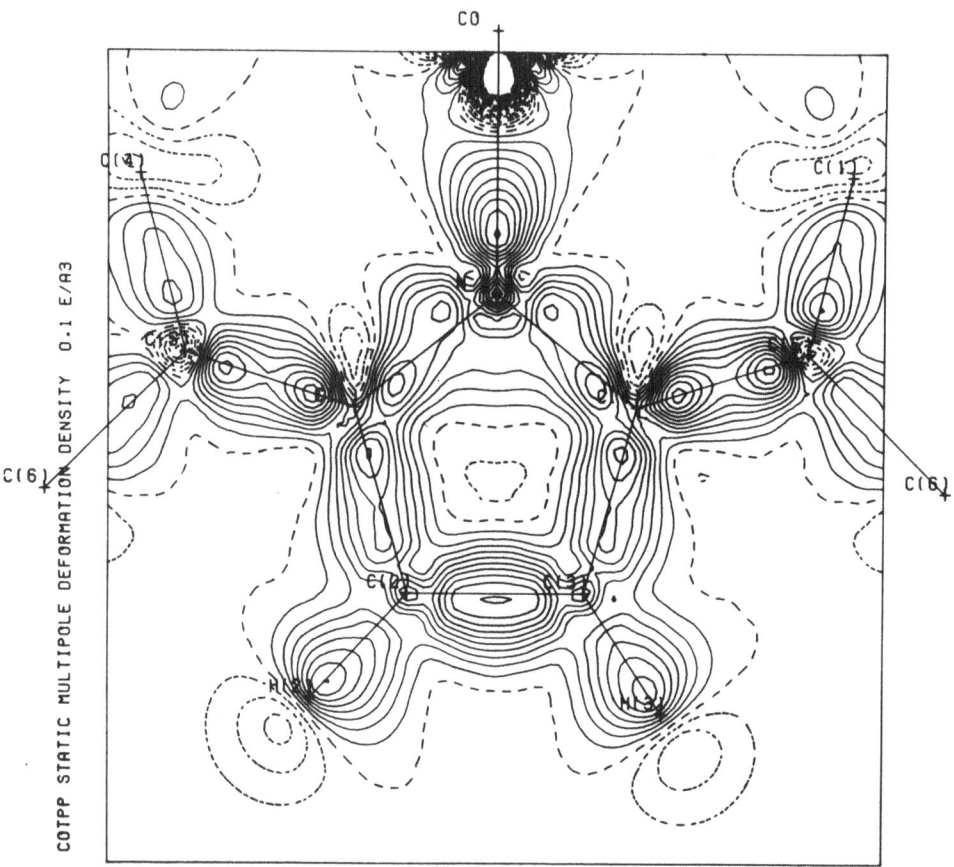

Figure 5. Static multipole model density of CoTPP plotted in the plane of the pyrrole ring.   Contours at 0.10 e/Å³.

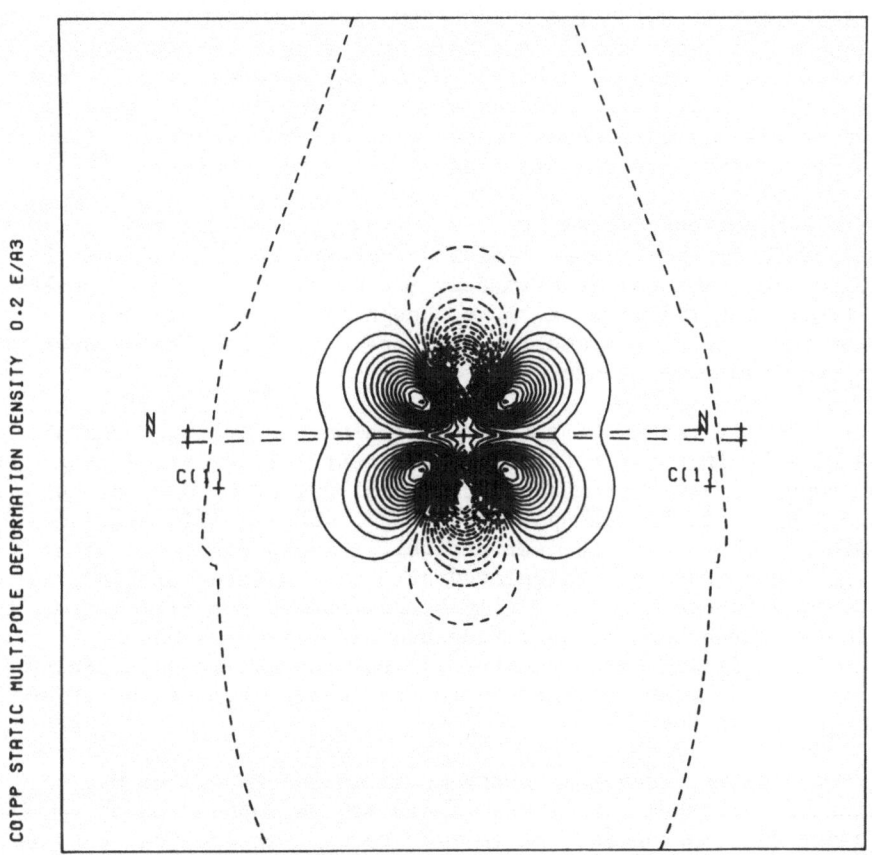

Figure 6. Static multipole model density of CoTPP plotted in a
          plane containing the cobalt atom, perpendicular to the
          porphyrin ring, and bisecting the NCoN angle.   Contours
          at 0.20 e/Å³.

In cobalt tetraphenylporphyrin, the observed orbital occupancies are consistent with a covalent interaction between the metal atom and the porphyrin molecule. An apparent occupancy of the $d_{x^2-y^2}$ orbital is expected due to σ donation by nitrogen, while π backbonding would decrease the populations of the d_{xz} and d_{yz} orbitals.

The extent to which the assumption that the atomic wavefunction can be described in the form of a single determinant influences the results is unknown. This may become a significant consideration in accurate studies since the inclusion of several excited states has been shown to be important in describing the ground state structure of some transition metal complexes.[18]

The limitations imposed by the restriction of the model to a minimal basis (with variable radial parameter) can be examined by looking at the residual density near the metal atom. Ghost peaks often occur near the metal atom, although since they generally do not show the molecular symmetry, they are more likely the result of unrecognized experimental errors.[19]

Finally, it should be asked "How reasonable are the orbital occupancies"? In these refinements, no restrictions are imposed on the populations other than on the total number of d-electrons. Yet in no case are the occupancies found to be significantly less than 0.0e or more than 2.0e per orbital. In addition, the local electric field gradients calculated at the iron site both in FeS_2 and $KFeS_2$, although not accurately determined by the diffraction experiment, agree both in magnitude and sign with Mössbauer quadrupole splitting measurements.[4,16] Perhaps most convincingly, the d-orbital occupancies agree with reasonable chemical descriptions of the systems.

In the future, several improvements in the technique may be anticipated. Constraints or restraints may be imposed on the refinement that the density agree with certain theoretical properties or reproduce other experimental quantities such as the electric field gradient.[20,21] Two-center terms might be included to account for covalency, in which case a constraint of N-representability may be necessary.[22]

ACKNOWLEDGEMENT

Support of this work by grants from the National Science Foundation and the National Institutes of Health is gratefully acknowledged. The author would also like to thank Professor P. Coppens for immeasurable encouragement and assistance.

REFERENCES

1. E. D. Stevens, J. Rys, and P. Coppens, J. Am. Chem. Soc.
    100:2324 (1979).
2. E. D. Stevens, Acta Crystallogr., Sect. B 36:1876 (1980).
3. M. Iwata and Y. Saito, Acta Crystallogr., Sect. B 29:822
    (1973).
4. B. Rees and A. Mitschler, J. Am. Chem. Soc. 98:7918 (1976).
5. E. D. Stevens, M. L. DeLucia and P. Coppens, Inorg. Chem.
    19:813 (1980).
6. B. N. Figgis, P. A. Reynolds, A. H. White and G. A. Williams,
    J.C.S. Dalton 371 (1981).
7. S. Ohba, K. Toriumi, S. Sato and Y. Saito Acta Crystallogr.,
    Sect. B 34:3535 (1978).
8. E. D. Stevens and P. Coppens Acta Crystallogr., Sect. A
    35:536 (1979).
9. R. F. Stewart Acta Crystallogr., Sect. A 32:565 (1976).
10. N. K. Hansen and P. Coppens Acta Crystallogr., Sect. A
    34:909 (1978).
11. M. E. Rose "Elementary Theory of Angular Momentum" Wiley &
    Sons, New York p. 61 (1957).
12. A. D. Rae Acta Crystallogr., Sect. A 34:719 (1978).
13. J. N. Varghese and R. Mason Proc. R. Soc. Lond, Ser. A 372:1
    (1980).
14. R. Mason this volume (1981).
15. E. D. Stevens "Electron and Magnetization Densities in
    Molecules and Crystals" P. Becker, ed., Plenum, New York p.
    823 (1980).
16. E. D. Stevens unpublished results.
17. E. D. Stevens J. Am. Chem. Soc. in press (1981).
18. M. Benard and A. Veillard Nov. J. Chim. 1:97 (1977).
19. M. Benard , P. Coppens, M. L. DeLucia and E. D. Stevens
    Inorg. Chem. 19:1924 (1980).
20. D. Schwarzenbach and N. Thong Acta Crystallogr., Sect. A
    35:652 (1979).
21. D. Schwarzenbach, Chapter 6.2, this volume (1981).
22. L. J. Massa, Chapter 2.2, this volume (1981).

NEUTRON SCATTERING BY PARAMAGNETS: WAVE FUNCTIONS OF ANTIBONDING

STATES IN TRANSITION METAL COMPLEXES

Ronald Mason

Ministry of Defence, London, UK and University
of Sussex, Brighton, UK

PREAMBLE

One broad theme underlies the papers in this Symposium:
diffraction intensities contain direct information on electronic
states in crystals with the x-ray diffraction experiment giving,
in principle, explicit valence electron densities and the polarised
neutron diffraction experiment giving magnetisation densities in
crystals. We only consider here the neutron scattering experiment
and how we have evolved experimental and theoretical techniques,
over the past five years, which give uniquely selective results
on spin densities in transition metal compounds of chemical
complexity (rather than, say, the simple metal oxides which were
among the earliest systems studied by magnetic scattering methods).

First, a note on some relevant theories and models. The
polarized neutron scattering experiments have now reached a point
where fairly complete sampling of reciprocal space is possible to
provide accurate magnetic structure factors for a large number of
Bragg reflexions from crystals with unit cell volumes of, say,
$1000°A^3$.[1] The magnetic structure factor is given by

$$\underset{\sim}{M}(\underset{\sim}{k}) = \sum_n f_n(\underset{\sim}{k}) \, \underset{\sim n}{M} \, \exp(i.\underset{\sim}{k}.\underset{\sim n}{r})$$

with $f_n(\underset{\sim}{k})$ the magnetic form factor for atom n having the easily
recognised form

$$f(k) = \frac{\int m(r) \exp(i.k.r.) \, dr^3}{\int m(r) \, dr^3}$$

with $m(r)$ being the magnetisation density of the atom. M_n is the magnetic moment of the nth atom positioned at r_n from the origin. When the magnetisation is due to electrons in a single unfilled shell, the form factor becomes

$$f(k) = \sum_\ell A_\ell(\hat{k}) <j_\ell(|k|)>$$

where

$$<j_\ell(k)> = \int_0^\infty U^2(r) \, j_\ell(kr) r^2 dr$$

$j_\ell(kr)$ is the spherical Bessel function of order ℓ and $U^2(r)$ the radial distribution function of electrons in the open shell. The $A_\ell(k)$ are coefficients, three in number for 'd' electrons and have been tabulated[2] for many transition metal ions, based on self-consistent wave functions. Finally we note that orbital magnetic contributions, if reasonably small, can be conveniently and adequately represented through a dipole approximation[3]

$$f(k) = 2S <j_0(k)> + L(<j_0(k)> + <j_1(k)>)$$

Several publications have used these equations, via least squares procedures, to comment on electron (spin) densities in complexes. Our earliest contribution was one which established that the magnetic structure factors for KNa_2CrF_6 contained evidence for the t_{2g}^3 orbital occupation for Cr^{3+}, expected on the basis of ligand field theory, and spin transfer $Cr \longrightarrow F$ was also demonstrated[4]. A fuller discussion[5] described the chromium as t_{2g} 2.66(5) e_g -0.06(5), and 0.020(5) spins in each $2p\pi$ orbital of fluorine and a negative spin (-0.02(1)) in the fluorine $2p\sigma$ orbital. I shall discuss these results - evidence for the importance of electron correlation - later. In the $CoCl_4^{2-}$ ion, in crystals of Cs_3CoCl_5, both metal configurations and overlap spin densities in the Co-Cl bonds were determined[6]. The most ambitious of what I shall call this "constrained multipole" approach lies in two analyses of the spin densities and metal electronic structures of phthalocyaninato-manganese[11,7] and - cobalt[11,8]. Spin populations for the cobalt ion are

$$3d_{xy}^{0.40(10)}, \quad 3d_{xz,yz}^{0.17(10)}, \quad 3d_z^2 \, 0.79(12),$$

$$3d_{x^2-y^2}^{-0.21(10)}, \quad 4s^{-0.14(6)}$$

with a total spin population of $-0.17(5)$ on the macrocycle nitrogen carbon atoms. Orbital electronic populations are then

$$3d_{xy} \ 1.60, \ d_{xz,yz} \ 1.83, \ d_{z^2} \ 1.21, \ d_{x^2-y^2} \ 0.21$$

with $4s^{0.14}$ or $^{1.86}$ − the choice being determined by electro-neutrality considerations. For the <u>manganese</u> complex − an organic ferromagnet − spin populations are

$$d_{xz}^{0.74(6)}, d_{xz,yz}^{1.17(6)}, d_{z^2}^{0.83(6)}, d_{x^2-y^2}^{-0.15(6)}, 4s^{-0.44(6)}$$

with $-0.31(3)$ spin on the ligand atoms. Electronic populations become

$$d_{xy}^{1.26}, \ d_{xz,yz}^{0.83}, \ d_{z^2}^{1.17}, \ d_{x^2-y^2}^{0.15}, \ 4s^{1.56}$$

It was shown that these orbital populations are in reasonable accord with predictions from angular overlap models.

    The last piece of theory we want for interpretation of the magnetic structure factors takes us to a form of the structure factor expressed in terms of multipole population densities positioned at atomic centres. The scattering (form) factor f_j associated with the centre j is given by the one-centre density representation as

$$f_j(\underset{\sim}{k}) = \int \rho(r,\theta,\phi) \exp(i.\underset{\sim}{k}.\underset{\sim}{r}) \ d\tau = f_j(S,\alpha,\beta)$$

ρ_j is the one centre electron (magnetisation) density associated with the centre j; (r,θ,ϕ) and (S,α,β) are the components of χ and k in spherical polar coordinates with respect to a local orthogonal frame of reference on the centre j.

    In a multipole expansion[9] the density is expanded as a linear sum of density fragments

$$\rho_{\ell m}(r,\theta,\phi) = M_\ell^m \ N_\ell^m \ Z_\ell^n(\theta,\phi) \ R(r)$$

$$\ell = 0,1,2,3,4\cdots$$
$$-\ell \leq m \leq \ell$$

$R(r)$ is the radial density distribution describing the density around the centre j, Z_ℓ^m are multipole functions (Surface or Tesseral Harmonics) given by

$$Z_\ell^m (\theta,\phi) = P_\ell^{|m|} (\cos\theta)\cos(|m|\phi) \qquad m \geq 0$$

$$= P_\ell^{|m|} (\cos\theta)\sin(|m|\phi) \qquad m < 0$$

P_ℓ^m are associated Legendre functions and N_ℓ^m a normalisation factor which requires

$$\int |\rho_{\ell m}| d\tau = 2 \text{ for } \ell \neq 0 \text{ and } \int \rho_{oo} d\tau = 1$$

M_ℓ^m is the population of the multipole fragment $\rho_{\ell m}$ and is normalised so that it becomes a measure of the number of electrons transferred from the negative to positive lobes of the surface harmonic.

Fourier transforms of multipole density fragments are

$$\mathcal{F}_{\ell m} (S,\alpha,\beta) = \int \rho_{\ell m}(r,\theta,\phi)\exp(-i\underset{\sim}{k}.r)d\tau$$

$$= i^\ell M_\ell^m N_\ell^n 4\pi<j_\ell(Sr)> Z_\ell^m (\alpha,\beta)$$

where

$$<j_\ell(Sr)> = \int_0^\infty R(r)j_\ell(Sr)r^2 dr$$

and

$$S = |\underset{\sim}{k}| = 4\pi\sin\theta\lambda^{-1}$$

and j_ℓ are spherical Bessel functions.

The generalised structure factor can be recast in multipole terms as

$$F(\underset{\sim}{k}) = \sum_{\substack{\text{cell}\\\text{centres n}}} \{ \sum_{\ell=o}^\infty \sum_{m=-\ell}^\ell i^\ell M_{n,\ell}^m N_\ell^m 4\pi<j_{n,\ell}>Z_\ell^m(\alpha,\beta)\exp(-i2\pi h.\underset{\sim}{x}_y-h^\dagger\underset{\sim}{\beta}_n)\}$$

x and β_n are the usual atomic coordinates and mean squares displacement tensors of the centre n.

For the most part this expression is adequate if terms up to hexadecapoles are included. There are often times when the number of non-zero multipoles is determined by local (crystallographic) symmetry arguments. A least squares convergence of the set of F(S) to the observed structure factors is via the optimisation of $M_{n,\ell}$; $\underset{\sim}{x}_n$ and $\underset{\sim}{\beta}_n$ having been established by a conventional neutron scattering experiment. The radial function R(r) is

constructed from Hartree-Fock or, perhaps, simple Slater functions;
flexibility can be achieved via Kappa refinement methods or by a
Taylor expansion technique.

We have shown[10] how it is possible to interpret multipole
populations in terms of chemical 'orbital' populations; by a
simple linear transformation 'orbital' populations determined
from one-centre orbital scattering factors are constrained
multipole refinements - for 'd' orbitals up to hexadecapole level
and for 'p' orbitals up to the quadrupole level. In determining
'orbital' populations one must remember that two-centre orbital
product terms are projected into one-centre terms, so any one-
centre density representation will have built into it the projected
two-centre density functions. 'Orbital' populations obtained by
single centre density functions are related to the diagonal
elements of the Dirac density matrix and should reflect the spin
state of the metal ion.

But we believe a more valuable interpretation of multipole
parametrisations of the spin density comes from a reconstruction
of the total spin density from the multipole density fragments
via the equation[11,12]

$$\rho(\underset{\sim}{r}) = \underset{\substack{\text{sphere} \\ \text{centres } j}}{\Sigma} \quad \Sigma_{\ell, m} \; \rho_{\ell m}^{j} \; (r, \theta, \phi)$$

where the sum is over a sphere which contributes to the density
of the cluster.

OBSERVED MOLECULAR SPIN DENSITIES

The spin density of the CrF_6^{3-} ion is shown in Figure 1 -
obtained by a multipole analysis of 92 magnetic structure factors
given by polarised neutron scattering from crystals of K_2NaCrF_6 (in
these crystals the Cr^{3+} have cubic site symmetry, the orbital
angular momentum is quenched (g = 1.998) so that magnetic structure
factors are directly related only to the spin density).

The features of Figure 1(a) are typical 'd' orbital spin
density around the Cr(III) ion, transfer of (positive) spin
density to fluorine $2p\pi$ orbitals and negative spin densities
concentrated around the Cr-F bond vectors. Transformation[10] of
the multipole populations into 'local orbital' populations resulted
in a $t_{2g}2.67(5)$ configuration on the metal, no significant population in the 'e' orbitals, a spin density of 0.31(8) in a 4s

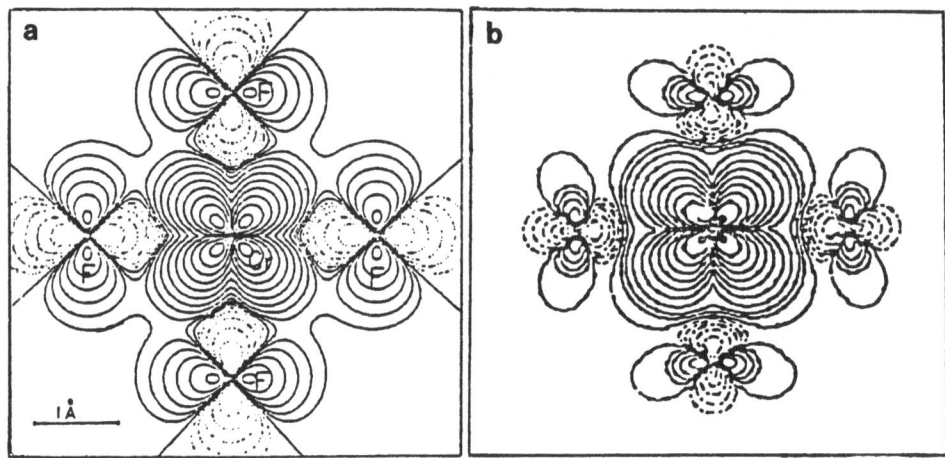

Fig. 1.   The spin density in the $[CrF_6]^{3-}$ ion in a plane containing
the Cr centre and four F centres obtained from a)
multipole analysis of the magnetic data b) Unrestricted
Hartree-Fock ab initio calculation at a double zeta
level.   Contour intervals are such that the nth contour
from the zero (first continuous) contour represents the
$\pm 2^{n-1} \times 10^{-3}$ eÅ^{-3} density level.

type orbital and a spin transfer to the fluorine ligands which
effectively provide for a redistribution of 0.02 e from the σ -
to the π-bond framework with a negative $p_z(p(\sigma))$ orbital population of 0.05 e (+0.03 e in the 2pπ(xy) fluorine orbitals).
Figgis et al[5] in their least squares (non-multipole) modelling
give very similar results: t_{2g} 2.66(5) eg -0.006(5), 4s 0.4(1),
F(2pπ) 0.02, and -0.02(1) in the Cr-F antibonding σ-orbitals.

The unrestricted Hartree-Fock calculation of the spin density
(ab initio;   double zeta level) is illustrated in Figure 1(b) -
it reproduces the three essential features of the observed map.
In particular, unlike simple calculations such as a Huckel model,
the negative 2pσ spin density is well accounted for and is
attributable to polarisation/correlation effects.

Figures 2(a) and 2(b) show similar experimental and theoretical densities in the Cl-Co-Cl plane of the $CoCl_4^{2-}$ ion in
Cs_3CoCl_5.

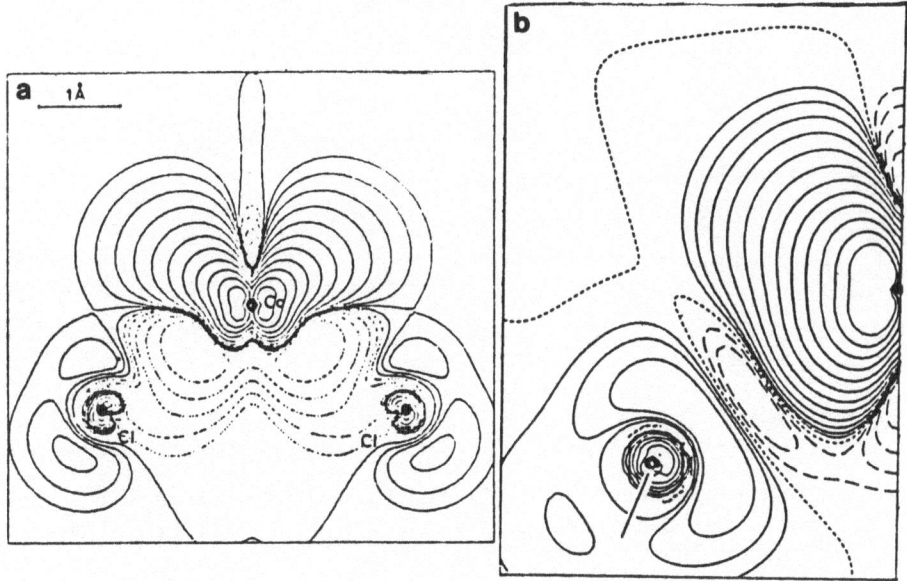

Fig. 2.   The spin density in the $(CoCl_4)^{2-}$ ion in Cs_3CoCl_5 a)
          obtained from a multipole analysis (ref. 10) of magnetic
          data, in the plane of the cobalt and two chlorine atoms.
          b) corresponding theoretical map.   Contour levels as in
          Figure 1.

The broad features are similar to those for CrF_6^{3-} but:

(i)   it is clear that the highest filled level in $CoCl_4^{2-}$
      is much more antibonding than it is in CrF_6^{3-};

(ii) as yet, the theory has a limited basis set and is not
      converged in the self-consistent process – electron
      correlation effects (negative spin densities) do not
      approach those in the observed map;   and

(iii) there is more spin delocalisation to the chloro–ligand
      than is the case for the chromium–fluorine bond (related,
      of course, to (i)).

We are still at an early stage of comparing experiment with
theory;   and, in particular, how relatively simple theory could
be formulated which will adequately represent polarisation/corre-
lation effects.   But it is clear that experiment is capable of
giving results for simple complex ions which have a quality which

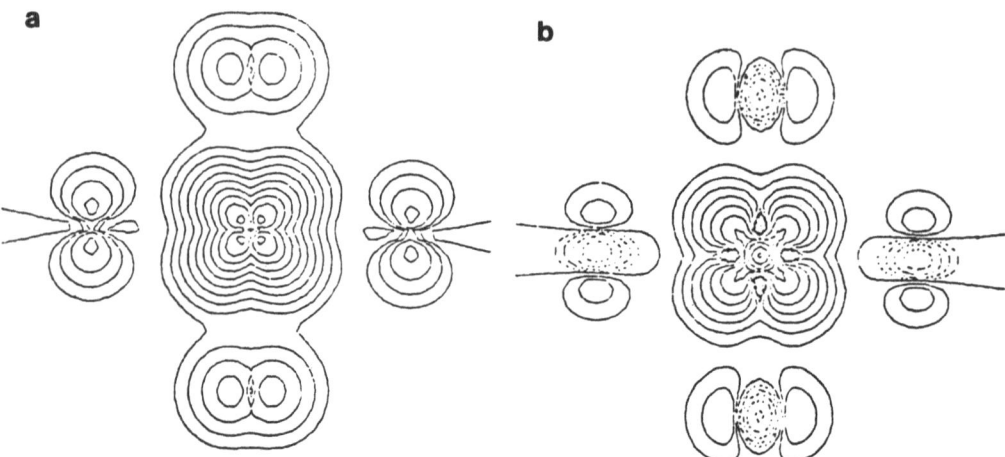

Fig. 3. Phthalocyaninato cobalt(II). Spin densities in the
        molecular plane and in a section 0.25Å above the plane.
        Contour levels as in Figure 1.

can be used to judge theory much more critically than has been
possible in the past.

    Two of the larger molecules we have studied to date have
been the phthalocyaninato-cobalt(II) and manganese(II) (vide
supra). Figure 3 shows the observed spin density in the molecular
plane of the cobalt(II) complex (a) and at 0.25Å above the plane
(b), while Figure 4 simulates the three-dimensional spin density.
Perhaps the simplest and most obvious conclusions to be drawn
from these maps is that, even for such a covalent molecule, the
highest filled orbital has such pronounced non-bonding character.
The negative spin densities in the nitrogen π-orbitals are of
considerable significance.

    We have compared the diffraction results with conclusions on
the intramolecular bonding drawn from spectroscopy - e.s.r. in
particular[7],[8]; and in the case of manganese phthalocyanine we
have been able to resolve the mechanism for intermolecular exchange
phenomena. But it remains the case that, with present experimental
facilities, the data on the phthalocyanines are more illustrative
of future possibilities than of providing accurate and quantitative
spin densities.

A final note seems appropriate; the polarised neutron
experiment seems destined to complement the conventional x-ray
method of determining valence electron densities. It has the
valuable property of being selective; but it tells one something
about doubly filled orbitals only by inference, and hence cannot
be said to be a truly direct probe of bonding. Nevertheless, it
is not unreasonable to claim that the paramagnetism of 'd' block
complexes, of value for more than 50 years as an indicator of
bonding, may have a new lease with the advent of the polarised
neutron scattering experiment.

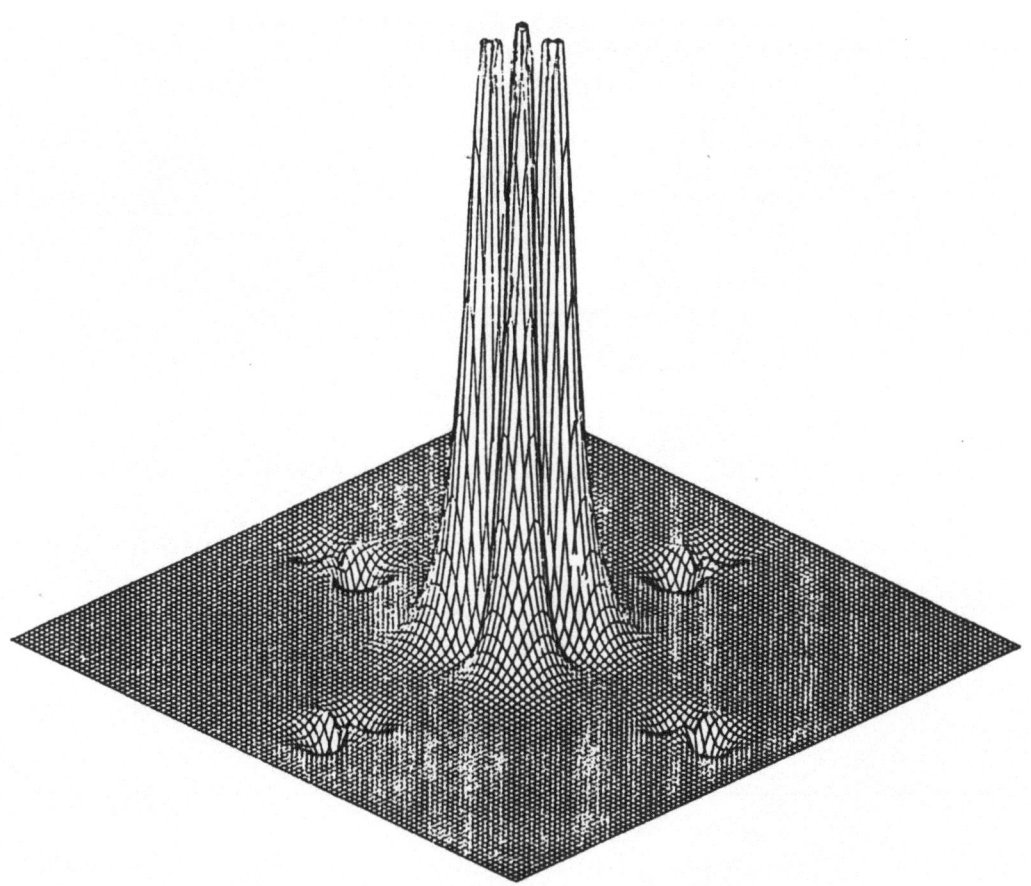

Fig. 4. The simulated three dimensional spin density in phthalo-
        cyaninato cobalt(II). Contour levels as before.

References

1. P. J. Brown, J. B. Forsyth, and R. Mason, Phil. Trans. Roy Soc (1980) B.
2. R. J. Weiss and A. J. Freeman, J. Phys. Chem. Solids 10:147 (1959).
3. W. Marshall and S. W. Lovesey, 'Theory of Thermal Neutron Scattering', Clarendon Press Oxford (1971).
4. B. N. Figgis, R. Mason, A.R.P. Smith, and G. A. Williams, J. Amer. Chem. Soc. 101:3673 (1979).
5. B. N. Figgis, P. A. Reynolds, and G. A. Williams, J. Chem. Soc. Dalton 2348 (1980).
6. B. N. Figgis, R. Mason, A.R.P. Smith, J. N. Varghese, and G. A. Williams J. Amer. Chem. Soc. 103:1300 (1981).
7. B. N. Figgis, G. A. Williams, J. B. Forsyth, and R. Mason, J. Chem. Soc Dalton (1981) (in press).
8. G. A. Williams, B. N. Figgis, and R. Mason, J. Chem. Soc. Dalton 734 (1981).
9. R. F. Stewart, J. Chem. Phys. 58:1668 (1973).
10. J. N. Varghese and R. Mason, Proc. Roy. Soc A372:1 (1980).
11. R. Mason and J. N. Varghese (unpublished work).

ELECTRON DENSITY DISTRIBUTION IN THE BONDS OF CUMULENES AND SMALL RING COMPOUNDS

Hermann Irngartinger

Organisch-Chemisches Institut der Universität
Heidelberg, Im Neuenheimer Feld 270
D-6900 Heidelberg, Bundesrepublik Deutschland

1.   Introduction

Already some time ago the chemists obtained detailed descriptions of covalent bonds and of the spatial distribution on the bonding electrons by theoretical calculations. They were however confined to smaller molecules. More recently measurements on larger organic molecules have been carried out and furnished experimental evidence of the electron density distributions in the bonds of organic molecules.[1] We have investigated experimentally two bonding systems: In the first group, the compounds have cumulenic double bonds.[2] The second set of molecules consists of the four-membered cyclobutadiene ring. Many calculations have been carried out regarding the stability and the structure of this molecule.[3]

We collected the X-ray reflection data of suitable crystalline derivatives at the temperature of liquid nitrogen. During the measuring period we have kept the temperature in the region of 100 and 110 K constant within 1 K. Each set of independent reflections has been measured three to four times and subsequently averaged. The positional and thermal parameters of the non-hydrogen atoms have been refined with high order data of the region $\sin \theta/\lambda = 0.7$ to 1.2 Å^{-1}. More than 30 independent reflections were available for one variable. Therefore all the maps of deformation densities which are given in this article have been obtained from X-X-data.

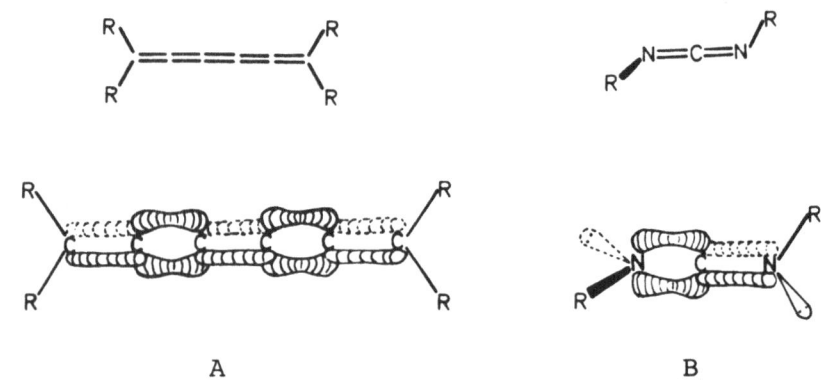

A   B

Fig. 1. Schematic representation of the π-bonds in the
        cumulenic systems of hexapentaene and carbodiimide

2.   Electron density distributions in cumulenes

The longest cumulenic system which can be obtained
as a stable crystalline derivative is the hexapentaene A
with five double bonds. The heterocumulenic system of a
carbodiimide B was particularly interesting, because in
addition to the bonding system we could also determine
the electron density of the lone pairs of the nitrogen
atoms. According to the theory of π-electrons the neighboring double bonds of cumulenes have perpendicular
orientation (Fig. 1).

2.1   Electron density distribution in the bonding
        system of hexapentaene.

The reactive cumulenic system of the hexapentaene 1
has been stabilized sterically on both sides by the
large groups of tetramethylcyclohexane rings.[4] As was
to be expected the cumulene chain is linear with bond
angles of 179° to 180° (Fig. 2), since the central carbon
atoms are sp hybridized.[5] The terminal $C(sp^2)-C(sp)$
double bond is shorter than 1.34 Å which is the common
bond length for $C(sp^2)-C(sp^2)$ double bonds. The adjacent
$C(sp)-C(sp)$ double bond is still shorter. The central
bond of 1 is of the same type and we expect the same
bond length. But this bond is definitely longer than the
two neighboring bonds. Therefore the hexapentaene shows
a bond alternation with long, short, long, short and
long double bonds. The reason for this surprising result
has to be sought in the interactions of the π-electrons
along the cumulene axis. Obviously mesomeric structures
with triple bond character have to be considered describing the ground state of this molecule.

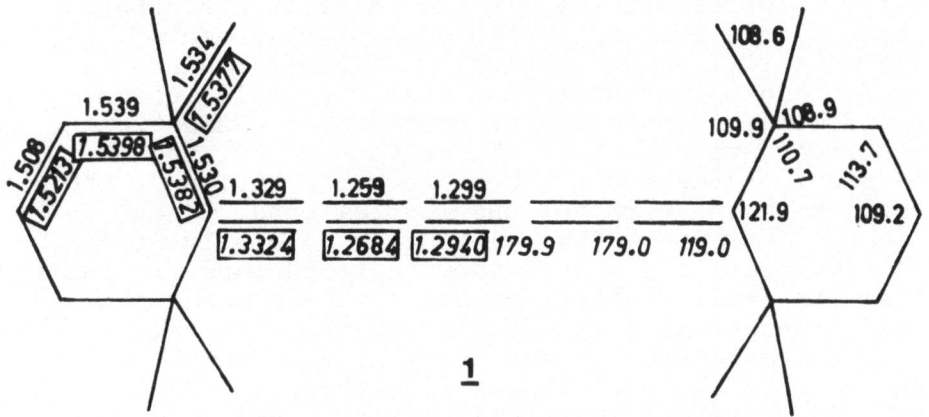

Fig. 2. Bond lengths (Å) and angles (°) of 1. The bond
        lengths within the frames are obtained from low
        temperature data (103 K). The remaining values
        are measured at room temperature.[5]

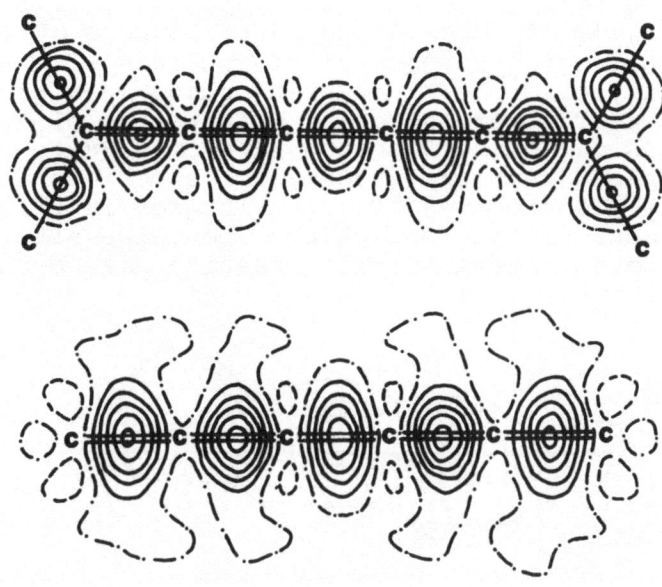

Fig. 3. The two perpendicular sections along the cumu-
        lenic axis of 1. The electron densities are
        averaged according to mm symmetry.[6] Contour
        interval is 0.1 e/Å³.

Fig. 3 represents the electron density distribution
of the bonding electrons in the cumulenic system of the
hexapentaene 1. The two sections are perpendicular to
each other and both contain the cumulenic axis. Assuming
a perpendicular orientation of neighboring π-electrons
to be correct, we expect in the section containing the
carbon atoms of the substituents no contribution of the
π-electrons in the terminal and central bond. However
the π-electrons of the two neighboring double bonds are
within this plane and should show contributions. In fact
the electron density distributions of these bonds show
stronger elongations perpendicular to the bond axis compared to the densities in the terminal and central bonds.
In the bonds of the cyclohexane ring the electron densities of the σ-electrons are appearing. The maxima of
the double bonds (0.5 e/$Å^3$) are higher than those of the
single bonds in the cyclohexane ring (0.4 e/$Å^3$). The
vertical section plotted in the second row shows the
contributions of the π-electrons in the terminal and
central double bonds. In contrast to the first section
the elongations of the electron densities become apparent in these bonds. The two remaining π-bonds are perpendicular to the second section. Therefore the density
lines hardly show any elongation.

These findings have been illustrated in an electron
isodensity diagram. A three dimensional picture of the
enveloping lines which connect all the points with an
electron density of 0.1 e/$Å^3$ is being shown in a projection.(Fig. 4) The perpendicular orientation of the
elongation axes in neighboring bonds becomes quite clear
along the complete cumulene system. The experimental
results on the electron density distribution in this
hexapentaene confirm the conceptions about the nature
of cumulenic bond systems. Similar results have been
obtained for a butatriene and an allene.[7]

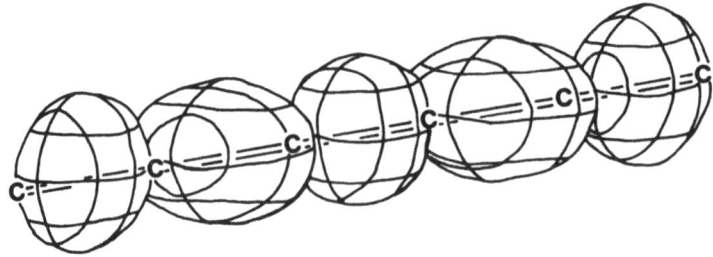

Fig. 4. Electron isodensity diagram of the cumulenic
        bond system of 1.

2.2 Electron density distribution in a carbodiimide
    derivative
    Bent bonds in non-cyclic structures

The bis-(para-methoxyphenyl)-carbodiimide 2 is an
ideal example to determine deformation densities.[8] It
has a large number of bonds with different bonding character and moreover several heteroatoms with lone pairs.
We obtained the bond lengths and angles given (Fig. 5)
by high order refinement with data collected at 103 K.
The total number of reflections was about 30,000. After
averaging the symmetry related reflections, about 10 000
reflections remain.[6]

The methoxy groups are coplanar with the corresponding phenyl rings. On the oxygen atom we observe an angular enlargement because of the repulsive forces between the phenyl ring and the methyl group.

The bond angles at the nitrogen atoms too are considerably enlarged by 9° because of 1...3 and 1...4
repulsion. The deviation of 11° from the linear arrangement of the cumulene system is surprisingly large.
Packing arguments which formerly have been proposed[9]
can not be responsible for this bending because deviations of the same order of magnitude have been found in
carbodiimides with various substituents.

Contrary to the carbodiimide 2 the derivative 3 is
substituted on both ends by aliphatic groups. Nevertheless this carbodiimide too is bent by 10° from linear
arrangement. Also in this case we recognize an angular
enlargement at the nitrogen atom produced by the repulsive contacts across the apex nitrogen atom.[8] The deviation from linear arrangement of carbodiimides is produced by bent bonds in a non-cyclic structural unit, as
will be explained in the following section.

2.2.1 Electron density distribution in the carbo-
      diimide group

The electron density in the left C=N double bond
of the first section in Fig. 7 shows a larger extension
normal to the bond axis than the density in the neighboring bond of the same kind because only the π-electrons
of the left bond are within this plane. The π-electrons
of the second C=N bond are oriented perpendicular to this
plane and consequently make no contribution to the density distribution within this section. In the second

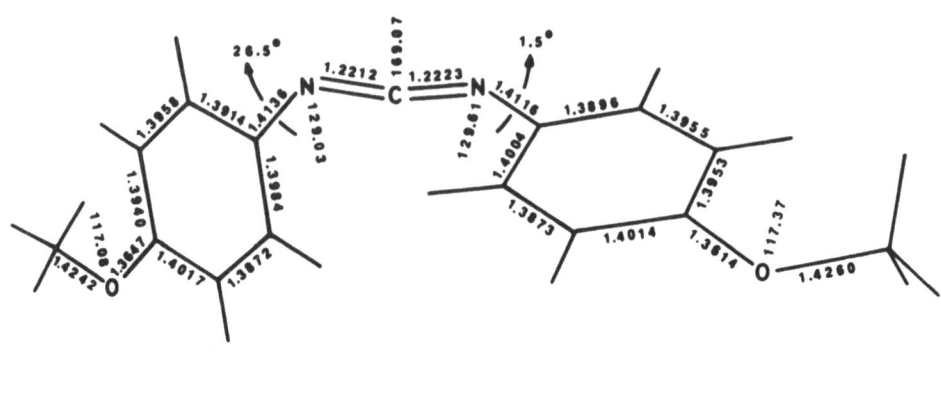

Fig. 5. Bond lengths (Å) and some bond angles (°) of 2
at -170°C. Standard deviations are 0.0005-
0.0008 Å and 0.07° respectively.

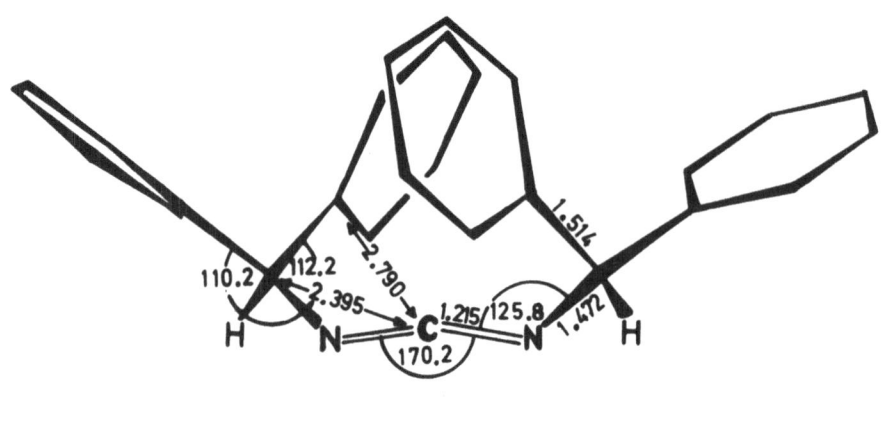

Fig. 6. Molecular dimensions of bis-(diphenylmethyl)-
carbodiimide.[8]

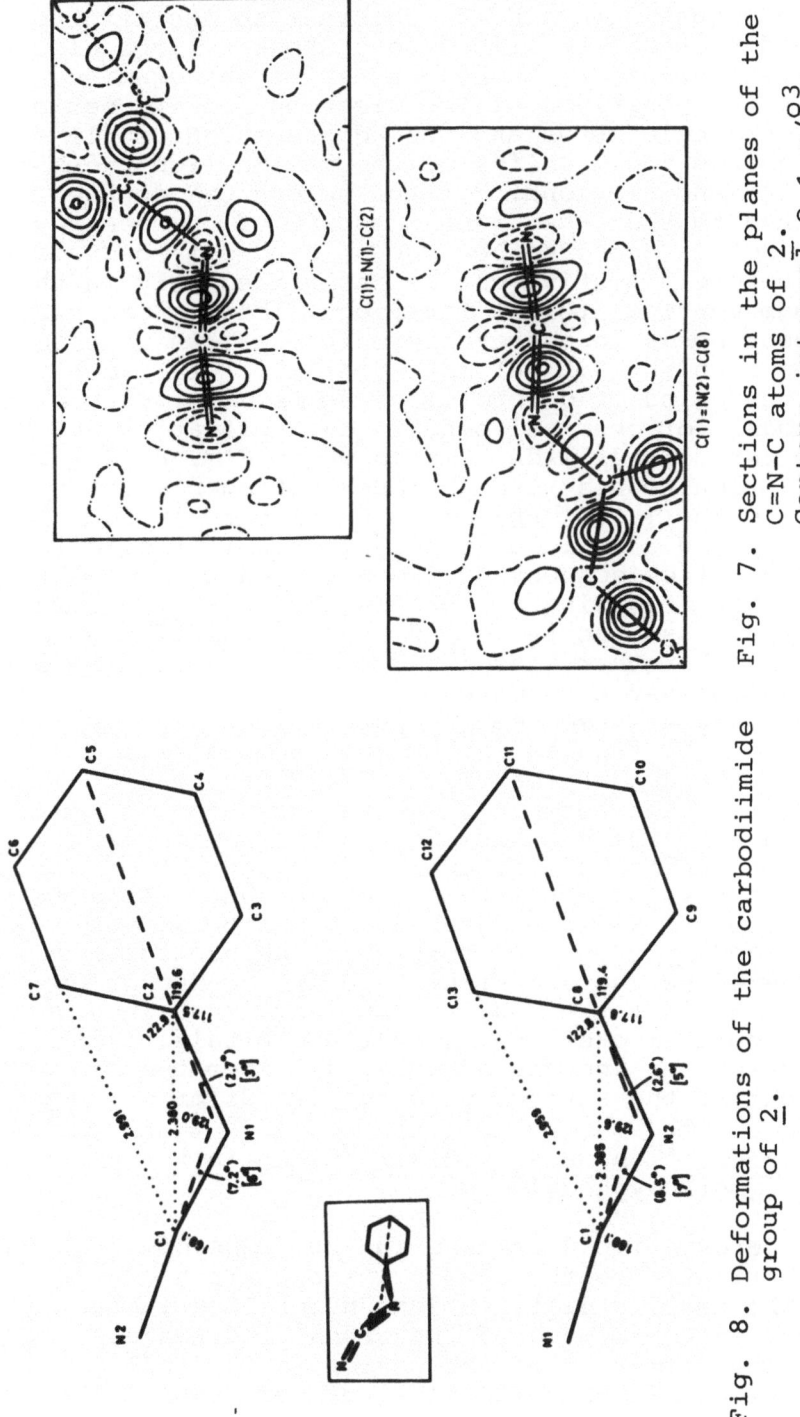

Fig. 7. Sections in the planes of the C=N-C atoms of 2. Contour interval 0.1 e/Å³

Fig. 8. Deformations of the carbodiimide group of 2.

section (Fig. 7) perpendicular to the first one the conditions are just reversed. In contrast to the first map
here the density of the right bond shows a large elongation perpendicular to the bond axis. If one examines
carefully the positions of the electron density maxima
of the C-N single bonds and the adjacent C=N double bonds
one recognizes small shifts of the maxima from the bond
axes to the endo regions. For the second C=N double
bond linked to the C-N=C group a shift of the density
peak can not be detected because this bond is bent down-
wards. Therefore a shift can be recognized only in the
second map vertical to the first one. This is one example of bent bonds in non-cyclic molecules, which recently
have been found by F.L. Hirshfeld et al.[10] In order to
explain this kind of deformation, he uses a flexible
spring model for chemical bonds. The application of this
model to the carbodiimide 2 is shown in Fig. 8. Mainly
the 1...3 repulsion across the apex nitrogen atom in
addition to the 1...4 contacts give rise to an enlargement of the C=N-C angle up to 129°. This is the angle
between the lines connecting the nitrogen atom to the
carbon atoms. According to the flexible spring model the
axes of the substituents--the diagonal line of the phenyl
ring and on the other side the second C=N double bond--
are tilted against the corresponding C-N bond axes. The
given tilt angles have been computed within the projection planes. The figures in the squared brackets are
taken from the density map (Fig. 7). They are the
approximate angles between the bond axes and the lines
connecting the atomic positions and the maxima of the
deformation densities. The overlapping regions of the
bonding orbitals do not lie on the bond axes. Apparently
the interorbital angle on the nitrogen atom is not as
much affected by repulsion forces as the nuclear positions. Therefore the substituents of the C-N=C group
are tilted against the C-N internuclear lines in order
to get an optimal overlapping of the orbitals. This
is the reason for the deviation of 11° from the linear
arrangement of the heterocumulenic carbodiimide group.

2.2.2 Electron density distribution in the methoxyphenyl group

The deformation densities in the plane of the phenyl
rings substituted to the carbodiimide 2 are shown in
Fig. 9. The density distributions determined independently for each C-C bond of the aromatic rings are very
regular. The variation of the maxima is 0.05 e/$\AA^3$, which
is a criterion of the reliability of our results. The

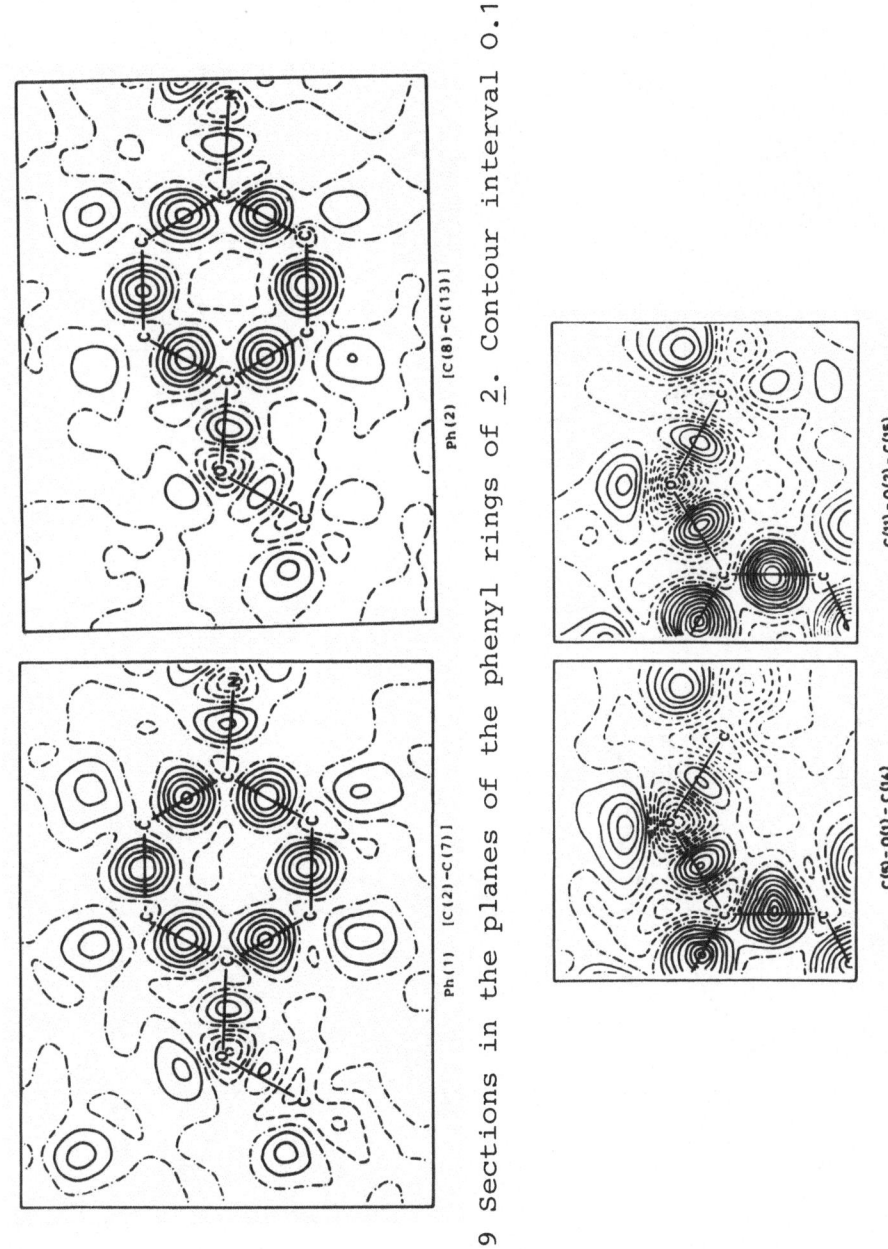

Fig. 9 Sections in the planes of the phenyl rings of 2. Contour interval 0.1 e/Å³.

Fig. 10. Sections through the methoxy groups of 2. Contour interval 0.05 e/Å³.

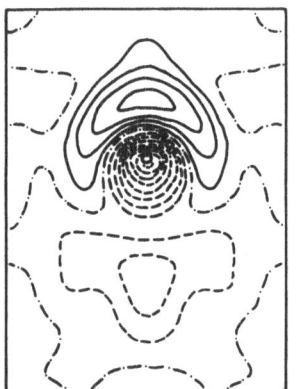
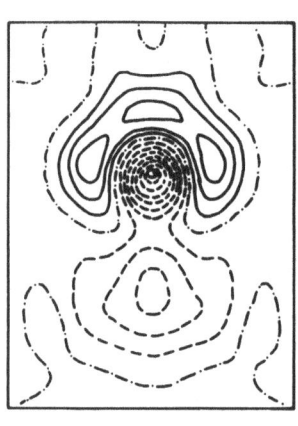

Fig. 12. Sections perpendicular to the C-O-C planes through the O-atoms and bisecting the C-O-C angles of 2. Averaged density (m symmetry). Contour interval 0.05 e/Å³.

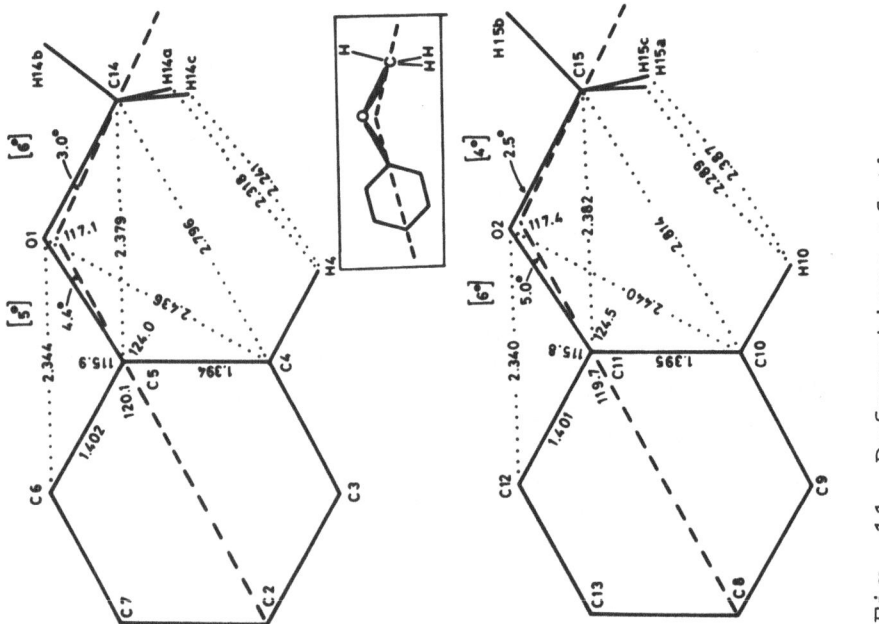

Fig. 11. Deformations of the methoxy groups of 2.

peaks above the oxygen atoms indicate the lone pair electrons.

Whereas the bond peaks of the phenyl rings are centered on the bond axes, the maxima of the methoxy groups are significantly shifted perpendicular to the bond axes into the direction of the bisecting line. Therefore here again we find bent bonds in non-cyclic structural units. In order to see these bent bonds in more detail Fig. 10 shows the sections through the three atoms of both methoxy groups. The bond peaks of all the C-O bonds are shifted by 0.05 to 0.1 Å perpendicular to the lines connecting the atoms into the endo direction of the methoxy groups.

By the repulsion forces across the apex oxygen atom, generated by short non-bonding distances, the nuclei of the carbon and oxygen atoms are bent away still sticking on the bonds which are imagined as flexible springs. (Fig. 11) Therefore the angle between the orbitals of the binding electrons are not as much affected as the bond axes which join two atomic positions. The three-fold symmetry axis of the methyl group and the diagonal line of the phenyl ring are tilted against the C-O axes. The figures in the squared brackets have been estimated from the density maps (Fig. 10).

So far we considered the electron density distributions in chemical bonds. The deformation densities of the lone pairs of the oxygen atoms are given in Fig. 12. The sections have been placed perpendicular to the planes of the methoxy groups of 2 along the bisecting lines of the C-O-C angles. In both cases density portions are appearing behind and above and below the oxygen atoms.

3.   Cyclobutadiene

3.1 Structures of some cyclobutadiene derivatives

After the demonstration of bent bonds in non-cyclic systems now we turn over to a cyclic system which has angular strain. For many decades the cyclobutadiene system was a fascinating subject for calculations and experimental investigations. Only recently have chemists succeeded in synthesizing cyclobutadiene and its derivatives or were able to prove its existence as an intermediate in reactions. Thereby the question arises again whether the four membered ring system of cyclobutadiene has a squared or a rectangular structure.[3] We had been

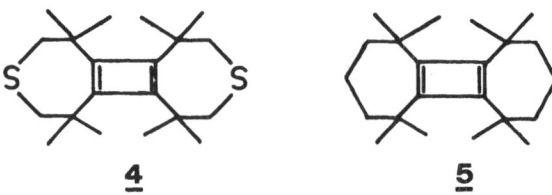

4   **5**

able to solve the first structure of a cyclobutadiene
derivative.[11] The cyclobutadiene ring of 4 is substituted by four tertiary butyl groups. Two of them are
linked by a sulfur atom to form seven membered rings on
both sides of the molecule. The bond length of 1.600 Å
for the single bond of the four membered ring of 4 and
the length of 1.344 Å for the corresponding double bond
certainly prove a rectangular structure.

We have also completed the structure determination
of the derivative 5 without heteroatoms to eliminate
the possibility that the sulfur atoms could have an influence on the bonding system of cyclobutadiene. We discovered two conformations of the molecule 5 in two
different crystal modifications. For the first conformation 5a (Fig. 13) we have found a chair-like form of the
seven membered rings which are connected to the central
four membered ring. The same conformation has
appeared in the sulfur containing derivative 4. This
conformation of cycloheptene has been drawn in the left
sketch of Fig. 13 and can be recognized in a side-on-
view of 5a (Fig. 13). The total molecule has a center
of symmetry at its center of gravity and shows only small
deviations from C_{2h} symmetry.[12] In the second modification 5b the cycloheptene ring has a twist conformation
(Fig. 13). The molecular symmetry is approximately D_2
and consequently the molecule is chiral.[13] The deviation
from planarity of the four membered ring is much smaller
(dihedral angle 2.8°) in 5b compared to tetra-tert.-
butyl-cyclobutadiene (9.8°).[14] The methyl groups in both
modifications of 5 are interlocked like gear wheels.
The intramolecular distances between these groups are
short. The repulsive forces give rise to a stretching
of the single bonds in the cyclobutadiene rings. These
non-bonding distances are a little shorter in 5a compared
to 5b. The single bonds of the cyclobutadiene rings are
stretched correspondingly (Fig. 14). As in the sulfur
containing derivative 4 the structure of the cyclobuta-

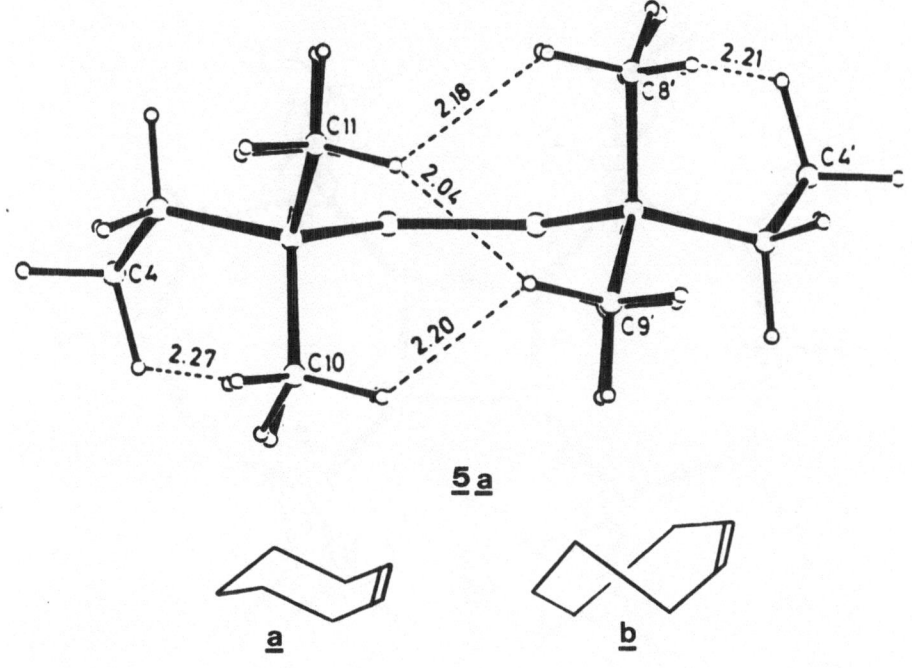

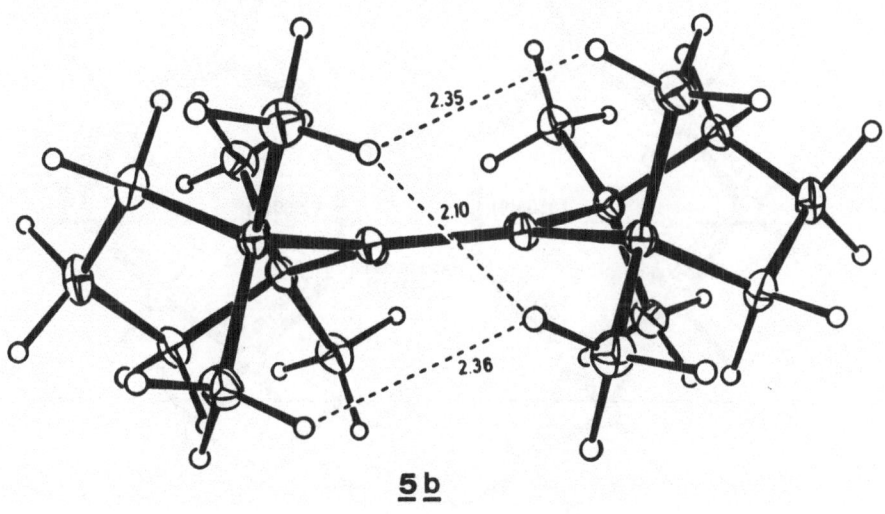

Fig. 13. Side-on-view of 5a and 5b parallel to the plane
of the four membered ring onto the edge of the
long bond. Interatomic distances (Å) at 115 K
for 5a and 110 K for 5b.

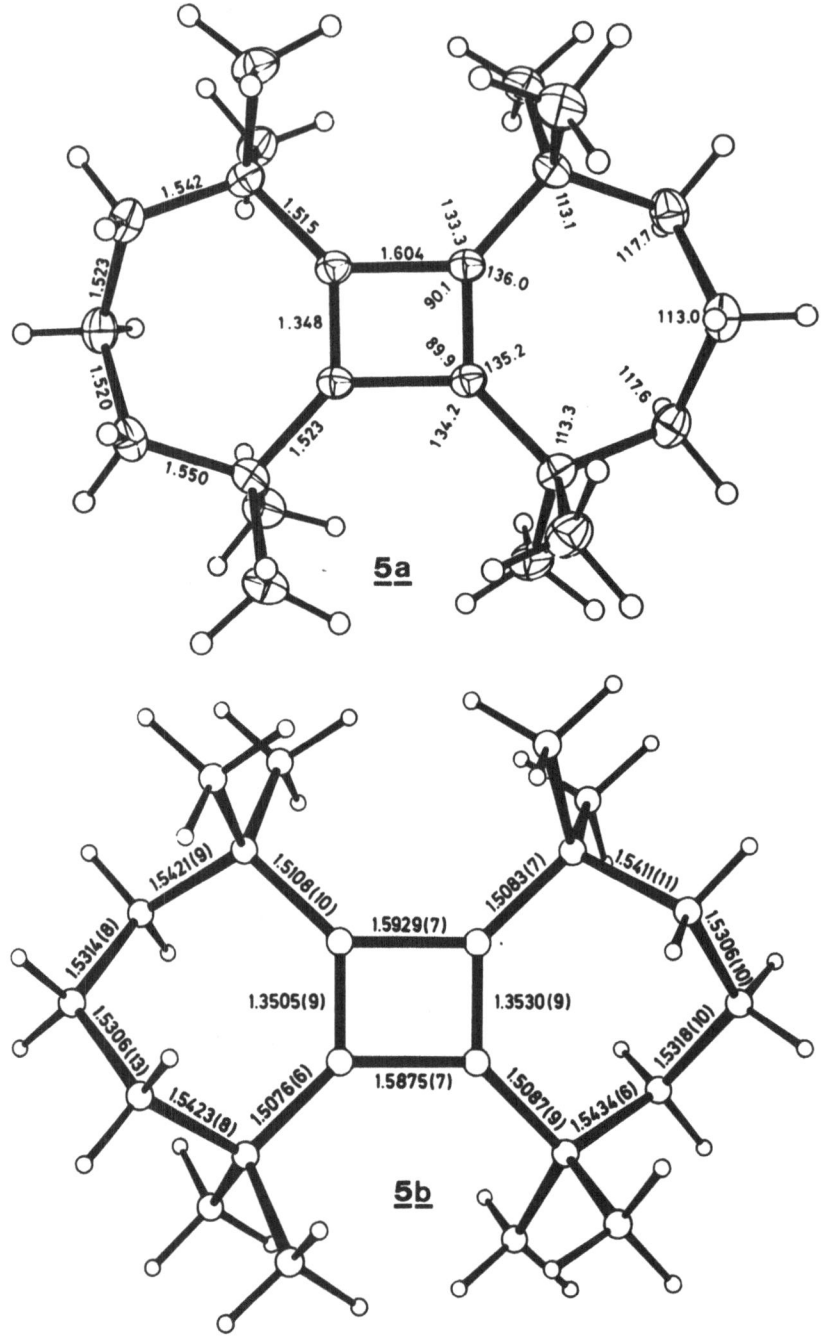

Fig. 14. Bond lengths (Å) and angles (°) of 5a (at
115 K)[12] and bond lengths of 5b (at 110 K)[13].

diene rings in 5a and 5b are far from a squared geometry. More information about the bonding system of cyclobutadiene is to be expected from the deformation densities.

3.2 Electron density distribution in cyclobutadiene

We have determined the deformation densities of the cyclobutadiene derivative 5b. The section in Fig. 15 has been made within the plane of the four membered ring. In the right map chemically equivalent bonds have been averaged. One can hardly recognize the difference with the left map in which the density of each bond has been determined independently. The density maxima of all four bonds are shifted outwards from the bond lines between the carbon atoms. In contrast the deformation densities of the bonds to the substituents are centered exactly on the midpoints of the bond axes. The deviation of the density peak in the double bond from the bond axis is 7°. The corresponding value of the single bond is 18°. So the angle between the deformation maxima is 115°. This value approaches the interorbital angle of 120° much better than the formal bond angle of 90°. The density peak of 0.5 e/$\AA^3$ on the maximum of the double bond is considerably higher than the corresponding maximum of barely 0:3 e/$\AA^3$ on the single bond. This is in accordance with the higher concentration of electron density in the double bond.

The sections in Fig. 16 are showing the density distribution in the planes perpendicular to this section of the four membered ring through the centers of the bonds. In the section through both short bonds the deformation density is clearly extended vertical to the molecular plane. The influence of the π-electrons is responsible for the shape of these density lines. The corresponding view parallel to the long bonds of cyclobutadiene reveals a much smaller deviation from a circular shape of the density lines which argues for a high σ-bond character. Already with room temperature data of the sulfur derivative 4 we observed the picture of bent bonds.[15] But the relative heights of the density peaks did not emerge correctly. According to these results and according to the determination of the bond distances, the cyclobutadiene derivative 5b has alternating single and double C-C bonds which are severly bent. This is in agreement with theoretical calculations of the density distributions.[15]

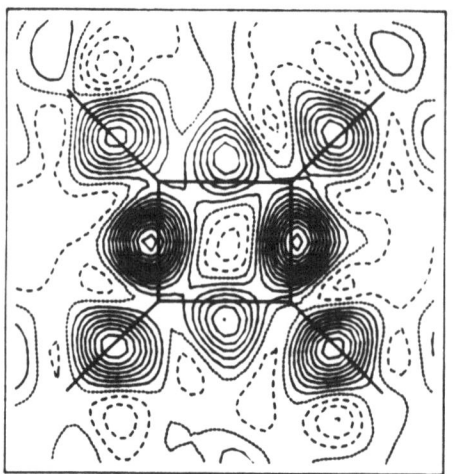

Fig. 15. Deformation densities in the section of the four
membered ring of 5b. Left: individually determined densities; right: averaged densities
(mm symmetry). Contour interval 0.05 e/$Å^3$.

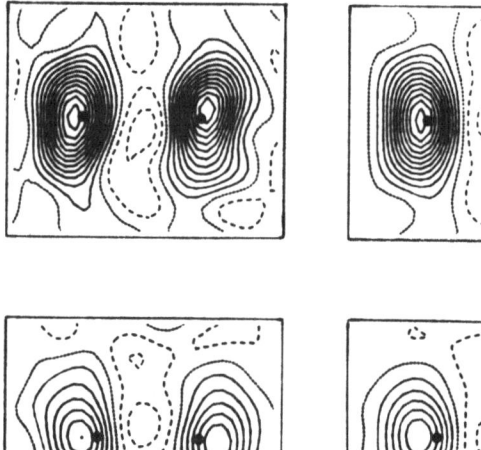

Fig. 16. Deformation densities in the sections perpendicular to the four membered ring of 5b through
the midpoints of the bonds (first row: double
bonds; second row: single bonds; right column:
averaged densities, mm symmetry).

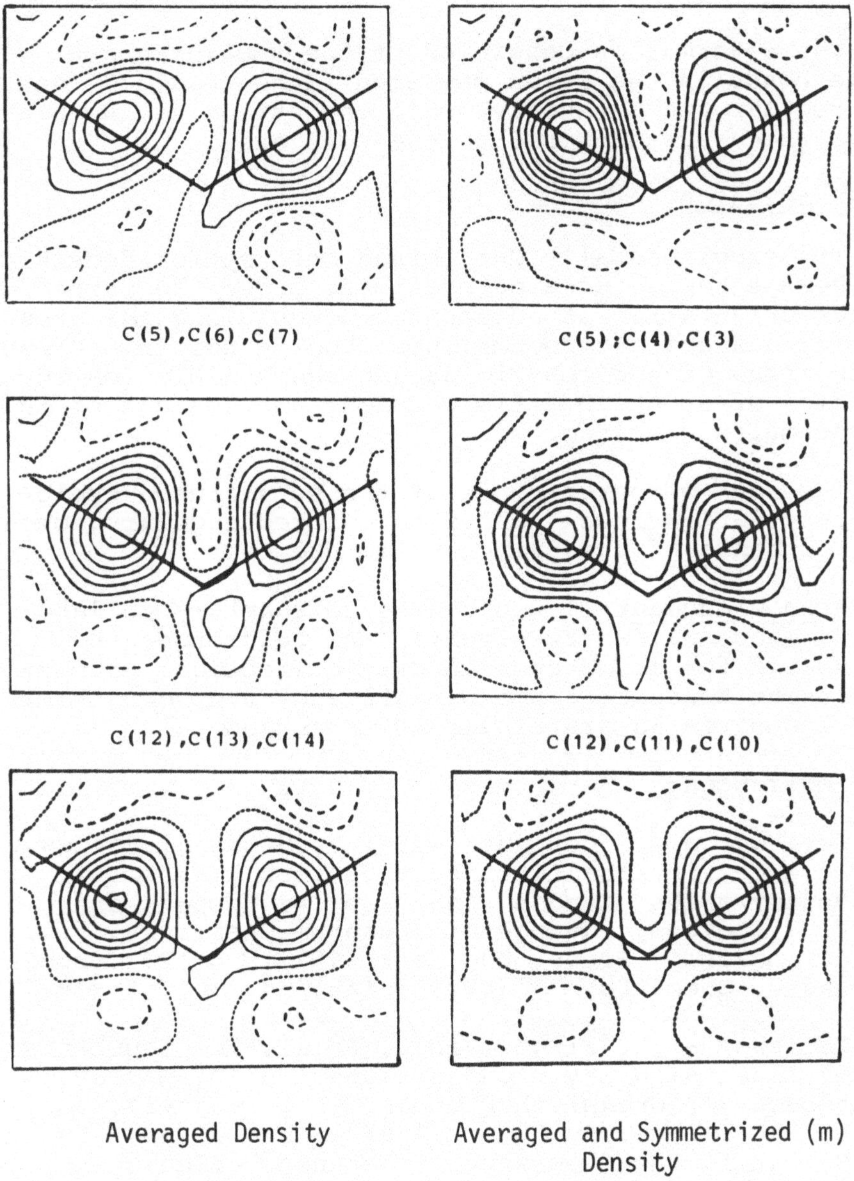

C(5),C(6),C(7)

C(5);C(4),C(3)

C(12),C(13),C(14)

C(12),C(11),C(10)

Averaged Density

Averaged and Symmetrized (m)
Density

Fig. 17. Deformation densities in the C-C-C bonds of the
seven membered rings of 5b. Contour interval
0.05 e/Å³.

The deformation densities of the remaining C-C
single bonds in the seven membered rings (Fig. 17) are
very regular and have only small variations of the peak
maxima which are centered on the bond axes.

4.   Conclusion

The experimentally determined deformation densities
in a hexapentaene and a carbodiimide prove the perpendicular orientation of adjacent π-bonds along the axes
of cumulenes and heterocumulenes. Bent bonds in the non-
cyclic group of a carbodiimide are responsible for the
deviation of about $10°$ from a linear arrangement of this
heterocumulenic system.

The four membered rings of the substituted cyclobutadiene derivatives 4 and 5 have a rectangular geometry. The bonds of this ring system are bent.

Acknowledgment: I would like to thank my collabo-
raters, whose names are given in the references and
H. Rodewald for assistence in data collection. Support
by Deutsche Forschungsgemeinschaft and Fonds der Chemischen Industrie is gratefully acknowledged.

References:

1.   Reviews: a) P. Coppens, MTP International Review of
      Science, Physical Chemistry, Series 2, Vol. 11.
      Butterworths, London (1975); b) P. Coppens and
      E.D. Stevens, Advances in Quantum Chemistry, Vol.
      10. Academic Press, New York (1977); c) P. Coppens,
      Angew. Chem. 89, 33 (1977); Int. Ed. 16, 32 (1977).

2.   Reviews: a) H. Fischer in S. Patai, The Chemistry
      of Alkenes. Wiley, London (1964); b) M. Murray, Me-
      thoden der Organischen Chemie (Houben-Weyl), Vol.
      5, 2a. Thieme, Stuttgart (1977); c) W. Runge in
      S. Patai, The Chemistry of Ketenes, Allenes and
      Related Compounds, Part 1, Wiley, Chichester (1980).

3.   Reviews: a) M.P. Cava and M.J. Mitchell, Cyclobuta-
      diene and Related Compounds. Academic Press, New
      York (1967); b) G. Maier, Angew. Chem. 86, 491
      (1974); Int. Ed. 13, 425 (1974); c) T. Bally and
      S. Masamune, Tetrahedron 36, 343 (1980).

4.   F. Bohlmann and K. Kieslich, Chem. Ber. 87, 1363
      (1954).

5. H. Irngartinger and H.-U. Jäger, Angew. Chem. 88,
    615 (1976); Int. Ed. 15, 562 (1976).

6. H.-U. Jäger, PhD-Thesis, Heidelberg (1977).

7. Z. Berkowitch-Yellin and L. Leiserowitz, J. Am.
    Chem. Soc. 97, 5627 (1975); Acta Cryst. B33, 3657
    (1977); Z. Berkowitch-Yellin, L. Leiserowitz and
    F. Nader, Acta Cryst. B33, 3670 (1977).

8. H. Irngartinger and H.-U. Jäger, Acta Cryst. B34,
    3262 (1978).

9. A.T. Vincent and P.J. Wheatley, J. Chem. Soc.
    Perkin 2, p. 687, 1567 (1972).

10. M. Eisenstein and F.L. Hirshfeld, Chem. Phys. 42,
    465 (1979).

11. H. Irngartinger and H. Rodewald, Angew. Chem. 86,
    783 (1974); Int. Ed. 13, 740 (1974).

12. N.H. Riegler, Diplomarbeit, Heidelberg (1979).

13. M. Nixdorf, part of a proposed PhD-Thesis.

14. H. Irngartinger, N. Riegler, K.-D. Malsch, K.-A.
    Schneider and G. Maier, Angew. Chem. 92, 214 (1980);
    Int. Ed. 19, 211 (1980).

15. H. Irngartinger, H.-L. Hase, K.-W. Schulte and
    A. Schweig, Angew. Chem. 89, 194 (1977); Int. Ed.
    16, 187 (1977).

Section 6
ELECTROSTATIC PROPERTIES

PSEUDOMOLECULAR ELECTROSTATIC PROPERTIES

FROM X-RAY DIFFRACTION DATA

Grant Moss

Medical Foundation of Buffalo, Inc., 73 High Street
Buffalo, New York 14203 and Chemistry Department
State University of NY at Buffalo, Buffalo, NY 14214

INTRODUCTION

The Hellman-Feynman[1] theorem establishes the link between quantum mechanics and electrostatics. In essence this theorem states that once the electron charge distribution in a molecule is known, the forces acting on the nuclei may be calculated using the formulae of classical electrostatics. In recent years a great deal of effort has been devoted to the study of molecular interactions from the standpoint of quantum mechanics.[2] In many cases the electrostatic term alone gives a good indication of the total interaction picture and has been used in the discussion of protonation, sites of reactivity, and formation of molecular complexes.[3-8] Such a simple picture is conceptually appealing since we have an intuitive grasp of how charges and dipoles interact. On the other hand, dispersion and exchange forces, having no direct classical analogue, present a greater problem to the imagination. The question we pose here is: Can we get this electrostatic information from X-ray diffraction data?

The past ten years or so have seen the growth of electron density studies both in theory and experiment. Comparisons of these two methods have shown that it is possible to obtain, with increasing reliability, features of the electron density from X-ray diffraction data that reinforce our conventional ideas of the processes involved in bond formation.[9] In some cases almost quantitative agreement has been obtained, the residual differences being attributed to systematic errors in the data and to the neglect of correlation and exchange in the theoretical calculations. As shown by Bernard[10] and Stevens[11] good agreement between theory and experiment can be obtained even for systems as large as metal

383

porphyrins. Generally comparisons are presented in the form of
deformation density maps, that is, the difference between (a) the
experimental or theoretical densities and (b) the density of the
promolecule, which is defined as a superposition of isolated
spherical atoms positioned at the corresponding nuclear sites.
While such maps are useful as a qualitative measure, a more quantitative basis for comparison is needed. Such measures have been
proposed by several authors. In the early work by Dawson[12] on
diamond, silicon, and germanium, the electron density features
were characterized in terms of the amount of charge shifted from
anti-bonding to bonding lobes. Kurki-Suonio and Salmo[13] have
determined the ionic state of several alkali halides and metal
oxides by direct integration of the observed structure factor
amplitudes over a spherical, atomic volume. Stewart[14] pointed out
that multipole fits to the experimental electron density offer the
possibility of further characterizing the electron density in terms
of atomic or molecular moments. Hirshfeld[15] has shown that such
quantitative measures, when applied to the static theoretical
densities of molecules containing the same functional groups,
yield an interpretation of bonding consistent with chemically
reasonable principles. Coppens and co-workers[16] have determined
net atomic charges and molecular dipole moments from multipole
refinements and direct integration techniques. More recently
Stewart[17] has developed expressions that allow electrostatic
properties, such as the electrostatic potential and the electric
field gradient, to be determined directly from X-ray diffraction
data.

This paper is limited to a discussion of electrostatic properties in molecular crystals, in which the intermolecular forces
are generally weak. Thus such effects as polarization and charge
transfer can be expected to be minimal. Even so, the properties
that are calculated will be pseudomolecular properties, that is,
properties of a molecule that has been extracted from the continuous crystal density. The degree to which these pseudomolecular
results are applicable to the calculation of intermolecular interactions must be given careful, case-by-case consideration. In
addition we also consider the related question of atomic properties
and their interpretation.

ATOMIC AND MOLECULAR MOMENTS AND THE ELECTROSTATIC POTENTIAL

The properties of any given system can be derived by evaluating the expectation value of the corresponding operator. X-ray
diffraction yields the electron distribution, expressed as a
series of observed Fourier coefficients, hence we are restricted
at the outset to properties for which the operator commutes with
the wavefunction. Further constraints on the properties that can
reasonably be extracted from X-ray data arise essentially from the

strengths and weaknesses of the X-ray experiment. Modeling of the electron density distribution yields a mathematical description that can then be cranked through the appropriate computational apparatus to yield a variety of molecular properties. However, not all of these calculations yield reliable results. The limited resolution of the X-ray experiment and generally lower accuracy of the high order data imply that the properties that emphasize the outer region of reciprocal space, i.e., the inner region of "atom space", are intrinsically less reliable than those that are determined by the lower order data. In the present work we are interested in the calculation of molecular outer moments (charge, dipole, etc.) and electrostatic potentials. Both kinds of properties may be expressed as Fourier series in which the effect of the high order data is downgraded.

In general a property χ of the one-electron density distribution may be written as the expectation value of the corresponding operator $\hat{\gamma}$. Assuming $\hat{\gamma}$ commutes with the wavefunction, the property χ is given by

$$\chi = \langle \hat{\gamma} \rangle = \int \hat{\gamma} \, \rho(\underline{r}) \, d\underline{r} \tag{1}$$

where the integration is performed over all space. For our present purposes the integration in equation 1 must be restricted to the volume of interest, that is, to the pseudomolecular volume M. The outer moments are thus defined by

$$\mu_{\alpha\beta\ldots\eta} = \int_M r_\alpha r_\beta \ldots r_\eta \, \rho(\underline{r}) \, d\underline{r} \tag{2}$$

where $r_\alpha = x, y, z$ for $\alpha = 1, 2, 3$.

An alternative formulation due to Buckingham[18] defines the outer moments of a charge distribution in terms of the Legendre polynomials. The Buckingham moments are easily derived from the outer moments. The components of the quadrupole moment, for example, are simply related to the second moments through the expressions

$$\Theta_{xx} = \mu_{xx} - \frac{1}{2} (\mu_{yy} + \mu_{zz})$$

$$\Theta_{xy} = \frac{3}{2} \mu_{xy} \quad \text{etc.} \tag{3}$$

By its definition the Buckingham quadrupole moment is zero for the promolecule hence it provides a measure of the deviations from spherical symmetry due to bond formation. It should also be noted that there are only five independent components of the quadrupole moment, whereas there are six independent second moments.

Thus the quadrupole moment can be derived from the second moments, but not vice versa. In the case of cylindrical symmetry, there will be only one independent component generally taken to be along the symmetry axis.

The second property of interest is the electrostatic potential which is defined by

$$V(\underline{r}_1) = \frac{1}{4\pi\epsilon_0} \int_M \frac{\rho(\underline{r})}{|\underline{r}-\underline{r}_1|} \, d\underline{r} \tag{4}$$

where integration is over the volume of interest. Since our interest is at distances removed from the molecule, it is convenient to follow Buckingham[18] and expand the electrostatic potential as a series in point multipoles, thus

$$V(\underline{r}_1) = \frac{1}{4\pi\epsilon_0} \left\{ \frac{\mu_0}{r_1} + \frac{\mu_\alpha r_\alpha}{r_1^3} + \frac{(3r_\alpha r_\beta - \delta_{\alpha\beta})\mu_{\alpha\beta}}{2r_1^5} \dots \right\} \tag{5}$$

where summation over repeated subscripts is implied. Thus at large distances the electrostatic potential is expressible in terms of the outer moments of the electron charge distribution, and the calculation of the electrostatic potential simply reduces to the determination of the outer moments from the X-ray data. The convergence of the multipole expansion is improved if the multipoles are distributed on several centers. Thus an expansion in terms of the moments of the constituent atoms is preferable to one based solely on the molecular moments. Practical evaluation of (2) therefore requires a suitable partitioning of the continuous crystal density into molecular and subsequently atomic fragments. The methods of partitioning are considered in the following section.

The usefulness of atomic charges in a variety of areas has, perhaps, given them an undeserved status. It is important to remember that atomic moments, unlike molecular moments, are not an observable. Without experimental guidelines their definition is thus arbitrary.

Finally we note that the need to partition the crystal density places obvious restrictions on both the systems to which these techniques can be applied and the use to which the results may be put. As our present interest is in the possibility of using the experimental electron density in the discussion of molecular interactions we restrict our attention to systems in which the intermolecular forces in the crystal are generally weak.

SPACE PARTITIONING OF THE CRYSTAL DENSITY

Becker[19] has given a comprehensive review of methods used to partition theoretical and experimental densities. There are two main techniques currently in use for the subdivision of the crystal density. Both are based on the principle of atomic superposition. An outline of the main features of the two methods is given in Table 1.

(a) Multipole Techniques

The multipole model has been described in detail by several authors.[20-21] Although there is great flexibility in the choice of the deformation function, $\delta\rho(\underline{r})$, Stewart[23] has shown that an nth order deformation function of the form $r^m \exp(-\alpha r)$ must be constrained to have $m \geq n$ in order that non-physical singularities

Table 1.   Multipole Refinement and Direct Integration Techniques in Molecular Crystals

---

$$\rho_{obs} = \Sigma\rho_{molecule}$$

$$\rho_{molecule} = \Sigma\rho_{atom}$$

$$\rho_{atom} = \rho_{spherical} + \Delta\rho$$

multipole*   direct integration

$$\Delta\rho_{atom} = \sum_{\ell=0}^{n} \delta\rho_\ell(\underline{r})$$

$$\Delta\rho_{obs} = \Sigma\Delta\rho_{atom}$$

$$\delta\rho_\ell(\underline{r}) = R(r)A(\theta,\phi)$$

$$= \rho_{obs} - \Sigma\rho_{spherical,atom}$$

$$\chi_{atom} = \chi_{sph} + \sum_{\ell=0}^{n} \int \delta\rho_\ell(\underline{r})\hat{\gamma}d\underline{r}$$

$$\chi_{atom} = \chi_{sph} + \int_{\substack{\text{volume} \\ \text{of atom}}} \Delta\rho_{obs}\hat{\gamma}d\underline{r}$$

---

*For a more general discussion of multipole models see the chapter by Coppens.

in the charge density and its derivatives be avoided. Likewise,
the angular function is largely a matter of taste by *caveat emptor*.
Stewart[21] and Coppens and Hansen[22] use a model based on the spherical harmonics. Hirshfeld[20] has chosen a set of functions of the
form $\cos^n \theta$, the angle θ being referred to a set of atom centered
polar axes. Both approaches yield equivalent results in the analysis of experimental data, although differences are observed between
the deconvoluted static deformation densities.[24] A possible difficulty of the multipole approach is that functions on one atom may
in fact expand to take up some of the density from a neighboring
atom. This problem has led Hirshfeld[25] to adopt an approach whereby
the least squares fit of the deformation functions is used simply
to "filter" the experimental density. The resultant deformation
model is then assumed to provide a good description of the total
molecular density, deconvoluted from the effects of thermal motion.
In a subsequent step, the stockholder partitioning step, the crystal
static deformation density is divided into atomic fragments by
numerical integration, in which the contribution from a given atom
to the deformation electron density at a particular point in space
is in direct proportion to the contribution from the corresponding
spherical atom density to the total procrystal density at that
point. I.e.,

$$\delta \rho_i = \omega_i \cdot \delta \rho_{total} \tag{5}$$

where

$$\omega_i = \frac{\rho_{spherical,i}}{\displaystyle\sum_{j=1}^{atoms} \rho_{spherical,j}} \, .$$

(b) Direct Integration Techniques

Direct integration of the experimental electron density distribution requires a careful definition of the molecular volume.
A detailed derivation of the method has been given elsewhere.[26]
In practice the electron density in the unit cell is calculated
on a fine grid, the grid is then scanned to determine whether a
particular grid point is to be considered to belong to a given
molecule. The definition of the molecular volume has been performed using two very different techniques to partition the molecular crystal.

(*i*) Discrete boundary partitioning. In this case the molecular volume is defined as a form of generalized Wigner-Seitz
cell.[27] The boundary surface is given by the condition

$$\frac{(r-r_A) \cdot r_{AB}}{R_A} = \frac{(r-r_B) \cdot -r_{AB}}{R_B} \tag{6}$$

where R_A and R_B are the van der Waals radii of two atoms A and B in different molecules. As shown in Fig. 1 this method of partitioning, when applied to the difference density, places the boundary surface in a physically reasonable region of very low electron density, and thus can provide corrections to the promolecule moments. Application to the derivation of atomic moments is conceptually more difficult, since it requires placement of a boundary surface through bond densities, and may lead to incorrect interpretation of the bonding. For example, discrete boundary partitioning of experimental difference densities has been used to determine the charge of the sodium atom in a number of salts.[28-30] In these studies the charge on the sodium was, on average, +0.2e. A theoretical study of Dannöhl *et al*,[31] however, found that for a procrystal composed of singly, but fully charged sodium and thiocyanate ions, direct integration of the difference density yielded a net charge of only +0.2e on sodium. Thus, in at least this test case, it did not prove possible to recover the input atomic charge by direct space integration.

(*ii*) Fuzzy boundary integration. This technique is applicable to either the total or difference densities. It is based on a direct application of Hirshfeld's[25] stockholder method to the experimental electron density. Thus each point on the integration grid is given a weight as defined in equation (5). In practice this method is used to derive corrections to the spherical atom moments from which molecular moments may be deduced. It should be noted that, by construction, the fuzzy boundary integration will correctly reproduce atomic and molecular moments of a promolecule density.

(c) Comparison of Multipole and Direct Space Techniques

The multipole and direct integration techniques are both empirical attempts to segment the unit cell into separate molecular entities. Both have their advantages and disadvantages. The principal benefit of a multipole refinement is that it filters out the noise inherent in the experimental data and, assuming the deconvolution of thermal motion to be valid, leads to an analytic representation of the static deformation density distribution, from which the calculation of molecular properties readily proceeds. At the atomic level the picture is more complex. The flexibility in the choice of deformation functions, while yielding essentially equivalent descriptions of the total density and hence of molecular properties, leads, in some cases, to wildly different estimates of the atomic properties. This fact has prompted

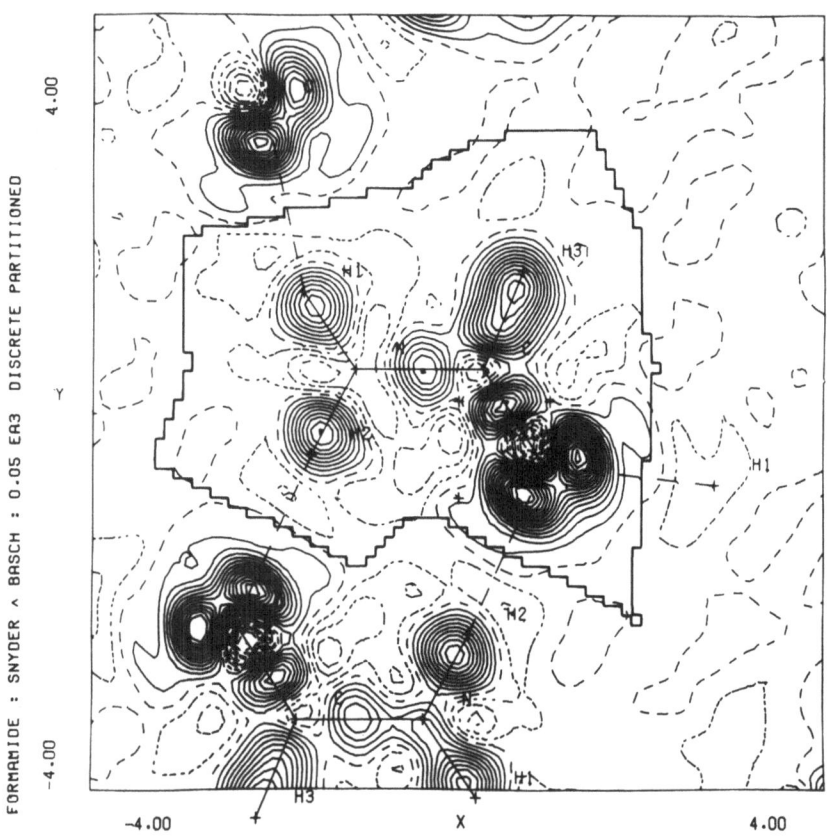

Fig. 1.   Discrete boundary partitioning of the difference density
           of a theoretical formamide crystal.[50] The heavy line
           shows the boundary surface in the plane of the molecule.
           Contours are at 0.05 e/$\overset{o}{A}^3$.

Hirshfeld[25] to adopt the extra step of the stockholder partitioning
mentioned above. Offsetting the advantage of the filtering, is
the problem of a large number of parameters and their often large
correlations. The correlation matrix, however, allows estimation
of the errors in the derived properties. Direct integration
techniques, on the other hand, treat the data at face value and
are therefore more susceptible to the effects of experimental
inaccuracies and incompleteness of the data. Atomic positional
and, in the case of a difference density integration, thermal
parameters must be supplied and, because of the bias of low order
data due to bond formation, these will generally come from the

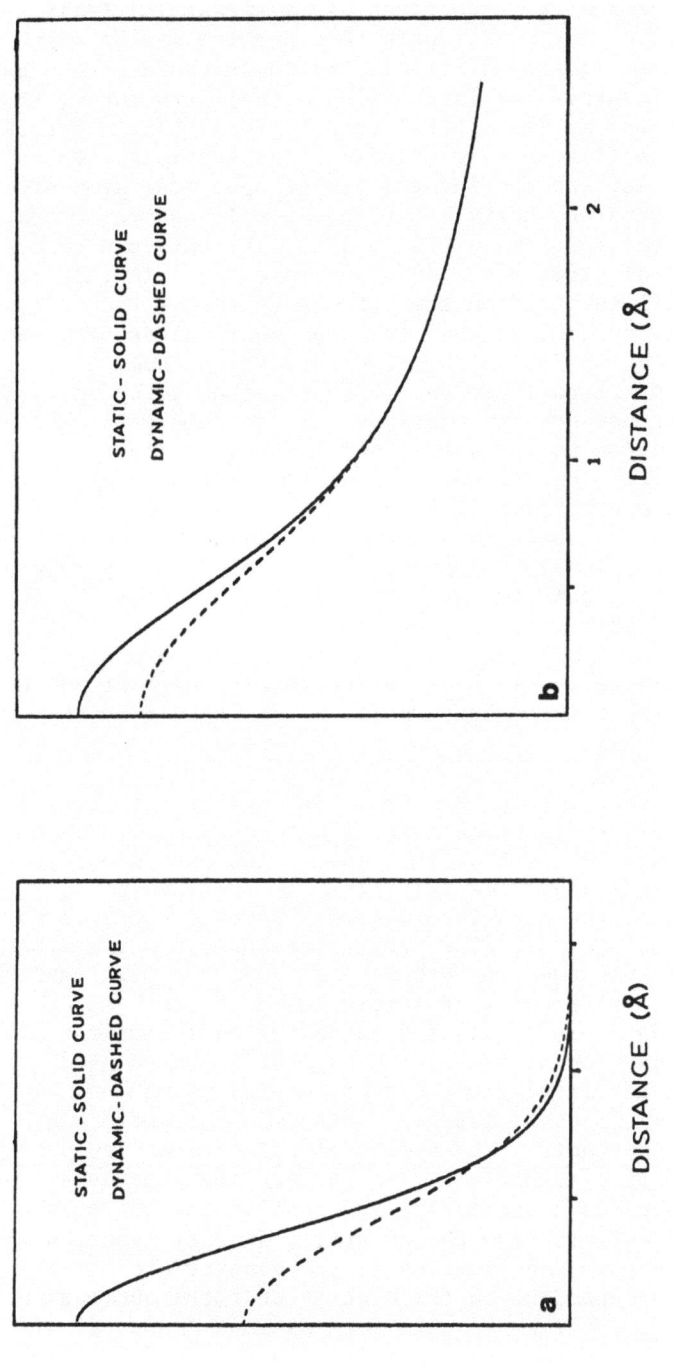

Fig. 2. (a) The effect of convoluting the static density function $\rho(r) = M \exp(-\alpha r^2)$ with the probability density function $p(r) = N \exp(-r^2/2\langle u^2 \rangle)$ where $\alpha = 4.0 \text{Å}^2$ and $\langle u^2 \rangle = 0.04 \text{Å}^{-1}$ - see text. (b) The electrostatic potential corresponding to the static and smeared densities of Fig. 2a.

less accurate high order data. In addition the structure factor
phases must be known and this restricts us, at least initially, to
centrosymmetric structures. Of course this problem may be over-
come by using a least squares multipole fit to determine the phases
of the experimental data. For large ($\sim$100 atoms) structures, however, this may present an impractical computational task. Errors
in the derived properties are also difficult to estimate. From
test calculations based on the integration of spherical atom structure factors, the fuzzy boundary technique reproduces atomic and
molecular moments to approximately 3 to 5%. (The only non-zero
moments for spherical atoms are even order moments.) In this case
the integration will suffer from the effects of series truncation
although these will be smaller than in a conventional density map,
because the contribution from high order data to the summation is
downgraded.[17] In practice it is generally found that for integra-
tions of the difference density, data beyond $\sin\theta/\lambda \sim 0.7 \text{Å}^{-1}$ make a
negligible contribution to the summation, and thus series truncation
will not be a problem. Due to the method of summation a correct
consideration of the experimental errors requires a knowledge of
the correlation between observed structure factors.[26] This correlation is generally assumed to be zero except for symmetry related
reflections generated from the same experimental observation. The
errors estimated in this way, however, appear to be unrealistically
small. Calculations of molecular dipole moments from refined
multipole parameters generally have errors in the range of 10-20%;
the error on the direct integration calculations should be of
comparable magnitude.

A major difference between multipole and direct integration
techniques is the manner in which thermal motion is treated.
Multipole refinements make use of the convolution approximation
to separate the static charge density from the thermal motion.
Direct integration of the structure factors, however, permits no
such separation, and the molecular properties derived may not be
comparable strictly for the two methods. A simple example, which
also illustrates the difficulty of obtaining inner moments from
X-ray diffraction data, is provided by considering a simple
Gaussian density function of the form $M \exp(-\alpha r^2)$, convoluted
with a Gaussian probability density function $p(r) = N \exp(-r^2/2 \langle u^2 \rangle)$,
where $\langle u^2 \rangle$ is the mean square isotropic displacement and M and N
are normalization constants. The effect of this thermal smearing
is illustrated in Fig. 2a, where it can be seen that the density
at the origin is altered drastically, and that on the whole, the
density is smeared out relative to the static density function.
The consequences of this modification to the density are: (i)
properties strongly dependent on the density near the nucleus,
and thus on the high order X-ray data such as inner moments, will
be very sensitive to the convolution approximation, and (ii) the
outer moments of the smeared density will be different from those
of the static density. An illustration of this second point is

provided by taking $\alpha = 4.0\overset{\circ}{A}^{-2}$ and $<u^2> = 0.04\overset{\circ}{A}^2$ which yield second moments $<r^2>$ of $0.375\overset{\circ}{A}^2$ and $0.495\overset{\circ}{A}^2$ for the static and smeared densities, respectively. This result implies that we must be careful when we compare results obtained with different experimental techniques. Direct integration yields thermally-averaged values for the molecular properties, while the multipole models offer, at least in principle, access to static values. The situation, however, is not so critical in the case of the electrostatic potential. In Fig. 2b, the electrostatic potentials corresponding to the static and smeared densities of Fig. 2a are plotted as functions of distance. At small distances $V_{dynamic}$ will always be smaller than V_{static}. At larger distances, our particular region of interest, $V_{dynamic}$ and V_{static} are equal.

The effect of thermal motion on the second moments of the molecular charge distribution is a possible source of difficulty, since the contribution from static spherical atoms will be quite different from that obtained using thermally smeared atoms. In the remainder of this paper we avoid this source of difficulty by adopting the quadrupole moment definition of Buckingham.[18]

A second major difference between the multipole and direct integration techniques, which has important consequences for the calculation of atomic properties, is the problem of extrapolation. Multipole models, by construction, have an implied extrapolation to infinite resolution. Hence the electron density close to the nucleus, a region not accessible to X-ray diffraction measurements, is modeled by the deformation functions as well. The contribution from this region to atomic properties may represent a considerable fraction of the total result. As an example we consider the simple exponential density function $\exp(-\alpha r)$. The ratio of the charge contained within a sphere of radius R, to the total charge, obtained by integrating the density function is given by

$$\frac{q_R}{q_{total}} = 1 - e^{-\alpha R} \sum_{s=0}^{2} \frac{(\alpha R)^s}{s!} . \tag{7}$$

The cut-off parameter R is related to the resolution of the experiment. It is frequently argued that features within approximately $0.3\overset{\circ}{A}$ of the nucleus cannot be determined by X-ray data. Taking this estimate for R and an exponent of $6.0\overset{\circ}{A}^{-1}$, a typical value for a simple exponential density function, yields a ratio of 0.26. Thus approximately 25% of the total charge is derived from a region about which the experimental data contains little or no information. In contrast the direct integration method deals only with the experimental set of observations. Implicit

in this method is the assumption that the promolecule correctly
describes the electron density in that region which is not accessible to X-ray diffraction due to the finite resolution of the experiment.

RESULTS

    The methods described above have been applied to several
systems. In this section the results of the fuzzy boundary direct
integration as applied to acetylene, formamide, and pyridinium
dicyanomethylide are discussed. In the following section the
results of a multipole analysis of pyrazine data and the subsequent
calculation of the electrostatic component of the interaction
energy are discussed. A summary of the crystallographic data is
given in Table 2.

(a) Direct Integration Analysis

    The direct integration of the experimental density was per-
formed using the fuzzy boundary partitioning scheme to determine
atomic and molecular moments. The unit cell was divided into
identical blocks of approximately $0.15\overset{\circ}{A}$ on edge, the charge in
each block being derived by Fourier summation. The charge of any
given block was divided among the constituent atoms via the stockholder partitioning scheme. These atomic contributions were then
used to calculate atomic, and subsequently molecular, deformation
moments, that is the moments due to deviations from isolated
spherical atoms.

    The results quoted in the Tables are deformation moments and
therefore do not include the spherical atom contribution. In the
case of atomic properties it is more convenient to use the deformation moments. For molecular properties we adopt the moment definition of Buckingham[18] in which case the contribution from the promolecule is zero.

Table 2.   Summary of Crystallographic Data

	Space Group	a	b	c	β	No. of Reflections	Temp. (K)	Reference
Acetylene	Pa3	6.091				164	141	32
Formamide	P2$_1$/n	3.604	9.041	6.994	100.5	922	118	35
Pyridinium Dicyanomethylide	P2$_1$/m	7.170	12.523	3.878	94.8	2390	118	24
Pyrazine	Pmnn	9.325	5.850	3.733		872	184	42

Acetylene. Accurate low temperature (141 K) data were collected by van Nes.[32] It is of interest to note that the molecule sits
on a three-fold axis and therefore cylindrical symmetry cannot, *a
priori*, be assumed.

Atomic and molecular deformation moments are given in Table 3.
For comparison, the equivalent quantities derived from a stockholder partitioning of a theoretical deformation density are also
listed. The wavefunction[33] (431-G) is of double-zeta quality, however, it does not include polarization functions. The partitioned
moments would therefore be expected to show smaller deviations
from spherical symmetry than those obtained from a more extended
basis.

Both the theoretical and direct integration results show a
transfer of charge to the carbon atoms. The atomic dipoles, defined
in the sense negative to positive (-...>+), indicate a shift of
charge into the bonds - the hydrogen atoms donating to the C-H bond

Table 3.   Atomic and Molecular Deformation Moments for Acetylene

(A) Direct Integration of Experimental Data

	μ_0	μ_x	μ_{xx}	μ_{yy}	μ_{zz}
C	-.142	.102	.045	.213	:213
H	.142	.099	.079	.060	.060
C_2H_2	.0	.0	1.816	.546	.546

(B) Stockholder Partitioning of Theoretical Density

	μ_0	μ_x	μ_{xx}	μ_{yy}	μ_{zz}
C	-.099	.131	.017	.037	.037
H	.103	.095	.068	.066	.066
C_2H_2	.004	.0	1.654	.164	.164

(C) Molecular Quadrupole Moment

	θ_{xx}
Direct Integration	1.27
Theoretical	1.49
Spectroscopic	3.0

Units in $+e\text{Å}^n$.
Molecular frame defined with x-axis from C to H.

and the carbon atoms to the C-C bond. The experimental second
moments show both atoms to be contracted relative to the spherical
atom density, the carbon atom being considerably more contracted in
the direction perpendicular to the molecular axis. Similarly the
theoretical second moments indicate a more pronounced contraction
of carbon perpendicular to the molecular axis, although in this
case, the carbon atom is less contracted than the hydrogen atom.

As stated earlier, the molecular second moments derived from
the direct integration of experimental data contain a significant
contribution from the thermally smeared spherical atoms. It is
therefore of more interest to consider the molecular quadrupole
moments as defined by Buckingham.[18] For a cylindrically symmetric
molecule, that is a molecule with an axis of symmetry that is at
least three-fold, there is only one independent component of the
quadrupole moment. Thus, in the case of acetylene, deviations from
cylindrical symmetry become apparent only in the third and higher
order moments.

The molecular quadrupole moment determined by direct integration has a value of 1.27 e$\mathring{A}^2$ which compares reasonably with an
experimental value of 3.0 e$\mathring{A}^2$ determined from spectroscopic measurements.[34] In the multipole refinement of van Nes,[32] the quadrupole
moment ranged from 7 to 9 eA^2 depending on the refinement model,
the variation being due to parameter correlation.

Although the molecular second moments are consistent with
cylindrical symmetry, examination of the higher order moments
reveals a degree of asymmetry. This asymmetry is also apparent in
the experimental difference density maps, but its effect is more
clearly seen in the electrostatic potential surrounding the molecule. In Figure 3 the electrostatic potential, calculated from
the direct integration atomic deformation moments, is plotted for
several levels of multipole approximation. Inclusion of the third
and fourth order deformation moments introduces a large asymmetry
particularly in the neighborhood of the C-C bond. At larger
distances, where the electrostatic and multipole approximations
are valid, the contribution from these terms falls off rapidly.

It is of interest to examine the origin of this asymmetry.
To this end a set of theoretical structure factors was calculated
using the 431-G wavefunction[33] mentioned above. To avoid problems
of overlap the isolated molecule was placed in a unit cell which
had cell dimensions double the experimental values. Integration
of this pseudocrystal data therefore shows the effects of the
crystallographic symmetry. This symmetry is reflected in the
reciprocal lattice, that is in the points at which the Fourier
transform of the continuous electron density distribution is
sampled. The electrostatic potential calculated from the results

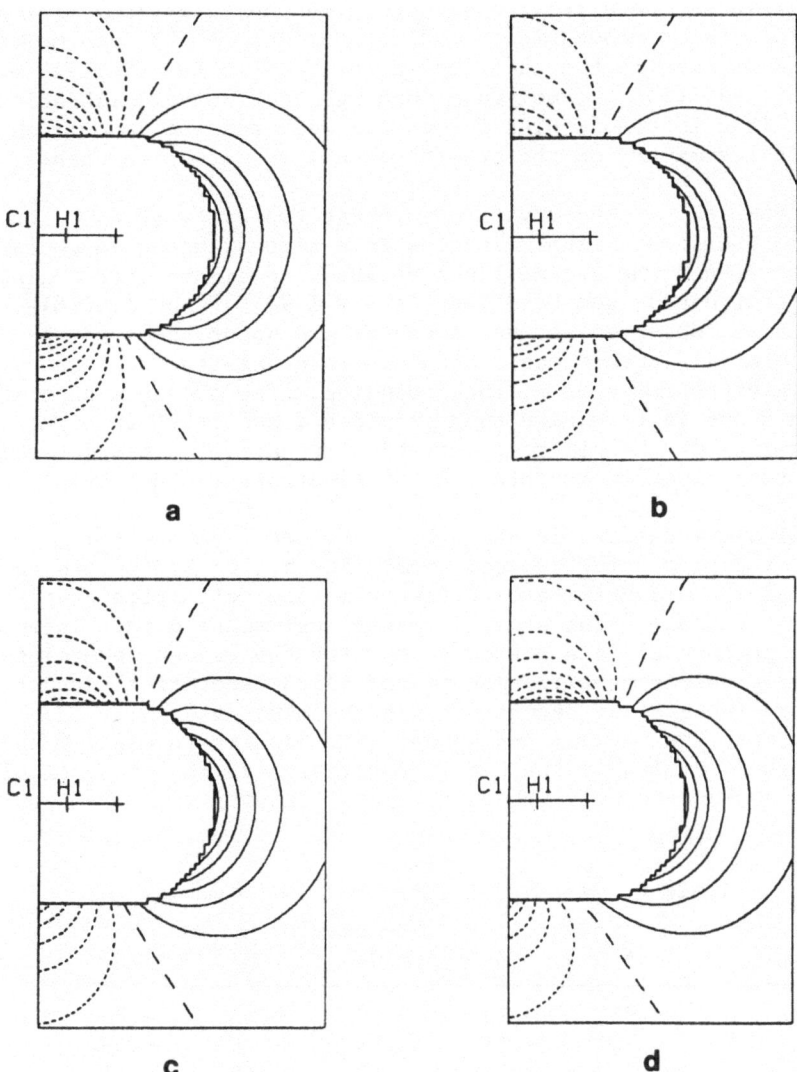

Fig. 3.   Deformation electrostatic potential of acetylene determined from direct integration of the experimental data for
different levels of the multipole expansion: (a) to first
order, (b) to second order, (c) to third order, and (d) to
fourth order.   Contour level is 2.0 Kcal/mole.

of the direct integration shows the same asymmetry as observed in
the analysis of the experimental data. Further confirmation that
this effect is of geometric origin is provided by the intermolecular
contact distances in the experimental unit cell, the shortest being
3.05Å. In addition the molecules undergo considerable librational
motion. Thus intermolecular forces are weak and the imposition of
cylindrical symmetry on the atomic moments is not unreasonable.

    Formamide. In the crystal there are two types of hydrogen
bonds.[35] The first joins molecules into dimers through a center
of symmetry, and the second links neighboring dimers into chains.
The respective hydrogen bond lengths are 1.94(5)Å and 1.90(6)Å.
No density is observed between acceptor and donor atoms in the
experimental difference maps, and this is consistent with an electrostatic interpretation of hydrogen bonding. The presence of three
hydrogen bonds in formamide thus provides a good test for the
applicability of the direct integration technique to the determination of the molecular moments and the electrostatic potential.

    In Table 4 the atomic charges determined from several different
methods of partitioning are compared. A considerable spread occurs
reflecting the essential arbitrariness in the definition of atomic
moments. In Table 5 the atomic moments derived by direct integration
of the experimental data are compared with the values determined
from a stockholder partitioning of the 431-G theoretical density.[33]
The direct integration results show a movement of density from
hydrogen into the C-H and N-H bonds. The dipoles on C and N show
an increased density in the C-N bond relative to the theoretical
result. A comparison of the experimental density with the results

Table 4. Atomic Charges (+e) in Formamide

	A	B	C	D	E
O	-.38	-.35	-.48	-.55	-.32
N	-.76	-.17	-.58	-.78	-.19
H2	.37	.13	.30	.40	.21
C	.26	.20	.50	.51	.01
H1	.36	.13	.28	.39	.16
H3	.15	.06	-.02	.03	.13

A: Mulliken charges from Snyder and Basch wavefunction[33]
B. Stockholder partitioning of Snyder and Basch wavefunction
C: Population analysis of Christensen et al[37] on an extended basis
    set (11,7,1) calculation
D: Kappa refinement by Stevens[35]
E. Present work; fuzzy boundary partitioning of experimental density

Table 5.   Atomic Deformation Moments for Formamide Referred to
           Local Coordinate Frames

(A) Fuzzy Boundary Direct Integration of Experimental Data

	μ_0	μ_x	μ_y	μ_z	μ_{xx}	μ_{yy}	μ_{zz}
O	-.32	-.047	.013	.012	.068	.072	-.120
N	-.19	-.088	.027	.004	.123	.150	.020
H2	.21	-.079	-.028	.003	.038	.153	.085
C	.01	-.035	-.051	-.075	.114	.137	.183
H1	.16	-.072	-.015	.020	.033	.135	.080
H3	.13	-.099	-.004	.028	.083	.101	.098

(B) Sotckholder Partitioning of Theoretical Density

	μ_0	μ_x	μ_y	μ_z	μ_{xx}	μ_{yy}	μ_{zz}
O	-.35	.183	.009	-.001	-.059	-.113	-.062
N	-.17	.013	.003	-.071	.081	.095	-.076
H2	.13	-.098	-.005	-.004	.070	.076	.048
C	.20	.013	-.022	.000	.066	.095	.197
H1	.13	-.099	-.003	.004	.070	.084	.047
H3	.06	-.064	.000	.000	.057	.069	.069

Units of $e\text{Å}^n$.

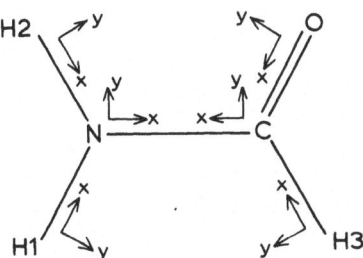

of an extended basis set calculation by Stevens et al[36] showed a
"substantial decrease in bond order of the C=O bond and an increase
of the C-N bond" in the solid compared to the gas phase.   The
uncertainty in the atomic parameters determined by direct integration precludes more than an observation of trends.   In the case of
the second moments all atoms appear to be contracted relative to
the spherical atom basis except oxygen, which exhibits a substantial
expansion perpendicular to the molecular plane.   This result is consistent with the observation of Stevens[35] that the oxygen lone pair
peaks are tilted out of the plane of the molecule.

The greater reliability of molecular as opposed to atomic moments is seen in Table 6 where the results of several determinations are compared. The theoretical values of Christensen et al[37] and Snyder and Basch[33] refer to a molecular geometry in which the pair of amide hydrogens are out of the plane of the molecule. In the case of the results derived from the X-ray data the non-zero out-of-plane component shows clear evidence for the effect of the out-of-plane hydrogen bond. The results are in generally good agreement with the spectroscopic value.[38]

The electrostatic potential in, and 2.0Å above and below the molecular plane in shown in Figure 4. These calculations are in reasonably good agreement with the theoretical calculations of Bonaccorsi et al.[3] The asymmetry of the electrostatic potential due to the out-of-plane dipole component is clearly evident. The results indicate a rather broad and deep potential well centered on the oxygen atom and clearly illustrate a preference for protonation at the oxygen site, in agreement with theoretical results.[3] Unlike exact calculations of the electrostatic potential from the wavefunction no potential minima can be found since at the distances at which such minima occur the multipole approximation breaks down.

Table 6. Molecular Dipole Moment (Debye) for Formamide

	μ_x	μ_y	μ_z	$|\mu|$	$\alpha°$
A	−4.02	−2.84	0.48	4.94	35.7
B	−3.25	−2.31	0.44	4.01	35.4
C	−2.88	−3.31	0.15	4.39	49.0
D	−3.56	−3.26	0.11	4.83	42.5
E	−2.85	−2.38	0.00	3.71	39.6
F	−3.06	−2.70	−0.49	4.07	41.9
G	−3.19	−2.93	−0.74	4.39	42.6

A: Discrete boundary direct integration: this work
B: Fuzzy boundary direct integration: this work
C: Kappa refinement[35]
D: Multipole refinement[15]
E: Microwave[38]
F: Christensen et al[37] (11,7,1) extended basis
G: Snyder and Basch[33] 431-G double-zeta basis

The molecular frame is defined with the x-axis from N to C and the oxygen atom in the x-y plane such that it has a positive y coordinate. The angle α is the angle made by the dipole with the negative x-direction.

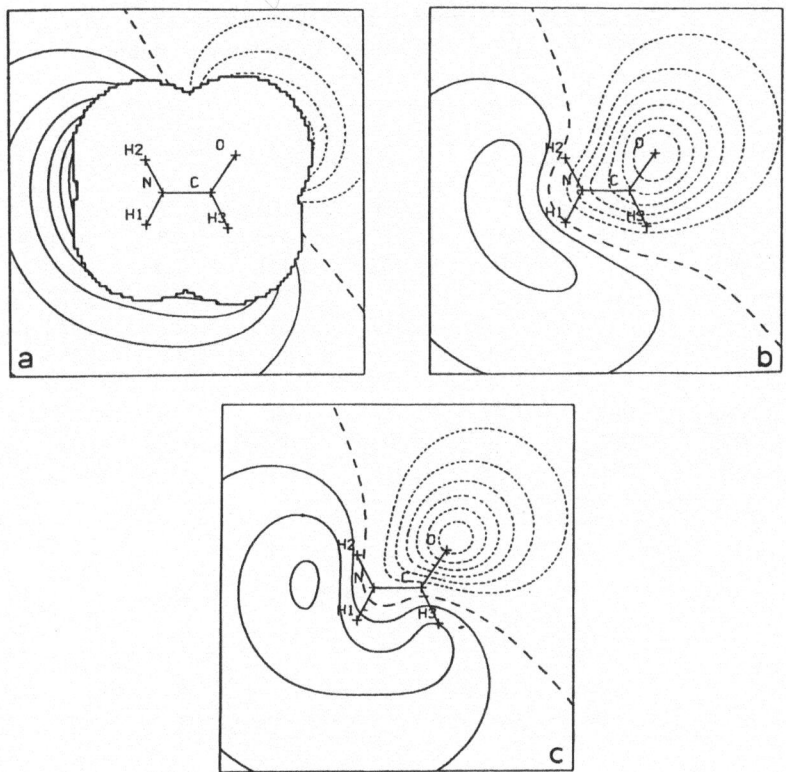

Fig. 4. Deformation electrostatic potential of formamide determined
by direct integration of experimental data, (a) in the
molecular plane, (b) 2Å above the plane, and (c) 2Å below
the plane. Contour level is 5 Kcal/mole.

Pyridinium Dicyanomethylide. This compound ($\underset{\sim}{1}$) has been
extensively studied using both room and low temperature neutron
data[39] and low temperature X-ray data.[24]
The X-ray data have been used in a comparative study of two deformation models.
A mirror plane bisects the molecule.
The cyano nitrogen is approximately 0.1Å
out of the plane of the pyridine ring.
Atomic moments derived from a direct
integration of the experimental data
are given in Table 7. Generally the
atomic charges are small except for
atoms H1 and N2. The results show the
pyridine nitrogen and the ylide carbon
atoms to be essentially neutral in
contrast to the conventionally assigned

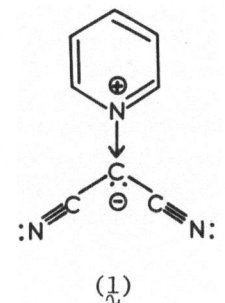

($\underset{\sim}{1}$)

Table 7. Atomic Deformation Moments for Pyridinium Dicyanomethyl-
          ide in Local Coordinate Frames from Fuzzy Boundary Direct
          Integration of Experimental Data

	μ_0	μ_x	μ_y	μ_z	μ_{xx}	μ_{yy}	μ_{zz}
H1	.140	-.091	.000	-.054	.064	.097	.007
H2	.093	-.065	-.018	-.019	.039	.073	.019
H3	.090	-.057	.000	-.002	.031	.060	.020
C1	-.015	.062	.000	.084	.096	.073	-.073
C2	-.057	.045	.019	-.057	.083	.062	-.072
C3	-.010	.019	-.029	-.020	.087	.060	-.060
C4	-.048	.002	.001	-.013	.097	.069	-.167
C5	.048	.082	-.027	-.014	-.038	.239	-.105
N1	.056	-.001	.000	-.005	.057	.066	-.047
N2	-.232	.006	-.017	-.012	-.048	.075	-.134

Units of $e\mathring{A}^n$.

unit formal charges. As in the case of acetylene and formamide,
the hydrogen atoms are contracted (positive second moments) and
there is a shift of density from the hydrogens into the bonding
regions. The atoms in the ring are contracted within the molecular
plane as in the ylide carbon atom. The nitrogen and carbon atoms
of the cyano groups appear to expand along the bond and contract
in the perpendicular, in-plane direction. All nonhydrogen atoms
expand out of the plane.

    It is interesting to compare the cyano groups with that in 2-
cyanoguanidine[40] and with the results of partitioning the theoretical[33] hydrogen cyanide molecule. Qualitative agreement is found
(see Table 8) for the nitrogen moments while the carbon atom moments
show more variation, probably due to the different bonding involved
in the three compounds. The sense of the atomic dipoles is, as
stated by Hirshfeld,[40] "in the same sense as the bond polarity,
ie. $^+C^-\ldots^+N^-$". The second moments show an expansion of the nitrogen

Table 8.   Comparison of Atomic Deformation Moments for Three Cyano Groups

		μ_0	μ_x	μ_y	μ_z	μ_{xx}	μ_{yy}	μ_{zz}
Carbon	A	.074	.066	.013	.000	-.072	.170	.127
	B	.048	.082	-.027	-.014	-.038	.239	-.105
	C	.051	.103	.000	.000	.074	.126	.126
Nitrogen	A	-.373	.134	.008	.000	-.180	-.046	-.075
	B	-.232	.006	-.017	-.012	-.048	.075	-.134
	C	-.184	.132	.000	.000	-.118	-.026	-.026

A: 2-Cyanoguanidine[40]:   Stockholder partitioning
B: Dicyano pyridinium methylide: present work; Fuzzy Boundary Direct Integration
C: Stockholder partitioning of Snyder and Basch wavefunction[33] for HCN
Units of $+e\overset{o}{A}^n$.
Local axes as defined in Table 7.

atom into the bond region. While these results are encouraging, further work needs to be done before any definite statements can be made concerning the possible transferability of functional group density features.

The molecular dipole moments determined from several analyses are compared in Table 9. The dipole arises from the large separation of the H1 and N2 atoms which have, by direct integration, charges of +0.14 and -0.23 e respectively. These results agree well with the solution value[41] and with the results of a multipole refinement.[24]

The electrostatic potential in and 2.0Å above the mean molecular plane is shown in Figure 5. In passing to the plane above the molecule the negative (attractive) potential region broadens considerably, extending partly over the pyridine ring. A corresponding section below the molecular plane shows the effect of the out-of-plane dipole component. The features above and below the molecular plane are in substantial agreement and suggest that within the approximations of the method the imposition of mirror symmetry for the molecular plane is reasonable.

SUMMARY

The results of the direct integration calculations suggest:

Table 9. Molecular Dipole Moment (Debye) for Pyridinium Dicyano-
          methylide*

	μ_x	μ_z	$|\mu|$
A	-9.1	0.0	9.1
B	-10.1	1.6	10.2
C	-8.2	1.2	8.3
D			9.2

A: Multipole refinement[24]
B: Fuzzy boundary direct integration: this work
C: Discrete boundary direct integration: this work
D: Solution value[41]
*The y-component of the molecular dipole is, by symmetry, zero.
The molecular frame has the same orientation as the local axes
at N1 (see Table 7).

(*i*) Molecular moments are reasonably well determined and
     agree well with the results of other techniques. The
     influence of the crystal environment is apparent;
     however, in the cases cited it is not a dominant
     effect, even in the case of formamide which is involved
     in three hydrogen bonds. The results provide a basis
     for the imposition of idealized symmetries to conform
     with the geometry of the isolated molecule. Such
     symmetries are generally imposed in a multipole re-
     finement, in order to reduce the number of parameters
     to a manageable level.

(*ii*) Atomic moments determined by the stockholder parti-
     tioning scheme as applied in the fuzzy boundary
     direct integrations, provide a qualitative inter-
     pretation of the charge redistribution involved in
     bond formation, which is in general agreement with
     intuitive chemical concepts.

(*iii*) The electrostatic potential is a useful tool for
     the interpretation of the electron density. Quali-
     tative agreement with theoretical results indicates
     that, given X-ray data of suitable quality, it is
     possible to obtain a reliable representation of the
     electrostatic potential.

(b) The Electrostatic Interaction of Two Pyrazine Molecules From
X-Ray Data

    The direct integration method places a heavy burden on data

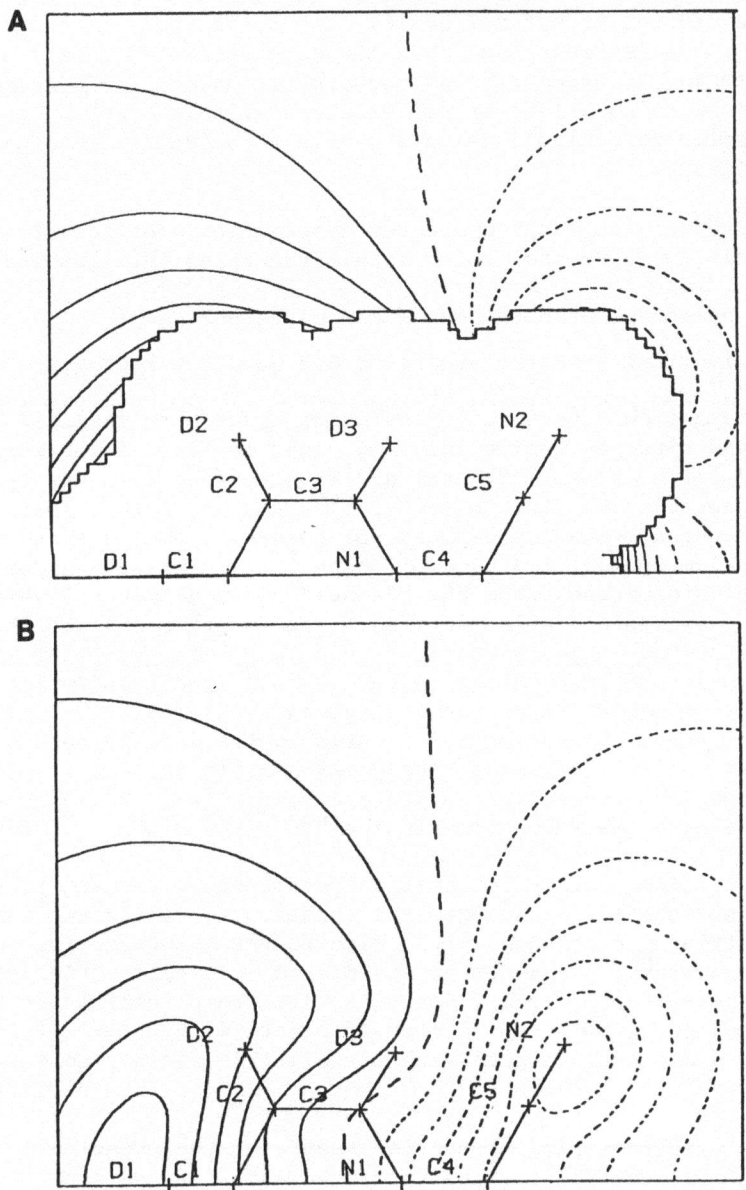

Fig. 5.   Deformation electrostatic potential of pyridinium dicyano-
          methylide determined by direct integration of experimental
          data, (a) in the plane and (b) 2Å above the plane.   Units
          as in Figure 4.

quality. In this section we investigate the Hirshfeld[20] multipole method as applied to the analysis of experimental data on pyrazine.[42] The data are not of optimal quality due to the poor mechanical properties of the crystals, and thus the data present a good test for the filtering afforded by least squares techniques. For comparison, theoretical calculations of the electrostatic interaction energy of two pyrazine molecules[43] using a double zeta quality wavefunction[44] are available.

To reduce the large number of deformation coefficients to a manageable level the following local symmetries were imposed:

(i) mm symmetry for both carbon and nitrogen; and

(ii) cylindrical symmetry about the C-H bond for hydrogen.

Deformation functions up to fourth order were included for carbon and nitrogen, while only functions to first order were retained for hydrogen. Several different refinements were performed; details are given elsewhere.[45] As is commonly found there was a strong correlation between the deformation dipole and positional parameters of hydrogen. Consequently the exponent of the hydrogen deformation functions was held fixed at several values, all other parameters being refined. The "best" model was then taken to be that which gave a C-H bond length in agreement with the NMR value.[46] The variation in C-H bond length and molecular quadrupole moment with exponent is given in Table 10. In all cases the agreement factor is essentially the same while the molecular quadrupole moment shows considerable variation. For the "best" model the molecular quadrupole moment is in reasonable agreement with the theoretical value.[43]

The refined atomic deformation parameters were used to calculate point deformation moments via equation 2. As in the case of the electrostatic potential, the electrostatic interaction energy at large separations may be expanded as a multipole series in which the charge of one molecule interacts with the potential of the second molecule, the first moment interacts with the electric field, the second moment interacts with the electric field gradient, and so on.[47]

The electrostatic interaction energy of two pyrazine molecules has been calculated using the atomic moments derived from the fixed exponent refinements. The results are shown in Figure 6 as a function of molecular orientation at a fixed center-of-mass separation of 15 atomic units ($\sim$7.9Å). For comparison, the theoretical result of Mulder and Huiszoon[43] is also plotted. The two results are in qualitative agreement. The variation with change in hydrogen deformation exponent is troubling since it was not possible to discriminate between the various models on the basis of the agreement

Table 10.   Variation of C-H Bond Length and Molecular Quadrupole
            Moment (*) in Pyrazine with Hydrogen Exponent

	α_H					
	4.0	4.5	5.0	5.25	5.5	6.0
C-H (Å)	0.967	0.988	1.030	1.076	1.134	1.203
θ_{xx}	-5.05	-5.38	-5.79	-6.17	-6.39	-5.79
θ_{yy}	2.84	4.24	6.35	8.35	10.26	9.97
θ_{zz}	2.21	1.14	-0.56	-2.18	-3.87	-4.18
R(F)%	4.73	4.73	4.73	4.73	4.72	4.70

Theoretical Molecular Quadrupole Moment[43]

$$\theta_{xx} \quad -8.92$$
$$\theta_{yy} \quad 10.39$$
$$\theta_{zz} \quad -1.47$$

*Atomic units

factors.   Although encouraging, these results indicate the need
for additional criteria to restrict the deformation model.   A
similar need has been discussed by Schwarzenbach and Ngo[48] in the
case of the electric field gradient at the nucleus.

CONCLUSIONS

      Two rather different methods, multipole refinement and direct
integration, have been applied to the analysis of X-ray diffraction
data in an attempt to obtain a more quantitative interpretation of
the molecular deformation density than is provided by difference
maps alone.   We asked, could we use these methods to derive molecular moments from which the molecular electrostatic potential, and
subsequently the electrostatic interaction energy, could be calculated.   We showed that it is possible to obtain at least qualitative
agreement with theoretical calculations, given data of sufficient
accuracy.   Molecular moments derived from both methods are in substantial agreement with other measurements.   In the case of pyrazine,
the agreement is very model dependent; however, the data are not of
optimal quality.   Introduction of an external constraint on the C-H
bond length improves agreement with theory.

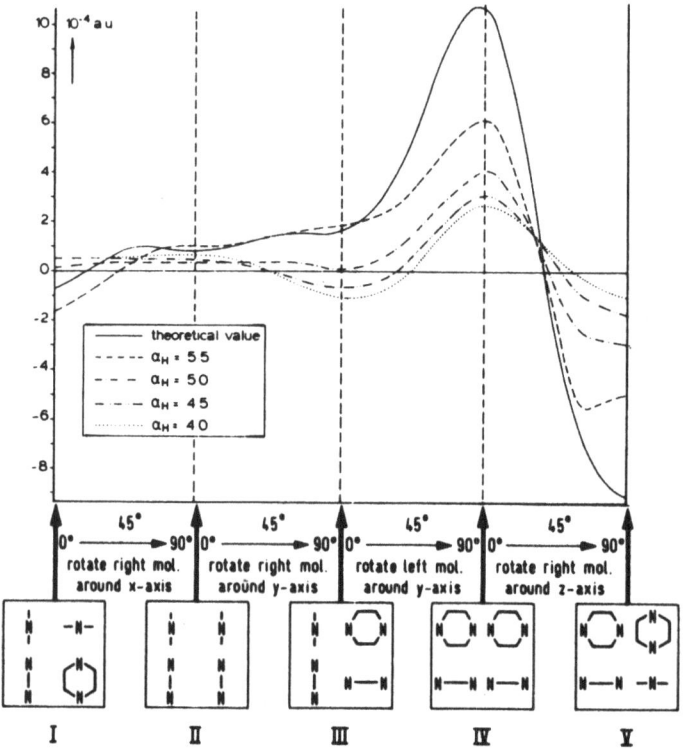

Fig. 6.   The electrostatic interaction energy of two pyrazine
          molecules as a function of orientation for a fixed center
          of mass separation of 15 atomic units.[45] The solid line
          is the theoretical result of Mulder and Huiszoon.[43]
          The dashed lines are the results obtained with multipole
          refinements of experimental data with fixed hydrogen
          exponents[45] - see Table 10 for details.

     For several molecules examined the electrostatic potentials
from direct integration and the electrostatic interaction energy
from multipole refinement are in qualitative agreement with theoretical calculations. Clearly there is a need for more accurate
data and for the incorporation of other experimental measurements
into the refinements. Further work also needs to be done to determine the range of validity of the multipole and direct integration
techniques.

     The atomic moments determined by stockholder partitioning, as
applied through the fuzzy boundary direction integration, provide
a convenient means for the interpretation of chemical bonding.
As shown in the case of formamide, the arbitrariness of the atomic
definition leads to a considerable spread in the moments derived

from different methods. Thus we are necessarily limited to a qualitative discussion of trends. In general the atomic moments conform
to our intuitive concepts of the redistribution of electron density
which occurs upon bond formation. Thus the atomic dipoles indicate
a shift of density into the bond regions. In the case of second
moments, the hydrogen atoms are consistently found to have positive
deformation moments, indicating a contraction relative to the free
atom density. This contraction is in agreement with theoretical
calculations by Stewart *et al*[49] on the hydrogen molecule.

ACKNOWLEDGEMENTS

        The author wishes to thank Professors D. Feil and P. Coppens
with whom most of this work was done, Drs. E. D. Stevens, G. T.
DeTitta, R. H. Blessing and B. K. Moss for their many discussions
and helpful comments, Ms. Gloria Del Bel who prepared the diagrams
and Ms. Brenda Giacchi who typed the manuscript.

        This work was supported by the following grants: National
Science Foundation Grant No. CHE79-05987; the National Resource
for Computation in Chemistry under a grant from the National
Science Foundation Grant No. CHE77-21305 and the Energy Sciences
Division of the U. S. Department of Energy Contract No. W-7405-
ENG-48; NSF Grant CHE79-11282 and DHEW Grant GM-26195.

REFERENCES

1. H. Hellman, "Quantenchemie", p. 285, Leipzig, Deuticke and
    Co. (1937); R. P. Feynman, Phys. Rev. 56:340 (1939).
2. For an up to date review of current methods see Israel J.
    Chem. 19 (1980).
3. R. Bonaccorsi, A. Pullman, E. Scrocco and J. Tomasi, Chem.
    Phys. Letters, 12:622 (1972).
4. Y. Ellinger, R. Subra, G. Berthier and J. Tomasi, J. Phys.
    Chem. 79:2440 (1975).
5. D. M. Hayes and P. A. Kollman, J. Amer. Chem. Soc. 98:3335
    (1976).
6. D. M. Hayes and P. A. Kollman, J. Amer. Chem. Soc. 98:7811
    (1976).
7. A. Agresti, F. Buffoni, J. J. Kaufman and C. Petrongolo,
    Mol. Pharmacol. 18:461 (1980).
8. E. Scrocco and J. Tomasi, "Advances in Quantum Chemistry",
    Vol. 11, p. 115, Academic Press (1978).
9. E. D. Stevens and P. Coppens, "Advances in Quantum Chemistry",
    Vol. 11, p. 1, Academic Press (1978).
10. M. Bernard, this symposium.
11. E. D. Stevens, this symposium.

12. B. Dawson, Proc. Roy. Soc. A298:264 (1967); *ibid.* A298:379 (1967); *ibid.* A298:395 (1967).
13. K. Kurki-Suonio and P. Salmo, Ann. Acad. Sci. Fenn. AVI:369 (1971).
14. R. F. Stewart, J. Chem. Phys. 57:1664 (1972).
15. F. L. Hirshfeld, in: "Electron and Magnetization Densities in Molecules and Crystals", p. 757, Plenum Press (1980).
16. P. Coppens, T. N. Guru Row, P. Leung, E. D. Stevens, P. J. Becker and Y. W. Yang, Acta Crystallogr. A35:63 (1979).
17. R. F. Stewart, Chem. Phys. Letters, 65:335 (1979).
18. A. D. Buckingham, Quart. Rev. 13:183 (1959).
19. P. Becker, in: "Electron and Magnetization Densities in Molecules and Crystals", p. 375, Plenum Press (1980).
20. F. L. Hirshfeld, Acta Crystallogr. B27:769 (1971).
21. R. F. Stewart, Acta Crystallogr. A32:565 (1976).
22. P. Coppens and N. K. Hansen, Israel J. Chem. 16:163 (1977).
23. R. F. Stewart, Israel J. Chem. 16:124 (1977).
24. F. Baert, L. Devos, P. Coppens and E. D. Stevens, submitted for publication to Acta Crystallogr.
25. F. L. Hirshfeld, Theor. Chim. Acta, 44:129 (1977).
26. G. Moss and P. Coppens, in: "Chemical Applications of Atomic and Molecular Electrostatic Potentials", p. 427, Plenum Press (1981); P. Coppens, G. Moss and N. K. Hansen, in: "Crystallographic Computing", p. 16.01, Indian Academy of Sciences (1980).
27. P. Coppens and T. N. Guru Row, Ann. New York Acad. Sci. 313:214 (1978).
28. J. W. Bats, Acta Crystallogr. B33:466 (1977).
29. J. W. Bats, Acta Crystallogr. B33:2035 (1977).
30. J. W. Bats, P. Coppens and A. Kvick, Acta Crystallogr. B33:1534 (1977).
31. H. Dannöhl, H. Mayer and A. Schwieg, Chem. Phys. Letters, 69:75 (1980).
32. G. J. H. van Nes and F. van Bolhuis, Acta Crystallogr. B35:2580 (1979).
33. L. C. Snyder and H. Basch, "Molecular Wavefunctions and Properties: Tabulated from SCF Calculations in a Gaussian Basis Set", Wiley, New York (1972).
34. W. Gordy, W. V. Smith and R. F. Trambarulo, "Microwave Spectroscopy", John Wiley (1953).
35. E. D. Stevens, Acta Crystallogr. B34:544 (1978).
36. E. D. Stevens, J. Rys and P. Coppens, J. Amer. Chem. Soc. 100:2324 (1978).
37. D. H. Christensen, R. N. Kortzeborn, B. Bak and J. J. Led, J. Chem. Phys. 53:3912 (1970).
38. R. J. Kurland and E. B. Wilson, Jr., J. Chem. Phys. 27:1769 (1957).
39. L. Devos, F. Baert and R. Fouret, Acta Crystallogr. B36:1807 (1980).
40. F. L. Hirshfeld, Israel J. Chem. 16:198 (1977).

41. C. Treiner, J. F. Skinner and R. M. Fuoss, J. Phys. Chem. 68: 3406 (1964).
42. G. DeWith, S. Harkema and D. Feil, Acta Crystallogr. B32:3178 (1976).
43. F. Mulder and C. Huiszoon, Mol. Phys. 34:1215 (1977).
44. J. Almlof, H. Johansen, B. Roos and U. Wahlgren. J. Electron Spectrosc. Relat. Phenom. 2:51 (1973).
45. G. Moss and D. Feil, Acta Crystallogr. A37:414 (1981).
46. P. Diehl, H. Bosinger, A. S. Tracy and R. Ader, Org. Magn. Resonance, 8:17 (1976).
47. J. O. Hirshfelder, C. F. Curtiss and R. B. Byrd, "Molecular Theory of Gases and Liquids", Wiley, New York (1964).
48. D. Schwarzenbach and Ngo Thong, Acta Crystallogr. A35:652 (1979); Ngo Thong and D. Schwarzenbach, Acta Crystallogr. A35:658 (1979).
49. R. F. Stewart, E. R. Davidson and W. T. Simpson, J. Chem. Phys. 42:3175 (1965).
50. G. Moss and P. Coppens, Chem. Phys. Letters, 75:298 (1980).

REFINEMENT OF CHARGE DENSITY MODELS USING CONSTRAINTS FOR ELECTRIC FIELD GRADIENTS AT NUCLEAR POSITIONS

D. Schwarzenbach and J. Lewis

Institut de Cristallographie, Université de Lausanne
1015 Lausanne - Dorigny
Switzerland

ELECTROSTATIC PROPERTIES OF THE CHARGE DENSITY

Once a charge density has been determined, the question obviously arises whether it can be used to compute physical properties of the crystalline substance. In this context, electrostatic properties are of prime interest since they are simple functions of the charge density[1]. In particular, electrostatic potentials, fields and field gradients can be readily computed, whereas the calculation of charges and dipole moments necessitates the definition of a cavity which is necessarily ambiguous.

The determination of the structure factors of a crystal structure is by no means straightforward. The structure factor $|F(\underset{\sim}{h})|$ is defined as the absolute value of the Fourier transform of the thermally averaged electron density in the unit cell:

$$|F(\underset{\sim}{h})| = |\int <\rho(r)>.\exp(2\pi i \underset{\sim}{h}.\underset{\sim}{r}).d^3\underset{\sim}{r}|.$$

Strictly speaking, it cannot be measured. It is derived from the measured integrated intensities $I(\underset{\sim}{h})$ using a set of models, assumptions and approximations which are independent of the accuracy of the physical measurement (extinction, dispersion, TDS). The resulting electron density is thus most probably affected by systematic errors whose importance we evidently try to minimize. The calculation of a physical property of the charge density which is also known from other sources than X-ray diffraction is therefore particularly interesting and may serve to judge the quality of the charge density.

The electric field gradient ∇E, i.e. the negative second
derivative of the electrostatic potential, is a symmetric tensor of
rank two. At the origin of a unitary coordinate system, it is
given by

$$\nabla E_{mn} = - \int \frac{3 x_m x_n - \delta_{mn} |\underset{\sim}{x}|^2}{|\underset{\sim}{x}|^5} \cdot \rho(\underset{\sim}{x}) \cdot d^3 \underset{\sim}{x},$$

$\rho(\underset{\sim}{x})$ being the total charge distribution, including the nuclei.
The trace is $\nabla E_{11} + \nabla E_{22} + \nabla E_{33} = 0$. The absolute value of the field
gradient $|\nabla E|$ at the site of a nucleus possessing a nuclear quadru-
pole moment can be measured by nuclear quadrupole resonance spectro-
scopy (NQR)[2]. Denoting the eigenvalues of ∇E by $|\nabla E_{zz}| \geq |\nabla E_{yy}|$
$\geq |\nabla E_{xx}|$, spectroscopists determine the following quantities:

(i)   the quadrupole coupling constant $|eQ.\nabla E_{zz}/h|$, eQ being the
        nuclear quadrupole moment and h Planck's constant;
(ii)   the asymmetry parameter $\eta = (\nabla E_{xx} - \nabla E_{yy})/\nabla E_{zz}$, $0 \leq \eta \leq 1$;
(iii)   up to three orientation parameters of the eigenvectors $\underset{\sim}{x}$, $\underset{\sim}{y}$
        and $\underset{\sim}{z}$ relative to a crystal fixed unitary coordinate system.

The sign of ∇E_{zz} is usually not known. In rare cases, however, it
has been measured using dynamic nuclear polarization (DNP), a
NQR - ESR double resonance method. In some structures, the measured
resonance lines cannot be assigned unambiguously to specific nuclei,
and the information on $|\nabla E|$ remains then incomplete. The most im-
portant uncertainty in the measured $|\nabla E_{zz}|$ value is due to the
nuclear moment Q which is only inaccurately known for some nuclei.

Unfortunately, the field gradients as measured with NQR cannot
be computed from X-ray structure factors without assumptions and
models supplementary to those needed to derive the structure factors
themselves. Specifically, we have to filter the experimental electron density with a multipolar deformation model, and it has to be
assumed that the functions of this model are able to represent the
relevant quadrupolar deformations. There are two major problems.
Firstly, we want to calculate the thermally averaged field gradient
as "seen" by a vibrating nucleus. This is not the same as the field
gradient at the mean nuclear site arising from the thermally
averaged charge density. The effects of thermal vibrations on the
charge density are usually described in terms of the convolution
approximation assuming rigid, but generally aspherical atoms[3].
The atomic electron density is thus at rest with respect to the
nucleus. The movements of the nearest neighbour atoms tend to be at
least partially correlated with those of the central atom, i.e.
their relative movements are small, and even zero in the case of an
ideally rigid molecule. On the other hand, the movements of distant

neighbours are hardly correlated with those of the central atom.
It is thus evident that the calculation of the thermally averaged
∇E requires the knowledge of the phonon spectrum of the crystal,
and that integrated X-ray intensities do not contain the necessary
information. We claim, however, that at least for the framework
structures treated so far, ∇E calculated from the static vibra-
tion-free charge density as obtained via a multipole refinement
should be a fair approximation to the thermally averaged ∇E [4].
The contribution to ∇E of the central atom is roughly 60 to 70%
and the remainder is nearly exclusively due to the nearest neighbour
atoms (see also conclusion 1 of this paper). This is in agreement
with the observation that experimentally determined field gradients
hardly change with temperature. In Li_3N, e.g., they are constant
between 220 and 350 K [5,6]. We have no experience with molecular
crystals where the thermal vibrations of hydrogen atoms with respect to a rigid molecule might not be negligible and could be
taken into account assuming a riding motion [4].

The second problem arises from the fact that a small quadrupolar deformation of the core electron distribution at a sufficiently small distance from the nucleus might have an important effect
on the gradient, and at the same time be hardly detectable in the
diffraction experiment. Theoretical calculations on N_2 indicate,
however, that the relevant quadrupole deformations should be well
within the resolution of the diffraction experiment [7].

The electric fields at the nuclei may also be of interest.
The Hellmann-Feynman electrostatic theorem implies that they should
vanish in a static structure at equilibrium. If we want to verify
this conclusion using the experimental charge density, we clearly
encounter the same problems as discussed above, with the added disadvantage that dipolar deformations of the atomic cores due to
chemical bonding seem in fact to exist [8,9].

ELECTROSTATIC CONSTRAINTS IMPOSED ON THE CHARGE DENSITY

If the calculated electrostatic property agrees with the
experimentally observed one, some confidence in the charge density
may be justified. The foregoing discussion shows on the other hand
that any disagreement can be rationalized with important arguments.
In this case, the assumptions underlying the calculations may
simply be assumed to be invalid. Alternatively, we may try to find
a charge density which correctly reproduces the measured property,
i.e. we reverse the original problem. We now consider the experimentally determined field gradient, or the zero-field condition,
as observations on the same level as the structure factors. We

only have to express the components of the field gradient or the field as functions of the parameters of the multipole model used to represent the charge density. The least-squares refinement of these parameters simultaneously with respect to structure factors and electrostatic quantities leads to a charge density which optimally ·reproduces all observations. Note that this procedure is not very different from the calculation of the standard deviations of ∇E and E from a refinement with respect to structure factors only. In both cases, the derivatives $\partial \nabla E_{mn}/\partial c_i$ and $\partial E_m/\partial c_i$ with respect to all positional and multipole parameters c_i are to be calculated. The variance of ∇E_{mn} is

$$\sigma^2(\nabla E_{mn}) = \underset{ij}{\Sigma\Sigma}(\partial \nabla E_{mn}/\partial c_i)(\partial \nabla E_{mn}/\partial c_j) \cdot \mathrm{cov}(c_i, c_j)$$

where $\mathrm{cov}(c_i, c_j)$ is the term (i,j) of the inverse least-squares matrix multiplied by the square of the goodness of fit. A large standard deviation indicates that the electrostatic condition can be satisfied without substantially altering the fit between observed and calculated structure factors, i.e. that there is little correlation between the quantities measured with the NQR and diffraction experiments. If the refinement includes ∇E_{mn} as observation, then the products of the derivatives multiplied by the weight assigned to ∇E_{mn} are contributed to the least-squares matrix. The above formula obviously produces then the inverse of the weight multiplied by the square of the goodness of fit.

The algorithms used in our laboratory are described elsewhere in detail[4]. The field gradient is computed as a sum of two terms

$$\nabla E = \nabla E(\text{procrystal}) + \nabla E(\text{deformation}).$$

The charge density of the procrystal is the commonly used superposition of neutral spherical atoms, including nuclei. The corresponding field gradient is calculated from the atomic form factor curves[10]. The gradient due to the deformation density is computed with a Fourier series, the Fourier coefficients being simple functions of the structure factors of the static deformation density. The calculation of its derivatives with respect to the multipole parameters is straightforward. The calculation of the electric fields is analogous. For purely historical reasons, we use Hirshfeld's deformation functions[11] to parametrize the deformation density. They are equivalent to 3 monopolar, 2 dipolar, 2 quadrupolar, 1 octopolar and 1 hexadecapolar sets of functions. Our program LSEXP includes the calculation of electrostatic potentials, fields E and field gradients ∇E at the atomic positions, as well as of the e.s.d's of E and ∇E, the introduction of E and ∇E as observations,

and the refinement of isotropic or anisotropic extinction[12]. It has
so far been applied to Li_3N [13], α-Al_2O_3 and $AlPO_4$ [14]. Static model
maps are computed with Fourier summations using the structure fac-
tors calculated with the deformation functions only. The e.s.d. of
the deformation density is estimated with the program DEFSYN[11].

LITHIUM NITRIDE, Li_3N

The structure of the ionic conductor Li_3N is hexagonal, P6/mmm.
There are three independent atomic sites, viz. N at 0 0 0, Li(1) at
0 0 1/2 and Li(2) at (1/3 2/3 0, 2/3 1/3 0). The site symmetries are
respectively 6/mmm, 6/mmm and $\bar{6}$m2. Diffraction data measured at $-40°C$
to a maximum $\sin\theta/\lambda$ of 1.07 Å^{-1} have been obtained from Schulz and
Schwarz[15]. The refinements of these authors indicate a fully ionic
structure $Li_3^+N^{3-}$. The quadrupole coupling constants of all atomic
sites have been measured by ^7Li, ^6Li and ^{14}N NQR [5,6,16]. Due to the
hexagonal site symmetries, each field gradient is completely characterized by one component ∇E_{33}, i.e. the gradient along the hexad axis.
The electric fields vanish at the nuclear positions. The signs of the
gradients are not determined experimentally, but high temperature
dynamic NQR results indicate opposite signs at Li(1) and Li(2) [17,18].

During the unconstrained charge density refinement with respect
to $|F|^2$, the scale factor, defined by $F_{obs} \sim$ scale x F_{calc}, in-
creased by 7% with respect to the procrystal structure. It showed
high correlations with the n = 0 monopolar functions and with the
temperature parameters of nitrogen. The scale obtained with a κ-re-
finement[19] agrees, however, within $\frac{1}{2}$% with the procrystal value.
This shows that the monopolar components of the Hirshfeld functions
are able to simulate the scattering of the core electrons at high
angles, the deformation model having enough parameters to describe
the intensities of the few low-order reflections. A similar effect
has been observed in a charge density refinement of quartz[20] using
a limited data set. Fixing the scale factor at a particular value
fixes the monopolar deformation terms. If we accept the common pro-
position that the core electrons are distributed as in the free
atoms, then the correct scale factor should be near the values obtained with the procrystal and κ-refinement. In order to judge the
effect on the calculated field gradients, two complete sets of eight
refinements were carried out, viz. one with variable scale and one
with the scale fixed at the procrystal value. Both sets agree well
and lead to identical conclusions concerning the gradients. Therefore, we report here only the results obtained with the fixed scale,
details being published elsewhere[13]. Note that we do not claim to
know the true value of the scale factor, the exact form of the
spherical deformation terms remaining thus uncertain.

The charge density was first refined with respect to structure factors only. The exponents α of the radial functions $r^n.\exp(-\alpha r)$ could not be refined due to correlation coefficients larger than 0.99 with other deformation parameters. This means that the α-s are superfluous for the parametrization of the rather flat deformation density. The field gradients were computed at convergence (Table 1). Their e.s.d's are quite large and the calculation of the quadrupole coupling constants from the structure factors appears thus to be impossible. The signs at Li(2) and N are, however, indicated to be probably negative. The sign at Li(1) is not indicated, but it should be positive since NQR indicates it to be opposite to the sign at Li(2). The extinction correction is important only for the reflection 001, and it has a minor effect on 002, 100 and 110. Its effect on $|\nabla E_{33}|$ of the Li-atoms is, however, not negligible.

Next, the field gradients were introduced as observations, assuming various sign combinations (Table 2). This considerably improved the convergence to the least–squares minimum, the correlations between the deformation parameters generally decreased, and the exponents α could be adjusted easily. As before, the signs at Li(2) and N appear to be negative. Static model maps resulting from refinements g (with the most probable signs + − − at Li(1), Li(2) and N) and d (with the signs − − ?) are shown in Figs. 1 and 2. Although the effect on the computed structure factors of changing the sign at Li(1) from + to − is negligible, the maps show a pronounced change of the form of Li(1). The + sign appears indeed to be more reasonable. Both figures show qualitatively the same features at N, but the effect of the smaller field gradient in Fig. 2 (-2.4×10^{14} esu) and the larger gradient in Fig.1 (-4.35×10^{14} esu) is clearly visible. The use of the field gradients in the refinement of the charge density thus modifies the results in the vicinity of the atoms. Note that the preferred signs + − − at Li(1), Li(2) and N agree with those computed from an ionic point charge model. The polarization of N corresponds to a large Sternheimer antishielding factor of the N^{3-} ion.

CORUNDUM, $\alpha- Al_2O_3$

The structure of corundum is rhombohedral, $R\bar{3}c$. There are two independent atomic sites, namely Al at 0 0 z and 0 at x 0 $\frac{1}{4}$. The corresponding site symmetries are 3 and 2. The electric field gradients at both sites and including their signs have been measured with ^{27}Al and ^{17}O NQR using DNP [21]. The gradient at Al is characterized by one component ∇E_{33} along c. The gradient at 0 is defined by three terms, namely ∇E_{zz}, η and the angle between the eigenvector

Table 1: Calculated field gradients ∇E_{33} in Li_3N. Units are 10^{14} esu; e.s.d-s are given in parentheses. Isotropic secondary extinction is neglected in (a) and included in (b).

	Li(1)	Li(2)	N	probable signs
$\lvert \nabla E_{33}(\text{obs})\rvert$	1.96	0.96	4.35	
a	−0.4(1.3)	−2.8(0.9)	−1.5(0.9)	? − −
b	+0.4(1.6)	−3.5(1.1)	−1.5(1.0)	? − −

Table 2: Reliability factors obtained by assuming various signs for the field gradients in Li_3N. The values of ∇E_{33} at N computed after the refinements c to f are respectively −2.5, −2.4, −1.8 and −2.0 x 10^{14} esu (e.s.d. 1.4).

Definitions: $GOF(\lvert F\rvert^2) = \{\Sigma weight \times (\lvert F_o\rvert^2 - \lvert F_c\rvert^2)^2/(n - m)\}^{\frac{1}{2}}$,
n = number of observations, m = number of variable parameters;
$R(\lvert F\rvert) = \Sigma\lvert F_o - F_c\rvert/\Sigma\lvert F_o\rvert$;
$R_w(\lvert F\rvert^2) = \{\Sigma weight \times (\lvert F_o\rvert^2 - \lvert F_c\rvert^2)^2/\Sigma weight \times \lvert F_o\rvert^4\}^{\frac{1}{2}} \cong 2.R_w(\lvert F\rvert)$;
y_{min} = smallest extinction factor, $F_o \cong y.F_{corr}$, type I crystal,
Lorentzian mosaic distribution.

Sign of ∇E_{33} at Li(1),Li(2),N			$GOF(\lvert F\rvert^2)$	$R(\lvert F\rvert)$	$R_w(\lvert F\rvert^2)$	n−m	y_{min}
a			1.373	0.00783	0.0137	100	1.0000
b			1.361	0.00767	0.0135	99	0.9479
c	+	−	1.247	0.00722	0.0123	98	0.9553
d	−	−	1.233	0.00732	0.0122	98	0.9514
e	+	+	1.309	0.00759	0.0129	98	1.0000
f	−	+	1.302	0.00793	0.0128	98	0.9947
g	+	−	−	1.251	0.00747	0.0124	99	0.9453
h	+	−	+	1.346	0.00769	0.0133	99	0.9774

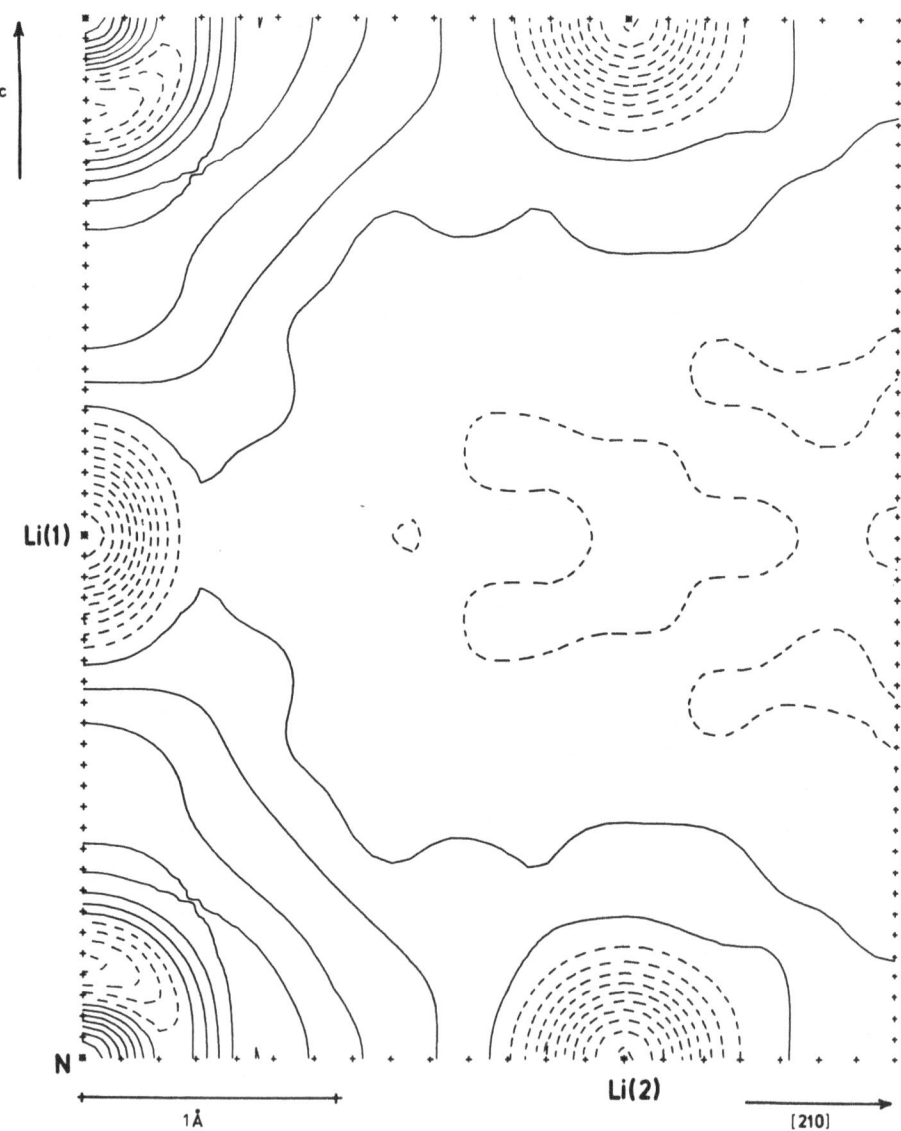

Fig. 1.   Static model deformation map of Li_3N in the plane $(\bar{1}2\bar{1}0)$
          obtained from refinement g, gradient signs + − −. Negative
          contours broken, contour interval 0.05 $e\text{Å}^{-3}$. E.s.d-s are
          0.01 $e\text{Å}^{-3}$ between the atoms and 0.1 $e\text{Å}^{-3}$ at the atomic
          sites.

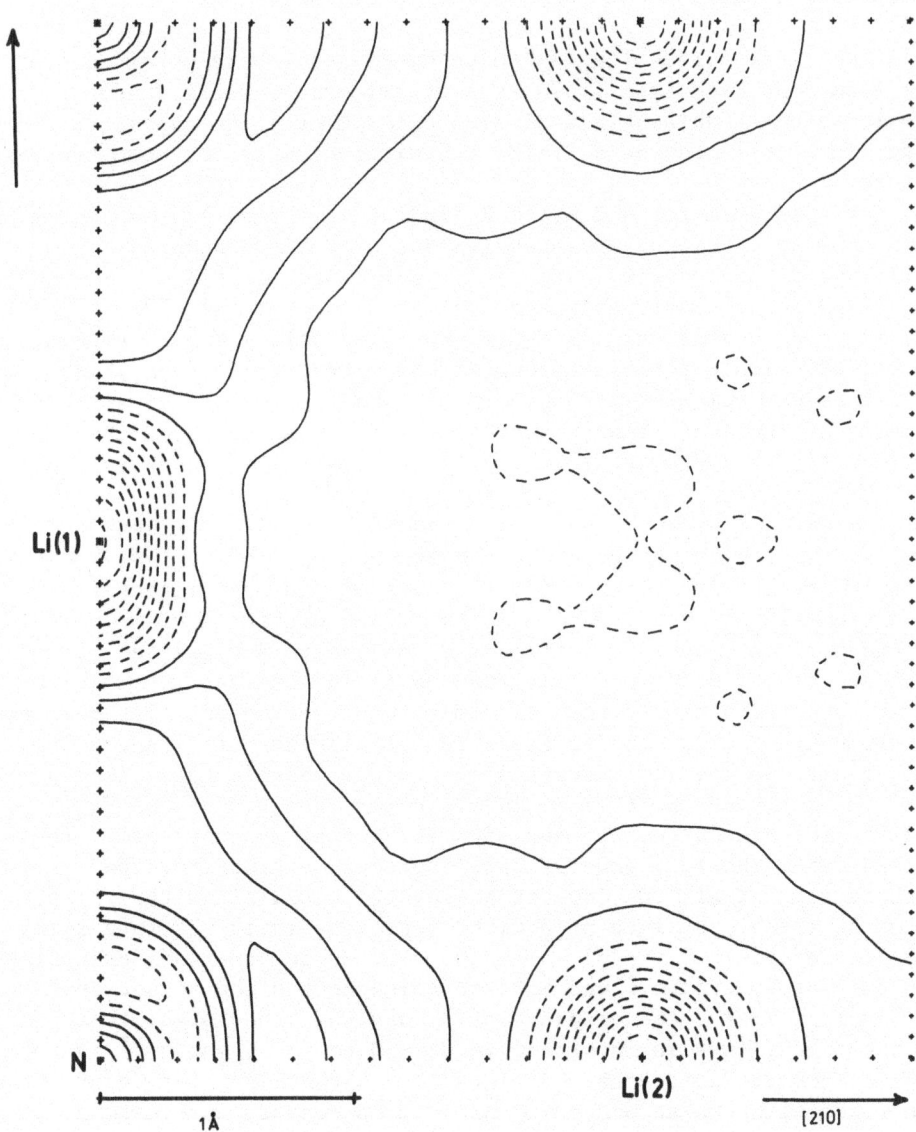

Fig. 2.   As Fig. 1, obtained from refinement d, gradient signs
          − − ?, in contradiction with the NQR data.

$\underset{\sim}{z}$ and $\underset{\sim}{c}$. The eigenvector $\underset{\sim}{x}$ of the smallest eigenvalue ∇E_{xx} (0) is parallel to the twofold axis. The gradients have been successfully calculated using a model composed of polarizable ions[22].

Spherical Verneuil grown corundum crystals were obtained from the firm Jean Sandoz SA (1470 Estavayer-le-lac, Switzerland). Two independent data sets including <u>all</u> symmetry equivalent reflections were collected on a SYNTEX P2$_1$ diffractometer at room temperature:

(i)   crystal diameter 0.2 mm, MoK$_\alpha$ radiation, $(\sin\theta/\lambda)_{max}$ = 1.19 Å^{-1},
      4402 reflections, 397 unequivalent structure factors;
(ii) crystal diameter 0.13 mm, AgK$_\alpha$ radiation, $(\sin\theta/\lambda)_{max}$ = 1.495 Å^{-1},
      8923 reflections, 804 unequivalent structure factors.

The internal consistency of equivalent reflections was generally excellent, except for the strongest low-order reflections which showed anisotropic extinction effects. The e.s.d-s of the structure factors of the Ag-data were about three times as large as those of the Mo-data, but the extinction effects were considerably smaller.

For each data set, four charge density refinements were carried out. An isotropic extinction correction was applied to the symmetry independent structure factors obtained from the mean intensities, <u>or alternatively</u> an anisotropic extinction correction was applied to the full data set. In both cases, the charge density was refined against structure factors only and then field gradients were calculated, <u>or alternatively</u> the refinement was carried out with respect to structure factors and gradient tensor elements simultaneously. The results of the refinements with respect to structure factors only are reported in Table 3. In <u>all</u> cases, the e.s.d-s of ∇E_{33}(Al) and ∇E_{zz} (0) are large, but the sign at 0 is always correctly predicted. From the refinements using isotropic extinction, the correct sign at Al, the correct angle $\phi(0)$, as well as the correct order of the eigenvalues at 0 with respect to their absolute size is obtained, i.e. $0 < \eta(0) < 1$. None of this information, excepting the sign at 0, can be found in the refinements using anisotropic extinction. There, the gradient at Al has the wrong sign, the order of the eigenvalues is incorrect ($\eta(0) < 0$ or > 1), and $\phi(0)$ is too small. This is due to the correlation of the quadrupolar deformation functions with the second order extinction tensor. Interestingly, the mosaic distributions of the two crystals are found to be similar. The main axes are aligned approximately along $\underset{\sim}{c}$, b^* and $\underset{\sim}{a}$ with half widths of about 4", 7" and 12". It can be shown that in this situation, the mean extinction correction of a set of symmetry equivalent intensities is a function of the angle between the reciprocal lattice vector and $\underset{\sim}{c}$. The scattering factors of the quadrupolar deformation functions of Al are also functions of this angle.

Table 3: Observed and calculated field gradients in $\alpha-Al_2O_3$. Units of ∇E are 10^{14} esu, e.s.d-s are given in parentheses. $\nabla E_{33}(Al)$ is the gradient along $\underset{\sim}{c}$. The eigenvectors $\underset{\sim}{x}$, $\underset{\sim}{y}$, $\underset{\sim}{z}$ of the observed gradient at 0 are defined by $|\nabla E_{xx}| \leq |\nabla E_{yy}| \leq |\nabla E_{zz}|$, $0 \leq \eta \leq 1$; $\underset{\sim}{x}$ is parallel to the $\underset{\sim}{a}$-axis for 0 at 0.306, 0, 1/4. The corresponding eigenvectors of the computed tensors are labelled as the nearest eigenvectors of the observed tensor, and <u>not</u> according to their eigenvalues. The asymmetry parameter is always defined as $\eta = (\nabla E_{xx} - \nabla E_{yy})/\nabla E_{zz}$. $\phi(0)$ is the angle between $\underset{\sim}{z}$ and $\underset{\sim}{z}$, $\phi(0) = +90°$ for $\underset{\sim}{z}$ parallel to $\underset{\sim}{b}*$. GOF and y_{min} (Lorentzian, type I) are defined as in Table 2. Refinements a to d were carried out with respect to structure factors only. The last column shows the GOF including ∇E as observations. The number of degrees of freedom is very different for isotropic and anisotropic extinction. Weights were $\sigma^{-2}(|F|^2)$.

Data		Extinction	$\nabla E_{33}(Al)$	$\nabla E_{zz}(0)$	$\eta(0)$	$\phi(0)$	GOF($|F|^2$)	y_{min}	GOF($|F|^2$) with ∇E
a	Mo	iso	-1.3(1.8)	+4.0(1.2)	0.3(0.4)	41(7)	2.897	0.76	3.011
b	Ag	iso	-8.0(1.0)	+3.9(1.0)	0.2(0.4)	44(6)	1.212	0.86	1.261
c	Mo	aniso	+7.7(2.7)	+17.4(2.0)	-0.3(0.2)	21(5)	5.982	0.77	6.030
d	Ag	aniso	+0.8(0.7)	+2.0(0.4)	1.5(0.4)	28(4)	1.135	0.82	1.142
observed			-2.186	+11.632	0.517	45.85			

Introducing the field gradients as observations again improved
the convergence of the least-squares calculations. In the anisotropic refinements, the goodness-of-fit (GOF) changed little, but
the extinction tensors became more isotropic.

Some of the resulting deformation maps are shown in Figs. 3
to 6. The agreement between the Mo- and the Ag-data is satisfactory,
but the latter are clearly superior. Only the maps (not shown here)
resulting from the anisotropic extinction refinements (c) of the
Mo-data with and without use of the field gradients are unclear
and show the features of the other maps through a maze of complicated contours. The corresponding GOF-s (Table 3) are in fact
rather high. In addition, the X-ray scale factor increased by 10%
with respect to all other refinements, including the procrystal
model (see Li_3N). For the Ag-data, anisotropic extinction leads
to similar maps as isotropic extinction. It seems thus that the
use of the anisotropic extinction formalism serves mainly to
destroy the field gradient information present in the structure
factors, and to increase the computing bill. For the Mo-data which
show more severe extinction effects, it is clearly harmful. This
is a surprising result, since the extinction is definitely anisotropic in both data sets.

Computed electric fields are typically $-1.4(0.4)$ and
$-1.0(0.3) \times 10^6$ esu along the triad axis at Al and the diad axis at
O respectively. The zero-field constraint has not been applied and
is not expected to modify the results appreciably.

$AlPO_4$, QUARTZ MODIFICATION

This structure is nearly identical to the one of α-quartz
SiO_2. The main difference consists of a doubling of the trigonal
$\c$-axis and the appearance of superstructure reflections with ℓ
odd. The field gradient at the site of Al (point symmetry 2) was
measured by [27]Al NQR [23]. The structure is non-centrosymmetric,
$P3_121$, and the phase problem of the structure factors is not entirely solved by the standard crystal structure determination. The
supplementary information available from the NQR experiment induced
us therefore to study the charge density of $AlPO_4$ rather than that
of SiO_2. In fact, introducing ∇E into the refinement improved the
convergence considerably. The lowest point in the shallow least-
squares minimum was reached using ∇E, but was not found with the
structure factors alone. The field gradient constraint modified the
phases of the superstructure reflections, resulting in a more pronounced difference between Al and P and in a seemingly more reasonable deformation map. The reader is referred to the full publication[14].

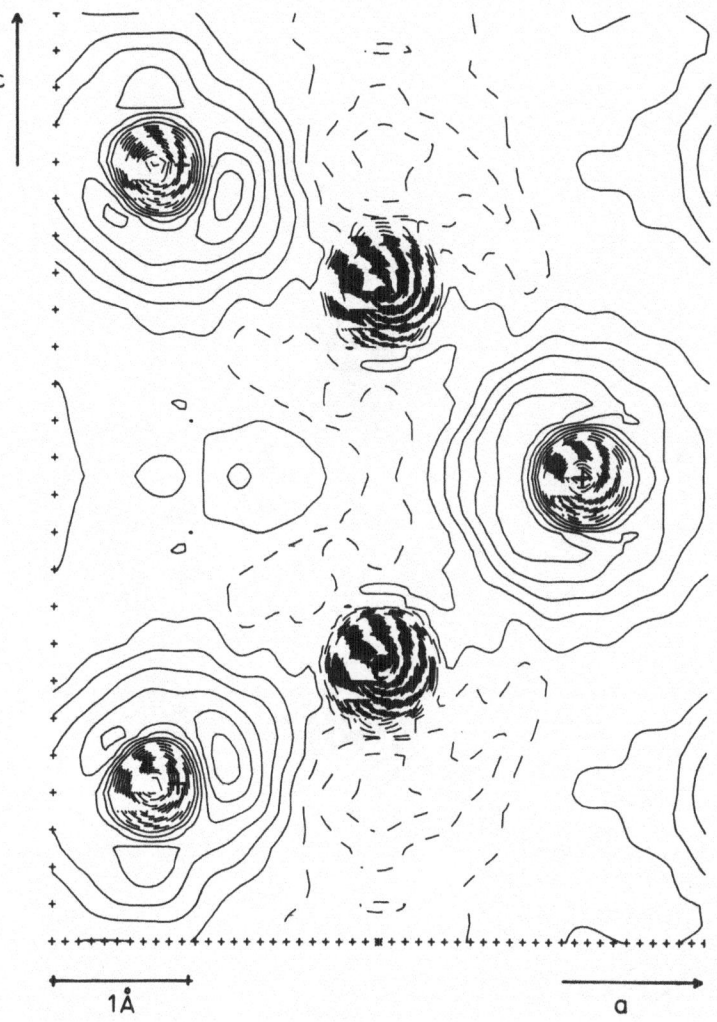

Fig. 3. Static model deformation map of α-Al$_2$O$_3$ in the plane (010)
through two Al and three O atoms, the terminal ones devia-
ting from the plane by 0.12 Å. The face shared by two AlO$_6$-
octahedra passes through the central O and is perpendicular
to the plot. The octahedra, i.e. the Al atoms, are related
by a diad axis. The lengths of the two unequivalent Al–O
bonds are 1.9715(2) Å towards the shared face and
1.8550(2) Å towards the unshared face.
Mo-data, isotropic extinction, refinement with respect to
structure factors only. Interval 0.05 eÅ^{-3}, negative contours broken. The e.s.d-s are indicated in Fig. 4.

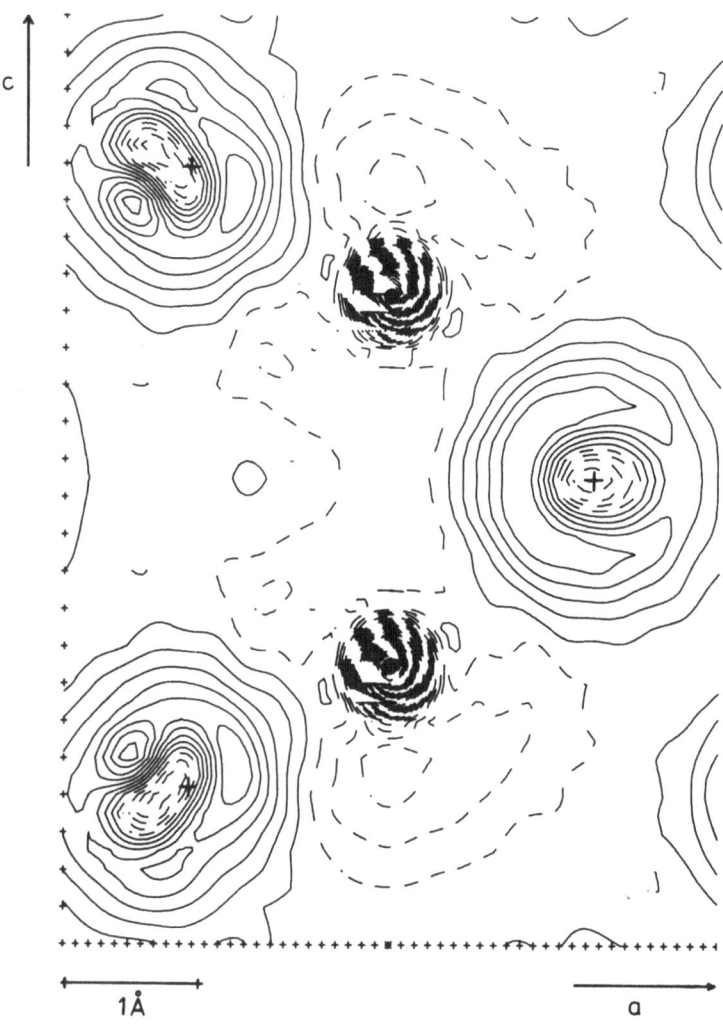

Fig. 4. As Fig. 3, Mo-data, isotropic extinction, refinement with respect to structure factors and field gradients. E.s.d-s are 0.03 e$\AA^{-3}$ between Al and 0, and 0.05 e$\AA^{-3}$ between Al and Al. They increase to 0.1 e$\AA^{-3}$ at distances of 0.42 and 0.35 $\AA$ from the sites of Al and 0, respectively.

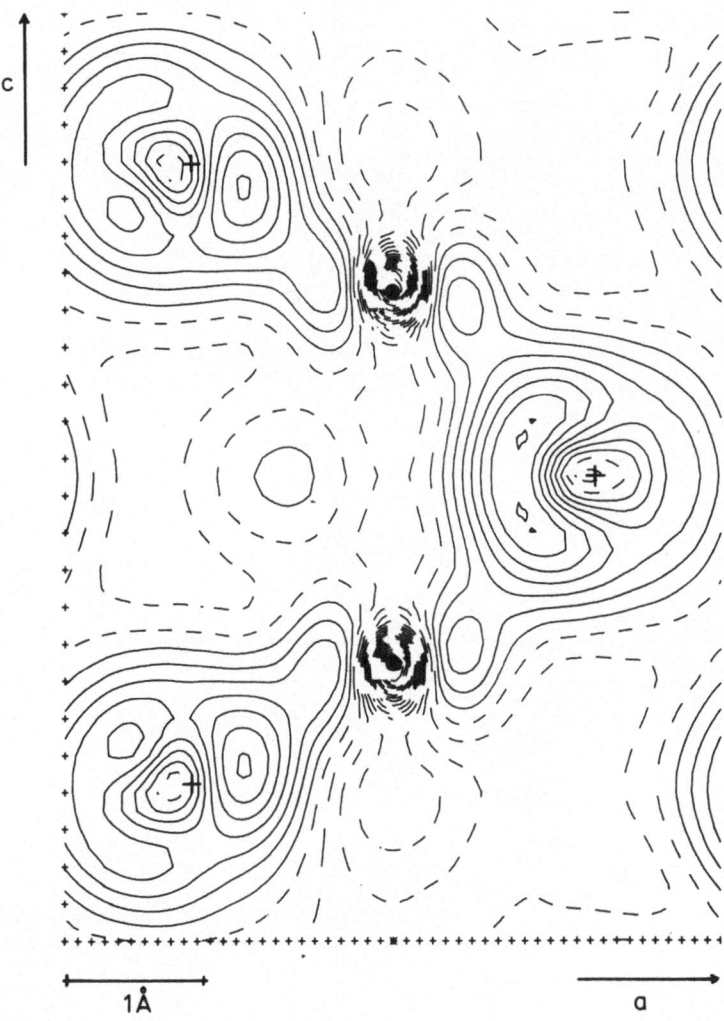

Fig. 5. As Fig. 3, Ag–data, isotropic extinction, refinement with
respect to structure factors only. E.s.d–s are 0.02 eÅ^{-3}
between Al and O, and 0.03 eÅ^{-3} between Al and Al. They
increase to 0.1 eÅ^{-3} at distances of 0.33 and 0.23 Å from
the sites of Al and O, respectively. The analogous refinement with anisotropic extinction yields a very similar
plot with e.s.d–s as in Fig. 6.

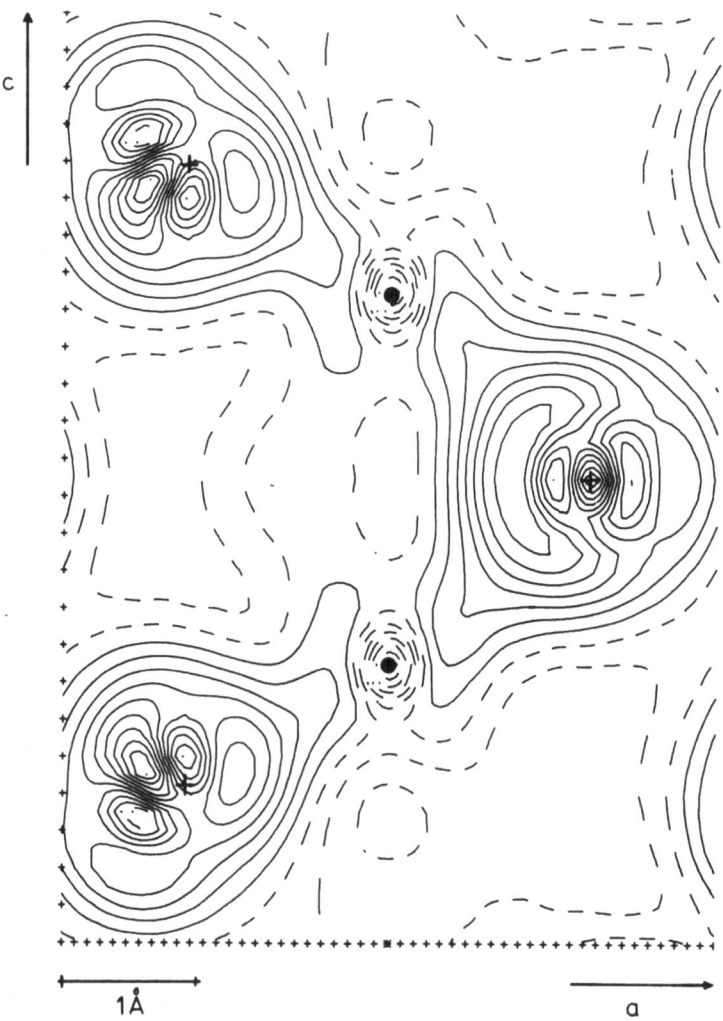

Fig. 6.   As. Fig. 3, Ag-data, <u>anisotropic</u> extinction, refinement
with respect to structure factors and field gradients.
E.s.d-s are 0.01 eÅ^{-3} between Al and O, as well as between
Al and Al. They increase to 0.1 eÅ^{-3} at distances of 0.26
and 0.17 Å from the sites of Al and O, respectively. The
analogous refinement with isotropic extinction yields a
nearly identical plot with e.s.d-s as in Fig. 5.

CONCLUSIONS

Our conclusions inferred from the extensive calculations on Li_3N and $\alpha\text{-}Al_2O_3$ are as follows:

(1) The most frequently published NQR-data are quadrupole coupling constants and asymmetry parameters. These quantities are computed with very large e.s.d-s and thus cannot be predicted from structure factors. X-ray and NQR data are complementary. The charge density far from the atomic centers is determined by the structure factors, the quadrupolar deformations near the atomic centers are determined by $|\nabla E_{zz}|$ and η.

(2) The orientational parameters of the gradient tensors, including the sequence of the eigenvalues with respect to their size, can be qualitatively retrieved from the X-ray data, subject to (4).

(3) The generally unknown sign of ∇E_{zz} can be computed. Note that for both compounds, the signs also agree with those obtained from simpler point charge calculations. The structure factors contain thus some of the information which is also present in the quantities of (2) and (3). This might be used to better determine the phases of the structure factors in a non-centrosymmetric structure such as $AlPO_4$.

(4) Refinement of secondary anisotropic extinction may destroy information on items mentioned in (2) and (3). We recommend to measure a full sphere of diffraction data. An isotropic extinction correction of the mean intensities appears then to be satisfactory. An anisotropic extinction correction applied to an incomplete sphere of data might give undesirable results.

(5) Refinement with respect to field gradients cannot serve to determine the X-ray scale factor. In fact, computed field gradients hardly depend on the assumed scale factor.

(6) The field gradient constraint greatly improves the convergence of the charge density refinement. Its use is recommended for this purely technical reason.

The calculations were carried out at the computer center of the Swiss Federal Institute of Technology at Lausanne (CDC CYBER NOS/BE). The project is supported by the Swiss National Science Foundation, grant no. 2.234-0.79.

REFERENCES

1. R. F. Stewart, Isr. J. Chem., 16:124 (1977).
2. S. Vega, Isr. J. Chem., 16:213 (1977).
3. K. Kurki - Suonio, Isr. J. Chem., 16:132 (1977).
4. D. Schwarzenbach and Ngo Thong, Acta Crystallogr., A35:652 (1979).
5. D. Brinkmann, W. Freudenreich and J. Roos, Solid State Commun., 28:233 (1978).
6. D. Brinkmann, M. Mali and J. Roos, in: "Proceedings Fast Ion Transport in Solids, Lake Geneva, Wisconsin", 483-486, North Holland, New York (1979).
7. R. F. Stewart, Chem. Phys. Lett., 49:281 (1977).
8. F. L. Hirshfeld and S. Rzotkiewics, Mol. Phys., 27:1319 (1974).
9. F. L. Hirshfeld, in: "Electron and Magnetization Densities in Molecules and Crystals, NATO Advanced Study Institutes Series B48: Physics", P. Becker, ed., 47-62, Plenum Press, New York (1980).
10. "International Tables for X-Ray Crystallography, vol. IV", J. A. Ibers and W. C. Hamilton, ed., Kynoch Press, Birmingham (1974).
11. F. L. Hirshfeld, Isr. J. Chem., 16:226 (1977).
12. P. J. Becker and P. Coppens, Acta Crystallogr., A31:417 (1975).
13. J. Lewis and D. Schwarzenbach, Acta Crystallogr., A37:in press (1981).
14. Ngo Thong and D. Schwarzenbach, Acta Crystallogr., A35:658 (1979).
15. H. Schulz and K. Schwarz, Acta Crystallogr., A34:999 (1978).
16. K. Differt and R. Messer, J. Phys. C, 13:717 (1980).
17. D. Brinkmann, M. Mali and J. Roos, Bull. Magnetic Resonance, 2:271 (1981).
18. R. Messer, H. Birli and K. Differt, J. Physique, Colloque C6, 41:28 (1980).
19. P. Coppens, T. N. Guru Row, P. Leung, E. D. Stevens, P. J. Becker and Y. W. Yang, Acta Crystallogr., A35:63 (1979).
20. Ngo Thong and D. Schwarzenbach, Abstracts 5th Europ. Crystallogr. Meet., 348, Copenhagen (1979).
21. E. Hundt, "Kernresonanz von ^{17}O und ^{27}Al in Rubin: Kopplungsparameter und Relaxation", PhD Thesis, University of Zürich, Switzerland (1972).
22. S. Hafner and M. Raymond, J. Chem. Phys., 49:3570 (1968).
23. E. Brun, P. Hartmann, F. Laves and D. Schwarzenbach, Helv. Phys. Acta, 34:388 (1961).

THE STUDY OF MOLECULAR ELECTRON DISTRIBUTION BY X-RAY PHOTOELECTRON

SPECTROSCOPY

William L. Jolly and Albert A. Bakke

Department of Chemistry, University of California
and the Materials and Molecular Research Division
Lawrence Berkeley Laboratory, Berkeley, CA 94720

INTRODUCTION

A molecule contains two kinds of electrons: valence electrons and atomic core electrons. The valence electrons occupy molecular orbitals and are more or less delocalized on the various atoms of the molecule, whereas the core electrons are highly localized on individual atoms of the molecule. In gas-phase X-ray photoelectron spectroscopy (XPS), one irradiates a sample chamber containing the molecules under study with an approximately monochromatic beam of X-rays and measures the kinetic energy distribution of the ejected electrons.[1-3] The spectrum is a plot of electron counting rate as a function of kinetic energy, E_K. Peaks are observed corresponding to electronic levels with binding energies (E_B) less than the X-ray photon energy, $h\nu$. By use of the Einstein relation, $h\nu = E_B + E_K$, one can readily calculate the binding energy corresponding to each peak. The X-ray has enough energy to eject not only valence electrons, but also some core electrons. The main use of XPS is in the determination of core electron binding energies, and this paper will be concerned with the information that one can glean from the core binding energies of molecules.[4]

It is well known that core electrons are not significantly involved in chemical bonding, and therefore one might well ask why chemists should have any interest in measuring their binding energies. The answer lies in the fact that the binding energy of a particular core level of an atom is a function of the chemical environment of the atom. On going from one molecule containing a particular element to another containing the same element, the core binding energy changes. In the case of a molecule containing atoms of the same element with markedly different bonding, the binding

431

energies corresponding to the different atoms will in general be different. For example, consider Figure 1, which shows the carbon 1s region of the X-ray photoelectron spectrum of the complex $(CH_2)_3CFe(CO)_3$, which has the following structure

The three main peaks on the right side of the spectrum correspond to the three different kinds of carbon atoms in the molecule.[5] The strong peak, at low binding energy, corresponds to the three CH_2 carbon atoms of the $(CH_2)_3C$ ligand. The weak peak, at intermediate binding energy, corresponds to the central carbon atom of the ligand. And the peak at relatively high binding energy, which has lost some intensity to the shake-up peaks shown on the left side of the spectrum, corresponds to the three CO carbon atoms. The question to be considered now is: What can one learn about molecular electron distribution from binding energies obtained from spectra such as this?

QUANTITATIVE EXPRESSIONS FOR BINDING ENERGIES

If, during the ejection of a core electron from an atom A in a molecule, all the other electrons in the molecule remained frozen in their original positions, the core binding energy would be precisely equal to the potential, $V(A)$, at the site of the core electron of atom A in the ground-state molecule. Of course, the other electrons do not remain frozen. As a consequence of the ionization, they tend to flow toward the core hole and thus reduce the energy required for electron ejection from atom A. The energy associated with this flow is the electronic relaxation energy associated with the core ionization, $E_R(A)$, and we may write

$$E_B(A) = V(A) - E_R(A), \tag{1}$$

or, for a difference (chemical shift) in binding energy,

$$\Delta E_B(A) = \Delta V(A) - \Delta E_R(A), \tag{2}$$

As a good approximation, the quantity $\Delta V(A)$ may be replaced by

$\Delta V(A)_{val}$, the change in potential due to the change in the distribution of the molecule's valence electrons (and of the cores of all the atoms except atom A)

$$\Delta E_B(A) = \Delta V(A)_{val} - \Delta E_R(A).$$

This is a good approximation because the potential at a core site in atom A due to the other core electrons of atom A is essentially independent of chemical environment.[6] The quantity $\Delta V(A)_{val}$ may be estimated with various degrees of accuracy, using procedures ranging in sophistication from ab initio quantum mechanical calculations to calculations based on estimated atomic charges. For example, by assuming that atom A has a radius r and a charge Q_A and that all other atoms are point charges Q_B, we may write

$$\Delta E_B(A) = \left\langle \frac{1}{r}\right\rangle \Delta Q_A + \Delta \sum_{B \neq A} (Q_B/R_{AB}) - \Delta E_R(A). \tag{3}$$

This equation has been used at various levels of approximation to correlate binding energies with atomic charges. At the highest level of approximation, all three terms of the equation are used. The relaxation energy term, $\Delta E_R(A)$, can be estimated from experimental Auger parameters,[7,8] from polarizability data,[8] or by a calculational method analogous to the transition-state method of Slater.[9] Because this relaxation term is often the smallest term in the equation, it is fairly common practice to assume that $\Delta E_R = 0$, corresponding to the following equation

$$\Delta E_B(A) = \left\langle \frac{1}{r}\right\rangle \Delta Q_A + \Delta \sum_{B \neq A} (Q_B/R_{AB}). \tag{4}$$

In fact, the second term in the equation is usually smaller than the first and, at the lowest level of approximation of equation 3, the second term is ignored, leading to the following approximate equation

$$\Delta E_B(A) = \left\langle \frac{1}{r}\right\rangle \Delta Q_A. \tag{5}$$

BINDING ENERGY-ATOMIC CHARGE CORRELATION

There are two main ways of using equations 3, 4, and 5. In one way, binding energies for a particular element are obtained for a series of compounds, the atomic charges of these compounds are estimated by some procedure, and then an appropriate function of the binding energy (depending on the equation used) is plotted against Q_A. For example, the correlation of the carbon 1s binding energies of some simple carbon compounds with carbon atomic charges is shown in Figure 2.[10] The quantity $\Delta E_B - V$ is plotted

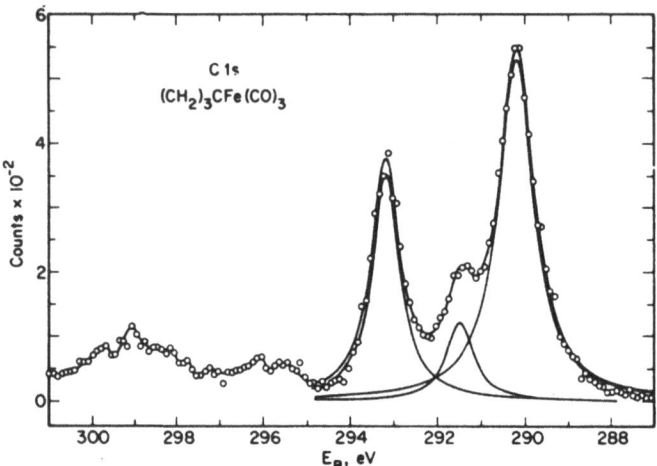

Fig. 1.   Carbon 1s spectrum of trimethylene-
          methaneiron tricarbonyl, deconvoluted
          using Lorentzian curves. The relative
          intensities of the three peaks are
          2.96:1:5.72. Reproduced with permission
          from ref. 5.

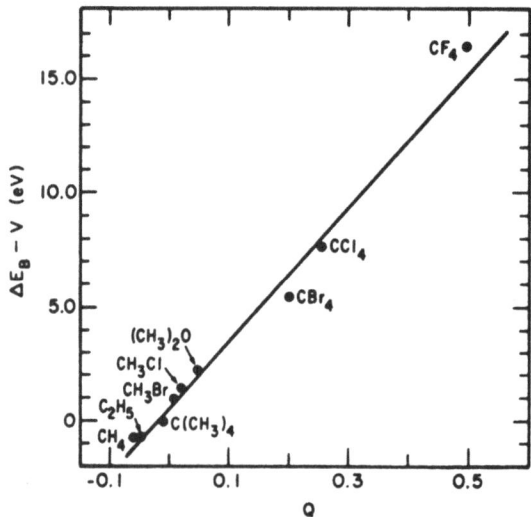

Fig. 2.   Plot of ΔE_B - V <u>vs</u>. Q for carbon
          1s binding energies. Charges
          calculated by CHELEQ method.
          Reproduced with permission from
          ref. 10.

vs. Q; this is equivalent to the use of equation 4. The charges were estimated using a simple electronegativity equalization procedure.[11] The main value of a plot such as that in Figure 2 is to verify that the binding energy data are consistent with the method used for estimating atomic charges. However, any reasonable method for estimating atomic charges usually gives fairly good straight lines in plots of this sort.[12] Thus the procedure is not very useful for determining the merits of an atomic charge estimation method. Occasionally, however, a plot of this sort can be used to determine the structure or the nature of the bonding of a compound. For example, if a compound has one of two conceivable structures, for which one estimates quite different atomic charges, then a binding energy determination will permit one to distinguish between the alternative structures.[13] Some methods of calculating atomic charge, when applied to molecules with possible contributions to bonding from the use of outer d orbitals, hyperconjugation, or back-bonding, require prior information regarding the extent of such contribution. In such cases the measured core binding energies, when combined with plots such as that in Fig. 2 (using only data for compounds with simple bonding), can be used to estimate the atomic charges and hence the degree of such extra bonding contributions.[4]

ATOMIC CHARGES FROM BINDING ENERGIES

The second main way of using equations 3 and 4 is in the derivation of "experimental" atomic charges from binding energy data. Suppose that we know the binding energies of all the atoms in a set of closely related molecules. We can write an equation (like equation 3 or 4) for each binding energy difference, thus obtaining a set of equations with the atomic charges as unknowns. If the molecules are carefully chosen, or if appropriate further approximations are made, the number of equations will be equal to or greater than the number of unknown atomic charges. Then it is possible to solve the equations to obtain atomic charges which exactly conform to the binding energy shifts or which correspond to a least-squares solution of the equations. We shall discuss several applications of this type.

Consider the binding energies for $Mn_2(CO)_{10}$ and $CH_3Mn(CO)_5$, shown in Table 1.[14] The $Mn_2(CO)_{10}$ molecule is a dimer of an electroneutral $Mn(CO)_5$ group. If we assume that equation 5 is approximately valid in this system, there are certain qualitative conclusions that can be drawn immediately. On going from $Mn_2(CO)_{10}$ to $CH_3Mn(CO)_5$, the manganese, carbon and oxygen binding energies all increase significantly, corresponding to more positive charges on these atoms in $CH_3Mn(CO)_5$ than in $Mn_2(CO)_{10}$. Thus we conclude that the methyl group in $CH_3Mn(CO)_5$ is negatively charged. These binding energies may also be treated quantitatively. The environments of

Table 1. Core Binding Energies of $Mn_2(CO)_{10}$ and $CH_3Mn(CO)_5$

	$Mn_2(CO)_{10}$	$CH_3Mn(CO)_5$
Mn $2p_{3/2}$	647.01	647.30
C 1s(CO)	293.28	293.62
O 1s	539.57	539.91

Table 2. Core Binding Energies of $(\eta^5-C_5H_5)M(NO)_2Cl$

M	O 1s	N 1s	C 1s	Cl $2p_{3/2}$
Cr	539.45	407.47	291.25	203.38
Mo	538.98	407.35	291.32	203.82
W	538.67	407.23	291.30	204.08

Table 3. Calculated Changes in Atomic Charge in
$CpM(NO)_2Cl$, M = Cr, Mo, W.

	Cr → Mo	Cr → W
ΔQ_N	-0.0019	-0.0060
ΔQ_O	-0.016	-0.026
ΔQ_{Cl}	0.022	0.034
ΔQ_M	0.014	0.030

the Mn, C, and O atoms in these compounds are so similar that it is probably a very good approximation to assume that $\Delta E_R = 0$ on going from one compound to the other. Thus we can employ equation 4. In calculating the potentials at the Mn, C, and O atom sites, we will not make a serious error if we assume that the potential due to the neighboring $Mn(CO)_5$ group in $Mn_2(CO)_{10}$ is zero and that the charge of the CH_3 group in $CH_3Mn(CO)_5$ is concentrated at the CH_3 carbon atom. Then we have three equations corresponding to the three binding energy shifts as well as the equation

$$-Q_{CH_3} = \Delta Q_{Mn} + 5\Delta Q_C + 5\Delta Q_O$$

From these we calculate $Q_{CH_3} = -0.137$, $\Delta Q_{Mn} = 0.001$, $\Delta Q_C = 0.020$, and $\Delta Q_O = 0.008$, in agreement with our first qualitative conclusion that the methyl group is negatively charged.

For a second example, consider the binding energies of the complexes $(\eta^5-C_5H_5)M(NO)_2Cl$ (M = Cr, Mo, W), shown in Table 2.[15] In this series, the carbon binding energy is essentially constant. However, on going from the chromium compound to the tungsten compound, the nitrogen and oxygen binding energies decrease and the chlorine binding energy increases. Apparently, on descending the transition metal family, π–donor bonding to the NO groups increases, and to compensate for the withdrawal of electron density from the metal atom, σ–donor bonding from the chlorine atom increases. Presumably the σ and π parts of the metal–cyclopentadienyl bonding are of comparable importance, and the opposing effects of changes in σ and π bonding cancel, resulting in essentially no change in the charge on the cyclopentadienyl group. Again, by solving four equations like equation 4, we can calculate the changes in the four atomic charges (ΔQ_N, ΔQ_O, ΔQ_{Cl}, and ΔQ_M) on going from the chromium compound to the molybdenum compound or from the chromium compound to the tungsten compound. The results, shown in Table 3, are in agreement with our first qualitative conclusions. It is interesting that the change in charge of the nitrogen atom is much smaller than the change in charge of the oxygen atom. We rationalize this result as follows. On descending the family, σ–donation by the NO group increases because of the large increase in nuclear charge of the metal, and the resultant improved dπ–π* overlap causes an increase in back–bonding. The stronger σ bonding between the nitrogen and metal atom shifts electron density from nitrogen to metal which compensates for the increase in π* electron density due to increased back–bonding. Thus the nitrogen atom charge is almost unchanged although the oxygen atom becomes more negative because of the increased back–bonding.

THE INADEQUACY OF THE SIMPLE ATOMIC CHARGE MODEL

The atomic charges which one can derive from binding energy

data by simultaneous solution of equations like equations 3 or 4
have qualitative or perhaps semi-quantitative significance, but they
generally cannot be relied upon to account quantitatively for other
physical properties which are sensitive functions of valence electron distribution.   It is well known, for example, that a reasonable
assignment of point charge atoms cannot account for the dipole
moments of molecules with lone pair electrons.   Ammonia and nitrogen
trifluoride (Figure 3) are the classic examples used to illustrate
this fact.[16]   Although the N-F bond polarity is at least as great
as the N-H bond polarity and although the F-N-F bond angle in NF_3
is more acute than the H-N-H bond angle in NH_3, the dipole moment
of NF_3 is much smaller than that of NH_3.   Of course this result is
due to the partial cancellation of the bond moments by the lone pair
moment in NF_3 and the reinforcement of the bond moments by the lone
pair moment in NH_3.   This same point can be illustrated quantitatively using the binding energy and dipole moment data for the following diatomic molecules: HF, F_2, ClF, Cl_2, HCl.   Using the fluorine 1s and chlorine $2p_{3/2}$ binding energies, we can write four
equations like equation 3, corresponding to two F 1s binding energy
shifts and two chlorine binding energy shifts.   The ΔE_R values can
be estimated by the transition state method, using CNDO/2 wave functions and the equivalent cores approximation.[9,17]   Because there
are essentially only three unknown charges (one each for HF, HCl,
and ClF), this system of equations is overdetermined and is solved
by a least-squares method.   The calculated charges are Q_{HF} = 0.294,
Q_{HCl} = 0.215, and Q_{ClF} = 0.062, corresponding to a standard deviation of 0.20 eV in the ΔE_B values.   Using these atomic charges, and
the assumption that the molecules can be represented by simple
point dipoles, the dipole moments can be calculated.   The calculated

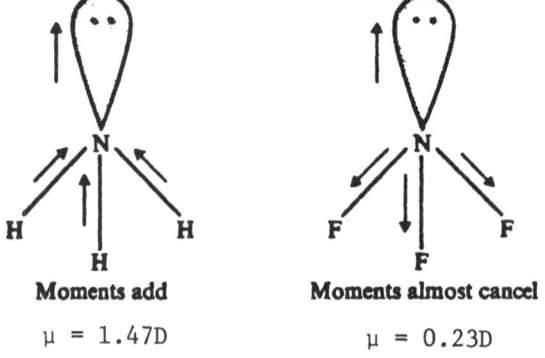

	Moments add	Moments almost cancel
	μ = 1.47D	μ = 0.23D

Fig. 3.   Explanation of the low dipole moment of NF_3.   Arrows
indicate the direction from + to −.

and experimental[18] dipole moments are as follows:

	μ_{calc}	μ_{expt}
HF	1.29D	1.83D
HCl	1.31	1.09
ClF	0.49	0.89

Obviously the simple model yields very poor dipole moments; even
the relative magnitudes of the HF and HCl moments are incorrectly
predicted. Clearly one must allow for the existence of lone pair
electrons in order to account for both core binding energy shifts
and dipole moments adequately.

A NEW POINT CHARGE MODEL

    Benson has suggested that the heats of formation and dipole
moments of molecules can be accounted for using a model in which
atoms have point charges, are polarizable, and (in the case of atoms
with lone pairs) have lone pair dipoles centered on their nuclei.[19]
This model, although more realistic than the simple atomic point
charge model, does not put any electron density in the bonding
regions between the atoms. Scheraga et al.[20] have shown that dipole
moments, lattice energies, and rotational barriers of molecules can
be fairly well accounted for assuming that molecules consist of
point charge atomic cores with bonding electron pair point charges
situated between the cores and appropriately positioned nonbonding
electron pair point charges. This model is probably unrealistic
because it puts too much electron density in the bonding and lone
pair regions. We have chosen a model involving point charge atoms
in which fractional negative charges are placed at points between
the bonded atoms and at points corresponding to lone pair electron
density, as indicated in the following structure for methyl fluoride.

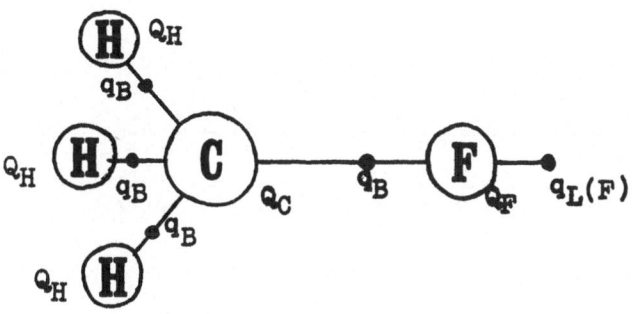

Table 4. Input Data for Calculation of Point Charges

Molecule	Core Binding Energies and Relaxation Energies, [a] eV						Dipole Moment, [b] debye
	C 1s E_B	C 1s E_R	F 1s E_B	F 1s E_R	Cl $2p_{3/2}$ E_B	Cl $2p_{3/2}$ E_R	
CH_4	290.91[c]	16.09					1.854
CH_3F	293.7[d]	16.06	692.92[d]	24.32			1.978
CH_2F_2	296.36[c]	15.99	693.65[d]	24.79			1.649
CHF_3	299.24[e]	15.62	694.62[d]	24.78			
CF_4	301.96[c]	15.47	695.57[c]	24.99			
CH_3Cl	292.48[f]	17.19			206.26[g]	9.56	1.895
CH_2Cl_2	293.9[h]	18.22			206.62[g]	10.16	1.62
$CHCl_3$	295.1[h]	18.79			206.86[g]	10.45	1.04
CCl_4	296.38[i]	19.00			207.04[g]	10.88	
$CClF_3$	300.31[i]	16.68	695.04[i]	25.48	207.83[g]	9.90	0.51
CCl_2F_2	298.93[i]	17.97	694.68[i]	25.02	207.47[g]	10.34	0.50
CCl_3F	297.54[i]	18.68	694.33[i]	26.12	207.20[g]	10.54	0.46
F_2			696.71[j]	25.24			
HF			694.22[c]	21.95			1.827
Cl_2					207.82[g]	9.88	
HCl					207.38[g]	8.38	1.093
ClF			694.54[k]	25.58	209.18[g]	8.81	0.888

For references, see p. 11

Table 4 References

[a] See refs. 9 and 17.
[b] Ref. 18.
[c] Thomas, T. D.; Shaw, Jr., R. W. J. Electron Spectrosc. Relat. Phenom. 1974, 5, 1081.
[d] Davis, D. W.; Shirley, D. A.; Thomas, T. D. in "Electron Spectroscopy," Shirley, D. A., ed., North-Holland Publishing Co., Amsterdam, 1972, p. 707.
[e] Davis, D. W. Ph.D. Thesis, University of California, Berkeley, Lawrence Berkeley Laboratory Report LBL-9900, May 1973.
[f] Perry, W. B.; Jolly, W. L. Inorg. Chem. 1974, 13, 1211.
[g] Aitken, E. J.; Bahl, M. K.; Bomben, K. D.; Gimzewski, J. K.; Nolan, G. S.; Thomas, T. D. J. Am. Chem. Soc. 1980, 102, 4873.
[h] Thomas, T. D. J. Am. Chem. Soc. 1970, 92, 4184.
[i] Holmes, S. A. M.S. Thesis, Oregon State University, Oct. 1974.
[j] Carroll, T. X.; Shaw, Jr., R. W.; Thomas, T. D.; Kindle, C.; Bartlett, N. J. Am. Chem. Soc. 1974, 96, 1989.
[k] Carroll, T. X.; Thomas, T. D. J. Chem. Phys. 1974, 60, 2186.

Table 4 lists the molecules which we have analyzed using this model, together with the core binding energies and dipole moments for these molecules. There are eleven ΔE_B(C 1s) values, nine ΔE_B(F 1s) values, and nine ΔE_B(Cl $2p_{3/2}$) values for each of which an equation like equation 3 can be written.[21] For each of the twelve dipolar molecules an equation equating the dipole moment to a function of point charges can be written. Thus 41 equations must be solved. To render the equations tractable, we have made certain assumptions and approximations. First, we have used values of ΔE_R estimated by the transition state method using CNDO/2 wave functions.[9,17] The E_R values are listed in Table 4. In the case of the chlorine relaxation energies, it is possible to compare these values with values obtained from Auger parameters;[7] the agreement is fairly good (standard deviation 0.29 eV in the ΔE_R values). Second, we assumed that the "bonding" point charges, q_B, are positioned between atoms such that the ratio of the distances to the two atoms is equal to the ratio of the covalent radii.[22] Third, we assumed that the "lone pair" point charges, q_L, are positioned on the back sides of the halogen atoms at distances equal to the average valence orbital radii, evaluated from Slater exponents.[23] Fourth, we assumed that all the q_B values are equal. Fifth, we assumed that all the q_L values for fluorine are equal, and that all the q_L values for chlorine are equal. Sixth, we assumed that the point charge at a hydrogen, fluorine, or chlorine atom is the same for all compounds in which the atom is directly bonded to the same element. Thus we forced all the fluorine atoms in the substituted methanes to have the same point charge. (Similarly for the hydrogen and chlorine atoms.) Seventh, in the 29 equations involving ΔE_R values, we used $\langle\frac{1}{r}\rangle$ values calculated from the Slater exponents of the valence shell orbitals. These are, for C, F, and Cl, 22.1, 35.4, and 20.3 eV/charge, respectively. The values are very similar to the corresponding Hartree-Fock-Slater $\langle\frac{1}{r}\rangle$ values.[24]

Noting that the sum of the Q's and q's for each molecule is zero, we find that the 41 equations contain nine adjustable parameters: q_B, $q_L(F)$, $q_L(Cl)$, three Q's for the substituted methanes, and three Q's for the diatomic molecules. A least-squares solution of these 41 linear equations in 9 unknowns yields q_B = -0.100, $q_L(F)$ = -0.252, $q_L(Cl)$ = -0.051, and the various Q values given in Table 5. The ΔE_B values and dipole moments are reproduced by these parameters with a standard deviation of 0.28 eV(debye). In performing the least-squares fit, the binding energies and dipole moments were all weighted equally because the uncertainties (in eV or Debye units) were similar in magnitude. If one wishes to convert the data in Table 5 to numbers corresponding to traditional "atomic charges," we recommend the following procedure: To the Q value for an atom add the corresponding q_L value (if the atom has lone pair electrons) and the atom's share of q_B in each of the bonds to the atom. We propose that q_B be split between the atoms of a bond inversely proportional to the covalent radii of the atoms. Thus one calculates atomic charges of 0.131 for H, 0.112 for C, and -0.081 for F in HCF_3.

As one might expect, the calculated Q and q values are sensitive to the choice of ΔE_R values. If we assume, for example, that all the ΔE_R values are zero (instead of using the estimated values in Table 4), the calculated parameters are quite different from those in Table 5, as shown by the following: q_B = -0.084, $q_L(F)$ = -0.831, $q_L(Cl)$ = -0.326, $Q_H(CH_4)$ = -0.048, $Q_F(CF_4)$ = 0.737, $Q_{Cl}(CCl_4)$ = 0.249, $Q_{Cl}(ClF)$ = 0.418, $Q_H(HF)$ = 0.087, and $Q_H(HCl)$ = -0.006. The negative value for $Q_H(HCl)$ is quite unrealistic and shows that it is important to use good ΔE_R values in this method.

If we assume that $q_L(F)$ = $q_L(Cl)$ = 0 (that is, if we ignore the lone pairs), the fit of the data is significantly poorer; the standard deviation rises from 0.28 to 0.42 eV(debye). Similarly, if we assume that q_B = 0 (that is, if we put no electron density between the atoms), the fit is considerably poorer; the standard deviation becomes 0.53 eV(debye). These results confirm our contention that a point charge molecular model should provide for lone pair and bonding electron density.

If we lift the restriction that Q_H, Q_F and Q_{Cl} are the same for all compounds in which the atom is bonded to the same element, the fit of the data is, of course, improved. For example, if we allow the Q_H values to vary linearly in the series CH_4, CH_3F, CH_2F_2, CHF_3 and CH_4, CH_3Cl, CH_2Cl_2, $CHCl_3$ (and make similar assump-

Table 5.   Calculated Q Values for Atoms

Molecule	Q_H	Q_C	Q_F	Q_{Cl}
CH_4	0.203	−0.412		
CH_3F	0.203	−0.185	0.228	
CH_2F_2	0.203	0.042	0.228	
CHF_3	0.203	0.269	0.228	
CF_4		0.496	0.228	
CH_3Cl	0.203	−0.187		0.029
CH_2Cl_2	0.203	0.038		0.029
$CHCl_3$	0.203	0.263		0.029
CCl_4		0.488		0.029
$CClF_3$		0.494	0.228	0.029
CCl_2F_2		0.492	0.228	0.029
CCl_3F		0.490	0.228	0.029
F_2			0.302	
HF	0.299		0.053	
Cl_2				0.101
HCl	0.287			−0.136
ClF			0.286	0.117

tions regarding Q_F and Q_{Cl}), the improvement in the standard deviation (to 0.22 eV(debye)) is just barely statistically significant.
However introduction of this extra freedom in the Q_H, Q_F and Q_{Cl}
values leads to some unrealistic trends (e.g., $Q_H(HCCl_3) < Q_H(CH_4)$);
hence we forced the Q value of an atom to be a constant for all
compounds in which it is directly bonded to the same element. It
is a typical feature of such least-squares calculations that the
introduction of more parameters improves the fit, but makes the
magnitudes of the parameters less realistic. A compromise must be
made.

ACKNOWLEDGEMENT

      This work was supported by the National Science Foundation
(Grant CHE-7917976) and the Division of Chemical Sciences, Office
of Basic Energy Sciences, U. S. Department of Energy, under Contract
No. W-7405-Eng-48. The authors are grateful to Dr. M. B. Hall for
a helpful discussion.

REFERENCES

(1) Siegbahn, K. et al. "ESCA Applied to Free Molecules," North-Holland: Amsterdam, 1969.

(2) Brundle, C. R.; Baker, A. D., ed., "Electron Spectroscopy: Theory, Techniques and Applications," Academic Press; London, 1977.

(3) Carlson, T. A. "Photoelectron and Auger Spectroscopy," Plenum: New York, 1975.

(4) Jolly, W. L. Coord. Chem. Rev. 1974, 13, 47; Chapter 3 in ref. 2, p. 119; Topics Curr. Chem. 1977, 71, 149.

(5) Koepke, J. W.; Jolly, W. L.; Bancroft, G. M.; Malmquist, P. A.; Siegbahn, K. Inorg. Chem. 1977, 16, 2659.

(6) Schwartz, M. E.; Switalski, J. D.; Stronski, R. E. in "Electron Spectroscopy," Shirley, D. A., ed., North-Holland: Amsterdam, 1972, p. 605

(7) Aitken, E. J.; Bahl, M. K.; Bomben, K. D.; Gimzewski, J. K.; Nolan, G. S.; Thomas, T. D. J. Am. Chem. Soc. 1980, 102, 4873; Thomas, T. D. J. Electron Spectr. Rel. Phen. 1980, 20, 117.

(8) Perry, W. B.; Jolly, W. L. Chem. Phys. Lett. 1973, 23, 529.

(9) Hedin, L.; Johansson, A. J. Phys. B Ser. 2, 1969, 2, 1336; Jolly, W. L. Discuss. Faraday Soc. 1972, 54, 13; Davis, D. W.; Shirley, D. A. Chem. Phys. Lett. 1972, 15, 185; Davis, D. W.; Shirley, D. A. J. Electron Spectr. Rel. Phen. 1974, 3, 137.

(10) Perry, W. B.; Jolly, W. L. Inorg. Chem. 1974, 13, 1211.

(11) Jolly, W. L., Perry, W. B. Inorg. Chem. 1974, 13, 2686.

(12) Carver, J. C.; Gray, R. C.; Hercules, D. M. J. Am. Chem. Soc. 1974, 96, 6851.

(13) Hendrickson, D. N.; Hollander, J. M.; Jolly, W. L. Inorg. Chem. 1969, 8, 2642.

(14) Avanzino, S. D.; Chen, H.-W.; Donahue, C. J.; Jolly, W. L. Inorg. Chem. 1980, 19, 2201.

(15) Chen, H.-W.; Jolly, W. L.; Xiang; S.-F.; Legzdins, P. Inorg. Chem. 1981, 20, 1779.

(16) Jolly, W. L. "The Principles of Inorganic Chemistry," McGraw-
      Hill: New York, 1976, p. 47.

(17) We used CNDO/2 wave functions and the relation $E_R = 0.5[V(A)_{val} - V(A^+)_{val}]$ to calculate the relaxation energies.

(18) McClelland, A. L. "Tables of Experimental Dipole Moments,"
      Vols. I and II, Freeman: San Francisco, 1963, 1974.

(19) Benson, S. W. Angew. Chem. Int. Ed. 1978, 17, 812.

(20) Shipman, L. L.; Burgess, A. W.; Scheraga, H. A. Proc. Nat.
      Acad. Sci. USA 1975, 72, 543; Burgess, A. W.; Shipman, L. L.;
      Scheraga, H. A. ibid., 1975, 72, 854; Burgess, A. W.; Shipman,
      L. L.; Nemenoff, R. A.; Scheraga, H. A. J. Am. Chem. Soc.
      1976, 98, 23; Scheraga, H. A. Acc. Chem. Res. 1979, 12, 7.

(21) The structural parameters for F_2, HF, Cl_2, HCl, ClF, CH_4, CF_4,
      and CH_2Cl_2 were obtained from Sutton, L. E., ed. "Tables of
      Interatomic Distances and Configurations in Molecules and
      Ions," Chemical Society: London, 1958. For CH_3F and CH_3Cl,
      Sutton, L. E., ed. "Tables of Interatomic Distances and Con-
      figurations in Molecules and Ions, Supplement," Chemical Soci-
      ety: London, 1965. For CH_2F_2, Hirota, E.; Tanaka, T. J. Mol.
      Spectrosc. 1970, 34, 222. For CHF_3, Cox, A. P.; Kawashima,
      Y. J. Mol. Spectrosc. 1978, 72, 423. For $CHCl_3$, Jen, M.;
      Lide, Jr. D. R. J. Chem. Phys. 1962, 36, 2525. For CCl_4,
      Bartell, L. S.; Brockway, L. O.; Schwendeman, R. H. J. Chem.
      Phys. 1955, 23, 1854. For CCl_2F_2, Takeo, H.; Matsumura, C.
      Bull. Chem. Soc. Jpn. 1977, 50, 636. For CCl_3F, Loubser, J.
      H. N. J. Chem. Phys. 1962, 36, 2808.

(22) Jolly, W. L. "The Principles of Inorganic Chemistry," McGraw-
      Hill, New York, 1976, p. 36. The covalent radii of H, C, F,
      and Cl are 0.30, 0.77, 0.58, and 1.00 Å, respectively.

(23) The "lone pair" points for fluorine and chlorine are positioned
      0.41 and 0.71 Å from the atoms, respectively.

(24) Lu, C. C.; Carlson, T. A.; Malik, F. B.; Tucker, T. C.;
      Nestor, C. W. Atomic Data 1971, 3, 1. The $\langle \frac{1}{r} \rangle$ values are:
      carbon (sp^3 hybridization) 23.0, fluorine (10% s character)
      36.7, chlorine (10% s character) 21.7 eV/charge.

ELECTRON DENSITY FUNCTIONS IN ORGANIC CHEMISTRY

Andrew Streitwieser, Jr., David L. Grier,
Boris A. B. Kohler, Erich R. Vorpagel and
George W. Schriver

Department of Chemistry
University of California
Berkeley, California 94720

A. CARBON-LITHIUM BONDING

(1) Electron density minima

A few years ago we computed the electron density function for methyllithium[1] and called attention to the low value of the electron density, $\rho(\equiv\rho(\underline{r}_1)$, the one-electron density function[2]) at its minimum along the C-Li internuclear axis. This low value did not seem to be consistent with significant C-Li covalency and we accordingly wrote that the carbon-lithium bond is essentially wholly ionic. Fig. 1 presents a perspective plot of ρ for $LiCH_3$ showing the low value of the shared electron density for C-Li compared to C-H. The electron density function has found frequent important use in under- standing bonding and in testing our bonding concepts.[2,3] The con- clusion that organolithium compounds are best interpreted simply as contact ion pairs not much different from lithium fluoride came as a complete surprise to us. We had been trained (and we even taught in the past) that the properties of organolithium compounds required substantial carbon-lithium covalency; but the electron density patterns computed for a number of organolithium compounds forced us to change our views. Not everyone agrees with us and our conclusion is still controversial. For example, one reviewer wrote of our conclusions, "...is surely wrong. There must be an error in the program."

Part of the problem, of course, is that ionic and covalent character are not physical observables and cannot be precisely defined. There is no ambiguity about ascribing complete covalency

447

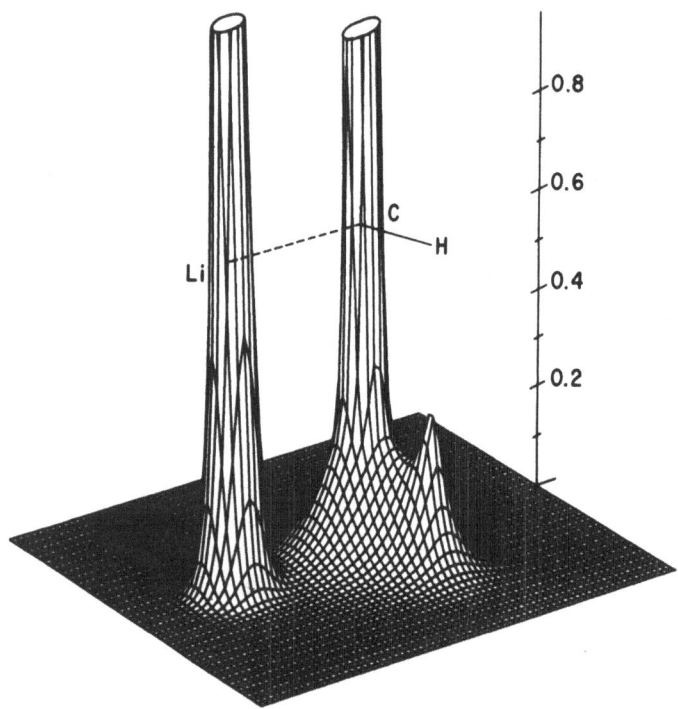

Fig. 1. Perspective plot of ρ for LiCH$_3$. For a Li–C–H grid
         plane ρ is plotted as the vertical axis. (Split shell
         plus polarization basis; Ref. 1).

to the bonding in a homonuclear diatomic and the interaction between
two widely separated ions is undoubtedly almost wholly "ionic", but
in a polar bond the division into fractions of ionic and covalent
character is necessarily ad hoc. There is less ambiguity about
qualitative terms. For example, the detailed analysis of LiF by
Bader and Henneker[3b] leaves no doubt that its bonding is almost
wholly ionic. The only ambiguity is whether the back-polarization
they found for fluorine π-electrons is simply an atomic polarization
resulting from a nearby charge or the manifestation of some
covalency. They conclude, "Thus, while the σ-density is ionic in
character, corresponding to almost complete transfer of charge to

the F, the π-density, in addition to being polarized, exerts small
but almost equal overlap forces on both nuclei in the manner of a
weak covalent bond."[3b] This careful result makes LiF an important
standard for comparison with other lithium compounds.

Graham, Marynick and Lipscomb[4] (GML) have recently made a careful study of the electron density function of methyllithium. At the
triple-zeta + double polarization level, surely close to the Hartree-
Fock limit, they found no important differences from our split shell
+ polarization (SSP) results.[1] They made the interesting suggestion
that ρ at the minimum between nuclei should be compared with reference systems at the same bond length, and pointed out that the
minimum density of CH_3Li is between those for Be_2H_2 and LiF at the
same bond distances. They concluded that C-Li is more covalent than
Li-F. This criterion can certainly be challenged--it compares, for
example, a "normal" methyllithium with a partially dissociated LiF;
nevertheless, the comparison can be instructive and we will consider
it further.

A more important limitation of the GML analysis is its dependence on the total ρ value at the minimum. It is well established
that shared electron density is a necessary condition for net covalent
bonding but it is not a sufficient condition. Such shared electron
density results in reduced electron kinetic energy along the bond
that allows a corresponding increase close to the nuclei; the concomitant change in potential energy is the principal factor in
covalent bonding.[2,5,6] Thus, integration of ρ for a "bonding region"[4]
is impossible because such a region does not exist. An important
example is He_2 (Hartree-Fock He_2, not the Van der Waals molecule!).
This molecule shows shared electron density--a ridge of ρ between
the nuclei--but is unbound. We can see why ρ does not drop to zero
between the nuclei. There exists a bonding MO, σ, that produces the
overlapping electron density. The corresponding antibonding MO, $\sigma*$,
is also filled and gives net repulsion but the node between the
nuclei cannot subtract electron density from σ! The result is the
familiar filled-shell repulsion and emphasizes that such repulsion
still exists in the presence of shared electron density. Obviously,
it is the electron density for the molecule as a whole that counts
but it is still useful for understanding to be able to dissect the
total density into meaningful parts. For this reason we have emphasized the use of valence electron densities since the core contribution to shared electron density does not contribute to bonding.
Although the "valence density" is not a physical observable it is
heuristically useful in the same sense as "atomic charges" and other
concepts that help reduce complex chemistry to manageable "theory".
Moreover, valence electrons are readily distinguishable from the
core in the context of MO theory as well as by electron density
partitioning.[7,8]

The point is emphasized by the comparison of ρ for LiF with

that of the unbound isoelectronic molecule HeNe, with the geometry
of both diatomics set to the C-Li bond distance in methyllithium,
2.021Å (Fig. 2). We note the close resemblance of both ρ functions
along the bond axis. One could argue that net covalent bonding would
require significantly more shared valence electron density at the
minimum. The values summarized in Table 1 show that the value of
ρ_{min} for LiF is indeed similar to that for HeNe. Moreover, subtracting the density associated with the two core MOs of LiF leaves
the valence density, ρ^{val}, shown in Fig. 2 as a dotted line. The
values summarized in Table 1 show that the core does contribute
significantly to ρ_{min} for LiF. The comparison with the larger basis
calculation of GML shows also that the 3-21G basis gives a satisfactory account of the properties of ρ in this region.

If charge transfer from Li to F were complete to form two
ions, the valence density would fall to zero and not show a peak at
Li. Fig. 2 shows that charge transfer is not complete but that
ρ^{val} at its minimum is very small. Fig. 3 is a comparable plot of
ρ for methyllithium. Charge transfer from Li to CH_3 in methyllithium
is less complete than for LiF but that ρ^{val} at its minimum is still
extremely small (Table 1). If we consider the value of ρ^{val} at the
position of ρ^{total}_{min}, we find a slightly higher value (Table 1) for
$LiCH_3$ than for LiF. This approach suggests that methyllithium is
indeed not quite so ionic as LiF but there is no doubt that any C-Li
covalency involved is exceedingly small. It may be further noted that
actual organolithium aggregates involve still longer C-Li bonds
(e.g., 2.23-2.27Å for methyllithium-TMEDA tetramer) and covalency
for such species must be still smaller (vide infra).

For non-linear molecules, it is not possible to display π and
σ densities on the same basis. For such systems we have found it
convenient to integrate along one axis--perpendicular to a chosen
plane--and present the resulting "Projection" function,[1,10,11] P,
in contour or perspective form. The projection functions generally
look similar to the ρ functions and most results change but little.
Applied to the present group of molecules, the position of P_{min} for
HeNe is close to that for ρ_{min}; for LiF and $LiCH_3$, P_{min} is a few
hundredths of an Angstrom closer to lithium (Table 2). The conclusions derived above based on ρ do not change if considered in terms
of P; we will see below examples where P is a more convenient function
to study.

(2) Electron Populations and Atomic Charges

An important application of electron density functions is their
use in deriving the Integrated Spatial Electron Populations, ISEP,
for a molecular region of interest. If the defined volume corresponds to what can be thought of as an atom in the molecule, an "atomic
charge" can be defined. Unfortunately, atoms in molecules are not
physical observables and there is correspondingly no absolute

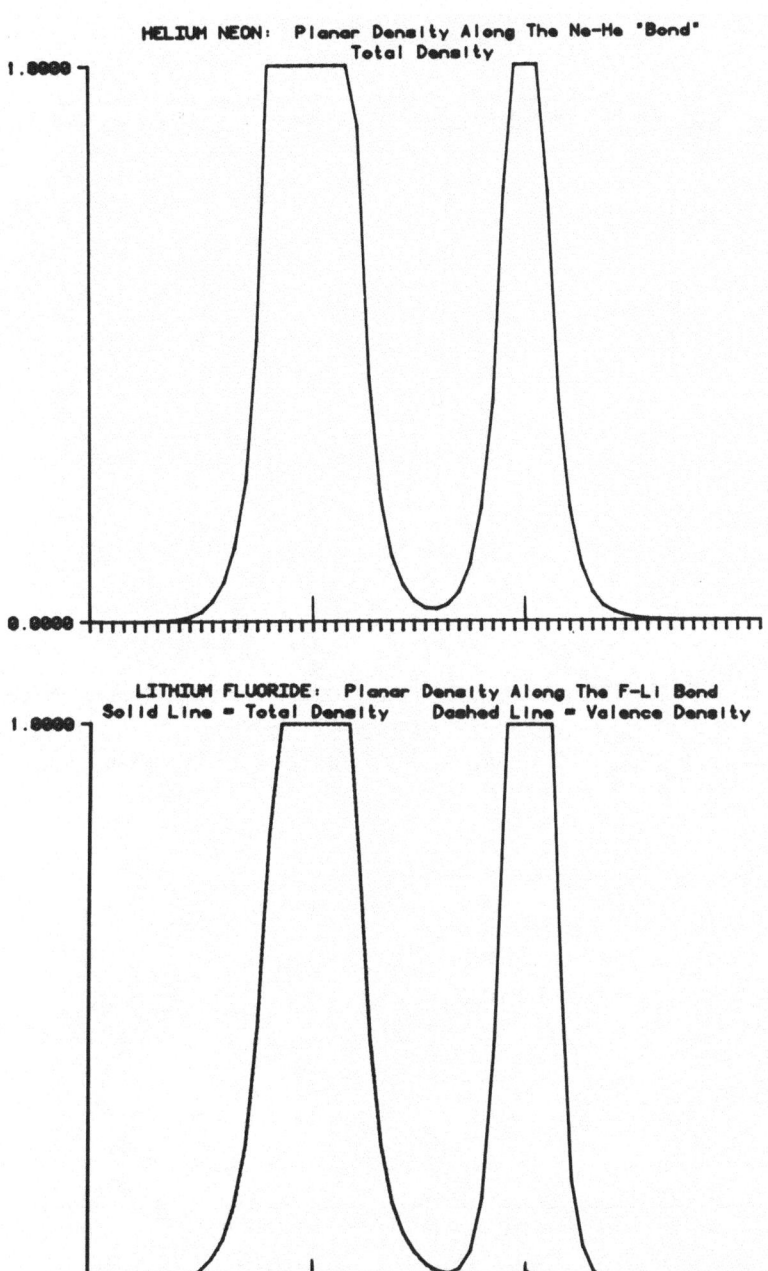

Fig. 2.   Electron density function ρ along the internuclear bond
axis for NeHe and FLi.   The light element is on the
right; 3-21G basis set.   Dotted lines for FLi show the
valence density.

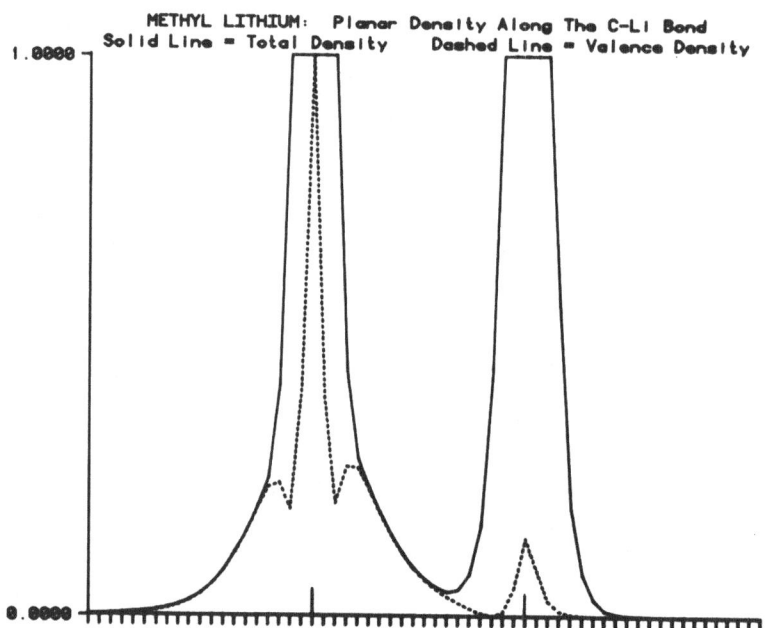

Fig. 3.   Electron density function ρ for methyllithium plotted
         as H_3CLi; 3-21G basis set. Dotted lines show the
         valence density.

Table 1.   Electron Density for R = 2.021Å
           3-21G Basis Set

	r_{min}, Å [a]	ρ_{min}^{Total}	ρ^{Val} at r_{min}	ρ_{min}^{Val} (r) [a]
LiF	0.77	0.023(.025) [b]	0.017	0.0002(0.33)
LiCH$_3$	.73	.042(.041) [b]	.032	0.0006(0.34)
HeNe	.88	.021		

[a] Distance from Li or He.
[b] Values of Ref. 4.

Table 2.   Electron Projection Function, $\underline{P}$
3-21G Basis Set

	r_{min} ^{a} Å	$\underline{P}_{min}$
LiF	0.72	0.051
LiCH$_3$	.66	0.113
	(.67)^b	(.110)^b
	(.67)^c	(.114)^c
HeNe	.89	.040

^{a}Position from light atom.
^{b}6-31G basis.
^{c}6-31G* basis.

definition of such an atom for any molecule other than a symmetrical
homonuclear species. Nevertheless, the concept of atomic charges in
molecules has such heuristic importance in chemistry that many
attempts have been made to define such atomic regions--too many to
cite completely. The methods fall into two distinct groups: those
based on isolated atoms and those based on the properties of ρ.

In the first method the molecule is divided into atomic spaces
using procedures based on the isolated free atoms. This approach was
pioneered in the "electron count" method of Politzer[12] and has been
applied successfully to a number of systems by him[13-17] and others.[18]
Related more recent methods have been proposed by Dean and Richards,[20]
Hirshfeld,[20] and by Iwata.[21] A fundamental problem with these methods
is that the effective size of an atom can vary in different molecules,
especially as a function of charge, and procedures based on isolated
neutral atoms may not be appropriate. This limitation is especially
critical for organolithium compounds since the effective size of
lithium in these compounds (effectively, Li$^+$) is much smaller than
in lithium atom or dilithium.

The second approach, based on properties of ρ, is exemplified
by the suggestion of Ransil and Sinai[22] that contours of ρ enclosing
one nucleus only be taken as defining "atomic" regions. This is
certainly a reasonable suggestion but excludes all "shared" electron
density. Bader, Beddall and Cade[23] have suggested for linear molecules a "natural partioning" using vertical planes centered at ρ$_{min}$
between the atoms. Bader and Beddall[24] have extended this definition
by defining a partitioning surface as the path of the gradient vector
of ρ passing through the point of minimum density between nuclei.
Bader and his group have shown that such so-called "virial regions"
are uniquely defined and have special properties.[25]   A comparison of

"natural" and "virial" partitioning is shown for LiF in Fig. 4. The
corresponding ISEP values give lithium charges of +0.84 and +0.93,
respectively. Note that the virial approach assigns to F some
electron density in regions behind Li. Both methods show extensive
but not complete transfer of one electron from Li to F. Our approach,
based on the electron Projection function, is analogous: a partitioning plane is chosen based on $\underline{P}_{min}$. Since this minimum is closer to
Li than ρ_{min}, the resulting ISEP values are in between those derived
from natural and virial partitioning. For LiF we have derived with
a similar function values of the lithium charge of +0.859 (4-31G)
and 0.874 (6-31G*).[26] The corresponding values for methyllithium
are +0.794 (4-31G) and +0.797 (6-31G*).[26] Electron transfer is still
high but is clearly slightly less than for LiF. Moreover, for basis
sets of at least split valence shell (SVS) quality, the results are
remarkably insensitive to basis set (see also Table 2).

Mulliken Populations give a much smaller Li charge that is
moreover highly basis set dependent; in LiF, +0.715 (4-31G) and
+0.639 (6-31G*), in LiCH$_3$, +0.451 (4-31G) and +0.416 (6-31G*).[26]
The limitations of Mulliken Populations have been pointed out often
and there is no need to repeat them here. One interesting observation, however, is made in Table 3 in which is summarized the exponent

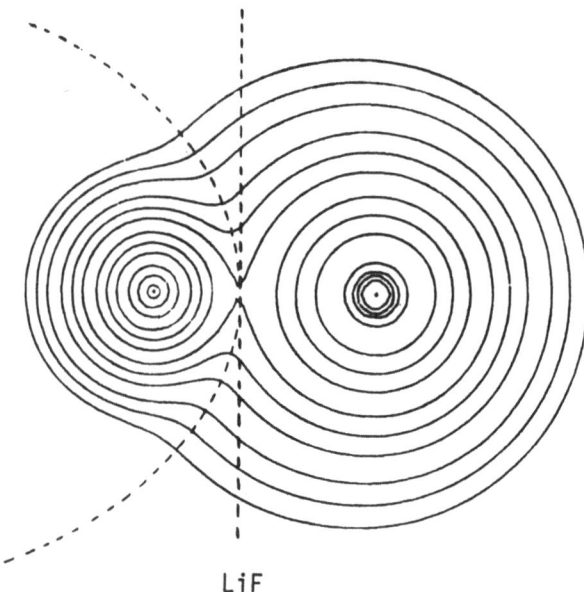

LiF

Fig. 4. Electron density contours for LiF. The vertical dashed
         line shows "natural partitioning"; the curved dashed
         line is that of "virial partitioning". (Adapted with
         permission from ref. 25b).

Table 3. Position of Outer Gaussian Maximum
(3-21G Basis Set[a])

	α Outer G	R_{max}, Å $(1/\sqrt{2\alpha})$	R_{X-Me} -0.77, Å
Li	0.0286	2.21	1.25
Be	.0774	1.35	.92
B	.1243	1.06	.79
C	.1959	0.85	.77
N	.2832	.70	.70
O	.3737	.61	.66
F	.4824	.59	.61

[a]Ref. 27.

of the outer Gaussian of the 3-21G basis set of Binkley, Pople and
Hehre.[27] This basis set has already been shown to be extremely
useful--it is of split valence shell quality but involves for computation no more integrals than the minimum basis STO-3G set and only
a somewhat larger SCF matrix. The outer 1G is the same as the 6-21G
set. For a 2s or the radial part of 2p gaussian, $\chi = xe^{-\alpha x^2}$. The
wave function rises from zero at the origin to a maximum at $x = 1/\sqrt{2\alpha}$
a.u. and falls exponentially to zero at larger x. The position of
the maximum is tabulated in Table 3 for first row elements, Y, and
compared with a bond radius taken for convenience simply as the C-Y
bond distance in CH_3Y with 0.77Å (carbon radius) subtracted. We note
that for most of the elements, the position of the maximum is close
to the "bond radius"; for these elements this outer gaussian is
important for describing the internuclear bonding region. For Be
and B the maximum occurs closer to carbon but is still in the "bonding region". For Li, however, the maximum occurs at the other side
of carbon; this gaussian function, although centered at lithium, is
actually used to describe electrons in the methyl group. Note that
in a Mulliken Population such electrons and corresponding overlap
populations, even though mostly in the C-H region, would be assigned
to Li! We will see below that Mulliken Populations can be reliable
and useful for many applications but they are absolutely meaningless
for lithium compounds and should never be used for such systems.

(3) a. Organolithium Aggregates

Organolithium compounds exist generally as aggregates in
solution. The dimer and tetramer of methyllithium have received
several theoretical studies.[28,29,30] These aggregates are often
considered in terms of electron-deficient multicenter bonding.

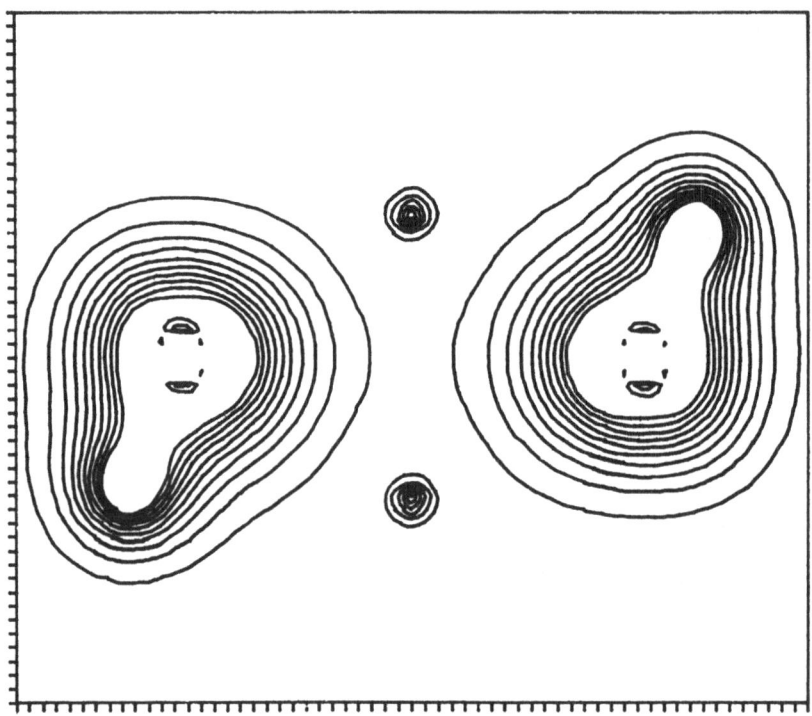

Fig. 5.   Valence electron density plot of methyllithium dimer
          (4-31/5-21 basis). Contour intervals are 0.02 e au^{-3}.
          The two lithiums are the circular features above and
          below the center of the figure.

Figure 5 shows a valence electron density plot for the molecular
plane of methyllithium dimer at the 4-31G (5-21G on Li) level. The
structure was optimized with the assumptions of C_{2h} symmetry, local
C_3 symmetry for each methyl and a Li-Li axis perpendicular to the
C-C axis. The resulting C-Li distance is 2.134Å, Li-Li is 2.160Å
and the HCLi angle is 113.3°; the resulting energy is -93.98764 a.u.
The figure is most readily interpreted as an ionic aggregate of two
lithium cations and two pyramidal methyl anions. The value of $\underline{P}$ at
the minimum is less than in methyllithium and there is in the middle
of the assembly no indication whatsoever of multicenter covalent
bonding. We note that four-center bonding MO derived from sigma-type
orbitals as in A would be expected to have enhanced density in the
center. A recent example was reported[31] in the X-ray analysis of
ethyllithium in which an electron density feature in the center of
a triangle of lithiums was taken as indicative of delocalized four-
center bonding. It seems more likely that the feature observed was
actually part of a carbanion lone pair. The involvement of Li 2p
orbitals as in B could lead to off-center density.

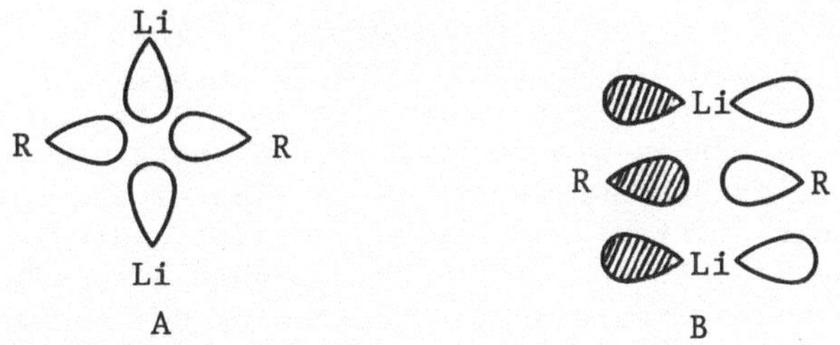

Li

R ◁ ▷ R

Li

A

Li

R ◁ ▷ R

Li

B

B.   THE n → π* TRANSITION IN FORMALDEHYDE

Formaldehyde and its excited states have been subject to a number of theoretical studies;[32] nevertheless, few studies have involved electron density analysis.[33,34] We have made such a study of the n → π* transition of formaldehyde using a DZ basis and configuration interaction with all single and double excited configurations for the experimental ground state geometry. The resulting natural orbitals were used for the electron Projection function analyses. This study was carried out in collaboration with Professor H. F. Schaefer.

A simplified orbital picture of the ground state and the singlet and triplet excited states is shown in Fig. 6. Because of the difference in electronegativities of C and O, the π MO of a carbonyl group has a large coefficient for the O p-orbital and a smaller coefficient for carbon. For the corresponding π* MO, the magnitudes

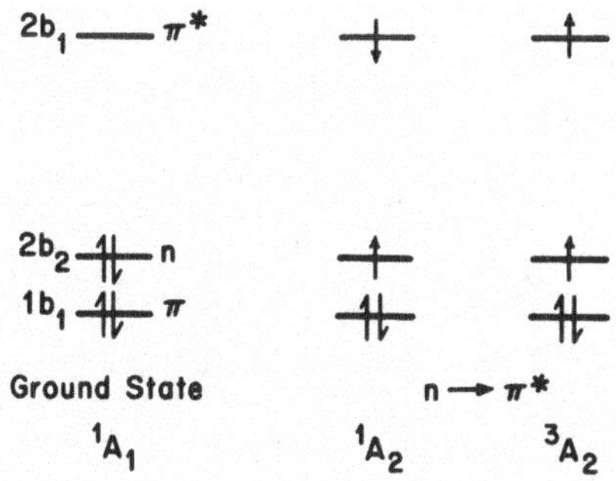

$2b_1$ —— π*

$2b_2$ n
$1b_1$ π

Ground State
$^1A_1$

n → π*
$^1A_2$ $^3A_2$

Fig. 6.   MO description of states of formaldehyde.

of these coefficients are approximately the same but reversed. Thus, a naive view of the n → π* transition transfers an electron from an essentially localized oxygen lone pair to a π-type molecular orbital dominated by carbon--a process that would appear to involve substantial oxygen-to-carbon charge transfer. The experimental dipole moment of the singlet n → π* state, $^1A_2$, is indeed lower in magnitude than the ground state but still has the same direction; that is, some oxygen-to-carbon charge transfer does occur but the amount is far less than the simple picture above would suggest. The electron density analysis shows what happens.

Projection function plots, $\underline{P}(^1A_1)-\underline{P}(^1A_2)$, are presented for two orthogonal planes, the molecular plane, Figure 7a, and the π-plane, Figure 7b. In both cases the total change over the entire figure vanishes. Both figures show loss of density primarily from an in-plane oxygen p-orbital but a gain primarily at the oxygen p_π-orbital rather than at carbon. These results are also evident in the ISEPs at oxygen using reasonable boundaries as summarized in Table 4. This table also compares experimental[35] and calculated dipole moments. Note that the calculated dipole moments refer to the same planar geometry as the ground state whereas the experimental quantities refer to the relaxed non-planar excited states. Further analysis by MOs show that the p_π lobe on C receives about 0.14 e from the b_2 MOs (n-system) but returns 0.04 e through the a_1 MOs (σ-system). Net charge transfer to carbon is relatively small. These results are rationalized on the basis that loss of an n-electron increases the effective electronegativity of oxygen resulting in withdrawal of additional electron density towards it; that is, in the excited state the doubly occupied π-MO is still further weighted towards O and largely compensates for the large C coefficient in the singly occupied π*-MO.

Table 4.   Integrated Spatial Electron Populations
           at Oxygen

MOs	Formaldehyde States		
	$^1A_1$	$^1A_2$	$^3A_2$
a_1	5.87	5.90	5.90
b_1	1.34	1.91	1.78
b_2	1.97	1.20	1.13
Dipole Moment,			
D, calculated	2.64	1.81	1.51
expt.[a]	2.33	1.56	1.33

[a]Ref. 35.

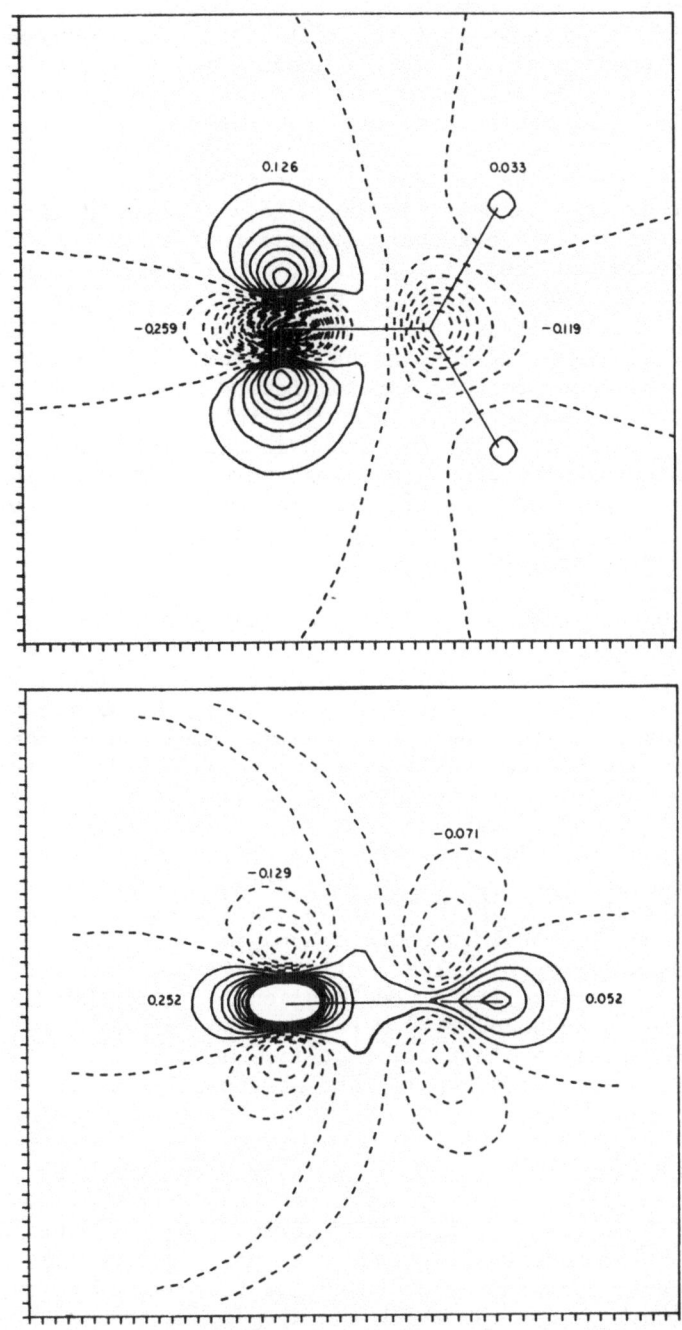

Fig. 7.   Difference Projection function plot for formaldehyde,
$^1A_1 - {}^1A_2$; DZ natural orbital CI.   Contour levels from
−0.18 (dotted) to .18 (full) by 0.02 e au^{-2}. (a) Molecular plane, (b) pi plane.   Numbers shown are ISEP changes
for reasonable boundaries not indicated.

Analysis also shows that on excitation, electron density is lost in the C=O "bonding region" in agreement with the expected decrease in bond order. This change is necessary to rationalize the non-planar geometry of the relaxed excited state.

Figure 8 shows the projection plot differences for the singlet and triplet excited states, $^1A_2$-$^3A_2$. Looking down at the molecular plane (Fig. 8a) we see net charge transfer of about 0.05 e from oxygen to carbon in the triplet state. This seems rather surprising considering that the $^3A_2$ state is more stable than $^1A_2$. Looking at the π-plane, however (Fig. 8b), shows that about 0.01 e is moved closer to the oxygen nucleus in the σ-system. This change provides additional shielding for the more diffuse π-electrons; that is, triplet oxygen is effectively less electronegative than singlet in the π-MOs. Charge transfer to carbon in the π-system is greater but a small number of electrons closer to the nucleus dominate the energetics. A more nucleophilic triplet carbon has been used to rationalize experimental differences in exiplex formation with electron-poor olefins.[36]

C. OXYGEN POPULATIONS OF SUBSTITUTED CARBONYLS

A perspective plot of the valence Projection function of acet-aldehyde is shown in Fig. 9 and clearly shows the arbitrary nature of any boundary assigned between C and O. Addition of core electrons does provide a definite saddle between C and O but the position varies with substituents and does not in any event correspond even approximately to empirical covalent bond radii. Fortunately, in physical organic chemistry we are generally concerned not with abso-lute electron populations but in population changes that occur with change in structure. Figure 10 shows contour plots of the difference Projection function between acetaldehyde and formaldehyde for the CHO plane, P(CH$_3$CHO)-P(H$_2$CO). The gently changing contours between C and O show that difference ISEPs are less sensitive to the precise loca-tion of an assigned boundary. Moreover, the 4-31 plot is very similar to the larger basis 6-31* plot; that is, the difference functions are not very sensitive to basis set, at least for split valence shell or better bases.

The positive contours at oxygen show unquestionably that the oxygen of acetaldehyde is more negative than that of formaldehyde. Similarly, the negative contours at carbon show the acetaldehyde carbon to be the more positive but some polarization-like structure is also clearly evident. The dominating feature, of course, results from the methyl group from which a hydrogen has been subtracted and is not meaningful. Because of the close structural change, the carbonyl carbon region is difficult to define. The carbonyl oxygen, however, provides an isolated region that can be discerned in an unambiguous manner. Our attention is directed to changes in oxygen ISEP with change in structure; thus, using reasonable choices for

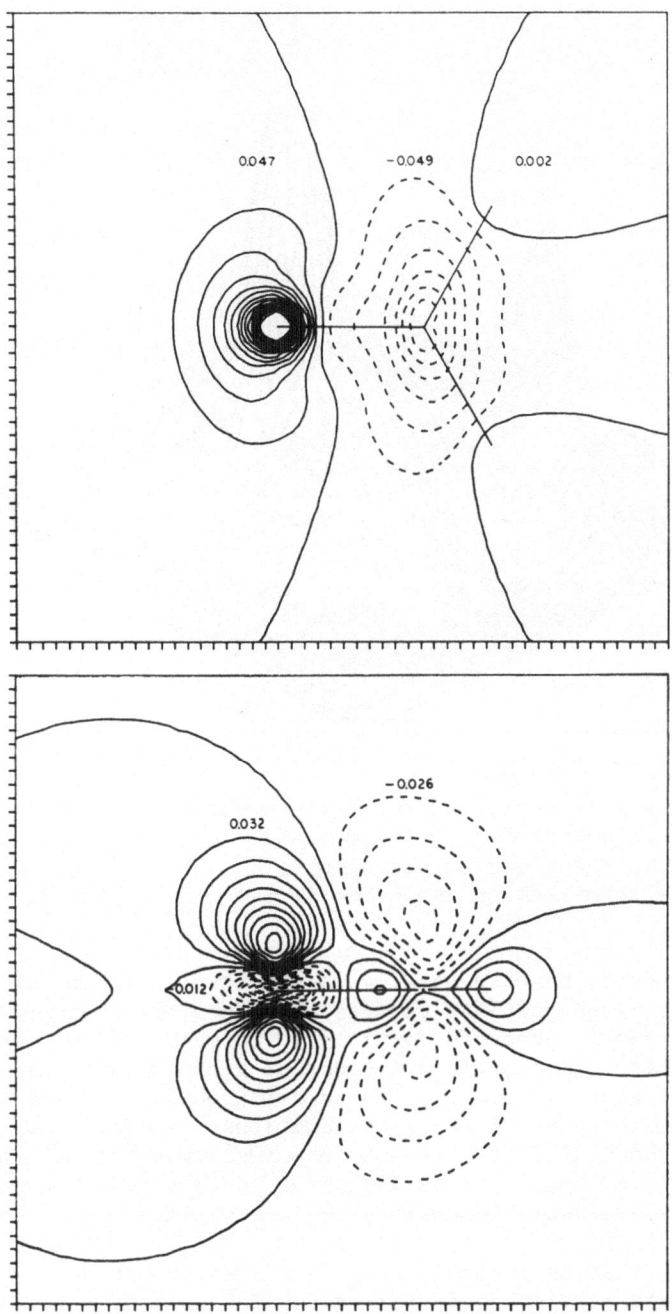

Fig. 8. Difference projection function plot for the singlet and
triplet n → π* excited states of formaldehyde, $^1A_2$ – $^3A_3$;
DZ natural orbital CI. Contour levels from –0.012 (dotted)
to +.030 (full) by 0.002 e au^{-2}. (a) Molecular plane,
(b) π-plane. Numbers shown are integrated electron
populations for different regions.

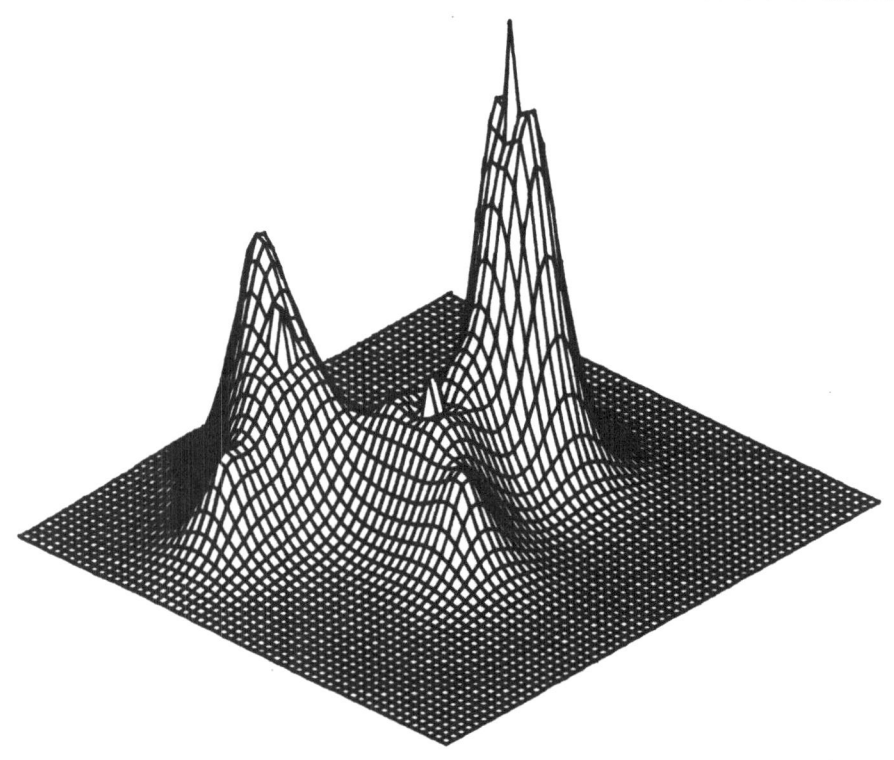

Fig. 9.   Perspective plot of the Valence Projection function of
          acetaldehyde for the CCHO plane. Oxygen is at the right
          rear, the aldehyde hydrogen is in the front right, and
          the methyl group is on the left.

the C-O boundary, based generally on the zero difference contour,
we have determined the ISEP changes at oxygen for a number of substituted carbonyl compounds compared to formaldehyde.[37] The 4-31
basis was used throughout and a constant geometry was taken for
the carbonyl group in order to avoid perturbations in the plots
associated with change in nuclear positions. In addition to the
total ISEP changes, contributions from MOs symmetric (σ) and antisymmetric (π) to the carbonyl plane were determined separately.
Some results are shown schematically in Fig. 11.

    We find that the methyl group is a π-donor but a weak σ-
acceptor. The isoelectronic BH_3^- group is a stronger π-donor and
a weak σ-donor whereas the NH_3^+ group is a π- and σ-acceptor. The
OH group, as expected, is a strong σ-acceptor but stronger π-donor
such that the net effect is strong valence donation. Fluorine, on
the other hand, is an equally strong σ-acceptor compared to OH but
a weaker π-donor; the net effect of F is to cause virtually no
valence population change at the carbonyl oxygen. The $NH_2(||)$

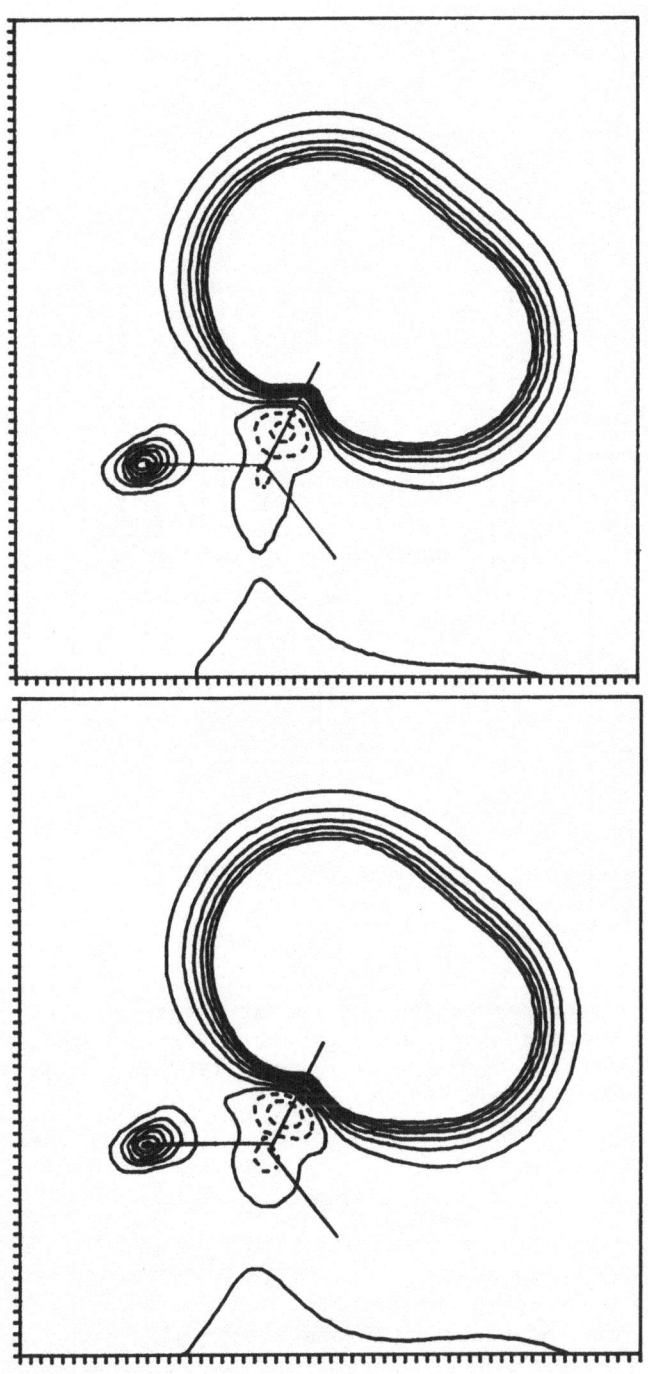

Fig. 10. Projection function of acetaldehyde minus that of form- aldehyde with superposed aldehyde group. Contour leve from −0.01 (dotted) to 0.03 by 0.005 e au^{-2}. (a, top. 4–31G basis, (b, bottom) 6–31G* basis.

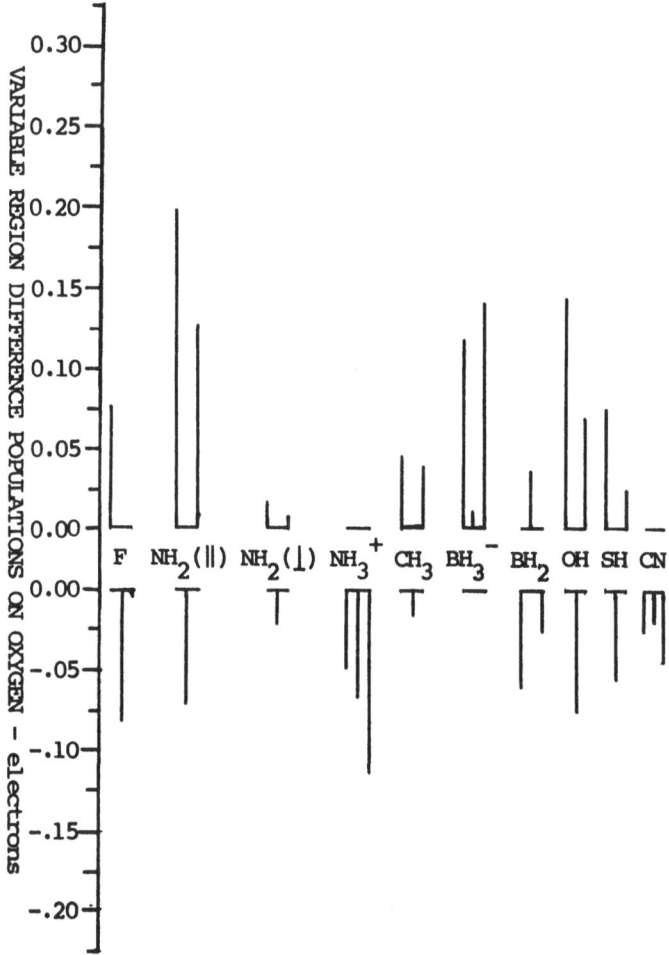

Fig. 11.   Integrated population changes at oxygen for YCHO compared
           to H_2CO with different Y groups. For each indicated Y
           group the π, σ and total valence ISEP change is shown
           graphically from left to right.

group which can conjugate has a far larger effect than the $NH_2(\perp)$
group in which the nitrogen lone pair cannot conjugate. The difference shows that much of the σ-effect is a reverse polarization
response to larger changes in the more polarizable π-system. Indeed, such reverse polarization in one electronic system as a
reaction to a more pronounced change in the other system is a fairly
general phenomenon in these results.

Figure 12 shows a comparison of the ISEP changes on O with the corresponding Mulliken Population changes. The correlation is only fair but does show large parallel trends. If we restrict attention to π-electron changes only, the corresponding correlation shown in Fig. 13 is excellent. The π-electron system is rather well defined for atoms both spatially and in terms of orbitals. Moreover, the π-overlap integrals between neighboring atoms are generally relatively small so that the arbitrary division of overlap into equal shares as in the Mulliken definition causes no real problem. Thus, relative π-Mulliken Populations appear to be valid representations of relative ISEP changes. For σ-electrons, however, as shown in Fig. 14, the parallelism is no longer satisfactory. Overlap integrals are now larger and atoms are less effectively defined by orbitals. The use of Mulliken Population changes in σ-electron systems must accordingly be made with due care.

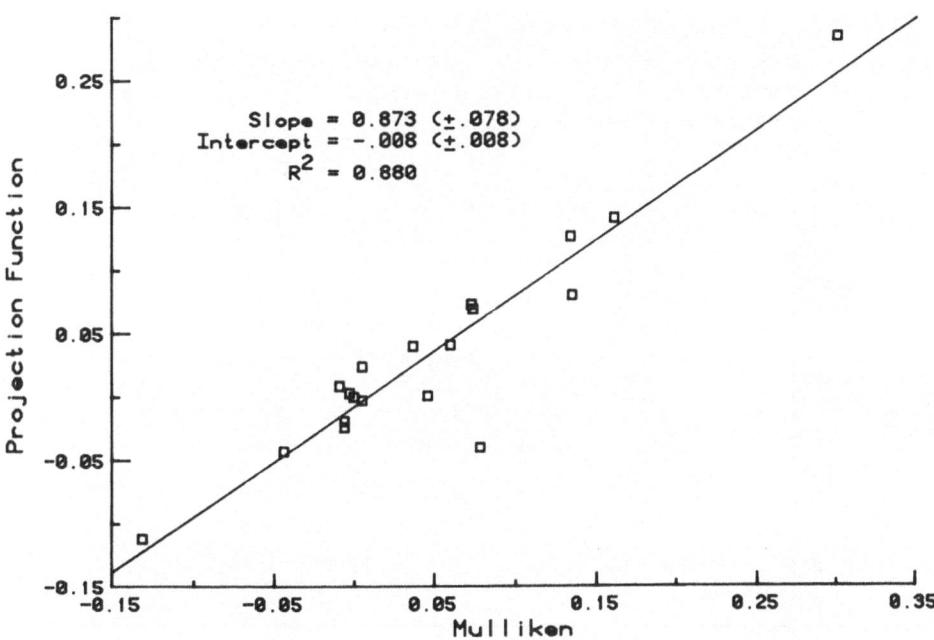

Fig. 12.   Comparison of the total ISEP difference with the corresponding Mulliken Gross Atom Population changes for the carbonyl oxygen of a number of ketones and aldehydes relative to formaldehyde.   4-31G basis.

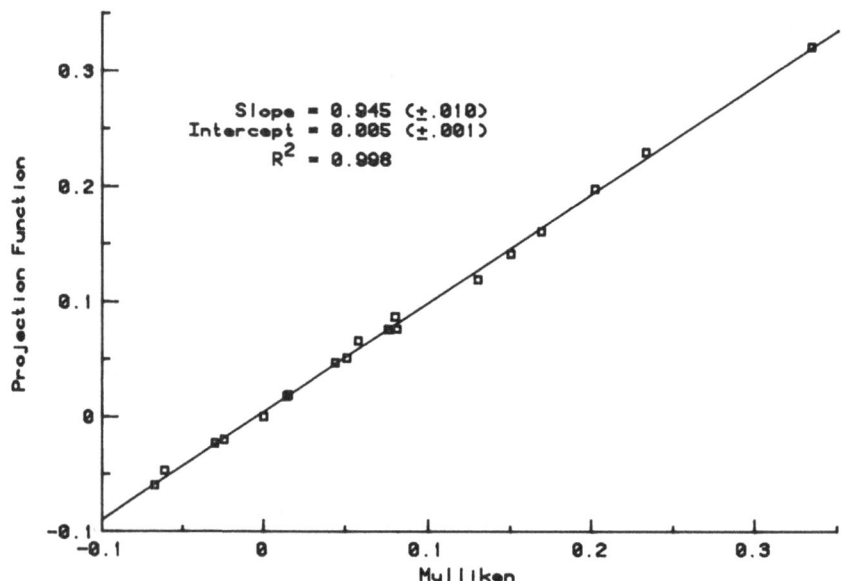

Fig. 13.   Comparison of the π ISEP difference with Mulliken π
            Population change for the carbonyl oxygens of a number
            of aldehydes and ketones.

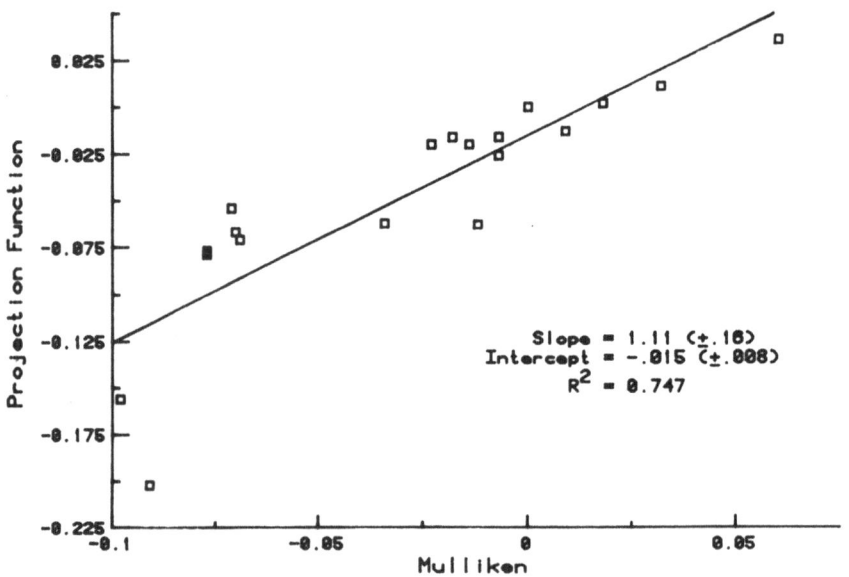

Fig. 14.   Comparison as in Fig. 13 but for σ-electrons only.

D.   INTEGRATED POPULATIONS IN SUBSTITUTED BENZENES

We have completed some preliminary studies of Projection
functions of substituted benzenes at the STO-3G level. Despite the
use of minimum basis sets, some important conclusions can be derived.
The projection function for benzene itself is shown in Fig. 15
separated into π and σ symmetry. The molecule has been divided into
6 regions by symmetry and integrations of each region show repro-
ducibility generally to better than 0.01 e.

Fig. 16 shows what happens when a dipole of 3.5D is placed outside of one hydrogen. The dipole mimics the pure field effect of a
nitro group and has no bonding connection to the ring. The $\underline{P}$
functions for this perturbed benzene were computed and the display
is that of the difference functions with unperturbed benzene subtracted. The π-system shows the expected strong polarization
towards the ipso-carbon with the excess electrons coming almost
wholly from the o- and p-carbons. Little change occurs at the m-
positions. The σ-changes are more complex. Alternations of density
indicative of successive polarizations occur throughout the molecule.
Integrations about carbon or CH regions for the σ system would
generally combine significant positive and negative contributions.
Net charge changes for the σ-system are accordingly less meaningful
and, indeed, would disguise the underlying polarization effects.

These patterns can be compared to a "real" substituent. Fig.
17 shows the π- and σ-difference $\underline{P}$ functions for fluorobenzene; i.e.,
$\underline{P}$ for PhF with benzene subtracted. The π-system shows increased
density at the o- and p-positions and a decrease at m-, in accord
with the usual expectations. One unexpected feature is that the
ISEP change at the o-positions is clearly substantially greater than
at ρ.

The σ-system shows the extensive net withdrawal expected for a
σ-acceptor substituent. However, we again find a complex pattern
of polarizations. One recurrent characteristic of these polarizations is that positive and negative features occur at the o- and m-
positions and tend to cancel when integrated over an entire region.
The net population changes are thus very small but these net ISEP
charges again obscure the underlying polarization effects. We note
generally that such polarizations are not revealed by any procedure
that integrates over whole regions--such as integrated "atomic"
charges or Mulliken Populations.

The underlying polarizations are clearly brought out by a plot
of the projection function along the benzene bonds as in Fig. 18.
The π-system shows clear positive and negative ISEP changes at atoms--
note that the zero values are close to bond midpoints. The σ-system
is totally different. Here the zero values are not only close to
bond midpoints but also at atom positions. The pattern is now one

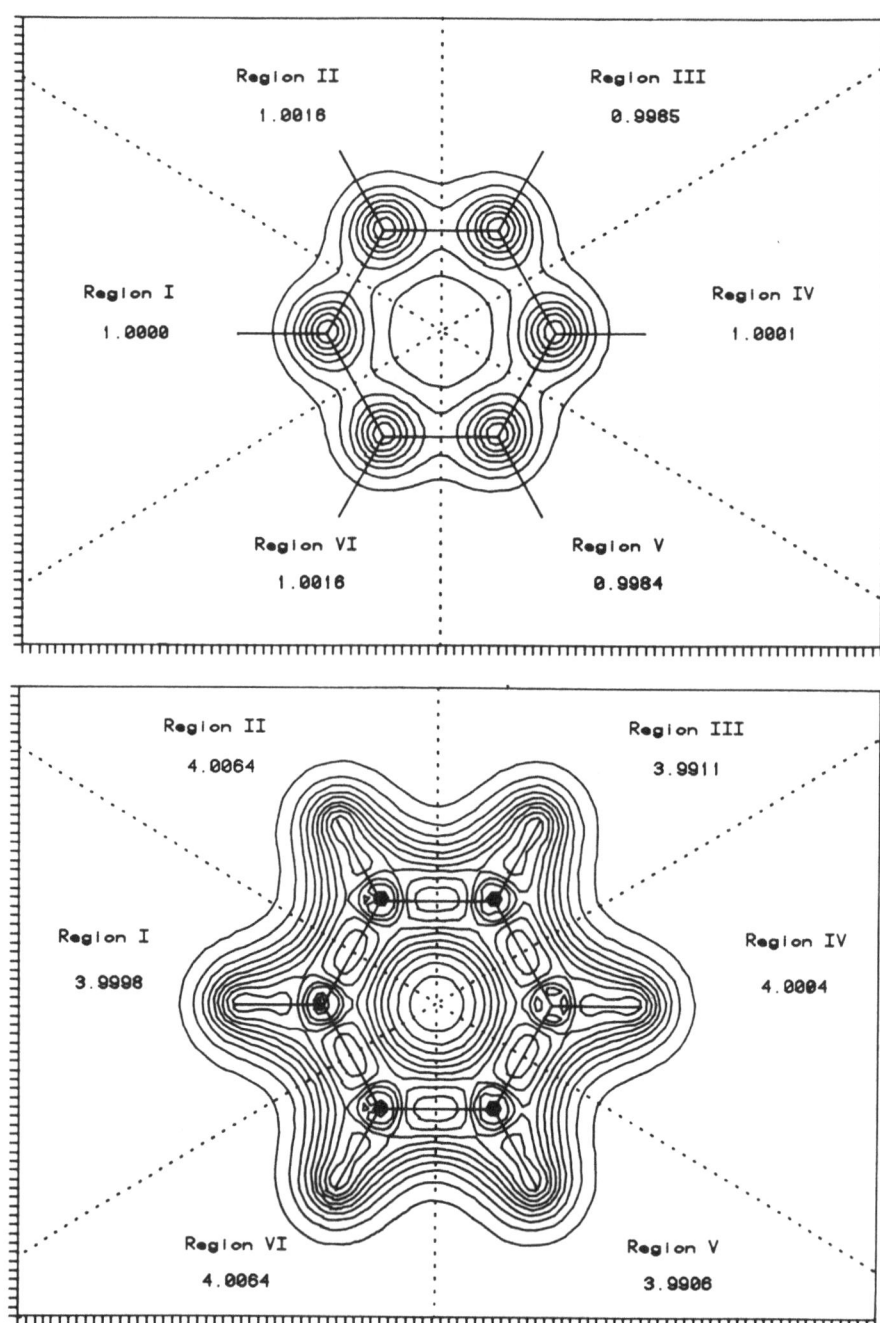

Fig. 15. Projection function plots for the molecular plane of
          benzene; STO-3G basis. Top: π-electrons. Bottom: σ-
          electrons. Numbers give the ISEP values for the regions
          indicated by dotted lines. Contour intervals 0.05 e au^{-2}.

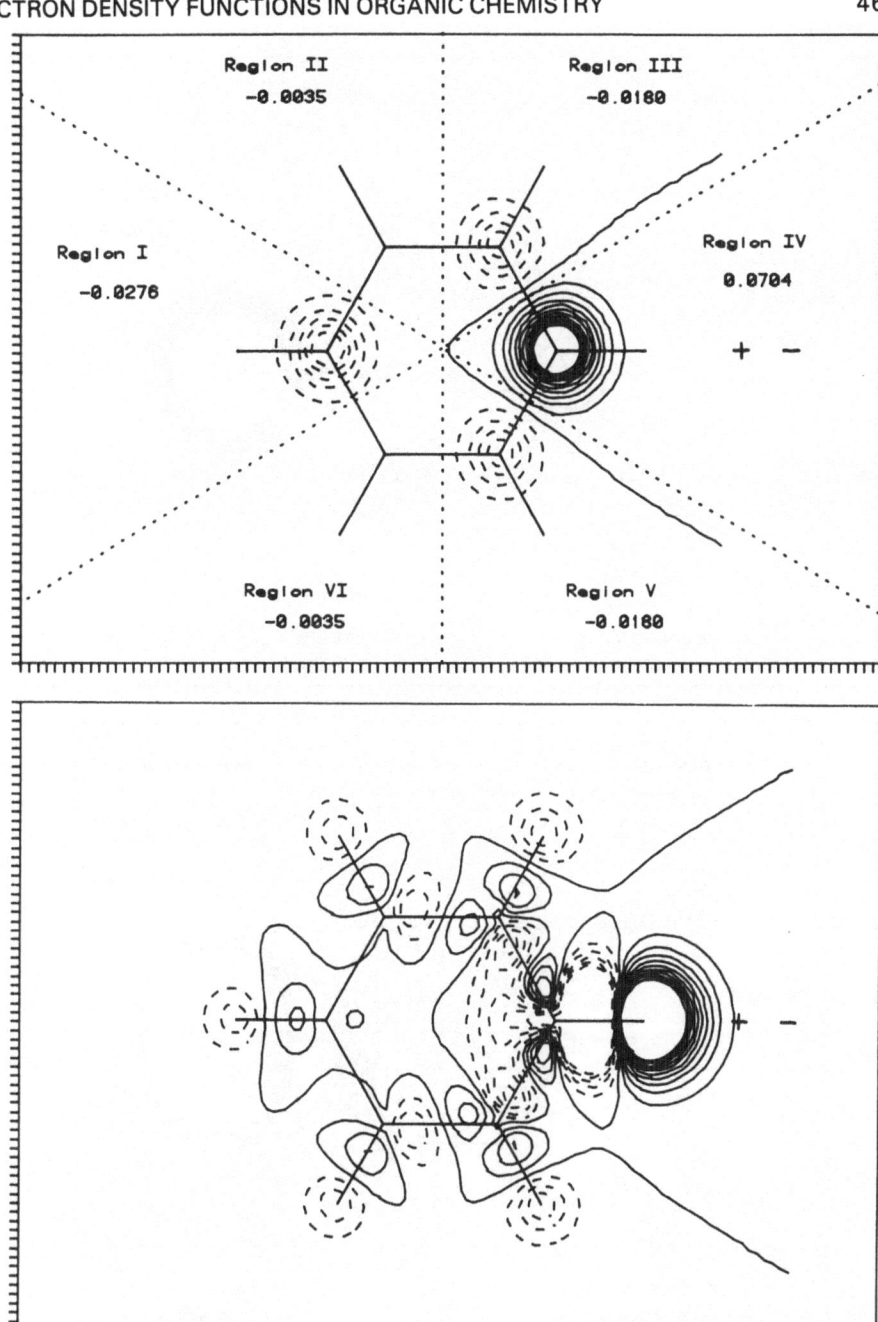

Fig. 16. Change in π- and σ- Projection functions of benzene by perturbation with a dipole of 3.5D placed as indicated. Top: π-electrons. Bottom: σ-electrons. The open contour in both figures is the zero contour; contour levels −0.01 to .02 by .002 e au⁻².

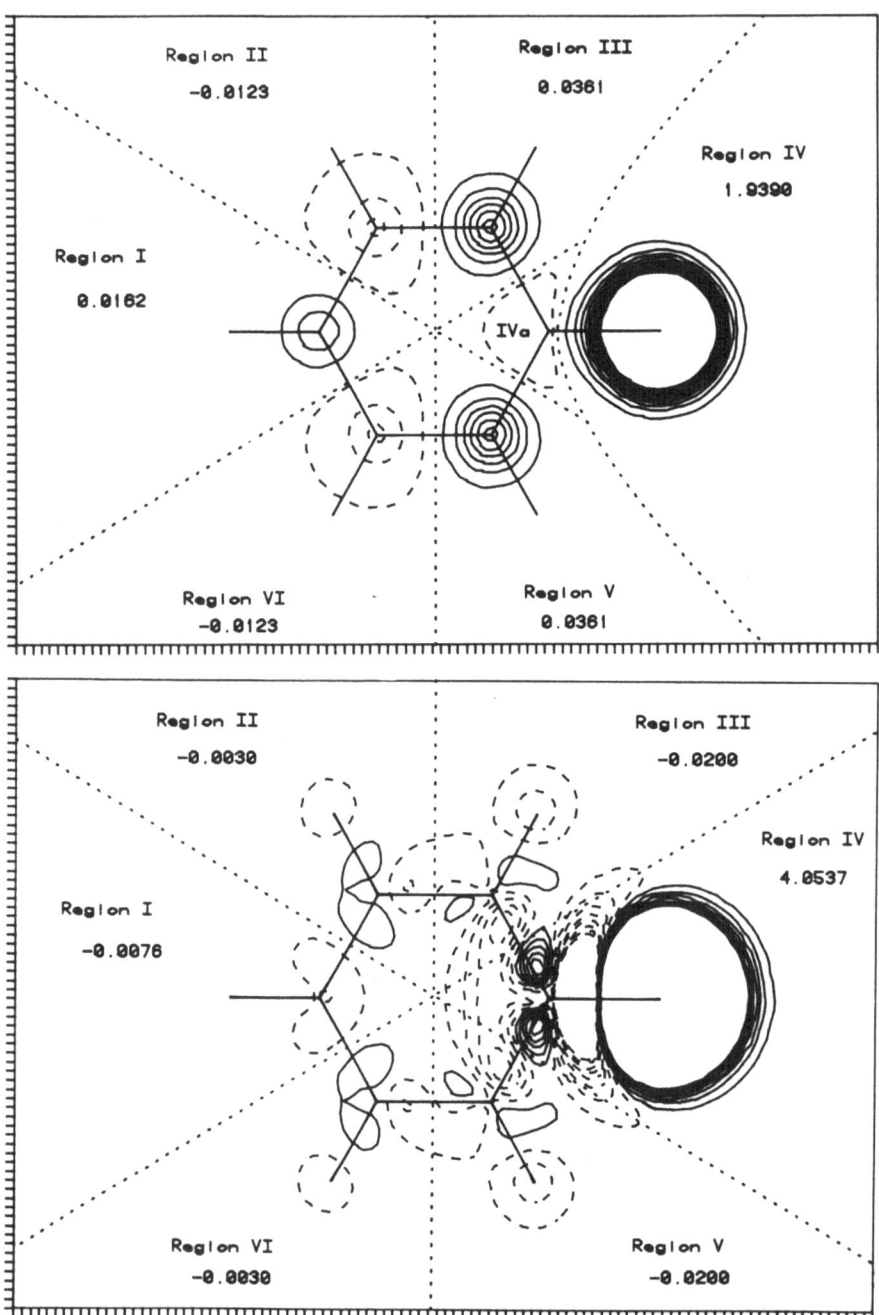

Fig. 17.   Pi and sigma Projection functions for fluorobenzene with
           those for benzene subtracted; e.g., $\underline{P}$(PhF) − $\underline{P}$(PhH).
           ISEP changes are indicated for regions defined by dotted
           boundaries; STO–3G basis. Contour levels −0.015 to .015
           by .002 e au^{-2}.

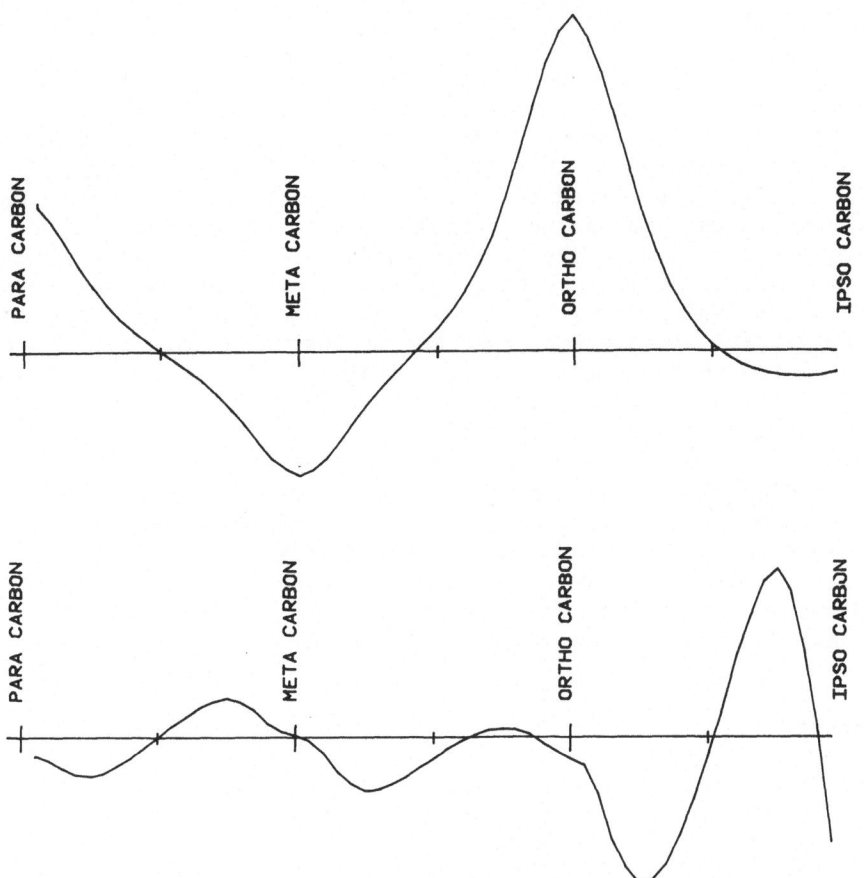

Fig. 18. Plots of the π and σ Projection function change,
P(PhF) - P(PhH), along the benzene C-C bond axes.

of polarization changes in C-C σ-bonding electrons rather than
charge transfer among atoms. This difference of π and σ electronic
systems is one of the new and significant results of such electron
density analyses. Although this result comes from a minimum basis
set computation whose details will undoubtedly change somewhat when
extended to larger basis sets, the important conclusions concerning
the general pattern of the results are unlikely to be seriously in
error.

E. CONCLUSIONS

These few examples show how electron density analysis can be used effectively to study important general concepts in the electronic theory of organic chemistry. This theory as it has developed over the past half-century is a qualitative theory of intramolecular charge transfer (the curved arrows) resulting from structural change. The theory has been based fundamentally on resonance concepts that require an evaluation of the relative weights of different resonance structures, although more recently perturbational MO concepts have played an increasingly important role. We expect that further electron density studies will sharpen our understanding of many of these qualitative theoretical concepts; already it appears that some of the traditional concepts and electronic "effects" are in error and require modification or discard.

Acknowledgment: This research was supported in part by the Office of Computer Affairs, University of California, Berkeley, by the National Resource for Computation in Chemistry and by the National Science Foundation. Dr. Kohler was supported in part by a scholarship of the Swiss National Fond number 82.611.1.78. We also thank S. Un and R. McDowell for technical assistance.

REFERENCES

1. A. Streitwieser, Jr., J. E. Williams, Jr., S. Alexandratos, and J. M. McKelvey, J. Am. Chem. Soc. 98:4778 (1976).
2. E. Steiner, "The Determination of and Interpretation of Molecular Wave Functions," Cambridge Univ. Press, London (1976).
3. (a) C. W. Kern, and M. Karplus, Chem. Phys. 40:1374 (1964); (b) R. F. W. Bader, and W. H. Henneker, J. Am. Chem. Soc. 87:3063 (1965); and many subsequent papers.
4. G. D. Graham, D. S. Marynick, and W. N. Lipscomb, J. Am. Chem. Soc. 102:4572 (1980).
5. K. Ruedenberg, Rev. Mod. Phys. 34:326 (1962).
6. W. Kutzelnigg, Angew. Chem. Intl. Ed. Engl. 12:546 (1973).
7. P. Politzer and R. G. Parr, J. Chem. Phys. 64:4634 (1976).
8. R. J. Boyd, J. Chem. Phys. 66:456 (1977).
9. H. Koster, D. Thoennes, and E. Weiss, J. Organometal. Chem. 160:1 (1978).
10. J. B. Collins, A. Streitwieser, Jr., and J. M. McKelvey, Comp. and Chem. 3:79 (1979).
11. A. Streitwieser, Jr., J. B. Collins, J. M. McKelvey, D. Grier, J. Sender, and A. G. Toczko, Proc. Natl. Acad. Sci. USA 76:2499 (1979).
12. P. Politzer, and R. R. Harris, J. Am. Chem. Soc. 92:6451 (1970).

13. P. Politzer and E. W. Stout, Jr., Chem. Phys. Letters 8:519 (1971).
14. P. Politzer, Theoret. Chim. Acta. 23:203 (1971).
15. P. Politzer and P. H. Reggio, J. Am. Chem. Soc. 94:8308 (1972).
16. P. Politzer and A. Politzer, J. Am. Chem. Soc. 95:5450 (1973).
17. P. Politzer, J. D. Elliott, and B. F. Meroney, Chem. Phys. Lett. 23:331 (1973).
18. R. S. Evans, and J. E. Huheey, Chem. Phys. Lett. 19:114 (1973).
19. S. M. Dean, and W. G. Richards, Nature, 256:473 (1975).
20. C. L. Hirshfeld, Theoret. Chim. Acta. 44:129 (1977).
21. S. Iwata, Chem. Phys. Lett. 69:305 (1980).
22. B. J. Ransil and J. J. Sinai, J. Chem. Phys. 46:4050 (1967).
23. R. F. W. Bader, P. M. Beddall and P. E. Cade, J. Am. Chem. Soc. 93:3095 (1971).
24. R. F. W. Bader and P. M. Beddall, J. Chem. Phys. 56:3320 (1972).
25. For recent reviews, see (a) R. F. W. Bader, Acct. Chem. Res. 8:34 (1975); (b) R. F. W. Bader and T. T. Nguyen-Dans, Adv. Quantum Chem. in press.
26. J. B. Collins and A. Streitwieser, Jr., J. Compt. Chem. 1:81 (1980).
27. J. S. Binkley, J. A. Pople, and W. J. Hehre, J. Am. Chem. Soc. 102:939 (1980).
28. M. F. Guest, I. H. Hiller, and V. R. Saunders, J. Organometal. Chem. 44:59 (1972).
29. N. C. Baird, R. F. Barr, and R. K. Datta, J. Organometal. Chem. 59:65 (1973).
30. T. Clark, P. V. R. Schleyer, and J. A. Pople, J.C.S. Chem. Comm. 137 (1978).
31. H. Dietrich, J. Organometal. Chem. 205:291 (1981).
32. A few of the more recent references are: R. R. Luchese, and H. F. Schaefer, J. Am. Chem. Soc. 100:298 (1978); J. A. Altman, I. G. Csizmadia, A. Robb, K. Yates, and P. Yates, J. Am. Chem. Soc. 100:1653 (1978); S. Bell, Mol. Phys. 37:255 (1979); J. D. Goddard, and H. F. Schaefer, J. Chem. Phys. 70:5117 (1979); S. K. Gray, W. H. Miller, Y. Yamaguchi, and H. F. Schaefer, unpublished.
33. R. J. Beunker and S. D. Peyerimhoff, J. Chem. Phys. 53:1368 (1970).
34. Ground State only: J. H. Dunning, Jr., and N. W. Winter, J. Chem. Phys. 55:3360 (1971).
35. D. E. Freeman, and W. Klemperer, J. Chem. Phys. 40:604 (1964); 45:52 (1966); J. R. Lombardi, D. E. Freeman, and W. Klemperer, J. Chem. Phys. 46:2746 (1967).
36. N. J. Turro, and G. L. Farrington, J. Am. Chem. Soc. 102:6051, 6056 (1980); N. J. Turro, "Modern Molecular Photchemistry," Benjamin-Cummings (1978).
37. D. L. Grier, and A. Streitwieser, Jr., J. Am. Chem. Soc. submitted.

INDEX